Deep Imaging in Tissue and Biomedical Materials

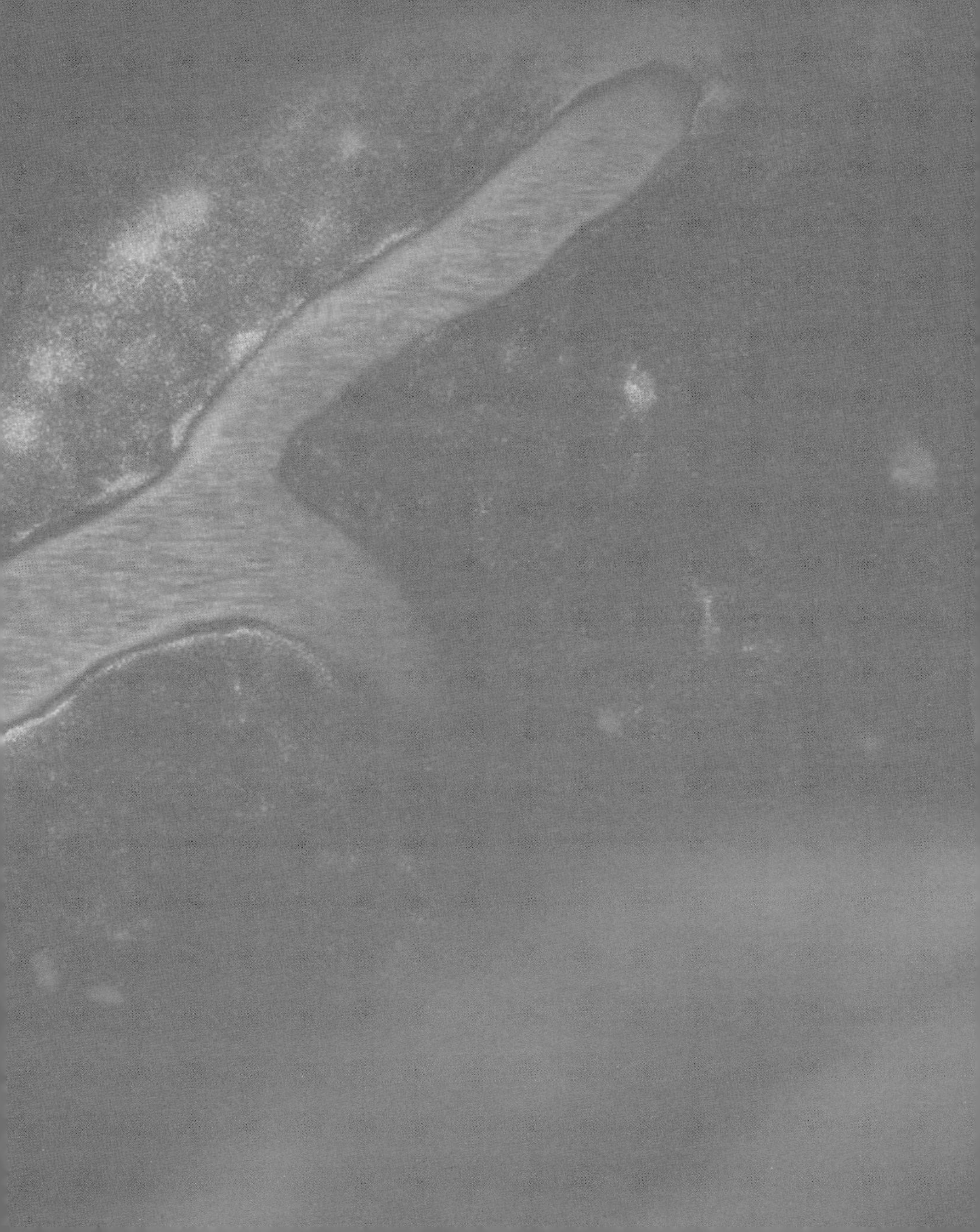

Deep Imaging in Tissue and Biomedical Materials

Using Linear and Nonlinear Optical Methods

edited by

Lingyan Shi

Robert R. Alfano

Pan Stanford Publishing

Published by

Pan Stanford Publishing Pte. Ltd.
Penthouse Level, Suntec Tower 3
8 Temasek Boulevard
Singapore 038988

Email: editorial@panstanford.com
Web: www.panstanford.com

British Library Cataloguing-in-Publication Data
A catalogue record for this book is available from the British Library.

Deep Imaging in Tissue and Biomedical Materials: Using Linear and Nonlinear Optical Methods

Cover image, taken by Lingyan Shi from Adrian Rodriguez-Contreras' Lab, shows a trans-cranial image of brain microvessels filled up with Texas red dextran 70 kDa and wrapped with Gcamp-6 GFAP astrocytes, using multiphoton fluorescence microscope.

ISBN 978-981-4745-88-8 (Hardcover)
ISBN 978-1-315-20655-4 (eBook)

Printed in Great Britain by Ashford Colour Press Ltd

Contents

Preface

The use of light for probing and imaging biomedical media is promising for developing safe, noninvasive, and inexpensive clinical imaging modalities with diagnostic ability. The advent of ultrafast lasers enables the applications of nonlinear optical processes for deeper imaging into biological tissues with higher spatial resolution. The primary goals of this book are to review the fundamentals in biophotonics and to introduce emerging novel optical imaging techniques for deep tissue imaging. This book is intended to serve as an introductory guide of optical imaging for students and a reference for engineers and researchers who seek a better understanding of deep imaging in tissues.

The book consists of 16 chapters and is divided into three parts. Part I consists of eight chapters. The first chapter, by Murugkar and Boyd, reviews the basic concepts of nonlinear optical imaging, including second harmonic generation, coherent Raman scattering, and self-phase modulation. The next chapter, by Galvez, reviews the fundamentals and physical phenomena of complex light beams, including Gaussian beam, Bessel beam, Airy beam, and Poincare beam. The third chapter, by Shi and coworkers, inspects the properties of Gaussian beam optics in multiphoton fluorescence imaging. The deep imaging in the optical windows in near-infrared (NIR) and short-wave infrared (SWIR) from 700 nm to 2500 nm is reviewed by Jacques. Next, Alfano and coworkers review the salient properties of light propagation in highly scattering media and tissue. The application of non-linear microscopy to life science is reviewed by Svindrych and Periasamy, which is followed by the chapter on smart biomarker of quantum dots for NIR fluorescence imaging by Jin, Sasaki, and Imamura. The last chapter of Part I, by Seddon, describes biomedical applications for deep probing in materials by using mid-infrared supercontinuum laser and new optical fibers.

Part II reviews the theories and properties of light propagating in tissue. The first chapter, by Bykov, Doronin, and Meglinski, overviews the theories and derived model for understanding light propagation in tissue-like media using Monte Carlo. The cumulant solution for light propagation in a turbid medium and its applications in deep imaging are reviewed in the following chapter by Xu, Cai, and Alfano. The final chapter of Part II, by Pu and coworkers, reviews the latest advancement of NIR scanning polarization imaging unit for prostate and presents an algorithm for diffusive image reconstruction using NIR banana pathways.

Part III presents recent technology developments in optical imaging and introduces the applications of different techniques for detecting disordered media and tissue. Karagoz and Altan introduce terahertz propagation in tissues and its limitation for thick tissue and present ways for use in smears and thin tissues for histology applications. The use of stimulated Raman scattering gain and loss microscopy in detecting brain tumor is then reviewed by Lewis and Orringer. The next chapter, by Cao and Cheng, introduces the technology of using new photoacoustics approach for deep imaging by detecting molecules' vibrational overtone of chemical bonds. Gannot and coworkers presents multiple acoustic and thermal methods for light–tissue interaction for detecting deeper structures. The final chapter, by Davy, Gigan, and Genack, describes the properties of transmission matrix that determine the net transmission pathways in biomedical and condensed media, and its use for deep imaging.

As will be seen, much has been accomplished and reviewed in the book, but much remains for the future. Therefore, not only is this book an introduction to students in the field, but it proposes directions for researchers to adapt their own or to explore new optical technologies for deeper and better imaging in life science. We wish to thank all the invited authors, who presented very interesting and knowledgeable chapters.

Lingyan Shi
Robert R. Alfano
New York
January 2017

Chapter 1

Overview of Second- and Third-Order Nonlinear Optical Processes for Deep Imaging

Sangeeta Murugkar[a] **and Robert W. Boyd**[b]

[a]*Department of Physics, Carleton University, Ottawa, Ontario K1S 5B6, Canada*
[b]*Department of Physics, University of Ottawa, Ottawa, Ontario K1N 6N5, Canada*

smurugkar@physics.carleton.ca, rboyd@uottawa.ca

1.1 Introduction: Nonlinear Optical Contrast in Biological Imaging

The field of optical microscopic imaging has rapidly evolved because of tremendous advances made in laser and detection technology. Various types of linear and nonlinear light–matter interactions have been harnessed for providing contrast in the microscopic images. Simple microscopy techniques such as bright-field and differential-interference-contrast reveal structural information at the cellular level owing to the refractive index contrast of the sample medium. Fluorescence microscopy offers higher chemical specificity and is the most popular contrast

Deep Imaging in Tissue and Biomedical Materials: Using Linear and Nonlinear Optical Methods
Edited by Lingyan Shi and Robert R. Alfano

ISBN 978-981-4745-88-8 (Hardcover), 978-1-315-20655-4 (eBook)
www.panstanford.com

mechanism used in biological studies. The contrast is achieved by means of targeted labeling of molecules using exogenous or endogenous fluorophores. However, external fluorophores are often perturbative since they may disrupt the native state of the sample, especially for small molecules whose size may be smaller than the fluorescent label itself. Besides, many molecular species are intrinsically nonfluorescent or only weakly fluorescent. It is also better to avoid external contrast agents for in vivo imaging applications since such contrast agents need concurrent development of appropriate delivery strategies and are often limited by problems of label specificity and induced toxicity. Vibrational microscopy techniques, on the other hand, are inherently label-free. They involve the excitation of molecular vibrations and offer intrinsic chemical specificity. Two such techniques include infrared absorption and Raman microscopy. Out of these, infrared microscopy has low spatial resolution owing to the long infrared wavelengths employed. In addition, water absorption of the infrared light is a major limitation for investigating live biological samples. Raman scattering, on the other hand, is based on the inelastic scattering of light by vibrating molecules and provides a molecular fingerprint of the chemical composition of a living cell or tissue. It offers a powerful label-free contrast mechanism and has been applied in various biological investigations. Linear contrast mechanisms based on fluorescence and Raman scattering typically employ continuous-wave visible light for excitation and sample scanning or laser scanning to generate an image. A confocal pinhole inserted at the detector facilitates a three-dimensionally sectioned image but unfortunately limits the sensitivity of detection.

In comparison, nonlinear optical microscopy or multiphoton microscopy employs near-infrared (near-IR) femtosecond or picosecond pulsed light to excite nonlinear optical processes that can only be accessed by application of two or more (multi) photons [1]. The nonlinear optical signal is generated only in the focal plane of the objective where the beam intensity is maximized. This results in the inherent three-dimensional sectioning capability without the need for a confocal pinhole. This also means that significantly greater sensitivity in signal detection is achieved since the confocal collection geometry is not necessary. The near-IR light used for excitation of the nonlinear optical signal enables deeper penetration in thick scattering samples and in addition is less biologically harmful. In addition, multiphoton microscopic imaging can take advantage of nonlinear optical processes involving endogenous contrast. This ability permits dynamic studies of live cells and tissue specimens in a label-free manner.

Two-photon-excited fluorescence (TPEF) or two-photon microscopy has been extensively applied for biological imaging over the past couple of decades [2, 3]. In TPEF, a single femtosecond pulsed laser beam is tightly focused in the specimen such that two low energy, near-IR photons are simultaneously absorbed by a fluorophore and then emitted as one photon at a higher frequency than the incident light. Along with TPEF, nonlinear optical imaging techniques such as second harmonic generation (SHG) and coherent Raman scattering (CRS) have experienced continued growth in the technology and applications over the past decade. In SHG, two photons with the same frequency interacting with a nonlinear optical material are effectively "combined" to generate new photons at twice the frequency of the incident light [4]. CRS, which refers to both stimulated Raman scattering (SRS) and coherent anti-Stokes Raman scattering (CARS), is a technique to enhance spontaneous Raman scattering. The spontaneous Raman effect involves incoherent excitation of molecular vibrations. It is inherently weak in nature with the typical photon conversion efficiencies for Raman being lower than 1 in 10^{18}. This results in long data acquisition times of 100 ms to 1 s per pixel and does not permit fast chemical imaging of a living system [5]. In contrast, CRS involves nonlinear Raman scattering such that molecular vibrations are driven coherently, in phase through stimulated excitation by two synchronized femtosecond (or picosecond) pulsed lasers. This translates to a reduction of the image acquisition time from many minutes or hours required in confocal Raman microscopy to only a couple of seconds using coherent Raman scattering microscopy [6, 7]. CARS and SRS have been shown to be promising techniques for the chemically selective imaging of lipids, proteins, and DNA in skin, brain, and lung tissue. TPEF and SHG enable selective imaging of auto-fluorescent proteins in tissue and of noncentrosymmetric structures such as fibrillar collagen, respectively [8]. We note that in many of these studies the illumination method of choice is one based on the nonlinear optical process of supercontinuum generation (SCG).

In this chapter we examine the basics of second-order and third-order nonlinear optical processes that are involved in biological imaging, with the emphasis on second harmonic generation, coherent Raman scattering and self-phase modulation (SPM). We begin with a classical description of nonlinear light–matter interaction in Section 1.2. This is followed by a summary of the classical and quantum mechanical approach to explain SHG signal generation in Section 1.3. The CRS process is described for CARS and SRS signal generation in Section 1.4 while the two-photon absorption process is described in Section 1.5. The process of SCG and the related process of self phase modulation (SPM) are described in Section 1.6.

1.2 Classical Description of Nonlinear Light–Matter Interaction

Linear and nonlinear optical effects arise from the interaction of the electric field associated with the incident light wave with the charged particles of the material. The rapidly oscillating applied electric field drives the motion of the bound electrons in the molecules of the material. The displacement of the electrons from their equilibrium positions induces an electric dipole moment $\tilde{\mu}(t)$.[1] The strength of this dipole moment is higher for weakly bound electrons as compared to electrons that are tightly bound to the nuclei of the molecules of the material. The macroscopic polarization is then obtained by adding the contributions of all N electric dipole moments induced in the material and is given by

$$\tilde{P}(t) = N\tilde{\mu}(t). \tag{1.1}$$

When the strength of the applied electric field is small compared to that of the electric field binding the electrons to the nuclei, i.e., in the weak field limit, the induced polarization can be written assuming the case of rapid response as

$$\tilde{P}(t) = \varepsilon_0 \chi \tilde{E}(t), \tag{1.2}$$

[1]The tilde is used to denote a quantity that varies rapidly in time.

where ε_0 is the electric permittivity in vacuum, χ is the susceptibility of the material and $\widetilde{E}(t)$ is the electric field associated with the incident light.

This type of linear dependence of the induced polarization in the material on the magnitude of the applied electric field is the source of all linear optical effects such as scattering, reflectance, absorption, and fluorescence. The linearity of the signal is defined through its linear dependence on the intensity of the incident light. It is important to note that although spontaneous Raman scattering is a linear effect, it can be described as a nonlinear interaction between the photon fields and the material as discussed later.

In the case of strong applied electric fields such as those associated with femtosecond or picosecond pulsed lasers, the induced polarization is not strictly linear and has higher orders terms as follows:

$$\widetilde{P}(t) = \varepsilon_0[\chi^{(1)}\widetilde{E}(t) + \chi^{(2)}\widetilde{E}^2(t) + \chi^{(3)}\widetilde{E}^3(t) + \cdots] \tag{1.3}$$

Here $\chi^{(1)}$ is the linear susceptibility, $\chi^{(2)}$ is the second-order susceptibility, $\chi^{(3)}$ is the third-order susceptibility, etc. Second-harmonic generation occurs as a result of the second-order response described by $\chi^{(2)}$, whereas the third-order response described by $\chi^{(3)}$, is responsible for coherent Raman scattering and two-photon excitation fluorescence.

1.3 Second Harmonic Generation

Second-harmonic generation is the prototypical nonlinear optical process [1]. Its discovery by Franken and coworkers in 1961 [9] is often taken as the birth of the field of nonlinear optics. The process of second-harmonic generation is illustrated in Fig. 1.1. A laser beam at frequency ω illuminates a nonlinear optical material with a nonzero second-order susceptibility $\chi^{(2)}$, and a beam of light at frequency 2ω is created. The transfer of energy from the input field to the output field can be depicted in terms of the energy-level diagram shown on the right-hand side of the figure. The solid line in this figure represents the atomic ground state while the dashed lines represent virtual levels. In this process, two photons from the input near-IR beam are lost and one photon in the output visible light beam is created. Second harmonic generation is an example of a parametric generation process since it involves the exchange of energy between incident and outgoing laser fields while molecules remain in the ground state after interaction. Other examples of parametric processes that involve nonlinear electronic polarization of the molecules under intense pulsed laser light excitation are third-harmonic generation and four-wave mixing. Coherent anti-Stokes Raman scattering (CARS) discussed in Section 1.4 is an example of a nonlinear four-wave mixing parametric process that involves the vibrational states of the molecules.

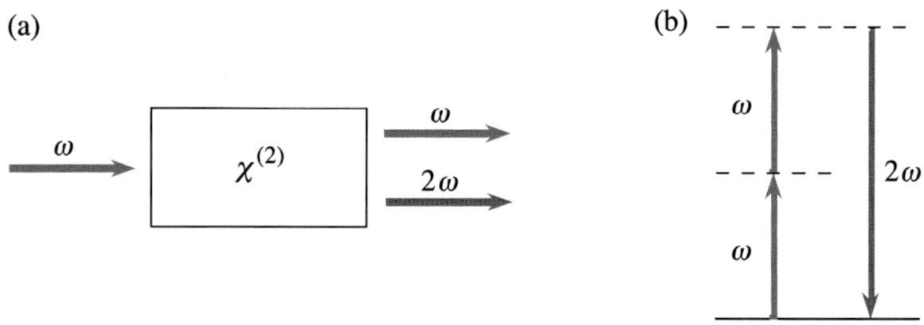

Figure 1.1 (a) Schematic representation of a second harmonic generation process (b) Energy level diagram for SHG. The bold line represents the ground electronic state of the molecule while the dashed lines are virtual excited states of the molecule.

The vector nature of the electric field and polarization needs to be taken into account in order to get a more complete description of the second-order nonlinear response. Besides, if the various orders of the nonlinear susceptibility are frequency dependent, the relationship between the electric field and polarization is best expressed in the frequency domain. For example, the second-order polarization can be expressed more generally as

$$P_i(\omega_n + \omega_m) = \varepsilon_0 \sum_{jk} \sum_{(nm)} \chi^{(2)}_{ijk}(\omega_n + \omega_m, \omega_n, \omega_m)\, E_j(\omega_n) E_k(\omega_m) \quad (1.4)$$

Here i, j, and k represent various Cartesian components of the field vectors and the notation (n, m) implies that the expression is to be summed over ω_n and ω_m but only those contributions that lead to the particular frequency $\omega_n + \omega_m$ given in the argument on the left-hand side of the equation are to be retained.

1.3.1 Quantum Mechanical Treatment of the Nonlinear Susceptibility

More accurate models of the nonlinear optical response are provided by quantum mechanical calculation [4, 10, 11]. For example, for the case of usual interest in which the applied and generated fields are detuned by at least several line widths from the closest material resonance, the second-order susceptibility can be expressed as

$$\chi_{ijk}^{(2)}(\omega_\sigma, \omega_q, \omega_p) = \frac{N}{\varepsilon_0 \hbar^2} \mathcal{P}_F \sum_{mn} \frac{\mu_{gn}^i \mu_{nm}^j \mu_{mg}^k}{(\omega_{ng} - \omega_\sigma)(\omega_{mg} - \omega_p)}, \tag{1.5}$$

where $\omega_\sigma = \omega_q + \omega_p$ and $\mathcal{P}$ is the number density of molecules, μ_{nm}^j represents the j-th Cartesian component of the electric-dipole moment matrix element connecting levels n and m, and ω_{mg} is the energy separation of levels m and g divided by $\hbar$. The symbol $\mathcal{P}_F$ is the full permutation operator, defined such that the expression that follows it is to be summed over all permutations of the frequencies ω_p, ω_q, and $-\omega_\sigma$. The Cartesian indices are to be permuted along with the related frequencies, and the final result is to be divided by the number of distinct permutations of the input frequencies ω_p and ω_q. In the general case in which ω_p and ω_q are distinct, this equation thus expands to six separate terms. Three of these six terms are illustrated in Fig. 1.2; the other three terms result from a simple interchange of ω_p and ω_q in these figures.

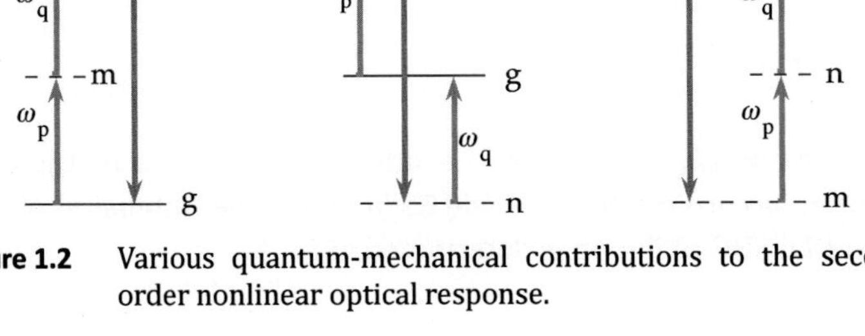

Figure 1.2 Various quantum-mechanical contributions to the second-order nonlinear optical response.

1.3.2 Wave Equation Description of SHG

The intensity of the radiation emitted in the harmonic generation process can be predicted by means of a propagation calculation [4]. Consider the wave equation in the form

$$\nabla^2 \widetilde{K} - \frac{n^2}{c^2}\frac{\partial^2 \widetilde{E}}{\partial t^2} = \mu_0 \frac{\partial^2 \widetilde{P}^{NL}}{\partial t^2}, \tag{1.6}$$

where μ_0 denotes the magnetic permeability of free space and $\widetilde{E}_j(z, t)$ represents the electric fields in the form of plane waves as follows:

$$\widetilde{E}_j(z,t) = A_j(z)e^{i(k_j z - \omega_j t)} + c.c., \tag{1.7}$$

where j = 1, 2 with $\omega_1 = \omega$ and $\omega_2 = \omega$, respectively, for the fundamental and second harmonic waves and *c.c.* stands for the complex conjugate. Here $A_j(z)$ represents the spatially slowly varying field amplitude of this wave. The nonlinear polarization is then given by

$$\widetilde{P}_2(z,t) = P_2(z)e^{i(2k_1 z - 2\omega t)} + c.c., \tag{1.8}$$

where $P_2 = P(\omega_2) = \varepsilon_0\chi^{(2)} = E_1^2$ represents the complex amplitude of the nonlinear polarization. Upon substituting the expressions for $\widetilde{P}_2$ and $\widetilde{E}_1, \widetilde{E}_2$ in the wave Eq. 1.6, a second-order equation in A_2 [1] is obtained. We make the assumption of the slowly varying envelope approximation (SVEA) which states that the change in amplitude A_2 of the second-harmonic wave is extremely small over a distance on the order of an optical wavelength. This approximation results in a simplified first-order equation in A_2. The resulting amplitude of the second-harmonic field after propagation through a distance L, is given by

$$A_2(L) = \frac{2i\omega\chi^{(2)}A_1^2}{n_2 c}\frac{e^{i\Delta k} - 1}{i\Delta kL}, \tag{1.9}$$

where $\Delta k = 2k_1 - k_2$. Since the intensity is related to the field strength according to $I = 2n\sqrt{\in_0/\mu_0}|A|^2$, the intensity of the generated SHG radiation is determined as

$$I_2(L) = \sqrt{\frac{\mu_0}{\epsilon_0}} \frac{2\omega^2}{n_1^2 n_2 c^2} |\chi^{(2)}|^2 I_1^2 L^2 \text{sinc}^2\left(\frac{\Delta k\, L}{2}\right), \tag{1.10}$$

where sinc(x) = sin(x)/x. The condition $\Delta k = 0$ is known as the condition of perfect phase matching, and is a requirement for efficient generation of second-harmonic radiation. When $\Delta k = 0$ the last factor in the above equation equals unity and the intensity of SHG signal scales as the square of the incident intensity I_1.

1.3.3 Symmetry Breaking and SHG Signal in Biological Imaging

Not all materials are capable of producing a SHG signal. A material must be noncentrosymmetric over macroscopic distances in order to produce appreciable radiation at the second-harmonic frequency [4]. The first biological SHG imaging was reported in 1986 by Freund and coworkers using rat tail tendon [12]. Since then SHG microscopy has been successfully applied to imaging subsurface tumor progression [13], collagen distributions and membrane potential [14–16]. SHG imaging is being developed as a powerful tool for investigating the changes to the tumor microenvironment during cancer invasion and metastasis [17]. Polarization-resolved SHG has been applied to yield information related to the molecular organization of collagen, while SHG imaging combined with various image analysis techniques has revealed changes in tissue architecture associated with connective tissue disorders, musculoskeletal diseases, organ fibrosis, cardiovascular pathologies, and cancers [18, 19].

1.4 Coherent Raman Scattering

Coherent Raman scattering includes both CARS and SRS and is related to spontaneous Raman scattering. Theoretical treatments of the coherent Raman scattering processes can be found both in the early research papers [20, 21] and in various textbook accounts [4, 7, 11]. We begin by describing the classical model of spontaneous Raman scattering. This sets the stage to arrive at a physical intuitive picture of coherent Raman scattering.

1.4.1 Classical Model of Spontaneous Raman Scattering

As discussed earlier in Section 1.2, the electric field corresponding to the incident visible or near-IR light beam induces a polarization in the material given by Eq. 1.4. Since the electrons are bound to the nuclei, the motion of the nuclei affects the displacement of the electrons. The purturbative effect of the nuclear modes can be expressed by connecting the electric dipole moment $\tilde{\mu}(t)$ with the electronic polarizability $\tilde{a}(t)$ as follows:

$$\tilde{\mu}(t) = \tilde{a}(t)\tilde{E}(t), \tag{1.11}$$

where

$$\tilde{\alpha}(t) = \alpha_0 + (\partial\alpha/\partial Q)\tilde{Q}(t) + \cdots \tag{1.12}$$

Here α_0 is the electronic polarizability in the absence of either the nuclear modes or the nonlinearities. $\tilde{Q}(t)$ describes the time-dependent nuclear coordinate and $(\partial\alpha/\partial Q)$ corresponds to the coupling strength between the nuclear and electronic coordinates.

Assuming the nuclear motion to be that of a classical harmonic oscillator,

$$\tilde{Q}(t) = 2Q_0\cos(\omega_v t + \varphi) = Q_0[e^{i\omega_v t + i\varphi} + e^{-i\omega_v t - i\varphi}] \tag{1.13}$$

Here Q_0 is the amplitude of the nuclear motion, ω_v is the nuclear resonance frequency, and φ is the phase of the nuclear mode vibration.

Assuming the form for the incident electric field as $E(t) = Ae^{-i\omega_1 t} + c.c.$, where A is the amplitude of the electric field, the dipole moment is as follows:

$$\tilde{\mu}(t) = \alpha_0 Ae^{-i\omega_1 t} + A(\partial\alpha/\partial Q)\,Q_0[e^{-i(\omega_1-\omega_v)t+i\varphi} + e^{-i(\omega_1+\omega_v)t-i\varphi}] + c.c. \tag{1.14}$$

It can be seen that the dipole moment oscillates at several frequencies. Here the first, second and third terms correspond to the elastically scattered Rayleigh light, the inelastic Stokes-shifted contribution at $\omega_s = \omega_1 - \omega_v$ and the anti-Stokes-shifted contribution at $\omega_{as} = \omega_1 + \omega_v$. Since the phase of the nuclear vibration of different molecules is uncorrelated, the phase of the radiated field from each dipole emitter is also uncorrelated. The spontaneous Raman scattered light is thus incoherent.

The amplitude of the Stokes electric field is directly proportional to the amplitude of the dipole oscillation $|\mu(\omega_s)|$ at the Stokes frequency ω_s. Using Eq. 1.14, we obtain the intensity of the Raman-scattered light by calculating the energy radiated by this dipole [7]. This procedure leads to the following expression:

$$I(\omega_s) = \frac{\omega_s^4}{12\pi\varepsilon_0 c^3} Q_0^2 |A|^2 (\partial\alpha/\partial Q)^2, \tag{1.15}$$

where c is the speed of light. It is evident from the above equation that the intensity of spontaneous Raman scattering is linearly proportional to the intensity of the incident beam $I_0 = |A|^2$ and scales with $(\partial\alpha/\partial Q)^2$ associated with the nonzero change in polarizability due to nuclear motion. The latter quantity is directly proportional to the Raman scattering cross section $\sigma(\omega_s)$. This simple classical model provides a useful physical picture for interpreting several attributes of the spontaneous Raman scattering process. However, a quantum-mechanical treatment is required to account for the spontaneous emission process. This latter approach reveals the similarity between the light–matter interaction involved in the spontaneous Raman scattering process and the interactions of a $\chi^{(3)}$ process. The intensity of the spontaneous Raman signal is found to be directly proportional to $\text{Im}\{\chi^{(3)}(\omega)\}$ [7].

1.4.2 Classical Model of Coherent Raman Scattering

In the classical picture of the coherent Raman scattering process, two incident light fields, $\tilde{E}_1$ and $\tilde{E}_2$ modeled as plane waves given by $\tilde{E}_1(t) = A_i e^{-i\omega_i t} + c.c.$, at frequencies ω_1 and ω_2 induce oscillations in the electron cloud surrounding the nuclei of the molecule. Although the incident frequencies are assumed to be much higher than the resonance frequency ω_v of the nuclear mode of the molecule, the combined effect of sufficiently intense optical fields can drive the nonlinear electron motion at the difference frequency of $\Omega = \omega_1 - \omega_2$. Since the electron motion is coupled to the nuclear motion through a nonzero $(\partial\alpha/\partial Q)_0$ the modulated electron cloud actively drives the nuclear mode to oscillate at frequency Ω with the amplitude given by

$$Q(\omega_\nu) = \frac{1}{m}(\partial\alpha/\partial Q)_0 \frac{A_1A_2^*}{\omega_\nu^2 - \Omega^2 - 2i\Omega\gamma}, \tag{1.16}$$

where γ is the damping constant and m is the reduced mass of the nuclear oscillator. It is seen from the above equation that the amplitude of the nuclear mode is largest when the difference frequency between the two incident fields matches the resonance frequency ω_ν of the nuclear mode.

When the molecule is set into oscillation, its polarizability will be modulated periodically in time. Thus, the refractive index of a collection of coherently oscillating molecules will be modulated in time in accordance with the following relation [4]:

$$\tilde{n}(t) = [1 + N\tilde{\alpha}(t)]^{1/2} \tag{1.17}$$

The temporal modulation of the refractive index will modify a beam of light as it passes through the medium. Thus, each input frequency develops sidebands shifted by ±Ω. The resulting macroscopic polarization is found using Eqs. 1.1, 1.11, and 1.12 to be [7]:

$$\tilde{P}(t) = N[\alpha_0 + (\partial\alpha/\partial Q)\,\tilde{Q}(t)]\{\tilde{E}_1(t) + \tilde{E}_2(t)\} \tag{1.18}$$

This results in the following expression consisting of nonlinear third-order polarization terms owing to the driven Raman mode.

$$\tilde{P}_{NL}(t) = P(\omega_{cs})e^{-i\omega_{cs}t} + P(\omega_2)e^{-i\omega_2 t} + P(\omega_1)\,e^{-i\omega_1 t} + P(\omega_{as})\,e^{-i\omega_{as}t} + c.c. \tag{1.19}$$

It contains four contributions as follows: $P(\omega_{cs})$ corresponds to coherent Stokes Raman scattering (CSRS) and $P(\omega_{as})$ corresponds to coherent anti-Stokes Raman scattering (CARS). The other two contributions $P(\omega_2)$ and $P(\omega_1)$ oscillate at fundamental frequencies ω_2 and ω_1 and are responsible for stimulated

Raman gain and stimulated Raman loss, respectively. The four coherent Raman effects have the same magnitude of the nonlinear susceptibility given by the following relation:

$$\chi_{NL}(\Omega) = \frac{N}{6m\varepsilon_0}(\partial\alpha/\partial Q)_0^2 \frac{1}{\omega_v^2 - \Omega^2 - 2i\Omega\gamma} \tag{1.20}$$

The classical model above provides an intuitive understanding for the basis of the coherent Raman scattering processes but does not recognize the quantized nature of the nuclear vibrations. A semi-classical description takes the quantum mechanical response of the material into account while treating the interacting fields as classical. The system's response to a particular combination of optical frequencies $\{\omega_1, \omega_2, \omega_3\}$ is described by the third-order susceptibility $\chi^{(3)}$. The magnitude of $\chi^{(3)}$ is highly dependent on the polarization states of the fields participating in the nonlinear interaction. It thus takes the form of a fourth rank tensor, $\chi^{(3)}_{ijkl}$ $\{-\omega_4; \omega_1, \omega_2, \omega_3\}$ where the polarization states of the fields indicated by the indices $\{i, j, k, l\}$ correspond to $\{\omega_4, \omega_1, \omega_2, \omega_3\}$. The negative sign in front of the frequency indicates that the field is emitted while the positive sign denotes an absorbed field. In the case of isotropic biological materials in CRS microscopy, only 21 out of the 81 elements of this tensor are nonzero and may be inter-related.

In the case of the CARS process, a semi-classical approach yields for example, nonresonant and resonant contributions to the resulting nonlinear susceptibility.

$$\chi^{(3)}(-\omega_{as}; \omega_1, -\omega_2, \omega_1) = \chi^{(3)}_{NR} - \sum_b \frac{A_b}{(\omega_1 - \omega_2) - \omega_{ba} + i\gamma_{ba}} \tag{1.21}$$

All the vibrationally nonresonant terms, including terms with two-photon electronic resonances, are lumped into $\chi^{(3)}_{NR}$. The second term corresponds to the vibrationally resonant contribution $\chi^{(3)}_R$ in which A_b is the effective amplitude associated with the ω_{ba} vibrational resonance. A further quantum mechanical treatment of the interacting fields and matter correctly predicts the Raman cross sections of the coherent Raman processes.

1.4.3 CARS Signal Generation in the Plane Wave Approximation

The CRS signal generation can be described by approximating the incident pump and Stokes beams as monochromatic plane waves. The two beams propagate collinearly along the z-axis through an isotropic Raman-active slab medium of thickness L. The third beam is degenerate with the pump beam in this four-wave mixing coherent Raman scattering process. When the energy difference between the pump and Stokes beams matches the energy of a particular vibrational transition, the molecular nuclei are driven to vibrate in phase at the beat frequency between the pump and Stokes beams. The degenerate pump beam scatters off this vibrational coherence such that the CARS signal is generated at the anti-Stokes frequency of $\omega_{as} = \omega_1 - \omega_2$ and has a scalar wave vector $k_{as} = 2k_1 - k_2$.

This situation is represented by the energy level diagram shown in Fig. 1.3a. The CARS signal does not completely vanish when the beat frequency between the incident pump and Stokes beams is tuned off-resonance. This nonresonant contribution in CARS is due to the material nonlinear electronic response enabled via virtual states as illustrated in Fig. 1.3b.

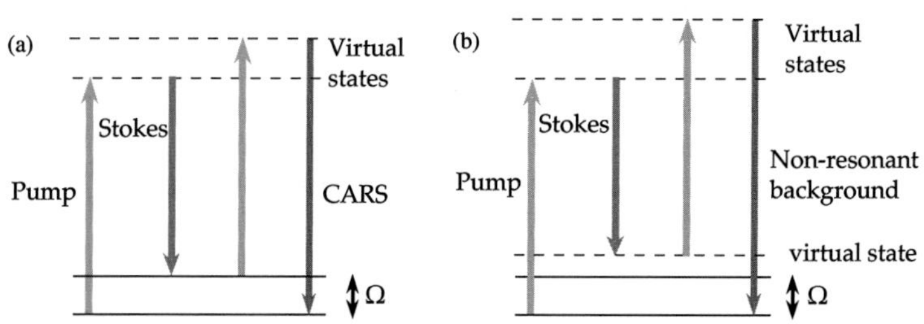

Figure 1.3 Energy Level Diagrams for CARS. (a) When the difference frequency between the pump and the Stokes beams matches the molecular vibration frequency, Ω, a resonant CARS signal is generated at the anti-Stokes frequency. (b) When the difference frequency between the pump and the Stokes beams is not resonant with any of the molecular vibration frequency in the material, a nonresonant background signal is still generated at the anti-Stokes frequency.

The intensity of the CARS signal is obtained by following a similar procedure used in Section 1.3.2. The wave equation 1.6

is solved using the appropriate expressions for the third-order nonlinear polarization and the CARS signal field $\tilde{E}_4(t)$ [7]. The intensity of the CARS signal is found as

$$I(\omega_{as}) \propto |\chi_e^{(3)}(\Omega)|^2 I_1^2 I_2 L^2 \mathrm{sinc}^2\left(\frac{\Delta kL}{2}\right). \tag{1.22}$$

Here I_1 and I_2 are the intensities of the pump and Stokes beams. The third-order nonlinear susceptibility is defined as $\chi_e^{(\Omega)} = \chi_e^{(3)}$ $(\omega_{as}; \omega_1 - \omega_2, \omega_1)$. $\Delta k = k_1 - k_2 + k_3 - k_4$ is the wave vector mismatch along the z-axis between the effective wave vector of the polarization field $(k_1 - k_2 + k_3)$ and the wave vector of the radiating field k_4. This phase matching condition arising from the rule of the conservation of momentum needs to be satisfied for the generated CARS signal to be significant. The first systematic study of CARS spectroscopy was carried out in 1965 by Maker and Terhune at Ford Motor Co. [19], while Duncan et al. demonstrated the first CARS microscope in 1982 [22]. In these earlier implementations of CARS, the CARS signal was detected in the phase matching direction using a noncollinear beam geometry. In 1999, Zumbusch et al. [23] demonstrated a practical implementation of CARS microscopy by using collinear laser beams that were tightly focused by means of a high numerical aperture (NA) objective. The large cone angles of the wave vectors of the excitation beams due to tight focusing relaxes the phase matching conditions in this case. Under the condition of tight focusing, the CARS signal has a large cone angle of k vectors in the forward direction. This is also the case in the backward (epi-) direction when any of the following three conditions are satisfied: (i) for an object with a size comparable to or smaller than the CARS wavelength, or (ii) for an interface between two media with different $\chi^{(3)}$ or for (iii) forward going CARS being scattered backward in thick specimens. This results from the constructive and destructive interference of CARS radiation from different parts of the sample.

Since the third order susceptibility for the CARS process contains both vibrationally resonant and nonresonant contributions as shown in Eq. 1.21, the following relation is obtained for the resulting intensity [7, 24]:

$$I_{CARS}(\Omega) \propto (|\chi_R^{(3)}(\Omega)|^2 + |\chi_{NR}^{(3)}|^2 + 2\chi_{NR}^{(3)}\,\mathrm{Re}[\chi_R^{(3)}(\Omega)])I_1^2 I_2 \tag{1.23}$$

As the last two terms in this equation suggest, there is a nonresonant background in the CARS intensity. This background poses serious problems for CARS microscopy. The last interference term distorts the CARS spectrum due to the constructive and destructive interference with the resonant vibrational contribution with increasing frequency. The CARS spectrum thus differs from the corresponding spontaneous Raman spectrum. Several methods have been developed to overcome the disadvantages of the CARS dispersive line shape. The detection sensitivity in CARS microscopy can be increased by suppressing the nonresonant background signals using techniques such as polarization-sensitive detection and frequency modulation CARS [5] or by implementing optical heterodyne detected CARS [25, 26].

As seen from Eq. 1.23, the CARS signal is proportional to the square amplitude of the nonlinear susceptibility $\chi^{(3)}$ which depends linearly on the density of Raman scattering molecules. Hence, the CARS signal varies quadratically as the number of molecules in the interaction volume. As a consequence, and in addition to the limit of detection set by the nonresonant background, low densities of scatterers such as those in the Raman “fingerprint region” of ~600–1700 cm^{-1} are not easily detected by CARS. On the other hand, CARS has been successfully applied to image high density of lipid molecules mainly in the high-wavenumber spectral region corresponding to 2800–3000 cm^{-1}.

1.4.4 SRS Microscopy

SRS microscopy probes the excited vibrational population directly instead of the vibrational coherence detected by CARS [4, 23]. Unlike CARS, which is a parametric generation process, SRS involves nonlinear dissipative optical processes in which molecules exchange energy with incident pump and probe laser fields. It consists of stimulated Raman Gain (SRG) and stimulated Raman loss (SRL) in which the energy difference between the pump and Stokes fields is transferred to the molecule for vibrational excitation. The energy level diagrams for SRL and SRG are given in Fig. 1.4.

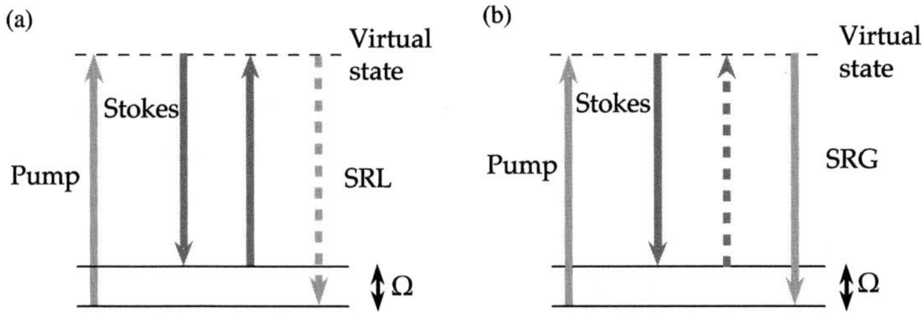

Figure 1.4 Energy level diagrams for SRS processes. (a) Stimulated Raman loss (SRL) is detected in the pump beam. (b) Stimulated Raman gain (SRG) is detected in the Stokes beam.

As seen in Eq. 1.19 earlier, the signal for the two different SRS processes is produced along with the CARS radiation at the anti-Stokes frequency. This occurs when the beat frequency of the pump and Stokes fields, $\Omega = \omega_1 - \omega_2$, matches the frequency ω_v of a Raman-active transition of the molecule. Two other third-order induced polarizations, $p(\omega_1)$ and $p(\omega_2)$ are generated at the fundamental pump and probe frequencies. In the plane wave approximation, the amplitude of the stimulated Raman wave at frequency ω_2 propagating forward through a slab of length L is given by

$$A_2^{stim}(L) \propto \chi_e^{(3)}(\Omega)\, A_1^* A_2 A_1 L \tag{1.24}$$

Since the Stokes field and the induced stimulated Raman signal have the same angular frequency, the two interfere at the detector. Hence, the net forward-detected SRG intensity at frequency ω_2 is

$$I(\omega_2) \propto |A_2(L) + A_2^{stim}(L)|^2 \tag{1.25}$$

The relative phase shift between the stimulated Raman field and the driving pump and Stokes fields determines the far field interference signal at the detector. The physical origin of this phase shift is understood by considering the radiating dipole model [7]. The net phase shift between the fields is the sum of the spectral phase shift and the geometric phase shift in the far field. The spectral phase shift between the induced dipole radiation and the driving pump field is $-\pi/2$ while it is $+\pi/2$

between the induced dipole radiation and the Stokes field. On the other hand, the geometric phase shift as a result of the mutual far-field interference of radiating dipoles is $-\pi/2$ in the far field. Hence, the stimulated Raman radiation is either 180° out of phase with the pump field or in phase with the Stokes field. Thus, the interference of the stimulated Raman field with the pump and Stokes fields produces a loss in the pump field intensity (SRL) and a gain in the Stokes field intensity (SRG) respectively. Within the semi-classical framework of the induced nonlinear polarization, the gain in intensity in the forward-detected Stokes field at frequency ω_2 is given by

$$\Delta I(\omega_2) \propto 2\mathrm{Im}\{\chi_e^{(3)}(\Omega)\}I_1I_2 \tag{1.26}$$

For the SRL, the stimulated Raman wave destructively interferes with the pump laser field ω_1 and results in an intensity loss as follows:

$$\Delta I(\omega_1) \propto -2\mathrm{Im}\{\chi_e^{(3)}(\Omega)\}I_1I_2 \tag{1.27}$$

The phenomenon of SRS was discovered by Woodbury and Ng [27] in 1962. Since then SRS spectroscopy has been widely applied for the highly sensitive detection of chemicals in the condensed and gas phases [28, 29]. The application of SRS as a method of contrast for imaging biological specimen was reported in 2008 [30–32]. The experimental setup for SRS microscopy typically involves the use of narrowband picosecond pulse pump and Stokes laser beams with (>1) MHz repetition rates. One of the excitation beams is modulated and the modulation transfer to the other beam is detected by means of phase sensitive detection. One of the challenges associated with ultrafast pulse excitation is the increased potential for photodamage. This is greatly reduced at longer wavelengths since the probability of two-photon absorption in the sample is significantly lower. Therefore, the SRL configuration is preferred in SRS microscopy since the higher excitation power is in the modulated Stokes beam at the longer wavelength [6]. The modulation transfer on to the filtered pump beam is detected utilizing a fast photodetector and a lock-in amplifier.

One of the notable features of SRS microscopy is that the signal strength, given by Eqs. 1.26 and 1.27, is linearly dependent on the nonlinear susceptibility $\chi^{(3)}$. Since this depends linearly on the density of Raman scattering molecules, the SRS signal varies linearly as the concentration of the analyte in the interaction volume analogous to spontaneous Raman scattering. This in addition to the absence of a nonresonant background in the SRS spectrum permits straightforward quantification and interpretation of the SRS image. Moreover, compared to spontaneous Raman microscopy, SRS gives orders-of-magnitude faster imaging speed due to the stimulated emission nature of the detected signal.

The phase-matching condition implies that the SRS signal can be detected in the forward direction in the pump or Stokes beams transmitted through the sample. However, in highly scattering tissue samples, a significant portion of this signal is back-scattered and can be collected using a large-area detector. Besides narrowband or single-frequency mode of SRS imaging, development of wavelength scanning hyperspectral SRS imaging and multiplex SRS imaging is an actively pursued area of current research in SRS microscopy [6].

1.5 Two-Photon Absorption

The two-photon absorption process is another third-order nonlinear light–matter interaction [4]. It involves a nearly simultaneous absorption of two laser photons to excite a molecule from the ground state to a higher energy electronic state. When the interacting molecule is a fluorophore, two-photon absorption is followed by subsequent fluorescence emission. This process is called two-photon excitation fluorescence (TPEF) [33]. Figure 1.5 illustrates the energy level diagrams for the single photon and two-photon excitation fluorescence processes Electronic ground and first excited states are denoted by thick horizontal black lines, whereas vibrational levels of the ground and first electronic excited states are denoted by thin horizontal black lines. Single photon excitation occurs through the absorption of a single laser photon. Two-photon excitation occurs through the absorption of two lower energy laser photons via short-lived intermediate states denoted by dashed horizontal line. After either excitation process, the fluorophore relaxes to the lowest energy level of the first excited electronic states via vibrational processes. Subsequent fluorescence emission processes are identical in both Fig. 1.5a,b.

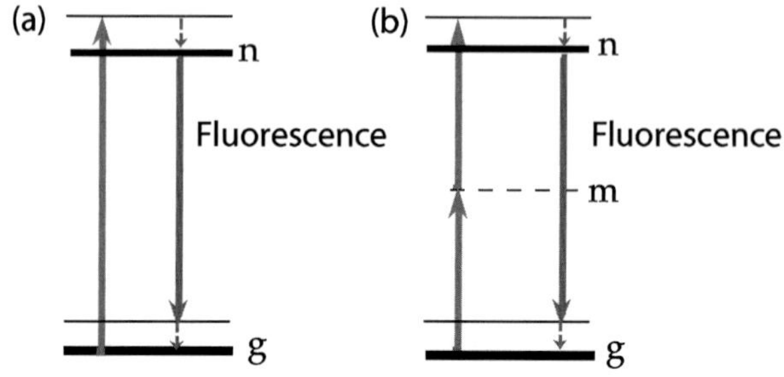

Figure 1.5 Energy level diagram for (a) single photon and (b) two-photon excitation fluorescence processes.

The transition rate for a two-photon absorption process, obtained from quantum mechanical calculations, is given by [4]

$$R_{ng}^{(2)} = \sigma_{ng}^{(2)}(\omega)I^2, \tag{1.28}$$

where $I = 2n\varepsilon_0 c\,|E|^2$ is the intensity of the light beam.

$\sigma_{ng}^{(2)}(\omega)$ is the two-photon absorption cross section and is given by

$$\sigma_{ng}^{(2)}(\omega) = \frac{1}{4n^2\varepsilon_0^2c^2}\left|\sum_m \frac{\mu_{nm}\mu_{mg}}{\hbar^2(\omega_{mg}-\omega)}\right|^2 2\pi\rho_f(\omega_{ng}=2\omega). \tag{1.29}$$

Here ρ_f is the density of final states and μ_{nm} and μ_{mg} are the electric dipole transition moments involved in the absorption of two photons from the ground state, *g*, to an excited state, *n*, via intermediate or virtual set of states, *m*.

The two-photon absorption cross section has units of cm^4 s, with 10^{-50} cm^4 s called a Göppert-Mayer or "GM" in honor of Maria Göppert-Mayer, who predicted the phenomenon of two-photon absorption in her doctoral dissertation in 1931 [34]. Since it is difficult to measure the two-photon absorption cross section directly, the two-photon "action" cross section is usually measured using TPEF. This is the product of the fluorescence quantum yield (φ_F) and the absolute two-photon absorption cross section [2]. It varies from ~10^{-4} GM for endogenous fluorophores such as NADH in biological tissue to ~50,000 GM for CdSe-ZnS quantum dots.

Two-photon absorption was first reported experimentally by Kaiser and Garrett in 1961 [35]. Denk et al. first demonstrated the potential of TPEF for laser scanning microscopy in 1990 [36]. Since this early work of the Webb group and Alfano et al. [37], TPEF imaging has rapidly evolved over the past two decades and has been extensively applied for deep biological imaging in various fields including neuroscience [38], embryology [39] and oncology [40]. It has been well-recognized, however, that tissue scattering places a fundamental limit on the imaging depth of the TPEF technique [41]. Three-photon fluorescence microscopy [42] has been shown to overcome this limitation by combining the advantages of long wavelength in the 1700 nm spectral excitation window and the higher-order nonlinear excitation [43].

1.6 Supercontinuum Generation

When a short intense light pulse propagates through a nonlinear optical medium, it often undergoes significant spectral broadening. This process has come to be known as supercontinuum generation (SCG). SCG is of interest both because of the rich physics that lies at its origin and because it has become a useful and versatile light source for use in studies in biophotonics.

1.6.1 Supercontinuum Generation in Bulk Media

The first dramatic example of SCG was reported in 1970 by Alfano and Shapiro [44], who observed broadening over much of the visible band when laser pulses of 4 ps duration and 530 nm wavelength were focused into samples of crystals and glasses. This process has come to be known as supercontinuum generation (SCG). In the work of Alfano and Shapiro, it was found that the SCG was usually accompanied by significant breakup of the laser beam into multiple filaments of the order of 20 microns in diameter. With the exception of SCG in optical fibers, it is believed that filamentation is usually required in order to obtain the high peak intensities required to produce SCG. In fact, although SCG is a very wide-spread phenomenon, the specific characteristics and underlying mechanism of SCG depend sensitively upon the conditions under which SCG is excited. Excellent reviews of the SCG process are available [45, 46], and can be consulted for treatments that go beyond that of the current treatment.

Although various physical processes can lead to SCG under different laboratory conditions, the most basic underlying process is that of self-phase modulation (SPM). This mechanism was first described theoretically by Shimizu [47]. Note also the more pedagogical treatment in the textbook of Boyd [4]. We assume that a quasi-monochromatic pulse of frequency ω_0 and intensity $I(t)$ acquires a nonlinear phase shift given by

$$\phi_{NL}(t) = n_2 I(t)\,\omega_0 L/c \tag{1.30}$$

in passing through a distance L of a material with nonlinear coefficient n_2. We can understand the origin of SPM as the smearing of the instantaneous frequency of the light from ω_0 to $\omega(t) = \omega_0 + \delta\omega(t)$, where

$$\delta\omega(t) = \frac{d}{dt}\,\phi_{NL}(t). \tag{1.31}$$

Let us assume for simplicity that the laser pulse has a Gaussian pulse shape, that the laser pulse duration (full width at half maximum) is T, and that the maximum nonlinear phase shift at the peak of the pulse is $\phi_{\mathrm{NL}}^{\max}$. We thus expect a spreading of the pulse spectrum by an amount of the order of $\delta\omega^{(\max)} = \phi_{\mathrm{NL}}^{\max/T}$. More formally, we can deduce the spectrum of the output pulse by means of the following equation:

$$S(\omega) = |\int_{-\infty}^{\infty} \tilde{A}(t) e^{-i\omega_0 t - i\phi_{NL} t} e^{i\omega t} dt|^2 \tag{1.32}$$

The nature of the resulting spectrum can be predicted by means of a simple mathematical construction described by Shen and Yang [45].

Recall that $\phi_{\mathrm{NL}}(t)$ has a shape that is Gaussian in time, and that the instantaneous frequency shift is $d\phi_{\mathrm{NL}}/dt$. Except at the points of inflection, there are two contributions to the integral that give the same frequency shift. These contributions will add either constructively or destructively depending on the relative phase of these contributions. The relative phase spans a total range of $\phi_{NL}^{\max}/$, and thus the total number of interference peaks is expected to be $\phi_{NL}^{\max}/2\pi$. In summary, we expect the spectrum of the light following SPM to possess a total width of $\delta\omega^{(max)} = \phi_{NL}^{\max}/T$ and to possess a total of $\phi_{NL}^{\max}/2\pi$ interference maxima. An example of such behavior is shown in Fig. 1.6.

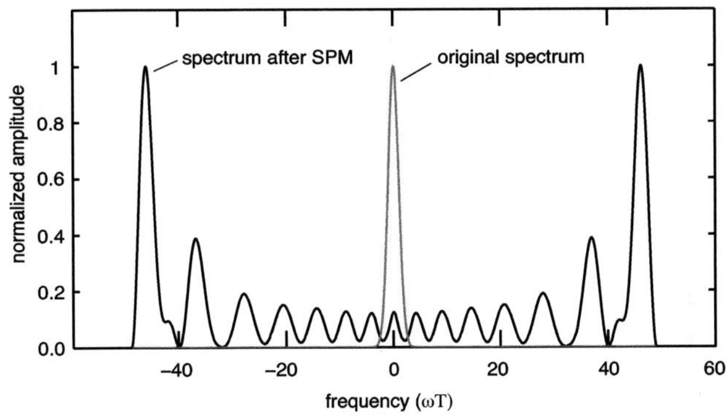

Figure 1.6 Modification of a laser pulse spectrum by the process of self-phase modulation (SPM). For this example, the peak intensity of the input pulse was adjusted to produce 30π radians of nonlinear phase shit.

As mentioned above, SCG is found to occur primarily when it is accompanied by the breakup of the laser beam into small-scale filaments. Nonetheless, Fork et al. [48] were able to observe extreme pulse broadening by means of the pure SPM process, without the accompanying formation of filaments. They made use of a jet of ethylene glycol sufficiently thin (500-μm-thick) that pulse reshaping could not occur within the medium. In addition, their excitation pulse (80 fs) was shorter than any pulse used up to that time, and consequently the incident laser intensity was large enough to excite SCG without the need for the increase of intensity by means of the self-focusing process. The emission spectrum extended from 0.19 to 1.6 μm. This work is also significant in that it demonstrates that SCG can occur as a consequence of the pure process of SPM.

The work of Fork et al. notwithstanding, the fact remains that SCG in bulk materials is routinely accompanied by the process of catastrophic self focusing. The explosive increase in beam pulse intensity associated with self focusing of course enhances the process of self-phase modulation, but it also gives rise to a variety of higher-order nonlinear optical processes. In fact, Yang and Shen [49], Rothenberg [50], Brabec and Krausz [51], and Gaeta [52] have all emphasized that the slowly varying amplitude approximation (SVAA) can break down under conditions of catastrophic self focusing. New effects described by these models that go beyond the SVAA include space-time focusing (different frequency components of the field diffract at different rates) and pulse steepening, which is a consequence of the intensity dependence of the group velocity. Gaeta [52] has emphasized that under conditions of catastrophic self focusing self steeping leads to the formation of an optical shock wave on the trailing edge of the optical pulse which thereby leads to the generation of an extremely broad output spectrum.

1.6.2 Supercontinuum Generation in Optical Fibers

Above we have described the process of SGC in bulk materials, and we have noted that the SCG process is typically profoundly modified by the process of self focusing. SCG can also be excited by propagation through optical fibers, and in this case the analysis is somewhat less complicated because there is no need to take account of the dynamics associated with the process of self focusing. Even here, however, there are several distinct situations whose properties tend to be quite different. We will need to distinguish between the case of conventional fibers and photonic crystal fibers, and also whether the excitation wavelength corresponds to normal or anomalous GVD.

We consider the case of supercontinuum generation in conventional optical fibers. By a conventional optical fiber, we here mean a fiber in which the fiber core has a larger refractive index than the cladding and where guiding thereby occurs by means of total internal reflection. One of the first studies along these lines was that of Lin and Stolen [53], who described a new broad-band continuum source pumped by a 20-kW, 10 ns visible-wavelength dye laser source, producing a spectral width of several thousand cm^{-1} and a total power of approximately 1 kW. However, entirely different physics can be observed if the generation process is excited by laser radiation with wavelengths greater than the "zero-gvd-wavelength" of silica glass. Under these conditions, which correspond to a wavelength of approximately 1.3 μm, the coefficient of group velocity dispersion (gvd) $\beta_2 = d^2k/\omega^2$ becomes negative, which permits the propagation of soliton pulses within the optical fiber. The properties of SCG are also significantly modified under excitation in the negative gvd regime. Several processes can occur to modify the spectrum of the generated radiation. One of these is the process of Raman-induced frequency shifts, which occurs as a result of intrapulse Raman scattering. In effect, the longer-wavelength components of the soliton become amplified at the expense of the shorter-wavelength components, leading to a migration of the spectrum to longer wavelengths. There is also the process of generation of dispersive waves. When the soliton condition is not perfectly met, wavelengths on the short-wavelength side of the carrier frequency tend to be not perfectly guided, leading to the generation of these short-wavelength components. Another competing process is that of soliton fission. The idea is as follows. The nonlinear Schrödinger equation, which describes the propagation of pulses within an optical fiber, possesses a family of solutions. In principle, these solutions are all stable, but the higher-order solutions are found to be highly susceptible to disruption from high-order nonlinear and dispersive effects. These higher-order solutions thus tend to decay into a sequence of low-order soliton solutions. One of these is intrapulse Raman scattering also known as Raman induced frequency shifts.

1.7 Conclusion

In this chapter, we have examined the basics of second-order and third-order nonlinear optical processes that are involved in biological imaging. We have considered the nonlinear light–matter interaction involved in second harmonic generation, coherent Raman scattering, two-photon absorption, and supercontinuum generation. We have discussed recent developments in nonlinear optical microscopy techniques based on these second-order and third-order nonlinear optical processes.

References

1. Boyd, R. W. (2008). Second- and higher-order harmonic generation, in *Handbook of Biological Nonlinear Optical Microscopy*, eds. Masters, B. R., and So, P. T. C., Chapter 6 (Oxford University Press, USA).
2. Zipfel, W. R., Williams, R. M., and Webb, W. W. (2003). Nonlinear magic: multiphoton microscopy in the biosciences, *Nat. Biotechnol.*, **21**(11), 1369–1377.
3. Hoover, E. E., and Squier, J. A. (2013). Advances in multiphoton microscopy technology, *Nat. Photon.*, **7**, 93–101.
4. Boyd, R. W. (2008). *Nonlinear Optics*, 3rd ed. (Academic Press, USA).
5. Evans, C. L. and Xie, X. S. (2008). Coherent anti-Stokes Raman scattering microscopy: chemical imaging for biology and medicine, *Annu. Rev. Anal. Chem.*, **1**(1), 883–909.
6. Zhang, D., Cheng, J. X., et al. (2014). Fast vibrational imaging of single cells and tissues by stimulated Raman scattering microscopy, *Acc. Chem. Res.*, **47** (8), 2282–2290.
7. Cheng, J. X., and Xie, X. S. (2010). *Coherent Raman Scattering Microscopy* (CRC Press, USA).
8. Yue. S., Slipchenko, M. N., and Cheng, J. X. (2011). Multimodal nonlinear optical microscopy, *Laser Photon. Rev.*, **5**(4), 496–512.
9. Franken, P. A., Hill, A. E., Peters, C. W., and Weinreich, G. (1961). Generation of optical harmonics, *Phys. Rev. Lett.*, **7**, 118–119.
10. Armstrong, J. A., Bloembergen, N., Ducuing, J., and Pershan, P. S. (1962). Interactions between light waves in a nonlinear dielectric, *Phys. Rev.*, **127**, 1918–1939.
11. Shen, Y. R. (1984). *The Principles of Nonlinear Optics* (Wiley, New York).
12. Freund, I., Deutsch, M., Sprecher, A. (1986). Connective tissue polarity, *Biophys J.*, **50**, 693–712.
13. Guo, Y., Savage, H. E., Liu, F., Schantz, S. P., Ho, P. P., and Alfano, R. R. (1999). Subsurface tumor progression investigated by noninvasive optical second harmonic, tomography, *Proc. Natl. Acad. Sci.*, **96**, 10854–10856.
14. Campagnola, P. J., Clark, H. A., Mohler, W. A., Lewis A, Loew, L. W. (2003). Second-harmonic imaging microscopy for visualizing biomolecular arrays in cells, tissues and organism, *Nature Biotech.*, **21**, 1356–1360.

15. Zipfel, W. R., Williams, R. M., Christie R., Nikitin A. Y., Hyman B. T., Webb W. W. (2003). Live tissue intrinsic emission microscopy using multiphoton-excited native fluorescence and second harmonic generation, *Proc. Natl. Acad. Sci.*, **100**, 7075–7080.

16. Nuriya, M., Jiang, J., Nemet, B., Eisenthal, K. B., Yuste, R. (2006). Imaging membrane potential in dendritic spines, *Proc. Natl. Acad. Sci. USA.*, **103**, 786–790.

17. Tilbury, K., and Campagnola, P. J. (2015). Applications of second-harmonic generation imaging microscopy in ovarian and breast cancer, *Perspect. Med.Chem.*, **7**, 21–32.

18. Campagnola, P. J., and Dong, C.-Y. (2011). *Laser Photon. Rev.*, **5**(1), 13–26.

19. Cicchi, R. et al. (2013). *J. Biophoton.*, **6**(2), 129–142.

20. Maker, P. D., and Terhune, R. W. (1965). Study of optical effects due to an induced polarization third order in the electric field strength, *Phys. Rev.*, **137**(3A), 801–818.

21. Begley, R. F., Harvey, A. B., Byer, R. L. (1974). Coherent anti-Stokes Raman spectroscopy, *Appl. Phys. Lett.*, **25**(7), 387–390.

22. Duncan, M. D., Reintjes, J., Manuccia, T. J. (1982). Scanning coherent anti-Stokes Raman microscope, *Opt. Lett.*, **7**, 350–352.

23. Zumbusch, A., Holtom G. R., Xie, X. S. (1999). Three-dimensional vibrational imaging by coherent anti-Stokes Raman scattering, *Phys. Rev. Lett.*, **82**, 4142–4145.

24. Min, W., Freudiger, C. W., Lu, S., and Xie, X. S. (2011). Coherent nonlinear optical imaging: beyond fluorescence microscopy, *Annu. Rev. Phys. Chem.*, **62**, 507–530.

25. Marks, D. L., and Boppart, S. A. (2004). *Nonlinear interferometric vibrational imaging, Phys. Rev. Lett.*, **92**, 123905.

26. Potma, E. O., Evans, C. L., and Xie, X. S. (2006). Heterodyne coherent anti-Stokes Raman scattering (CARS) imaging, *Opt. Lett.*, **31**, 241.

27. Woodbury, E. J., and Ng, W. K. (1962). Ruby laser operation in the near IR, *Proc. Inst. Radio Eng.*, **50**, 2347–2348.

28. Owyoung, A., and Jones, E. D. (1977). Stimulated Raman spectroscopy using low-power cw lasers, *Opt. Lett.*, **1**, 152.

29. Kukura, P., Mc Camant, D. W., and Mathies, R. A. (2007). Femtosecond stimulated Raman spectroscopy, *Annu. Rev. Phys. Chem.*, **58**, 461.

30. Nandakumar, P., Kovalev, A., Volkmer, A. (2009). Vibrational imaging based on stimulated Raman scattering microscopy, *New J. Phys.*, **11**, 033026–35.

31. Ozeki, Y., Dake, F., Kajiyama, S., Fukui, K., and Itoh, K. (2009). Analysis and experimental assessment of the sensitivity of stimulated Raman scattering microscopy, *Opt. Express*, **17**, 3651–3658.

32. Freudiger, C. W., Min, W., Saar, B. G., Lu, S., Holtom, G. R., He, C., Tsai, J. C., Kang, J. X., and Xie, X. S. (2008). Label-free biomedical imaging with high sensitivity by stimulated Raman scattering microscopy, *Science*, **322**, 1857.

33. So, P. T. C., Dong C. Y., Masters, B. R., and Berland, K. M. (2000). Two-photon excitation fluorescence microscopy, *Annu. Rev. Biomed. Eng.*, **2**, 399–429.

34. Göppert-Mayer, M. (1931). Uber elementarakte mit zwei quantensprüngen, *Ann. Phys.*, **9**, 273–294.

35. Kaiser, W., Garrett, C. G. B. (1961). Two-photon excitation in CaF_2: Eu^{2+} *Phys. Rev. Lett.*, **7**, 229.

36. Denk, W., Strickler, J. H., and Webb, W. W. (1990). Two-photon laser scanning fluorescence microscopy, *Science*, **248**, 73–76.

37. Alfano, R. R., Demos, S. G., Galland, P., Gayen, S. K., Guo, Y., Ho, P. P., Liang, X., Liu, F., Wang, L., Wang, Q. Z., and Wang, W. B. (1998). Time resolved and NLO imaging for medical applications, *Adv. Optical Biopsy, Ann. N. Y. Acad. Sci. N. Y.*, **838**, 14.

38. Kerr, J. N. D., and Denk, W. (2008). Imaging in vivo: watching the brain in action, *Nat. Rev. Neurosci.*, **9**, 195–205.

39. Olivier, N., et al. (2010). Cell lineage reconstruction of early zebrafish embryos using label-free nonlinear microscopy, *Science*, **329**, 967–971.

40. Williams, R. M. et al. (2010). Strategies for high-resolution imaging of epithelial ovarian cancer by laparoscopic nonlinear microscopy, *Trans. Oncol.*, **3**, 181–194.

41. Kobat, D., Horton, N. G., and Xu, C. (2011). In vivo two-photon microscopy to 1.6-mm depth in mouse cortex, *J. Biomed. Opt.*, **16**, 106014.

42. Hell, S. W., Bahlmann, K., Schrader, M., Soini, A., Malak, H. M., Gryczynski, I., and Lakowicz, J. R. (1996). Three-photon excitation in fluorescence microscopy, *J. Biomed. Opt.* **1**, 71–74.

43. Horton, N. G., Wang, K., Kobat, D., Clark, C. G., Wise, F. W., Schaffer, C. B., and Xu, C. (2013). In vivo three-photon microscopy of subcortical structures within an intact mouse brain, *Nat. Photon.*, **7**, 205–209.

44. Alfano, R. R., and Shapiro, S. L. (1970). Observation of self-phase modulation and small-scale filaments in crystals and gases, *Phys. Rev. Lett.*, **24**, 591.
45. Shen, Y. R. and Yang, G. (2006). In *The Supercontinuum Laser Source*, ed. Alfano, R. R., Chapter 2 (Springer, NY).
46. Dudley, J. M., Genty, G., and Coen, S. (2006). Supercontinuum generation in photonic crystal fiber, *Rev. Mod. Phys.*, **78**, 1135; see especially section II for a historical literature review.
47. Shimizu, F. (1967). Frequency broadening in liquids by a short light pulse, *Phys. Rev. Lett.*, **6**, 1097.
48. Fork, R. L., Shank, C. V., Hirlimann, C., Yen, R., and Tomlinson, W. J. (1983). Femtosecond white-light continuum pulses, *Optics Lett.*, **8**, 1.
49. Yang, G., and Shen, Y. R. (1984). Spectral broadening of ultrashort pulses in a nonlinear medium, *Opt. Lett.*, **9**, 510.
50. J. E. Rothenberg (1992). Space-time focusing: breakdown of the slowly varying envelope approximation in the self-focusing of femto-second pulses, *Opt. Lett.*, **17**, 1340.
51. Brabec, T., and Krausz, F. (1997). Nonlinear optical pulse propagation in the single-cycle regime, *Phys. Rev. Lett.*, **78**, 3282.
52. Gaeta, A. L. (2000). Catastrophic collapse of ultrashort pulses, *Phys. Rev. Lett.*, **84**, 3582.
53. Lin, C., and Stolen, R. H. (1976). New nanosecond continuum for excited-state spectroscopy, *Appl. Phys. Lett.*, **28**, 216.

Chapter 2

Complex Light Beams

Enrique J. Galvez

Department of Physics and Astronomy, Colgate University,
13 Oak Drive, Hamilton, New York, 13346, USA
egalvez@colgate.edu

2.1 Introduction

Optical beams derived from lasers have been useful in many applications for a long time. The invention of lasers has led to our ability to control very precisely many parameters of the light, such as wavelength and time. In many cases the spatial mode and polarization of the light has remained constant. However, most recently, the technology has evolved to allow us greater control over the light, enabling us to specify the phase, amplitude, and polarization within a light beam. Complex light refers broadly to beams of light where one or more parameters are changed within the beam of light. In particular, spatial modes have received much attention, generated either directly from lasers or by manipulation of the wavefront via diffractive optical elements. These studies have led us to Laguerre–Gauss beams, Bessel beams and Airy beams. The control of their depth of focus and width has found application in high-resolution confocal and multiphoton microscopes at near or below the diffraction limit.

Deep Imaging in Tissue and Biomedical Materials: Using Linear and Nonlinear Optical Methods
Edited by Lingyan Shi and Robert R. Alfano

ISBN 978-981-4745-88-8 (Hardcover), 978-1-315-20655-4 (eBook)
www.panstanford.com

In this chapter, I delve into complex light by treating the most important beams of light that are useful for research in many areas of science and technology. It includes a description of the functional forms of the modes and their parameters, which lead to better resolution and deeper penetration. I will first cover the fundamentals of Gaussian beams, which are naturally generated by a laser, and which constitute the workhorse of laser-based research on multiphoton and confocal microscopes, and in manipulation of matter via optical forces. I will then continue with interesting optical realizations that carry a new range of physical phenomena, such as optical vortices and orbital angular momentum in the case of Laguerre–Gauss beams; or diffraction-less propagation and self-reconstructing beams in the case with Bessel and Airy beams; or space-variant polarization and singularities in polarization in the case of vector and Poincaré beams. The different sections also discuss and cite the various methods for producing these beams. It is hoped that they will stimulate new research in imaging, manipulation, and new areas where complex light has yet to make an impact, leading to the discovery of new phenomena and applications.

2.2 Gaussian Beams

In this section, we discuss optical beams that are described by a Gaussian intensity profile. This is the case of laser beams. Thus it is important to begin with a discussion of the fundamental Gaussian beam and its properties. We then follow with sections describing important high-order Gaussian beams, which are also beams that can be derived from the fundamental one.

2.2.1 Fundamental Gaussian Beams

We begin by deriving the equation of the fundamental Gaussian beam. This treatment is also covered in standard textbooks on lasers [40, 45]. In classical electromagnetism, light is represented by an electromagnetic wave, with electric and magnetic fields that obey Maxwell's equations. Since the electric and magnetic fields are related to each other, it is usual to express light in terms of one of them: the electric field

$$\vec{E} = E_0 \Psi \hat{e}, \tag{2.1}$$

where E_0 is the scalar magnitude of the field of the beam, Ψ is a normalized wave function, and $\hat{e}$ is a unit vector. The scalar part of the field must satisfy the wave equation

$$\nabla^2\Psi - \frac{1}{c^2}\frac{\partial^2\Psi}{\partial t^2} = 0, \tag{2.2}$$

where

$$\nabla^2 = \frac{\partial}{\partial x^2} + \frac{\partial}{\partial y^2} + \frac{\partial}{\partial z^2}$$

is the Laplacian and c is the speed of the light. For a harmonic wave, we can introduce a trial solution that separates out the time dependence:

$$\Psi(x,y,x,t) = U(x,y,z)e^{-iwt} \tag{2.3}$$

where U depends only on the spatial coordinates, and w is the angular frequency of the light. Then by replacing Eq. 2.3 into Eq. 2.2, we arrive at the Helmholtz equation:

$$\nabla^2 U + k^2 U = 0 \tag{2.4}$$

where $k = w/c$ is the wave number. We will be restricting ourselves to optical beams that are likely produced by lasers. These are collimated beams that have a finite transverse extent, and travel mainly in one dimension, say along the positive z axis of a Cartesian coordinate system. Thus we can require a solution to be similar to a plane wave, with a term $\exp(ikz)$. Therefore, we propose that the solution has to be of the form

$$U(x,y,z) = U_0(x,y,z)e^{ikz} \tag{2.5}$$

Replacing Eq. 2.5 into Eq. 2.4 and simplifying we get

$$\frac{\partial^2 U_0}{\partial x^2} + \frac{\partial^2 U_0}{\partial y^2} + \frac{\partial^2 U_0}{\partial z^2} + 2ik\frac{\partial U_0}{\partial z} = 0. \tag{2.6}$$

A beam is a wave of finite transverse extent, so diffraction causes the beam to spread as the light travels. To account for this diffraction we allow U_0 to depend on z. We can picture this the following way: the intensity of the light for fixed values of the transverse coordinates, x and y, is going to decrease as z increases. Thus we can say that U_0 changes slowly with z, and thus neglect the term $\partial^2 U_0/\partial z^2$ compared to the other ones and remove it from Eq. 2.6. The resulting equation is

$$\frac{\partial^2 U_0}{\partial x^2}+\frac{\partial^2 U_0}{\partial y^2}+2ik\frac{\partial U_0}{\partial z}=0. \tag{2.7}$$

It is called the paraxial wave equation.

We proceed to find U_0 by requiring that it have the form

$$U_0(x,y,z)=Ae^{\frac{ik(x^2+y^2)}{2q(z)}}e^{ip(z)}. \tag{2.8}$$

Replacing Eq. 2.8 into Eq. 2.7 we get separate equations for $q(z)$ and $p(z)$. The solution for $q(z)$ is

$$q(z)=z-iz_R, \tag{2.9}$$

where z_R is a constant known as the Rayleigh range. A more appropriate way to write this equation is

$$\frac{1}{q(z)}=\frac{1}{z+\frac{z_R^2}{z}}+\frac{i}{\frac{z^2}{z^R}+zR}. \tag{2.10}$$

The solution for $p(z)$ is

$$e^{ip(z)}=\frac{w_0}{w}e^{-i\varphi(z)}, \tag{2.11}$$

Inserting Eqs. 2.10 and 2.11 into Eq. 2.8 yields

$$U_0(x,y,z)=A\frac{w_0}{w}e^{-\frac{x^2+y^2}{w^2}}e^{\frac{ik(x^2+y^2)}{2R}}e^{-i\varphi}, \tag{2.12}$$

where w is the beam spot or half width of the beam, which depends on z:

$$w=w_0\sqrt{1+\frac{z^2}{z_R^2}}, \tag{2.13}$$

and where the constant w_0 is called the beam waist. The Rayleigh range z_R is related to the waist by:

$$z_R = \frac{\pi w_0^2}{\lambda} \tag{2.14}$$

The above relation is the traditional and useful way to relate z_R and w_0, but a more intuitive relation is

$$\frac{z_R}{w_0} = \pi \frac{w_0}{\lambda}, \tag{2.15}$$

relating the ratio of z_R and w_0 to the ratio of w_0 and λ. A new term that appears in the solution, *R*, is the radius of curvature of the wavefront, which also depends on *z*:

$$R = z + \frac{z_R^2}{z} \tag{2.16}$$

Equation 2.12 also has a new phase φ, given by

$$\varphi = \tan^{-1}(z/z_R), \tag{2.17}$$

and known as the Gouy phase.

Rewriting the solution to the wave function, we get

$$\Psi(x,y,z,t) = A\frac{w_0}{w} e^{-\frac{x^2+y^2}{w^2}} e^{i(kz-wt)} e^{\frac{ik(x^2+y^2)}{2R}} e^{-i\varphi} \tag{2.18}$$

The first three terms specify the amplitude of the wave, and the next three terms contain the phase of the wave embedded in the exponential terms. It is important to understand the meaning of these terms, as they describe the properties of the beam. Below we analyze the most important features of them in detail.

2.2.1.1 The beam spot *w*

The function *w* of Eq. 2.13 represents the half width of the beam, also known as the beam spot. Figure 2.1 shows a graph of *w* as a function of *z*. When $z = 0$ the beam spot is w_0, also known as the waist, and is the smallest value that *w* can have under this description. In the positive and negative *z* directions *w* increases. Thus, the point $z = 0$ is either some type of starting point. Indeed, the central point of most lasers is a waist. Once the light comes out of a laser, its width w increases due to diffraction, which is accounted by the solution. Figure 2.1 also describes the width of a beam when it is focused by a lens. Below we will discuss how to control the size of the waist when we focus a beam of light. When $z = z_R$ the beam spot is $w = \sqrt{2}\,w_0$. This point is an important parameter in the propagation of the light: it marks the transition from the beam spot being nearly constant to one that increases linearly with z.

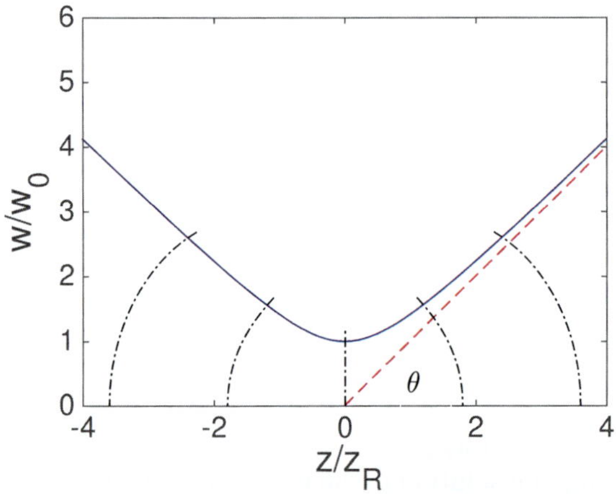

Figure 2.1 Graph of the beam spot of the beam (solid blue) and wavefronts (black, dash-dot). The red dashed line corresponds to the graph of $w = \theta z$ (see text).

When we focus a beam of light with a converging lens, and think of the light as a set of parallel rays, we imagine those lines to cross exactly at the focal point. However, when $|z| < z_R$ the ray description of the propagation of the light breaks down, and the wave aspect sets in. When $z >> z_R$ the waist becomes linearly dependent on the distance from the waist: $w \simeq (w_0/z_R)z$.

This asymptote is shown by the red dashed line in Fig. 2.1. Because in this regime the beam spot expands nearly linearly, we can express it as

$$w = \theta z, \tag{2.19}$$

where θ is called the divergence angle of the beam, which, using Eq. 2.14, can be written as

$$\theta = \frac{\lambda}{\pi w_0} = \sqrt{\frac{\lambda}{\pi z_R}}. \tag{2.20}$$

The divergence angle is also associated with the numerical aperture of the beam, similar to the concept used in imaging. This treatment is valid as long as the beam is paraxial, and justifies the approximations made. This analysis is for divergence angles below about 0.5 rad, or about 30°. More intuitively, this condition means that the beam width has to be much larger than the wavelength.

We can also see that the divergence of the beam is consistent with diffraction. For example, the divergence of the first diffraction minimum from a circular aperture of radius a when it is illuminated by a plane wave is given by

$$\theta = \frac{1.22\lambda}{2a}.$$

In our case we could visualize the waist of our optical beam as a fuzzy Gaussian aperture of radius w_0.

In "ordinary" situations we have that the following relation holds

$$\lambda \ll w_0 \ll z_R. \tag{2.21}$$

Let's do a numerical example: if we have a HeNe laser beam (λ = 632.8 nm) that is focused to a spot w_0 = 0.5 mm, the Raleigh range comes out to be $z_R = \pi w_0^2/\lambda$ = 1.2 m. In examining Eq. 2.15, we see that as w_0 decreases, z_R decreases more rapidly, so that as w_0 approaches λ, z_R approaches w_0. Efforts to produce three-dimensional imaging with confocal microscopes relies on creating the conditions where both w_0 and z_R are small, and filtering the light with apertures.

2.2.1.2 Beam intensity

The explicit expression for the amplitude is

$$U_0 = \sqrt{\frac{2}{\pi}} \frac{1}{w} e^{-r^2/w^2}, \tag{2.22}$$

which in the transverse plane has the typical Gaussian profile, as shown in Fig. 2.2.

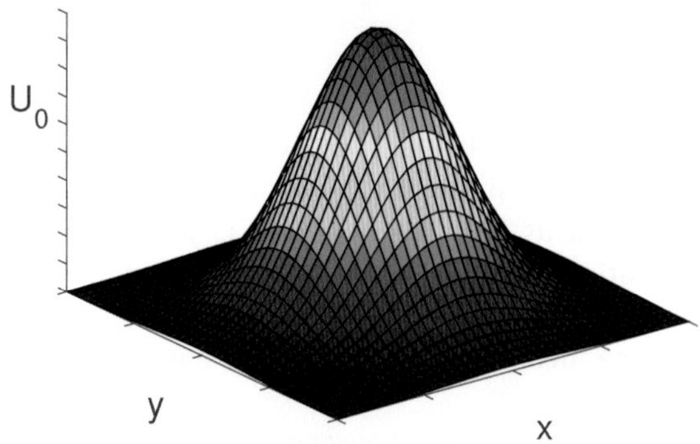

Figure 2.2 A 3-D view of the magnitude of the amplitude of the fundamental Gaussian beam as a function of the transverse coordinates.

The irradiance of the beam is defined as the energy per unit area per unit time delivered by the beam on a surface. It is given by

$$I = \frac{c\varepsilon_0}{2}(E_0 U_0)^2, \tag{2.23}$$

If we integrate the irradiance, we get the total power of the beam:

$$P = \int_0^\infty I 2\pi r dr = \frac{c\varepsilon_0}{2} E_0^2, \tag{2.24}$$

It is usual to know the total power of a beam because it can be measured easily with a power meter. By combining the previous equations, we can express the irradiance of the beam in terms of the total power of the beam:

$$I = \frac{2P}{\pi w^2} e^{-2r^2/w^2}. \tag{2.25}$$

Should we send the beam through a circular aperture of radius a, the transmitted power will be

$$P_a = P(1 - e^{-2a^2/w^2}). \tag{2.26}$$

If we set $a = w$, we will get 86% of the light through, and if we set $a = \pi w$, we will get 99% of the light through.

At $r = 0$ the dependence of the amplitude with z is given by $1/w$. From the analysis of w in the previous section we gather that for $z < z_R$ the intensity is nearly constant, but for $z >> z_R$, since $w \propto z$, the intensity decreases as $1/z^2$.

2.1.1.3 Wavefront

The wavefront is normally referred to the surface that contains the points of equal phase. The first phase term of Eq. 2.18, $\exp[i(kz - \omega t)]$, represents the phase of a plane wavefront perpendicular to z. All alone, it would represent a plane wave. The next phase term, $\exp(ik(x^2 + y^2)/2R)$ represents a correction to the planar phase front, reflecting the expansion of the beam. It has the same variation as that of a spherical surface of radius R. However, per Eq. 2.16, R depends on z: at $z = 0$ the radius of curvature is $R = \infty$ (i.e., a vertical plane, as shown in Fig. 2.1). When $z >> z_R$ then the radius of curvature is $R(z) \simeq z$, and the wavefront is a spherical surface traveling away from $z = 0$. A special point is $z = z_R$, where the radius of curvature is $R = 2z$. This distance is also known as the confocal parameter. It is important for many problems in laser-beam optics such as in designing laser resonators. In these cases the laser medium is put in between two spherical mirrors of radius of curvature equal to their separation.

2.1.1.4 Gouy phase

The last phase term in Eq. 2.18 is almost always overlooked or not even mentioned in basic treatments in optics, but it is important in understanding complex light beams. It is the Gouy phase, given by Eq. 2.17. This phase arises due to diffraction. The warping of the wavefront from $z = 0$ to $z >> z_R$ comes at a cost: an additional phase is accrued by the light wave. The accumulated Gouy phase can be as high as π from end to end when focusing a beam (and higher for high-order beams, as discussed below). However, it is an overall phase, so if the beam is not interfering with another beam coherent with it, this phase has no effect.

Figure 2.3 shows a graph of the Gouy phase, which depends on the propagation coordinate. This phase changes rapidly between $z = -z_R$ and $z = +z_R$, and reaches asymptotes $\pm\pi/2$ when $|z| >> z_R$.

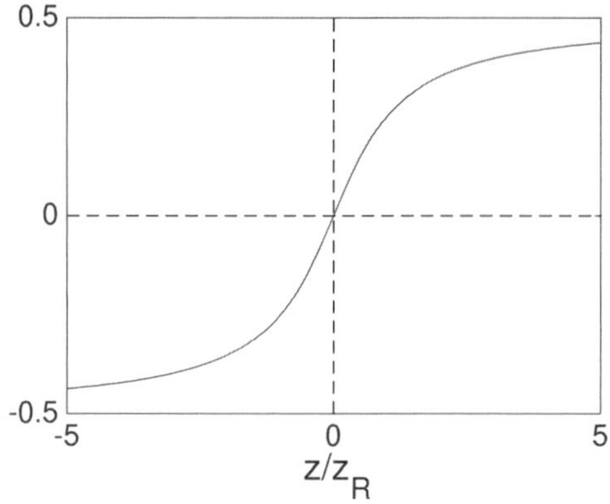

Figure 2.3 Graph of the Gouy phase of Eq. 2.17 in units of π.

2.1.1.5 Momentum

A photon of light has an energy given by

$$E = h\nu, \tag{2.27}$$

where h is Planck constant and ν is the frequency of the light. A beam of light with a wavelength of 500 nm consists of photons, each with an energy of 4×10^{-19} J. Thus the number of photons traveling in a beam of power P is

$$N = P/E, \tag{2.28}$$

which for 500 nm light translates into 2.5×10^{15} per second. At the same time, a photon also carries a linear momentum of

$$p = E/c, \tag{2.29}$$

where c is the speed of light. The momentum for a photon of wavelength 500 nm is 1.3×10^{-27} kg m s^{-1}, which is quite small to make a visible effect on a macroscopic object. However, on a microscopic object light can exert a significant force. A particle that absorbs all of the light incident upon it will experience a force

$$F = Np, \tag{2.30}$$

which is 3.3 pN for the photon of our example. This force is greater than the weight of a 5 μm plastic sphere: 0.67 pN.

The realization of these magnitudes and the widespread use of lasers has led to the development of optical tweezers, which can trap and move micron-sized objects in a small sample, such as a microscope slide. Moreover, when light is focused on transparent objects they refract the light, changing its direction, and thus receiving a recoil force. This situation is so fortunate that a transparent object immersed in a liquid of distinct index of refraction can be trapped at the waist of a focused beam of light [4, 41]. This is shown schematically in Fig. 2.4: two rays aimed above the center of a transparent sphere refract. In doing so, the initial momentum of the rays, p_i, leaves the sphere with a final momentum p_f. The change in the momentum of the light, Δp, results in a recoil momentum of the sphere p_{recoil} that pushes the sphere toward the focus of the light. This is the principle of optical tweezers [33]. The previous argument is the one that is most appropriate for objects of size greater than about a micrometer. For smaller-size spheres, a better approach is one that treats the sphere as a radiating dipole, and its interaction with the light leads to a potential energy gradient that pulls objects to the focus of the light [49].

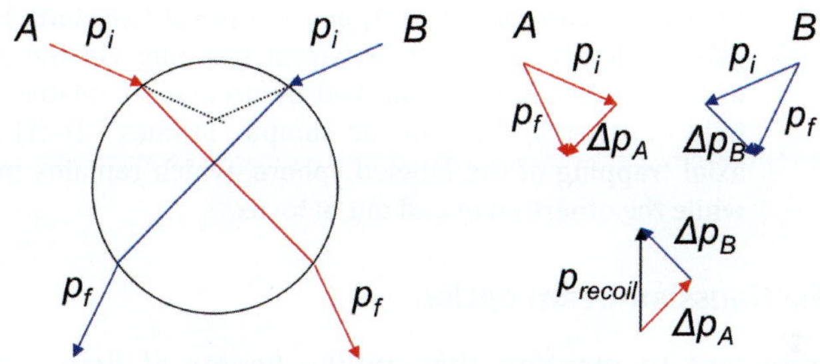

Figure 2.4 Diagram of light rays, refracted by a sphere showing how the change in momentum of the light translates into a recoil of the sphere toward the focus of the rays.

Figure 2.5 shows images of a sphere trapped in an optical tweezer setup. Frames (a), (b), and (c) show the motion of the labeled sphere through a sample with other spheres. The sideways motion can go as fast the trap can overcome the viscous drag force [47].

$$F_{drag} = 6\pi\eta a v, \tag{2.31}$$

where a is the radius of the sphere (2.5 μm), v the velocity of the sphere (5 μm/s) and η the viscosity (of water, 0.001 Pl), which results in a magnitude of 0.24 pN. The three-dimensional aspect of the trap is observed in the frames (d), (e), and (f), where the sample was moved in and out of focus while the trapped sphere remained in focus. Gaussian beams have had a huge impact in biomedical research via the manipulation of subjects under the microscope, in the device known as optical tweezers [32, 41].

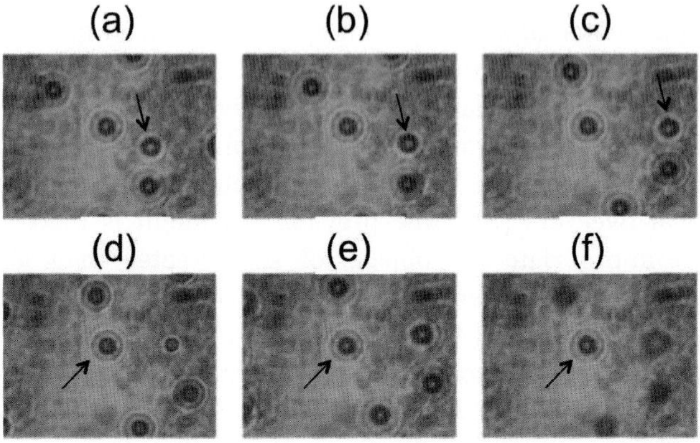

Figure 2.5 Images of latex spheres trapped in optical tweezers. Frames (a)–(c) shows the effect of lateral trapping via the motion of the trapped sphere (marked by an arrow) relative to the field of other spheres in the sample. Frames (d)–(f) shows axial trapping of the labeled sphere, which remains in focus while the others go in and out of focus.

2.1.1.6 Gaussian-beam optics

It is important to mention that, unlike beams of light carrying images, Gaussian beams do not follow the simple rules of geometric optics such as the thin lens formulas. This is because Gaussian beams straddle a regime between a ray and wave. The correct treatment of propagation of Gaussian beams is given by the ABCD theory of Gaussian-beam propagation [40]. Because $q(z)$ and $p(z)$, of Eqs. 2.9 and 2.11, respectively, define the wave function of the light, we must follow the changes that those parameters incur during propagation. In this regard, it is instructive to realize that Eq. 2.10 can be rewritten as

$$\frac{1}{q} = \frac{1}{R} + i\frac{\lambda}{\pi w} \tag{2.32}$$

which relates the real and imaginary parts of q to R and w. Thus to find the values of R and w along an optical system we only need to know how to find q as a function of z.

For propagation through free space between points z_1 and z_2, the transformation for q is given by

$$q_1 - z_1 = q_2 - z_2. \tag{2.33}$$

Thus, combining Eqs. 2.32 and 2.33 can lead to a method of obtaining R and w at all points starting from a given set of initial points. The effect of a lens on a Gaussian beam results in a change in the radius of curvature of the wavefront:

$$\frac{1}{R_2} = \frac{1}{R_1} - \frac{1}{f}, \tag{2.34}$$

where R_1 and R_2 are the radii of curvature of the wavefront before and after the lens, respectively, and f is the focal length of the lens. The sign convention for R is: positive when the center curvature is to the left, and negative when it is to the right. This is shown in Fig. 2.6 for a particular case where $R_1 > 0$ and $R_2 < 0$. It is easy to see that if the incoming beam on a converging lens of focal length f has a large radius of curvature (say $R_1 = \infty$), then the radius of curvature of the light after the lens is $R_2 = -f$. This seems to imply that the waist past the lens will be a distance R_2 from the lens. However, this is only true if the Rayleigh range after the lens is $z_{R,2} << R_2$. Otherwise, we need to propagate q to the waist (i.e., where R is infinite and w is minimum). If we combine propagation with the effect of the lens, we find that the location of the waist is given by [42]

$$z_2 = \frac{R_2}{1 + [R_2 \lambda / (\pi w_2^2)]^2} \tag{2.35}$$

where w_2 is the beam spot of the beam just after the lens, which is the same as the beam spot just before the lens, w_1. The new waist is given by

$$w_{02} = \frac{w_2}{\{1 + [\pi w_2^2 / R_2 \lambda)]^2\}^{1/2}}, \tag{2.36}$$

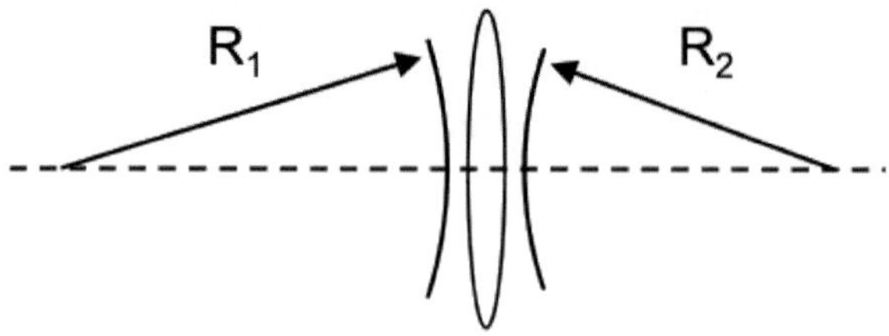

Figure 2.6 Change in the radius of curvature of the wavefront by passage through a lens.

Notice that when

$$R_2 \ll \pi w_2^2/\lambda. \tag{2.37}$$

we get

$$z_2 \simeq R_2 \tag{2.38}$$

and

$$w_{02} \simeq \frac{\lambda R_2}{\pi w_2}. \tag{2.39}$$

Note that Eq. 2.37 is readily satisfied in situations where we want to focus a beam with a short focal length lens. For example, a visible laser beam with a width of 1 mm and a visible wavelength of 632 nm, results in $\pi w_2^2/\lambda \sim 5$ m. If $R_1 >> f$, then we only need $f \sim R_2$ to be less than a meter or so. In many applications, we need to get the smallest waist possible. From Eq. 2.39 we get that what we need to do is to have the largest value of the input width w_2 and the smallest value for R_2. If the incoming beam has $R_1 >> f$, then $R_2 \simeq f$, which implies that we need a lens with the shortest focal length possible. For other cases we need to apply the full formalism.

2.2.2 Hermite–Gaussian Beams

In the previous section we found a solution to the paraxial wave equation by introducing a solution with an initial mathematical form. However, the final solution was "rigged" from the start: it was forced to be symmetric about the transverse plane. A more general solution would allow solutions with distinct dependencies on x and y coordinates. Thus, a more general trial solution is

$$U_0(x,y,z) = Ag(x,z)h(y,z)e^{\frac{ik(x^2+y^2)}{2q(z)}}e^{ip(z)}. \tag{2.40}$$

The solution of the paraxial wave equation yields independent functions g and h that are Hermite polynomials [45]:

$$U_{m,n}(x,y,z) = \frac{A}{w}\mathrm{H}_m\left(\frac{\sqrt{2}x}{w}\right)\mathrm{H}_n\left(\frac{\sqrt{2}y}{w}\right)e^{-\frac{x^2+y^2}{w^2}}e^{\frac{ik(x^2+y^2)}{2R}}e^{-i\varphi(z)}. \tag{2.41}$$

The constant term in Eq. 2.41 that normalizes the square of the amplitude is given by

$$A = \left(\frac{2^{1-N}}{\pi n!m!}\right)^{1/2}, \tag{2.42}$$

and H_m and H_n are Hermite polynomials with (positive) integer indices m and n. Hermite–Gauss modes are often denoted as $\mathrm{HG}_{m,n}$. The order of the mode is given by

$$N = n + m. \tag{2.43}$$

It is illustrative to see the expressions for the lowest-order Hermite polynomials $\mathrm{H}_m(\upsilon)$:

$$\mathrm{H}_0(\upsilon) = 1 \tag{2.44}$$

$$\mathrm{H}_1(\upsilon) = 2\upsilon \tag{2.45}$$

$$\mathrm{H}_2(\upsilon) = 4\upsilon^2 - 2 \tag{2.46}$$

$$\mathrm{H}_3(\upsilon) = 8x^3 - 12x \tag{2.47}$$

The product of the Hermite polynomials and the Gaussian term, in Eq. 2.41, result in the beam profiles shown in Fig. 2.7. Note that the zero-order ($N = 0$) is the fundamental Gaussian mode. The indices m and n are also the number of zeros of the mode in the x and y coordinates, respectively. Thus, one can easily understand and predict the general shape of the mode from the values of m and n. One can also see that the size of the mode increases with the order. In laser resonators, modes of the same order are degenerate. Often laser resonators have an aperture within the laser cavity to eliminate these (unwanted) high-order modes and lase only in the fundamental mode.

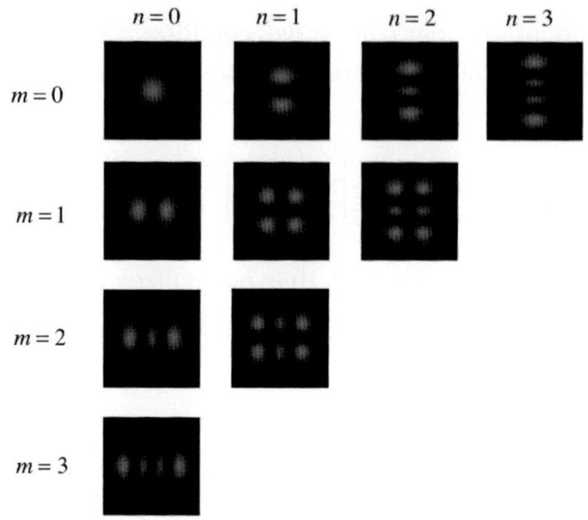

Figure 2.7 Graph of the irradiance of the lowest-order Hermite–Gauss modes.

The phase terms of the mode are independent of the transverse coordinates. They are the same as those in Eq. 2.18 except for the Gouy phase, which in Eq. 2.41 has a new factor [9]:

$$\varphi(z) = (N + 1)\tan^{-1}(z/z_R). \tag{2.48}$$

If a beam is a superposition of modes, then propagation introduces a relative Gouy phase between the modes. Such a phase will affect the shape of the composite mode as the light propagates, especially if it is focused by lenses. Not all the points in the mode have the same phase, because the Hermite functions have positive and negative values. For example, inspection of the HG_{10} mode shows a node at $x = 0$, which correlates with H_1 in Eq. 2.45. Because H_1 has different sign for $x > 0$ and $x < 0$, the two lobes of HG_{10} are 180° out of phase. In fact, this is true with all adjacent lobes in Fig. 2.7. Figure 2.8 shows the amplitudes for two HG modes, illustrating the phase difference between adjacent lobes.

Hermite–Gaussian modes are best produced directly from an open-cavity laser by inserting thin crossed wires within the cavity. The laser modes that are generated will have nodes along the wires because scattering losses will discourage all other modes. This is easily done in any type of open-frame laser, such as gas lasers. The advantage of the method is that it is very efficient. The disadvantage is that higher-order modes do not lase with as high gain as low-order modes, so as the order of the generated mode goes up, the laser power in that mode goes down. Laser cavities with asymmetric frames or astigmatic gain regions can also be used to create mode superpositions and tailor a particular beam shape. Hermite–Gauss modes are of interest mostly for their connection to Laguerre–Gauss modes (described next) because the latter can be expressed as superpositions of Hermite–Gauss modes.

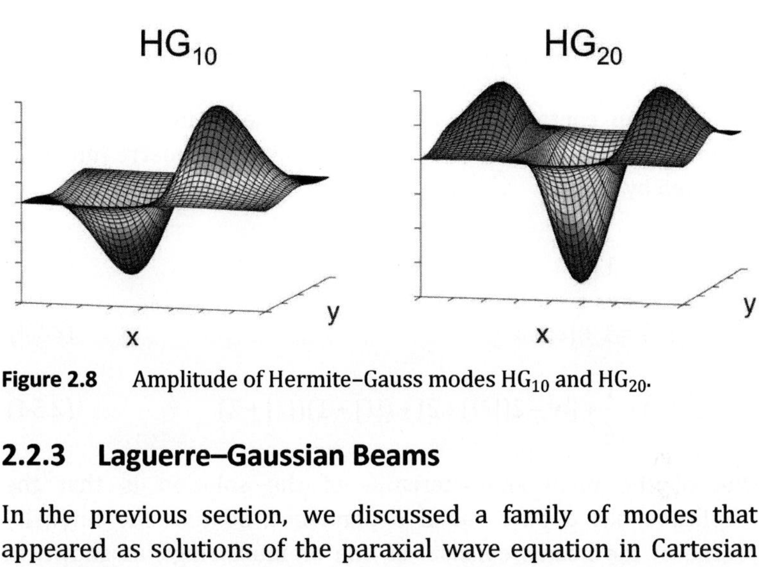

Figure 2.8 Amplitude of Hermite–Gauss modes HG_{10} and HG_{20}.

2.2.3 Laguerre–Gaussian Beams

In the previous section, we discussed a family of modes that appeared as solutions of the paraxial wave equation in Cartesian coordinates. If we choose a different coordinate system, such as cylindrical coordinates, we will get a new set of solutions. These solutions give rise to Laguerre–Gauss modes. These modes have very interesting properties, so we will subdivide this section into several parts.

2.2.3.1 Fundamentals

A first step is to set up the paraxial wave equation in cylindrical coordinates, and then allow the solutions to have independent solutions in terms of the transverse coordinates $r = (x^2 + y^2)^{1/2}$ (the distance from the origin) and $\phi = \tan^{-1}(y/x)$ (the angle about the transverse horizontal axis, x). The solution of the resulting equation is given by [2]

$$U_{p,\ell}(r,\theta,z)=\frac{A}{w}\left(\frac{\sqrt{2}r}{w}\right)^{|\ell|}\mathrm{L}_{\mathrm{p}}^{|\ell|}\left(\frac{2r^2}{w^2}\right)e^{-r^2/w^2}e^{ikr^2/[2R]}e^{i\ell\phi}e^{-i\varphi}, \tag{2.49}$$

where p and ℓ are integer indices, with p being only positive, and so the order of the mode is given by

$$N=2p+|\ell|. \tag{2.50}$$

The normalization constant is given by

$$A=p!\left(\frac{2}{\pi p!(|\ell|+P)!}\right)^{1/2}. \tag{2.51}$$

The solution contains the associated Laguerre function L_p^ℓ that depends only on r. The lowest-order associated Laguerre functions are given by

$$\mathrm{L}_0^{|\ell|}(v)=1 \tag{2.52}$$

$$\mathrm{L}_1^{|\ell|}(v)=1+|\ell|-v \tag{2.53}$$

$$\mathrm{L}_2^{|\ell|}(v)=\frac{1}{2}+[v^2-2(|\ell|)+2]+(|\ell|+1)(|\ell|+2) \tag{2.54}$$

One of the main characteristics of the solution is that the amplitude has a pure radial dependence, so that the intensity profile of the beams consists of one or more rings, as shown in Fig. 2.9. Indeed, the index p specifies the number of radial nodes. Because of the term of the form $r^{|\ell|}$, only when $\ell = 0$ there is a central maximum. For other values of ℓ the central spot is dark. When $\ell \neq 0$ and $p = 0$ the beams have a characteristic single-ringed "doughnut" shape, with the radius of the doughnut given by

$$r_\ell=\sqrt{\frac{|\ell|}{2}}w, \tag{2.55}$$

and a maximum amplitude at that radius that decreases monotonically with increasing $|\ell|$. The dark interior of Laguerre–

Gauss modes also makes them amenable to imaging. In particular, it has been used in STED microscopy for high-resolution imaging [28, 50].

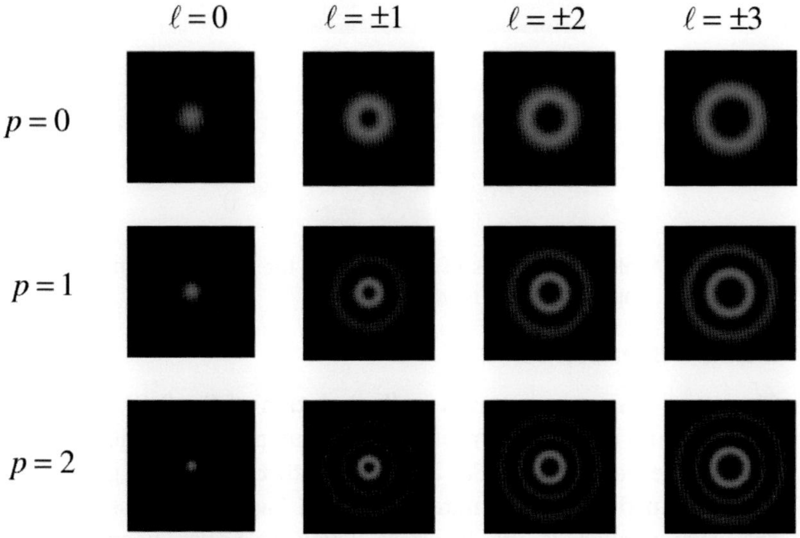

Figure 2.9 Graph of the irradiance of the lowest-order Laguerre–Gauss modes.

The most interesting term of Eq. 2.49 is the one where the phase varies as $\ell\phi$. It specifies that the phase of the wave depends on the angular coordinate ϕ, and which winds ℓ-times-2π per revolution about the center of the beam. The constant ℓ is known as the topological charge. This is illustrated in the color-coded phase maps of the modes, shown Fig. 2.10. The center of each of the phase plots has a phase singularity or optical vortex. Since that point has all the phases, then the light interferes with itself destructively, explaining the dark spot at the center of the beams. This aspect of Laguerre–Gauss modes has been the subject of much research [1].

The central spot containing the optical vortex is particularly sensitive to perturbations, so that when the beam is not in a pure mode but in a superposition, it is likely that the exact position of the optical vortex will be affected. When $|\ell| > 1$, any perturbation will split the vortex of charge ℓ into ℓ singly-charged vortices [6].

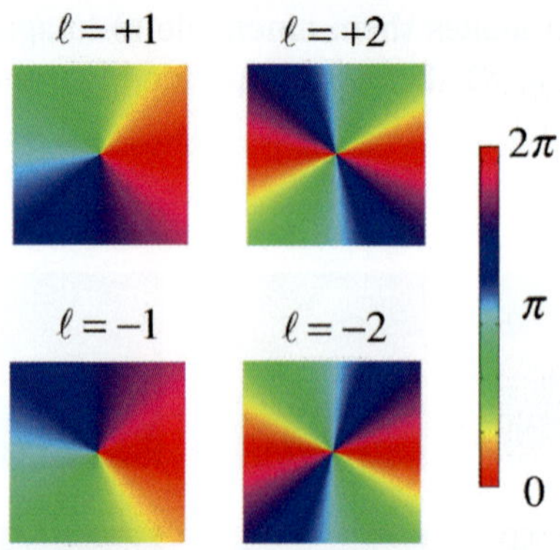

Figure 2.10 False-color maps of the phase of the first-and second-order Laguerre–Gauss modes with $p = 0$.

2.2.3.2 Interference

There are other phase terms in the solution of Eq. 2.49, such as the curvature of the wavefront and the Gouy phase. The latter now follows the definition of Eq. 2.48, but with the order N given by Eq. 2.50. In this way, Hermite–Gauss and Laguerre–Gauss modes are very similar. If we interfere Laguerre–Gauss beams, one gets patterns such as those shown in Fig. 2.11. Frames (a) and (d) show respectively the modeled and measured the interference patterns for the case of two modes with $\ell = 1$ and $\ell = 0$ interfering collinearly. As can be seen, it is an asymmetric pattern. For higher values of ℓ the pattern consists of $|\ell|$ maxima symmetrically distributed about the center of the beam. If the relative phase between the modes is varied, the pattern rotates about the center of the beam. We can understand this the following way: the mode has a phase

$$\psi_\ell = \ell\phi \tag{2.56}$$

while the other mode has a constant phase ψ_0. We get constructive interference when

$$\psi_\ell - \psi_0 = 2\pi n, \tag{2.57}$$

where n is an integer. This will occur at angles

$$\phi_n = \frac{2\pi n + \psi_0}{\ell}. \tag{2.58}$$

If the phase of the zero-order mode changes, then the angle at which the n-th maximum occurs will rotate about the center of the beam. Note that for higher values of ℓ the pattern as a whole will rotate at a slower rate. If we change the sign of ℓ the pattern will rotate in the opposite sense for the same change in the relative phase.

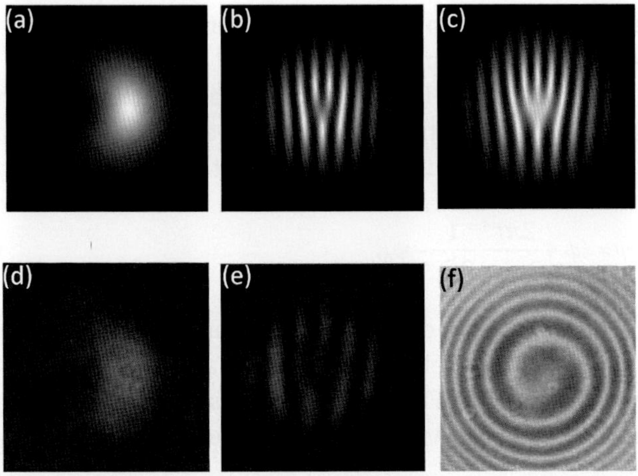

Figure 2.11 Modeled interferograms of $\ell = 1$ and $\ell = 0$ modes interfering (a) collinearly, and (b) at an angle; and (c) of $\ell = 2$ and $\ell = 0$ interfering at an angle; measured interference pattern of $\ell = 1$ and $\ell = 0$ beams (d) collinearly (taken by P. Crawford), and (e) at an angle (taken by K. Beach); and (f) collinear interference of $\ell = 1$ with an expanded $\ell = 0$ beam (taken by H. Sztul).

Non-collinear interference gives rise to peculiar fringe patterns. Because of the optical vortex in these modes, which is a phase dislocation, the fringe interference pattern has a distinctive forked pattern. If the zero order mode forms an angle β, then its phase varies along the plane that contains both beams. Let us call that the coordinate x, and assume that the initial phase is ψ_0 at $x = 0$. The phase difference between the two modes on a screen in the x–y plane centered on the beam's intersection is given by [7]

$$\Delta\psi = \ell \tan^{-1}\left(\frac{y}{x}\right) - \frac{2\pi}{\lambda} x \sin\beta - \psi_0, \tag{2.59}$$

where we have expressed the variable ϕ in terms of the x and y coordinates. The graph of $\Delta\psi = 2\pi n$ gives rise to a forked pattern. The number of tines in the fork is the value of $|\ell| + 1$, as shown in Fig. 2.11b,c where the interference pattern has been modeled for the cases $\ell = 1$ and $\ell = 2$, respectively. Thus, the interference pattern can be used to diagnose the topological charge of the beam. Figure 2.11e shows the measured forked pattern. If the charge is negative, then the fork points in the opposite direction (down for the cases of the figure).

If the $\ell = 0$ beam is collinear but has a radius of curvature that is different from the one of the beam carrying the optical vortex, then the phase difference between the two modes is

$$\Delta\psi = \ell\phi + \frac{2\pi r^2}{\lambda}\frac{1}{2R_0} - \psi_0, \tag{2.60}$$

where we have assumed for simplicity that $R_\ell \sim \infty$. If we require $\Delta\psi = 2\pi n$, we get a relation between r and ϕ corresponding to ℓ-intertwined spirals. Figure 2.11f shows the measured pattern for the case with $\ell = 1$. As ψ_0 is changed, the spiral pattern rotates.

2.2.3.3 Angular momentum

If we ignore phase due to the curvature of the wavefront and the Gouy phase, then the phase of the wave is given by the z and ϕ dependencies, or

$$\psi(r,\phi,z) = kz + \ell\phi. \tag{2.61}$$

This phase function follows the form of a helix of pitch λ. Points of equal phase, known as the wavefront of the beam, make ℓ-intertwined helices. This is shown in Fig. 2.12 for the cases with $\ell = +1$ and $\ell = +2$. The angular phase of the mode brings with it a new property of light beams: orbital angular momentum [2]. Fundamental light beams are known to carry linear momentum associated to their planar wavefront. Helical beams, such as Laguerre–Gauss with $\ell \neq 0$ carry angular momentum. This angular momentum depends on the rate of change of the phase with the angular coordinate, so it increases with $|\ell|$.

Each photon in a Laguerre–Gauss mode carries an angular momentum [2]

$$L = \ell\hbar, \tag{2.62}$$

where $\hbar = h/2\pi$, with h being Planck constant. Because the waveform is tilted on one side of the beam and tilted the other way on the opposite side,

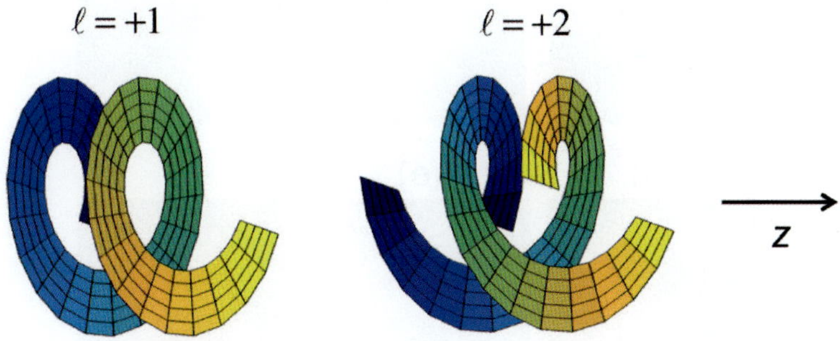

Figure 2.12 A pictorial view of the wavefronts of beams with $\ell = +1$ (left) and $\ell = +2$ (right).

Laguerre–Gauss beams can exert torques on objects, especially irregular ones. To a fully absorptive particle the torque is given by [29, 41]

$$\tau = \frac{\ell P}{\omega}, \tag{2.63}$$

where P is the absorbed power and w is the angular frequency of the light. A sphere of radius a will rotate with an angular frequency given by

$$\Omega = \frac{\tau}{8\pi\eta a^3}. \tag{2.64}$$

Figures 2.13(a–c) show the rotation of a partially absorptive sphere effected by light in an LG_0^2 mode. The rotation of the irregular features of the sphere demonstrates its rotation.

When the objects are transparent but irregular, say cylindrical, the asymmetry of the objects changes the mode, as seen in the next section, which involves an exchange of angular momentum between the light and the object. In that case, too, the object will experience a torque. In addition to rotating objects in place, when the objects are small, they get trapped by the intensity gradient along the ring of the mode, and because the wavefront is slanted along the ring, the objects will move along it, as seen in Figs. 2.13(d–f).

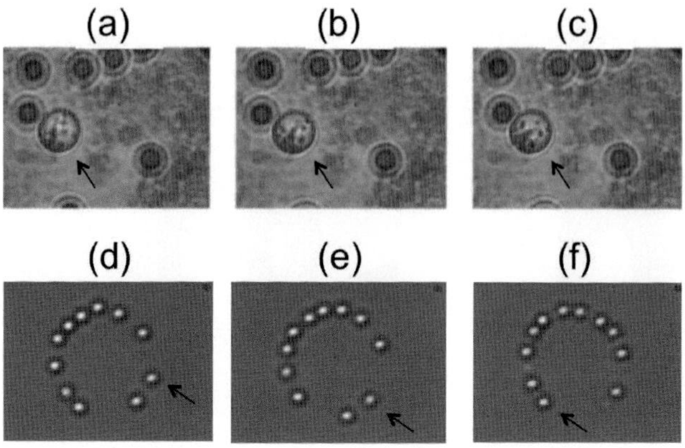

Figure 2.13 (a)–(c) Images of a 10 μm sphere (labeled) rotated in place by light with $\ell = 2$ (taken by N. Zehlev); (d)–(f) images of latex spheres moved along the ring of light with $\ell = 20$ (courtesy of M. J. Padgett). The arrow marks the same sphere.

2.2.4 Relations between Mode Families

Because Hermite–Gauss and Laguerre–Gauss functions each form a complete set, any mode can be expressed as a superposition of modes of either set, and in particular, the modes of one family can be decomposed into a superposition of modes of the other family. Moreover, the modes of a given order N of one family have a decomposition in terms of a superposition of modes of the same order in the other family [9].

The connection between first-order modes is very illustrative. According to Eq. 2.41, the first-order Hermite–Gauss modes are given in abbreviated form by

$$\mathrm{HG}_{10} = B_1 xGW \tag{2.65}$$

$$\mathrm{HG}_{01} = B_1 yGW, \tag{2.66}$$

where

$$B_1 = \frac{2\sqrt{2}}{\sqrt{\pi}w^2} \tag{2.67}$$

is a constant,

$$G = e^{-r^2/w^2}, \tag{2.68}$$

is the Gaussian envelope, and

$$W = e^{-ikr^2/(2R)-i\varphi}, \tag{2.69}$$

is an overall phase term. In a similar way, we can define the first-order Laguerre–Gauss modes:

$$\mathrm{LG}_0^{+1} = A_1 r e^{i\phi} GW \tag{2.70}$$

$$\mathrm{LG}_0^{-1} = A_1 r e^{-i\phi} GW, \tag{2.71}$$

where

$$A_1 = \frac{2}{\sqrt{\pi} w^2} \tag{2.72}$$

is the normalization constant.
Notice that $A_1 = B_1/\sqrt{2}$. The Laguerre–Gauss modes can also be expressed as

$$\mathrm{LG}_0^{+1} = A_1(x+iy)GW \tag{2.73}$$

or equivalently,

$$\mathrm{LG}_0^{+1} = \frac{1}{\sqrt{2}}(\mathrm{HG}_{10} \pm i\mathrm{HG}_{01}), \tag{2.74}$$

The previous equation means that the superposition of the two first-order Hermite–Gauss modes with a relative phase of $\pi/2$ give rise to a first-order Laguerre–Gauss mode. This type of relation can be easily realized in the laboratory, but more interestingly, it can be used, in combination with the Gouy phase, to convert first order modes of one family to the other. To understand this we first have to realize a relation between rotated Hermite–Gauss modes:

$$\mathrm{HG}_{10}(\theta) = \mathrm{HG}_{10}\cos\theta + \mathrm{HG}_{01}\sin\theta, \tag{2.75}$$

where θ is the angle that mode HG_{10} is rotated. In particular, when $\theta = \theta/4$ one gets

$$\mathrm{HG}_{10}(\pi/4) = \frac{1}{\sqrt{2}}(\mathrm{HG}_{10} + \mathrm{HG}_{01}). \tag{2.76}$$

That is, the superposition of the two Hermite–Gauss modes in phase gives rise to the mode rotated by $\pi/4$. Note the similarity between Eqs. 2.74 and 2.76. If we have a way to introduce a phase of $\pi/2$ between the two Hermite–Gauss modes, we convert a rotated Hermite–Gauss mode into a Laguerre–Gauss mode. This can actually be done with optical components using the Gouy phase and cylindrical lenses; a device known as a mode converter [9]. The idea is that a cylindrical lens introduces a Gouy phase in one dimension but not the other, so modes aligned with the axis of the cylindrical lens will experience a Gouy phase that is different from the mode aligned with the axis perpendicular to the cylindrical lens. This is shown in Fig. 2.14. The combination of two spherical and two cylindrical lenses results in creating a phase of $\pi/2$ between the modes parallel and orthogonal to the cylindrical axes. By inputting a mode oriented at $\pi/4$, per Eq. 2.76, it is decomposed into the modes parallel and perpendicular to the astigmatic axis of the mode converter. The latter introduces a $\pi/2$ phase between them via the Gouy phase. The resulting mode is then a Laguerre–Gauss mode.

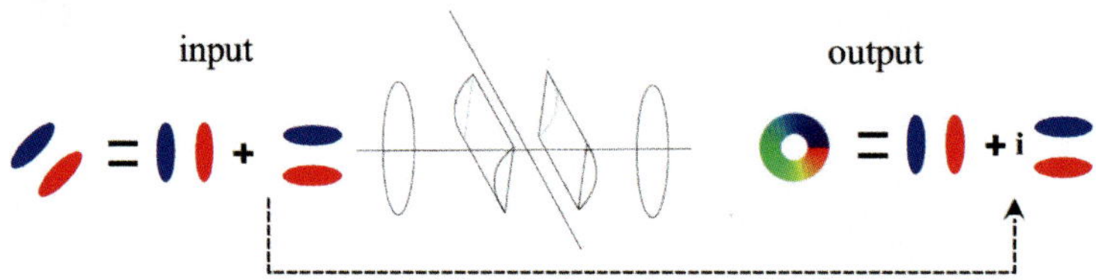

Figure 2.14 A $\pi/2$ mode converter.

Hermite–Gauss modes do not carry orbital angular momentum. Because the mode converter converts a Hermite–Gauss mode into a Laguerre–Gauss mode, the light exchanges angular momentum with the apparatus. In the case of optical tweezers, irregular objects can exchange angular momentum with the light, and thus receive a recoil torque.

The two first-order Hermite–Gauss modes can also be put in terms of the two first-order Laguerre–Gauss modes:

$$\mathrm{HG}_{10} = \frac{1}{\sqrt{2}}(\mathrm{LG}_0^{+1} + \mathrm{LG}_0^{-1}) \tag{2.77}$$

$$\mathrm{HG}_{01} = \frac{-i}{\sqrt{2}}(\mathrm{LG}_0^{+1} - \mathrm{LG}_0^{-1}). \tag{2.78}$$

The zero orders are the same for the two families: $\mathrm{LG}_0^0 = \mathrm{HG}_{00}$. These relations between families can be extended to higher orders [9]. For example,

$$\mathrm{LG}_0^{+2} = \frac{1}{2}\mathrm{HG}_{02} - \frac{i}{\sqrt{2}}\mathrm{HG}_{11} - \frac{1}{2}\mathrm{HG}_{20} \tag{2.79}$$

$$\mathrm{LG}_0^{-2} = \frac{1}{2}\mathrm{HG}_{02} + \frac{i}{\sqrt{2}}\mathrm{HG}_{11} - \frac{1}{2}\mathrm{HG}_{20} \tag{2.80}$$

$$\mathrm{LG}_1^0 = \frac{1}{2}\mathrm{HG}_{02} + \frac{1}{2}\mathrm{HG}_{20} \tag{2.81}$$

The converse relationship can be deduced from the previous. One final case of importance is that of the cosine form of the Laguerre–Gauss mode:

$$\mathrm{LG}c^{|\ell|} = \frac{1}{\sqrt{2}}(\mathrm{LG}_0^{\ell} + \frac{1}{2}\mathrm{LG}_0^{-\ell}) \tag{2.82}$$

which is a symmetric mode with $2|\ell|$ lobes symmetrically arranged about the center of the beam. A relative phase between the two only rotates the pattern.

Finally, there exists a distinct but related family of modes, known as Ince–Gaussian beams [5]. They provide a continuous transition between Laguerre–Gauss and Hermite–Gauss modes.

2.2.5 Laboratory Methods of Production

Low-order Hermite–Gauss modes can be produced directly by lasers bearing apertures or obstacles to encourage maximum gain for a particular mode and too much loss for all the other modes [9]. This is not so easy for Laguerre–Gauss modes because it is not straightforward to enhance one topological charge ℓ and discourage the mode with $-\ell$. Thus, Laguerre–Gauss modes are mostly produced externally via mode transformation. In the previous section, we mentioned a method that uses mode converters to transform

Hermite–Gauss modes into Laguerre–Gauss modes. That method is a good one for a few set of low-order modes.

2.2.5.1 Spiral phase plate

Higher-order Laguerre–Gauss modes can be produced via spiral phase plates, sometimes called vortex lenses. These devices are also available commercially. They consist of a glass substrate that has a dielectric deposited over it in the shape of a helical ramp. This ramp retards the light through its angle-dependent thickness, shown schematically in Fig. 2.15a. The step is of a height that leads to delaying the wave from either side of the step by an integer number of wavelengths. In particular, If the medium of the ramp has an index of refraction n, and the step has a height h, then at either side of the step the light acquires a phase $\phi_+ = 2\pi nh/\lambda$ and $\phi_- = 2\pi h/\lambda$ (assuming air is the surrounding medium). For a Laguerre–Gauss mode of order ℓ the difference in phase at either side of the step is $\phi_+ - \phi_- = 2\pi\ell$. Therefore, the height of the step is

$$h = \frac{\ell\lambda}{n-1}. \tag{2.83}$$

There are experimental challenges to creating the sharpest step, but the greatest challenge is to produce a pure mode. Let us assume that the phase plate has no imperfections, then sending a Gaussian beam through it will produce a mode with topological charge ℓ, but not in a $p = 0$ mode, but in a superposition

$$\mathrm{LG}^{\ell}_{\mathrm{sup}} = \sum_{p=0}^{\infty} c_p \mathrm{LG}^{\ell}_{p}, \tag{2.84}$$

where c_p is a coefficient that will be greatest for $p = 0$ and decrease in magnitude for higher values of p. The resulting mode may even resemble a singly ringed $p = 0$ mode, but its radius will not correspond to what one may predict via the equations outlined in sections above. The reason is quite simple: the light going through the center of the spiral phase plate has all phases, so it undergoes destructive interference. A few wavelengths away past the plate, the light beam already has a dark spot, but the light that initially was at the center of the input mode has to go somewhere, so diffraction takes care of it. The final mode is a multi-ringed mode with topological charge ℓ. In many applications, this is still acceptable, but in applications where one requires a pure mode, then the light that enters the phase plate has to already have the proper ringed-shaped intensity. That is, for producing the purest mode we need to apply phase and amplitude modulation.

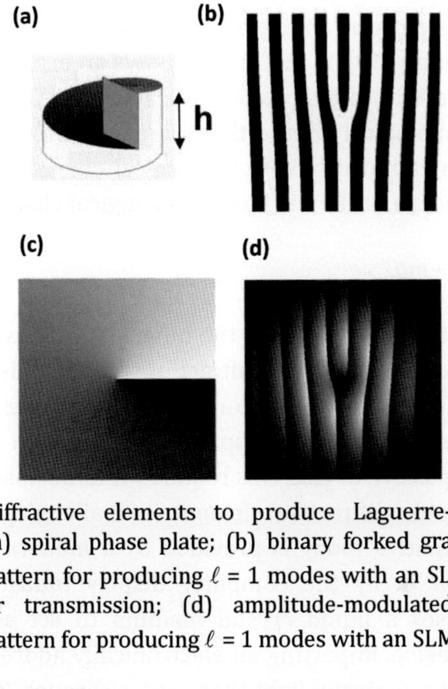

Figure 2.15 Diffractive elements to produce Laguerre–Gauss modes: (a) spiral phase plate; (b) binary forked grating; (c) phase pattern for producing $\ell = 1$ modes with an SLM on reflection or transmission; (d) amplitude-modulated phase-blazed pattern for producing $\ell = 1$ modes with an SLM by diffraction.

The analogous device for a Hermite–Gauss mode is a stepped phase plate. The adjacent lobes of all Hermite–Gauss modes are π out of phase, so a plate that shifts the phase by the right amount can also be used. In first-order modes where the there are only two lobes, a simple piece of glass of the proper thickness, or tilted for fine adjustment, can be used to generate them from an incoming fundamental Gaussian mode. The method still brings the diffractive effects as with the spiral phase plate: the lack of amplitude modulation creates additional lobes.

2.2.5.2 Holographic diffraction

By far the most efficient and versatile method of producing high-order modes is through the use of diffractive optical elements.

These come in two forms: passive and active. Passive gratings can be holographic, engraved, or binary. Active ones can use liquid-crystal-based electronic devices, such as a spatial light modulator (SLM) to dial the desired pattern on demand.

The simplest diffractive device is the binary grating, an amplitude grating that either blocks or transmits light. It is generated simply by taking a black and white photograph of the diffraction pattern and using the developed negative as the diffraction grating [30]. Figure 2.15b shows an example of a binary grating pattern with a dislocation charge $q = 1$. Diffracted orders will have a topological charge that increases with the order. The incident beam can also have a topological charge ℓ_{inc}. In general the m-th diffractive order will have a topological charge.

$$\ell_{\text{diff}} = \ell_{\text{inc}} + mq. \tag{2.85}$$

The most important problem of the binary grating is its efficiency. It is partially opaque and the diffraction efficiency is low. Despite this, many early research on complex light problems was done using this method. A better approach is to use a phase-blazed grating. For a passive device, this requires it to be holographic [36]. The photographic diffractive element is challenging to fabricate. The easy alternative today is provided by a modern electro-optic device brought in by the computer display industry: the SLM. This device uses a liquid crystal medium to act as a pixelated birefringent device, imparting an electronically addressable space-variant retardance on the light that passes through it. Thus, these devices allow one to dial-in the desired light mode by phase sculpturing it. They can also be highly efficient. They can be used in two distinct modes: on reflection/transmission, and on diffraction. The grayscale encoded pattern for the former is shown in Fig. 2.15c. It is the pattern that gets programmed onto the SLM, and where the gray level becomes a phase shift imparted onto the light, and where white is, say, phase zero and black is phase 2π. The pattern of the figure was generated by encoding in grayscale the phase of Eq. 2.60 modulo-2π, with $R = \infty$, $\ell = 1$, $\psi_0 = 0$ and r and ϕ specifying SLM pixels. This is the electronic form of the spiral phase plate, where the phase delay is imparted by the birefringence of the SLM instead of a path through a refractive medium of varying thickness.

The spiral patterns encoded onto the SLM have the same shortcomings of the spiral phase plate. The alternative is to generate the modes on diffraction. To resolve the problem of efficiency, the pattern of the grating is blazed, and to resolve the problem of mode purity, the grating is amplitude modulated [13]. An example of such a pattern is shown in Fig. 2.15d for the case of $\ell = 1$. In this case, Eq. 2.59 was used, modulo-2π, and multiplied by a modulation factor related to the shape of the mode. The combination of amplitude phase modulation allows one to engineer a pure mode. The SLM has other advantages: the phase can also be manipulated to include deliberate mode superpositions and to simulate other optical elements such as lenses and astigmatic elements [34].

2.3 Non-Diffracting Optical Beams

The previous sections were devoted to beams of light that are solutions of the wave equation, and which suffer from diffraction: as the beam propagates, they evolve by expanding and phase-shifting via the Gouy phase. New and distinct types of beams have been studied and developed that, through a finite region of space, preserve their size and shape unaltered. That is, beams that temporarily do not diffract. Some may even say that they are a three-dimensional interference pattern, but one that remains intact for a design distance. We also saw that the concept of a pure mode is meaningful within a mathematical framework, because a pure mode in one family, say Laguerre–Gauss, is a superposition (and consequently an interference beam) of pure modes of another family, such as Hermite–Gauss. Even a fundamental Gaussian beam can be thought of an infinite superposition of plane waves. The case of non-diffracting light beams is fascinating for its simplicity and elegance, and for their unique properties. In this section we will discuss two distinct types of non-diffractive beams: Bessel beams and Airy beams. In both cases, the ideal function is of infinite extent, as are plane waves, but implementations of truncated functions still show the same ideal properties for finite and but not insignificant regions of space.

2.3.1 Bessel Beams

The best way to understand Bessel beams is by starting with how they can be made, shown in Fig. 2.16a. A beam of light with a planar wavefront is incident on an axicon. This is a conical optical element: a prism of revolution. It is flat on one side (facing the incoming light) and conical on the other side. It is normally characterized by its apex or opening angle α. Figure 2.16a shows a section of the axicon that includes its axis; a view where the axicon looks like a prism. A more convenient angle is the base angle, shown in the figure, and related to the apex angle by

$$\gamma = \pi/2 - \alpha/2, \tag{2.86}$$

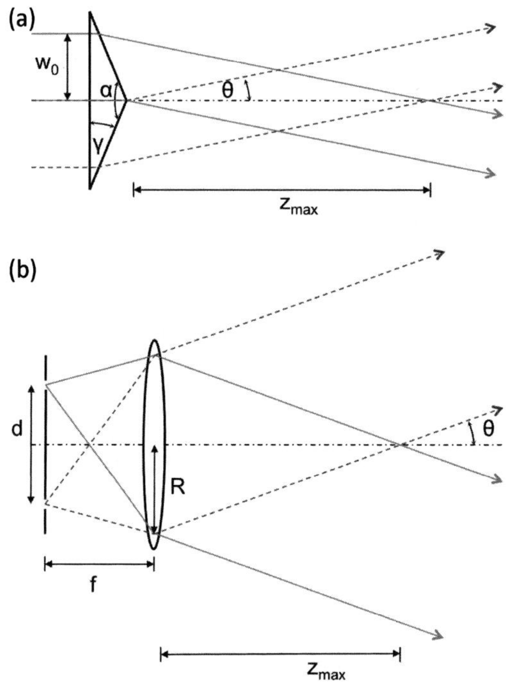

Figure 2.16 Methods of production of Bessel beams: (a) with an axicon of base angle γ; and (b) with an annular slit of diameter d, and a lens of focal length f.

If the incoming rays are parallel to the axis, they are not refracted by the first surface. Refraction at the second surface yields, via Snell's law, the relation

$$n\sin\gamma = \sin(\theta + \gamma), \tag{2.87}$$

where θ is the angle that the refracted rays form with the z-axis, and n is the index of refraction of the axicon. Incoming rays above the axis are refracted downward, and similarly, incoming rays below the axis are refracted upward. This is of course the description of just the rays contained in the plane of the figure. All the refracted rays of a given input impact parameter w_0 (distance from the axis) form a three-dimensional cone after the axicon. For small angles, the refracted angle is given by

$$\theta = (n - 1)\gamma. \tag{2.88}$$

The group of rays from the top overlaps with the ones from the bottom over a distance

$$z_{\max} \simeq \frac{w_0}{\theta}. \tag{2.89}$$

This quantity is also the depth of focus of the beam.

The beam that results can be thought of being made of a superposition of plane waves that have a wave number k forming an angle θ with the axis.

These waves have a longitudinal component

$$k_z = k\cos\theta \tag{2.90}$$

and a radial (transverse) component

$$k_r = k\sin\theta. \tag{2.91}$$

A solution for the electric field within the region of overlap is [21]

$$E(r,\phi,z) = A_0 e^{ik_z z} J_0(k_r r) \tag{2.92}$$

where A_0 is a constant, the second term is the phase propagator associated with a plane wave traveling along the z-axis, and $J_0(k_r r)$ is the Bessel function of order 0. Notice that the magnitude of the mode depends only on r, and not at all on z. It means that in the region where the rays overlap, the magnitude of the field in the transverse plane remains constant. Thus, the width of the beam does not change throughout the overlap region.

For this reason, these beams are also known as non-diffracting. Figure 2.17 shows the intensity of the transverse profile. Note that a large fraction of the energy is concentrated in the central maximum of radius

$$r_0 = \frac{2.405}{k_r}, \tag{2.93}$$

which is the first zero of the Bessel function. For example, a glass axicon with a base angle $\gamma = 2.25°$ yields $\theta = 0.022$ rad, or $z_{max} = 46w_0$. If $w_0 = 7.5$ mm, then $z_{max} \sim 340$ mm. The radius of the central maximum for $\lambda \sim 600$ nm is 10 μm. A Gaussian beam with a waist of the same value will expand due to diffraction and have a half-width of 6.5 mm at a distance $z = z_{max}$. This Bessel beam has huge depth of focus compared to that of the Gaussian beam: $2z_R = 1$ mm.

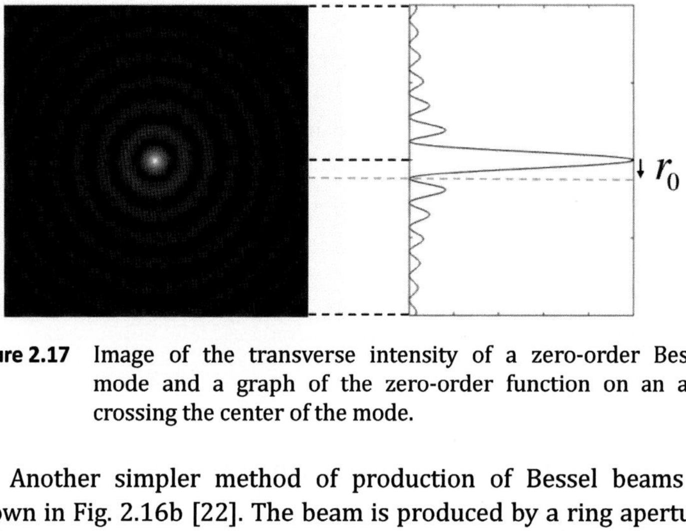

Figure 2.17 Image of the transverse intensity of a zero-order Bessel mode and a graph of the zero-order function on an axis crossing the center of the mode.

Another simpler method of production of Bessel beams is shown in Fig. 2.16b [22]. The beam is produced by a ring aperture and a lens. The ring has a diameter d and an aperture width Δd. Due to single-slit diffraction, the beam expands further after the ring. If the lens is placed a focal length f away from the aperture, it needs to have a minimum radius

$$R = \frac{d}{2} + \frac{\lambda f}{\Delta d}. \tag{2.94}$$

The second term accounts for the slit diffraction of the light up to the first minimum. The non-diffractive distance for this case is (for small angles)

$$z_{\max} = \frac{R}{\theta}, \tag{2.95}$$

and where

$$\theta = \frac{d}{2f}. \tag{2.96}$$

The method, although inefficient (because most of the incoming light is blocked), is simple and inexpensive. To produce a Bessel mode with the same θ and $z_{\max}$ as the previous example would require a ringed aperture with a Non-Diffracting Optical Beams diameter of 13.2 mm, and a lens of 300 mm focal length with a 15 mm diameter. One can also understand this method of production in terms of Fourier analysis: the Bessel mode is the Fourier transform of the ring, a transform effected by the lens. Bessel beams can also be produced diffractively using either holograms [51], or spatial light modulators [18].

Beyond $z_{\max}$ the mode disappears, as there is no longer beam overlap or interference. Because these can be thin and non-expanding, they are particularly suited for imaging via scanning light-sheet microscopy [23, 44]. Bessel beams are "self-healing" [38]. That is, an obstacle placed at a point along the beam does not cast a lasting shadow. Using simple geometry, we get that when a Bessel beam encounters an object of transverse side a, the mode repairs itself after a distance

$$z_{\text{rep}} = \frac{a}{2\theta}. \tag{2.97}$$

The explanation is simple: the mode is produced by plane waves (rays) traveling at oblique angles, as shown in Fig. 2.16. Beyond z_{rep} the rays no longer hit the object, and so they recreate the beam as if the object was not there. In actual implementations there are deviations from the ideal modes described above, which can yield a z-dependent amplitude A_0 and other phase terms that may vary with z and r [35]. However, the pattern retains its constant Bessel shape as a function of z.

A Bessel beam with an azimuthal phase can be generated diffractively [18, 52], by using a high-order Laguerre–Gauss beam as input to an axicon [3], or by applying a phase delay around the aperture of the ring [53]. In this case the Bessel mode is given by

$$E_\ell = A_0 e^{ik_z z} e^{i\ell\phi} J_{|\ell|}(k_r r) \tag{2.98}$$

where now the Bessel function is a high-order one, with the order given by ℓ, the topological charge. The center will no longer be dark because it will carry an optical vortex. Other solutions to the wave equation yield non-diffracting solutions include Mathieu beams, which are elliptical realizations of Bessel beams [16].

2.3.2 Airy Beams

The Airy function is a one-dimensional function that is asymmetric about its coordinate, say x, which is oscillatory for $x < 0$ (extending out to infinity) and decaying exponentially for $x > 0$. It is a solution of a differential equation that appears in several physical situations, and especially in optics, around caustics [10]. In our case, implementations have involved a two-dimensional function that is the product of two one-dimensional Airy functions in orthogonal directions. If for now we concentrate on one direction, the Airy function appears as a solution of a paraxial one-dimensional wave equation [46], similar to Eq. 2.7:

$$\frac{\partial^2 U}{\partial x^2} + 2ik\frac{\partial U}{\partial z} = 0, \tag{2.99}$$

where $k = 2\pi/\lambda$. For $z = 0$ the solution $U(x, z)$ is simply the Airy function

$$U(x,0) = \mathrm{Ai}\left(\frac{x}{x_0}\right), \tag{2.100}$$

where x_0 is a scale factor. For $z > 0$ the solution is:

$$U(x,z) = \mathrm{Ai}\left(\frac{x}{x_0} - \frac{z^2}{4k^2 x_0^4}\right) e^{i[xz/(2kx_0^3) - z^3/(12k^3 x_0^6)]}. \tag{2.101}$$

Note that the amplitude of the solution remains intact: only its argument changes, varying quadratically with z. That is, the pattern curves toward $x > 0$ as it propagates. More precisely, the beam describes a parabolic trajectory of the form

$$x = \frac{z^2}{4k^2 x_0^3}. \tag{2.102}$$

This motion is often referred as "acceleration," although the light does not follow a curved trajectory. It is the superposition (interference pattern) that curves. This trajectory follows the momentum which tilts the mode further as it propagates [48]. This one-transverse-dimensional version of the Airy beam has applications in light-sheet microscopy, producing a continuous sheet of non-diffracting light, [54] and at the same time relieving the need to scan the beam, as done in other forms of the technique [23, 44].

A realization of this beam is recreated by truncating the function at the input ($z = 0$). It is also recreated in 2 dimensions, which results in a product of Airy functions of the x and y coordinates. However, the main features remain the same. The complete implementation has the modified form [14]

$$\begin{aligned} U(x,y,z) = {} & \mathrm{Ai}\left(\frac{x}{x_0} - \frac{z^2}{4k^2 x_0^4} - \frac{iaz}{kx_0}\right) \mathrm{Ai}\left(\frac{y}{y_0} - \frac{z^2}{4k^2 y_0^4} - \frac{iaz}{ky_0}\right) \times \\ & e^{ax/x_0 - az^2/(2k^2 x_0^4)}\, e^{i[(a^2 + x/x_0)z/(2kx_0^2) - z^3/(12k^3 x_0^6)]} \times \\ & e^{ay/y_0 - az^2/(2k^2 y_0^4)}\, e^{i[(a^2 + y/y_0)z/(2ky_0^2) - z^3/(12k^3 y_0^6)]}, \end{aligned} \tag{2.103}$$

where $0 < a < 1$, and where $1/a$ is related to the transverse extent to the beam, similarly to the role of w for a Gaussian beam. There are two types of exponentials for each coordinate. The real exponentials produce an attenuation of the oscillatory part of the Airy beam in each dimension, x and y. The imaginary exponentials add coordinate-dependent phase factors. The latter result is the wavefront leading to the curvature of the beam.

Figure 2.18 shows three ways to view a two-transverse-dimension Airy beam. In Fig. 2.18b we see the transverse image of the mode. The lobes correspond to the oscillations of the Airy function along the two coordinates. They show clearly the asymmetry of the function. The intensity as a function of x for

$y = 0$ and $z = 0$ is shown in Fig. 2.18a. A different view is shown in Fig. 2.18c, which shows a contour-type graph of the intensity as a function of x and z for $y = 0$. The curvature of the graph clearly shows the acceleration of the beam. This feature appears due to the superposition of oblique plane waves, similar to the case of Bessel beams. For that reason, Airy beams also exhibit the self-healing effect.

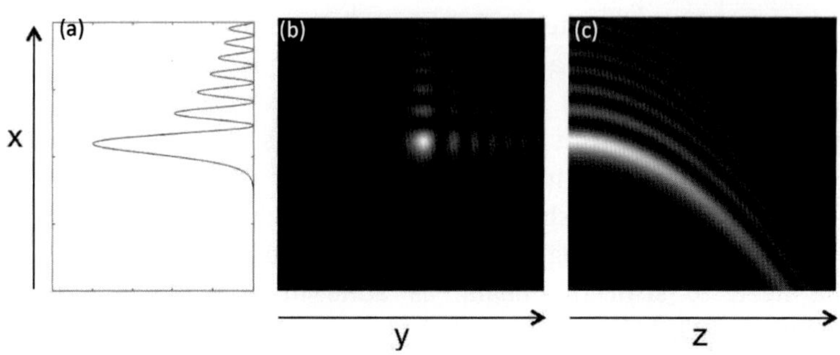

Figure 2.18 Airy modes: (a) Intensity along the transverse coordinate x; (b) grayscale graph of the intensity in the transverse plane x–y; (c) grayscale graph of the transverse intensity along axes x and z for a one dimensional Airy beam.

2.4 Beams with Space-Variant Polarization

The previous sections covered scalar beams, where every point in the beam had the same polarization. That is, the vectorial part of the beam is the same for all points. Here we discuss beams of light where the state of polarization is not constant throughout the beam. We address these types of beams in three sections. A first section gives a brief introduction to polarization of light, and the following sections present two types of beams: vector and Poincaré beams.

2.4.1 Polarization

Electromagnetic waves consist of transverse oscillations of electric and magnetic fields, $\vec{E}$ and $\vec{B}$, respectively. The two vectors are orthogonal to each other, and their magnitudes are related by

$$E = cB. \tag{2.104}$$

Thus, it is convenient to express the light only in terms of the electric field, with the understanding that the magnetic part is proportional to it and orthogonal in direction. Another reason to consider is that matter normally interacts more strongly with the electric field component of the light. Although not discussed here, there are situations where the magnetic field and its interactions play an important role.

In a plane-wave description of the light, the electric field of a propagating electromagnetic wave oscillates linearly in the plane that is transverse to the propagation. In linear polarization, the electric field oscillates in a single plane that contains the propagation direction. If we denote the state of linear polarization along the x-direction by $\hat{e}_x$, and similarly, for the linear polarization along y-direction by $\hat{e}_y$, then a general expression for the electric field is a linear superposition of these two components [42]:

$$\vec{E} = E_0(\cos\alpha\,\hat{e}_x + \sin\alpha e^{-i2\delta}\,\hat{e}_y), \tag{2.105}$$

where α and δ are angular variables that specify the relative amplitude and phase between the two components. When $2\delta = 0$, π, the polarization is in general elliptical. That is, the tip of the electric field describes an ellipse in the transverse plane. The orientation of the semi major axis and its relation to the semi minor axis are non-trivially related to α and δ. For this reason, we take a different approach to represent the fields.

If $2\delta = \pm\pi/2$ and $\alpha = \pi/4$, the linear states are of the same magnitude but out of phase by $\pi/2$, and the electric field rotates, with the tip of the vector describing a circle. This is known as circular polarization. Depending on whether the phase is positive or negative, the electric field rotates in one sense or the other, and so we call right and left circular polarization when the electric field rotates clockwise and counter clockwise, respectively, as a function of time in a given transverse plane when we are looking into the beam of light. The states of right (R) and left (L) circular polarization can be described in terms of the linear states as:

$$\hat{e}_R = \frac{1}{\sqrt{2}}(\hat{e}_x - i\hat{e}_y) \tag{2.106}$$

and

$$\hat{e}_L = \frac{1}{\sqrt{2}}(\hat{e}_x + i\hat{e}_y), \tag{2.107}$$

respectively. The complex factor is $i = \exp(i\pi/2)$ encodes the phase between the two components. Because $\hat{e}_R$ and $\hat{e}_L$ are orthogonal, we can express a general state of polarization in terms of them:

$$\hat{e} = \cos\chi\hat{e}_R + \sin\chi^{-2\theta}\hat{e}_L, \tag{2.108}$$

This state represents an ellipse with an ellipticity [17]

$$\varepsilon = \tan(\pi/4 - \chi) = \pm\frac{b}{a}, \tag{2.109}$$

where b and a are the semi-minor and semi-major axes of the ellipse, respectively; and an orientation of the semi major axis θ. That is, in this representation the ratio of amplitudes of the two components determines the ellipticity and their relative phase specifies the orientation.

The Poincaré sphere of Fig. 2.19 is a convenient geometrical construction that allows us to represent all the states of polarization by points on the surface of the sphere [12]. Angle χ is half of the polar angle, and θ is half the azimuthal angle, or longitude on the sphere, as shown in Fig. 2.19.

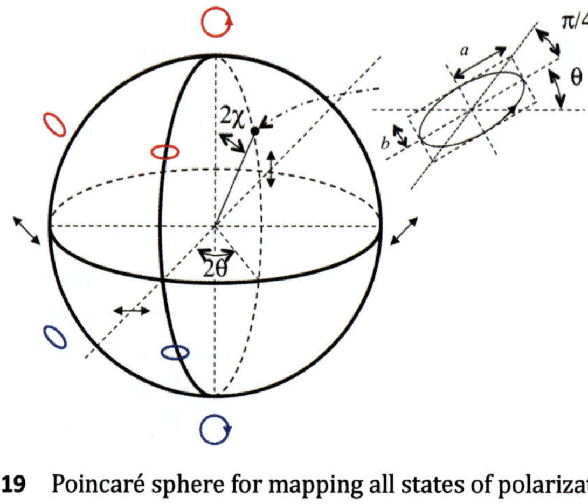

Figure 2.19 Poincaré sphere for mapping all states of polarization.

The north and south poles are the states of right and left circular polarization, respectively. All the points along the equatorial line are states of linear polarization with orientations that depend on the azimuth angle. Points in between, in the northern and southern hemispheres are states of elliptical polarization with right and left-handedness, respectively. The full array of states is shown in Fig. 2.20.

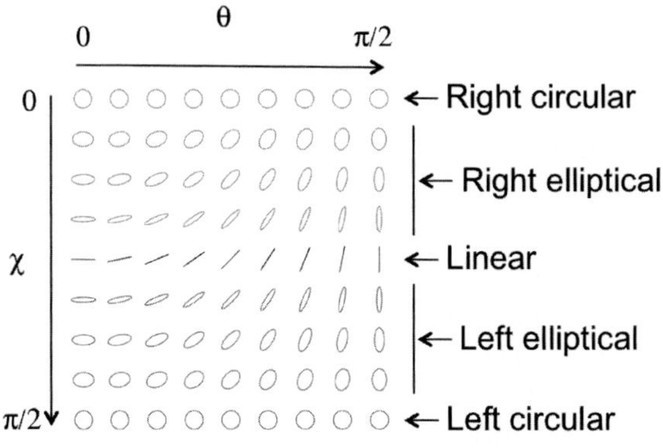

Figure 2.20 Mapping of states of polarization via angles χ and θ.

2.4.2 Vector Beams

A class of these beams has only linear polarization and are known as vector beams or modes. The most important vector mode is the radial mode. It is a mode where the state of polarization is linear but its orientation is radial, as shown in Fig. 2.21a. This mode can be understood easily as a superposition of two first-order Laguerre–Gauss modes in orthogonal states of circular polarization:

$$U = \frac{1}{\sqrt{2}}(\mathrm{LG}_0^{+1}\hat{e}_R + \mathrm{LG}_0^{-1}e^{-i\gamma}\hat{e}_L), \tag{2.110}$$

where γ is a phase. Replacing the expressions for the Laguerre–Gauss modes of Eqs. 2.70 and 2.71 into Eq. 2.110, and setting $\gamma = 0$ gives

$$U = \frac{1}{\sqrt{2}}A_{1r}\mathrm{GW}(e^{+i\phi}\hat{e}_R + e^{-i\phi}\hat{e}_L), \tag{2.111}$$

where G and W are defined by Eqs. 2.68 and 2.69. Notice that the amplitudes get factored out, so the state of polarization is linear. Inside the parentheses, we have a superposition of circular states with a relative phase between them. Thus, the orientation of the resulting linearly polarized state, which is given by half the relative phase, depends only on the transverse angle: $\theta = \phi$. The vector mode can also be produced by a superposition of first-order Hermite–Gauss modes:

$$U = \frac{1}{\sqrt{2}}(\mathrm{HG}_{10}\hat{e}_x + \mathrm{HG}_{01}\hat{e}_y). \tag{2.112}$$

This can be understood by analyzing the x and y dependencies of the expressions for the modes in Eqs. 2.65 and 2.66, and realizing that $x = r\cos\phi$ and $y = r\sin\phi$.

This state has been of interest because when focused by a lens, a longitudinal electric field appears at the center of its waist [24]. This mode is attractive because the waist is smaller than the one produced with uniform polarization [20]. When $\gamma \neq 0$, the orientation of the polarization is given by $\theta = \phi + \gamma/2$. When $\gamma = \pi$, the orientation is perpendicular to the radial dimension, or tangential, as shown in Fig. 2.21b.

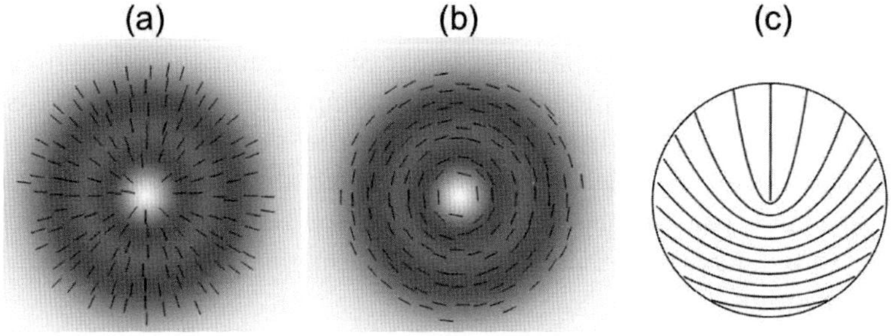

Figure 2.21 Radial (a) and tangential (b) vector modes, where orientation of the linear state of polarization varies across the beam profile, denoted by lines. (c) Pattern of fast axis orientations in a spatially variable half-wave plate that gives rise to vector modes.

These vector modes can be made in several ways. One is by direct superposition of Laguerre–Gauss or Hermite–Gauss modes with an interferometer (see Refs. [26] and [55] for a review of previous methods). Another method involves using a specially designed optical element that acts as a half wave plate with a fast axis orientation that varies with the angular coordinate:

$$\alpha_{\text{fast-axiz}} = q\phi \tag{2.113}$$

with $q = 1/2$. The pattern followed by the fast axes of the element is shown in Fig. 2.21c. Linearly polarized light entering a half-wave plate gets its orientation rotated by twice the angle that it forms with the fast axis. Thus, this device rotates incoming uniformly polarized light by an amount that depends on ϕ. The relative angle of the device as a whole with the incoming polarization specifies the orientation of the light at each angle. In the case of Fig. 2.21c, incoming vertical polarization gives rise to radial polarization; and incoming horizontal polarization gives rise to tangential polarization. Devices like these can be made stitching standard birefringent materials cut into pie sectors, with each pie sector having its fast axis forming an angle given by Eq. 2.113 [20]. A continuously varying fast axis in a single element has been designed using liquid crystals, and is already available commercially. A particularly interesting device is known as a "q-plate" allows the fast-axis angle of the device (q in Eq. 2.113) to be tunable [37]. This also allows one to create beams like the ones discussed in the next section.

The superpositions of Eqs. 2.110 and 2.112 are non-separable. That is, they are described by an equation that does not allow one to factor the modes and polarizations separately. It is a type of superposition that gives rise to modes with space-variant polarization. The range of distinct modes that can be produced can be mapped onto geometrical representations [39], allowing an investigation of the variety of modes that can be produced and their underlying mathematical structure.

2.4.3 Poincaré Beams

When we do a superposition of spatial modes of different order and polarization, we get a mode where both orientation and ellipticity vary from point to point. To describe this in the simplest terms let us start with the Laguerre–Gauss and circular polarization bases, similar to Eq. 2.110 but now with the modes LG_0^ℓ and LG_0^0. This yields

$$U = \frac{1}{\sqrt{2}}(\text{LG}_0^\ell \hat{e}_R + \text{LG}_0^0 e^{-i\gamma}\hat{e}_L), \tag{2.114}$$

The amplitude and phase of the two spatial modes do not have the same values at each point, and so they will yield in general elliptical states at any given point. If we use the expressions for each mode, we can rewrite Eq. 2.114 as

$$U = \frac{1}{\sqrt{2}} AGW(\cos\chi \hat{e}_R + \sin\chi e^{-i\gamma}\hat{e}_L), \tag{2.115}$$

where now

$$\chi = \tan^{-1}\left(\frac{A_0}{A_\ell r^\ell}\right) \tag{2.116}$$

and

$$\theta = \ell\phi/2 + \gamma/2, \tag{2.117}$$

A is a constant and G and W are defined by Eqs. 2.68 and 2.69. Notice that the relative amplitude of the two states of polarization depends only on the radial coordinate, and the relative phase depends only on the angular coordinate. Let us consider the case $\ell = +1$. Since at $r = 0$ the amplitude of LG_0^{+1} is 0, while that of LG_0^0 is maximum, then at $r = 0$ the state of polarization is left circular. As r increases, the amplitude of LG_0^{+1} increases while the amplitude of LG_0^0 decreases, and at a certain radius r_v they will be equal and the polarization is linear. In the range $0 < r < r_v$ the polarization will be left-handed elliptical, with ellipticity that depends on r. For $r > r_v$ the amplitude of LG_0^{+1} will be greater than that of LG_0^0, and so the state of polarization will be right-handed, and at $r \rightarrow \infty$ the polarization will be right circular. The mode LG_0^{+1} has a phase that varies with ϕ whereas LG_0^0 has a constant phase. Therefore, the relative phase at a point (r, ϕ) is $\phi + \gamma$, and therefore the orientation of the elliptic-linear states is $\theta = \phi/2 + \gamma/2$. In all, this new mode has an ellipticity that varies with radial distance and an orientation that varies with φ. It contains a mapping of all the states on the Poincaré sphere onto the transverse beam mode, and for that they are called Poincaré beams. It yields the pattern of Fig. 2.22a. If we use $\ell = -1$ mode, we get the pattern of Fig. 2.22b. These two patterns are known as the lemon and the star; names derived from an analysis of their topological structure: line singularities [11, 19, 25], such as those found in fingerprints [43].

We can make two types of changes to Eq. 2.114 and get yet other polarization patterns. If we change the mode to $\ell = 2$, we get a radial mode but with an ellipticity that varies with r (see that Eq. 2.117 yields $\theta = \phi$ as in the radial vector mode). Conversely, we can change the set of states of polarization used to get an asymmetric polarization pattern, as shown in Fig. 2.22c for the case of the linear x and y linear polarization states. This pattern is also of interest because it has a polarization gradient across its profile.

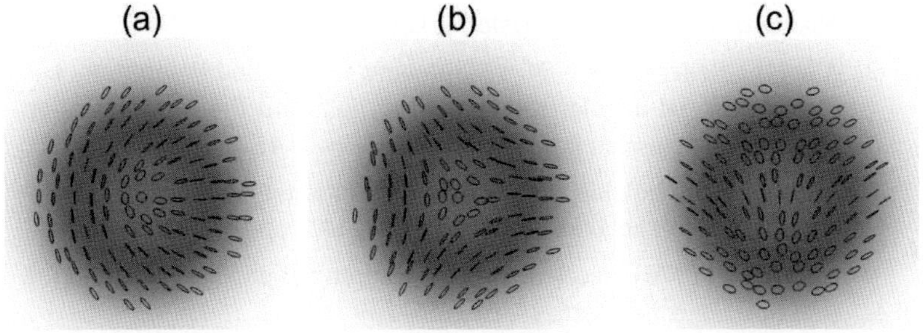

Figure 2.22 Lemon (a) and star (b) Poincaré modes obtained using Laguerre–Gauss modes respectively with $\ell = 1$ and $\ell = -1$ in a state of right-hand circular polarization in superposition with fundamental mode in a state of left-hand circular polarization. (c) Same construction as for (a) but using states of linear polarization instead of circular states.

These types of modes can be produced with various laboratory designs that use superposition of collinear modes [27], or transmission through optical elements with dislocations, such as the q-plates [15], solids with stress birefringence [8] or optical fibers [31].

2.5 Discussion and Conclusions

The control of amplitude, phase, and polarization of light has yielded an interesting set of optical beams presented in this chapter. They carry an array of unique properties, such as optical vortices, angular momentum, self-reconstruction, and vectorial dislocations, to name a few. They have already been at the center of many applications, and hold promise of more. Although these works have approached these beams from a classical perspective, their properties and applications extend to the quantum-physical domain, where they also provide settings for fundamental tests and applications. The field of complex light is growing into a one that provides an array of optical tools for research and applications. The present chapter presents an overview of what has been done, with the promise that more is yet to come.

References

1. Allen, L., Barnett, S. M., and Padgett, M. J. (2003). *Optical Angular Momentum* (Institute of Physics, Bristol).
2. Allen, L., Beijersbergen, M. W., Spreeuw, R. J. C., and Woerdman, J. P. (1992). Orbital angular momentum of light and the transformation of Laguerre-Gaussian laser modes, *Phys. Rev. A*, **45**, pp. 8185–8189.
3. Arlt, J., and Dholakia, K. (2000). Generation of high-order Bessel beams by use of an axicon, *Opt. Commun.*, **177**, pp. 297–301.
4. Ashkin, A. (1970). Acceleration and trapping of particles by radiation pressure, *Phys. Rev. Lett.*, **24**, pp. 156–159.
5. Bandres, M. A., and Gutierrez-Vega, J. C. (2004). Ince-Gaussian beams, *Opt. Lett.*, **29**, pp. 144–146.
6. Baumann, S. M., Kalb, D. M., MacMillan, L. H., and Galvez, E. J. (2009). Propagation dynamics of optical vortices due to Gouy phase, *Opt. Express*, **17**, pp. 9818–9827.
7. Bazhenov, V. Y., Soskin, M. S., and Vasnetsov, M. V. (1992). Screw dislocations in light wavefronts, *J. Mod. Opt.*, **39**, pp. 985–990.
8. Beckley, A. M., Brown, T. G., and Alonso, M. A. (2012). Full Poincarè beams II: Partial polarization, *Opt. Express*, **20**, pp. 9357–9362.
9. Beijersbergen, M. W., Allen, L., van der Veen, H. E. L. O., and Woerdman, J. P. (1993). Astigmatic laser mode converters and transfer of orbital angular momentum, *Opt. Commun.*, **96**, pp. 123–132.
10. Berry, M. V., and Balazs, N. L. (1979). Nonspreading wave packets, *Am. J. Phys.*, **47**, pp. 264–267.
11. Berry, M. V., and Hannay, J. H. (1977). Umbilic points on Gaussian random surfaces, *J. Phys. A*, **10**, pp. 1809–1821.
12. Born, M., and Wolf, E. (1999). *Principles of Optics* (Cambridge).
13. Bowman, R., D'Ambrosio, V., Rubino, E., Jedrkiewicz, O., Trapani, P. D., and Padgett, M. J. (2011). Optimisation of a low cost SLM for diffraction efficiency and ghost order suppression, *Eur. Phys. J.*, **199**, pp. 149–158.

14. Broky, J., Siviloglou, G. A., Dogariu, A., and Christodoulides, D. N. (2008). Self-healing properties of optical Airy beams, *Opt. Express*, **16**, pp. 12880–12891.

15. Cardano, F., Karimi, E., Marrucci, L., Lisio, C., and Santamato, E. (2013). Generation and dynamics of optical beams with polarization singularities, *Opt. Express*, **21**, pp. 8815–8820.

16. Chavez-Cerda, S., Padgett, M. J., Allison, I., New, G. H. C., Gutierrez-Vega, J. C., O'Neil, A. T., Vicar, I. M., and Courtial, J. (2002). Holographic generation and orbital angular momentum of high-order Mathieu beams, *J. Opt. B*, **4**, pp. S52–S57.

17. Collett, E. (1993). *Polarized Light: Fundamentals and Applications* (Taylor and Francis).

18. Davis, J. A., Carcole, E., and Cottrell, D. M. (1996). Nondiffracting interference patterns generated with programmable spatial light modulators, *Appl. Opt.*, **35**, pp. 599–602.

19. Dennis, M. R. (2002). Polarization singularities in paraxial vector fields: Morphology and statistics, *Opt. Commun.*, **213**, pp. 201–221.

20. Dorn, R., Quabis, S., and Leuchs, G. (2003). Sharper focus for a radially polarized light beam, *Phys. Rev. Lett.*, **91**, pp. 233901–1–4.

21. Durnin, J. (1987). Exact solutions for nondiffracting beams. I. the scalar theory, *J. Opt. Soc. Am. A*, **4**, pp. 651–654.

22. Durnin, J., Miceli Jr., J. J., and Eberly, J. H. (1987). Diffraction-free beams, *Phys. Rev. Lett.*, **58**, pp. 1499–1501.

23. Fahrbach, F. O., and Rohrbach, A. (2012). Propagation stability of self-reconstructing Bessel beams enables contrast-enhanced imaging in thick media, *Nat. Commun.*, **3**(632), pp. 1–8.

24. Fontana, J. R., and Pantell, R. H. (1983). A high-energy, laser accelerator for using the inverse Cherenkov effect, *J. Appl. Phys.*, **54**, pp. 4285–4288.

25. Freund, I. (2001). Polarization flowers, *Opt. Commun.*, **199**, pp. 47–63.

26. Galvez, E. J. (2013). *Vector Beams in Free Space* (Cambridge University Press, Cambridge), pp. 51–69.

27. Galvez, E. J., Khadka, S., Schubert, W. H., and Nomoto, S. (2012). Poincarè-beam patterns produced by non-separable superpositions of Laguerre-Gauss and polarization modes of light, *Appl. Opt.* **51**, pp. 2925–2934.

28. Harke, B., Keller, J., Ullal, C. K., Westphal, V., Schonle, A., and Hell, S. W. (2008). Resolution scaleng in sted microscopy, *Opt. Express*, **16**, pp. 4154–4162.

29. He, H., Friese, M. E. J., Heckenberg, N. R., and Rubinsztein-Dunlop, H. (1995). Direct observation of transfer of angular momentum to absorptive particles from a laser beam with a phase singularity, *Phys. Rev. Lett.*, **75**, pp. 826–829.

30. Heckenberg, N. R., McDuff, R., Smith, C. P., and White, A. G. (1992). Generation of optical phase singularities by computer Bibliography generated holograms, *Opt. Lett.*, **17**, pp. 221–223.

31. Jayasurya, Y. V., Inavalli, V. V. G. K., and Viswanathan, N. K. (2011). Polarization singularities in the two-mode optical fiber output, *Appl. Opt.*, **50**, pp. 131–137.

32. Jones, P. H., Marago, O. M., and Volpe, G. (2015). *Optical Tweezers Principles and Applications* (Cambridge University Press, Cambridge).

33. Lang, M. J., and Block, S. M. (2003). Resource letter: Lbot-1: Laser-based optical tweezers, *Am. J. Phys.*, **71**, pp. 201–215.

34. Leach, J., Wulff, K., Sinclair, G., Jordan, P., Courtial, J., Thompson, L., Gibson, G., Karundi, K., Cooper, J., Laczik, Z. J., and Padgett, M. (2006). Interactive approach to optical tweezers control, *Appl. Opt.*, **45**, pp. 897–903.

35. Lin, Y., Seka, W., Eberly, J., Huang, H., and Brown, D. (1992). Experimental investigation of bessel beam characteristics, *Appl. Opt.*, **31**, pp. 2708–2713.

36. Mair, A., Vaziri, A., Weihs, G., and Zeilinger, A. (2001). Entanglement of the orbital angular momentum states of photons, *Nature*, **412**, pp. 313–316.

37. Marrucci, L., Manzo, C., and Paparo, D. (2006). Optical spin-to-orbital angular momentum conversion in inhomogeneous anisotropic media, *Phys. Rev. Lett.*, **96**, pp. 163905 1–4.

38. McGloin, D., and Dholakia, K. (2005). Bessel beams: Diffraction in a new light, *Contemp. Phys.*, **46**, pp. 15–28.

39. Milione, G., Sztul, H. I., Nolan, D. A., and Alfano, R. R. (2011). Higher-order Poincarè sphere, stokes parameters, and the angular momentum of light, *Phys. Rev. Lett.*, **107**, pp. 053601 1–4.

40. Milonni, P. W., and Eberly, J. H. (1988). *Lasers* (John Wiley & Sons, New York).

41. Padgett, M. J., Molloy, J., and McGloin, D. (2010). *Optical Tweezers: Methods and Applications* (CRC Press Taylor and Francis Group, Boca Raton).

42. Pedrotti, F. L., and Pedrotti, L. S. (1993). *Introduction to Optics* (Prentice Hall).

43. Penrose, R. (1979). The topology of ridge systems, *Ann. Hum. Genet. Lond.*, **42**, pp. 435–444.

44. Planchon, T. A., Gao, L., Milkie, D. E., Galbraith, J. A., Galbraith, C. G., and Betzig, E. (2011). Rapid three-dimensional isotropic imaging of living cells using bessel beam plane illumination, *Nat. Meth.*, **8**, pp. 417–423.

45. Siegman, A. E. (1986). *Lasers* (University Science Books, Mill Valley).

46. Siviloglou, G. A., Broky, J., Dogariu, A., and Christodoulides, D. N. (2008). Ballistic dynamics of Airy beams, *Opt. Lett.*, **33**, pp. 207–209.

47. Smith, S. P., Bhalotra, S. R., Brody, A. L., Brown, B. L., Boyda, E. K., and Prentiss, M. (1999). Inexpensive optical tweezers for undergraduate laboratories, *Am. J. Phys.*, **67**, pp. 26–35.

48. Sztul, H. I., and Alfano, R. R. (2008). The Poynting vector and angular momentum of Airy beams, *Opt. Express*, **16**, pp. 9411–9416.

49. Tlusty, T., Meller, A., and Bar-Ziv, R. (1998). Optical gradient forces in strongly localized fields, *Phys. Rev. Lett.*, **81**, pp. 1738–1741.

50. Torok, P., and Munro, P. R. T. (2004). The use of Gauss-Laguerre vector beams in STED microscopy, *Opt. Express*, **12**, p. 3605.

51. Turunen, J., Vasara, A., and Friberg, A. T. (1988). Holographic generation of diffraction-free beams, *Appl. Opt.*, **27**, pp. 3959–3962.

52. Vasara, A., Turunen, J., and Friberg, A. T. (1989). Realization of general non diffracting beams with computer-generated holograms, *J. Opt. Soc. Am. A*, **6**, pp. 1748–1754.

53. Vasilyeu, R., Dudley, A., Khilo, N., and Forbes, A. (2009). Generating superpositions of higher-order Beasel beams, *Opt. Express*, **17**, pp. 23389–23395.

54. Vettenburg, T., Dalgarno, H. I. C., Nylk, J., Llado, C. C., Ferrier, D.E. K., Cizmar, T., Gunn-Moore, F. J., and Dholakia, K. (2014). Light sheet microscopy using an Airy beam, *Nat. Meth.*, **11**, pp. 541–544.

55. Zhan, Q. (2009). Cylindrical vector beams: From mathematical concepts to applications, *Adv. Opt. Photonics*, **1**, pp. 1–57.

Chapter 3

Gaussian Beam Optical Parameters in Multi-Photon Excitation Fluorescence Imaging

Lingyan Shi,[a,b,c,d] **Adrián Rodríguez-Contreras,**[c,d] **and Robert R. Alfano**[d]

[a]*Department of Chemistry, Columbia University, New York, NY 10027, USA*
[b]*Department of Biomedical Engineering, The City College of New York 160 Convent Ave, New York, NY 10031, USA*
[c]*Department of Biology, The City College of New York 160 Convent Ave, New York, NY 10031, USA*
[d]*Institute for Ultrafast Spectroscopy and Lasers, Departments of Physics The City College of New York 160 Convent Ave, New York, NY 10031, USA*

lshi365@gmail.com

This chapter inspects and compares the critical optical properties of Gaussian laser beam in theory, including the beam spot size, depth of focus, and intensity profile, in two-photon and multiphoton fluorescence imaging for quantifying nanometer-scale solutes.

Deep Imaging in Tissue and Biomedical Materials: Using Linear and Nonlinear Optical Methods
Edited by Lingyan Shi and Robert R. Alfano

ISBN 978-981-4745-88-8 (Hardcover), 978-1-315-20655-4 (eBook)
www.panstanford.com

3.1 Introduction

Gaussian (G) beam is widely used in optics. The profile of the G beam is a key element for imaging in multiphoton microscopy, such as two-photon (2P) and three-photon (3P) applications. A 140 fs pulsed G beam is used in the near-infrared region for 2D and 3D imaging of objects, particularly cellular structures in the brain and other biological systems.

Propagation for G beam in the media has been reviewed in depth in Chapter 1 and by Yariv [1]. This chapter examines the salient properties of G beam in the nonlinear optical process of two-photon fluorescence (2PF) that include the beam spot size, depth of focus (DOF), intensity profile in the radial (r, x, y) direction, and propagation in the z direction.

3.2 Gaussian Beam Model

The intensity I_G of a G beam is given by

$$I_G(r,z) = I_0\left(\frac{w_0}{w(z)}\right)^2 \exp\left(-\frac{2r^2}{w^2(z)}\right), \tag{3.1}$$

where r is the transverse radius, z is the path length, I_0 is the incident intensity, w_0 is the narrowest spot size (also known as the beam waist) that equals $1/e$ of Gaussian electric field E_G, and $w(z)$ is the spot size given by [1]

$$w^2(z) = w_0^2\left(1 + \left(\frac{z}{z_0}\right)^2\right), \tag{3.2}$$

where

$$z_0 = \frac{\pi w_0^2}{\lambda} \tag{3.3}$$

is the Rayleigh length (Fig. 3.1) and λ is the wavelength. Length $2z_0$ is known as the DOF. At the beam waist, where $r = w_0$ and $z = 0$ (Fig. 3.1), the electric field of G beam drops to $1/e$ of the maximum electric field E_0, and the intensity drops to $1/e^2$ of I_0.

The intensity along the z axis (r = 0) can be obtained from Eq. 3.1 as follows:

$$I_G(0, z) = I_0 \frac{1}{1+(z/z_0)^2}, \tag{3.4}$$

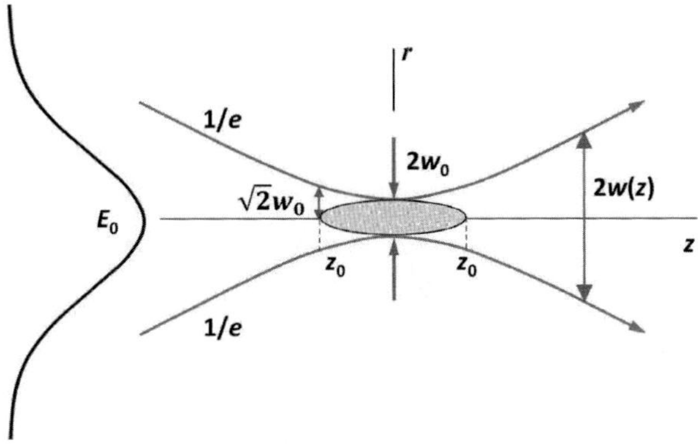

Figure 3.1 Illustration of the key spatial electric field profile of a Gaussian beam. E_0, the maximum incident electric field at the origin; r, the transverse radius; z, path length; z_0, the Rayleigh length; $w(z)$, spot size; w_0, the narrowest spot size (beam waist). The electric field amplitude of Gaussian beam is given by $|E_G(r,z)| = E_0 \frac{w_0}{w(z)} \exp\left[\frac{-r^2}{w(z)^2}\right]$.

which is an elliptical shape with the spanning $2z_0$. The intensity reaches the maximum I_0 at z = 0, and it reduces to ½ I_0 at z_0 (Fig. 3.2a). The DOF is the critical distance for G beam intensity in the z direction and is dependent on the magnification and numerical aperture (NA) of the microscope objective lens. The DOF ($2z_0$) is given by Eq. 3.3 as

$$2z_0 = \frac{2\pi w_0^2}{\lambda} \tag{3.5}$$

and

$$w_0 = \frac{\lambda}{\pi \mathrm{NA}} \tag{3.6}$$

The NA is given by

$$\mathrm{NA} = n\frac{D}{2f}, \tag{3.7}$$

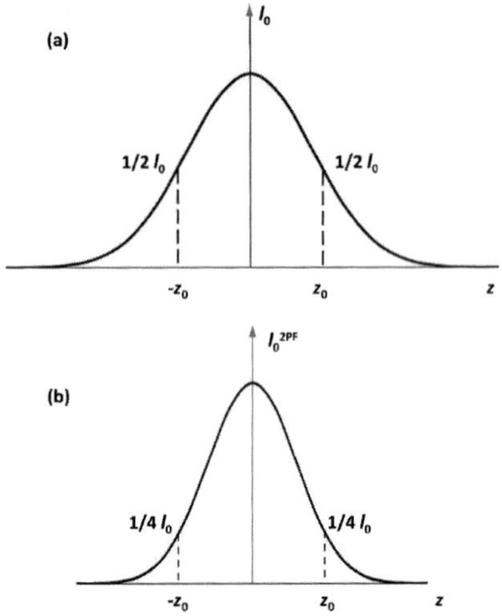

Figure 3.2 Axial intensity of Gaussian beam as a function of z at $r = 0$. (a) At $z = 0$, the intensity decreases to 1/2 at z_0. (b) 2PF intensity profile of 2PF, in which t, the intensity decreases to 1/4 at z_0.

where n is the index of refraction, D the diameter of the lens, and f the focal length. Therefore, the relationship between DOF and NA of the objective lens can be obtained from Eqs. 3.5–3.7

$$2z_0 = \frac{2\lambda}{\pi \mathrm{NA}^2} \tag{3.8}$$

The beam waist (w_0) and DOF ($2z_0$) are two most salient features to locate dyes or diffraction limited bright particles such as quantum dots (QDs) or fluorescent beads in tissue compartments (e.g., inside and outside a blood vessel). At the DOF the focal spot area is blurred to twice of that at the beam waist w_0. Smaller beam waist and a smaller DOF are desired for a better image resolution in the x–y plane and z direction, and for a smaller point spread function (PSF).

3.3 Parameters in Multiphoton Imaging

The intensity I_{2p} of a G beam in 2PF is proportional to the square of the intensity I_G, i.e.,

$$I_{2P} \propto \eta \sigma_2 I_G^2(r,z), \tag{3.9}$$

where η is the fluorescence efficiency and σ_2 is the two-photon dye cross-section with unit GM, where 1 GM = 10^{-50} $cm^4s/photon$ [2]. The intensity I_{2p} is used to locate the excitation value of 2PF. It spreads over $2z_0$ at the z axis, reduces to 1/4 of I_0 at z_0 (Fig. 3.2b), and falls off fast to $1/e^4$ I_0 at the beam waist ($r = w_0$ and $z = 0$).

An important feature of 2PF imaging of G beam is that its emission is in the regions that are occupied by G beam. The G beam scans first in the x–y plane and then along the z direction. The resulting 2PF intensity is displayed in the x–y plane at $z = 0$, and the intensity of the G beam I_G^2 is plotted in y–z (or x–z) plane. The length of the y–z plot is $2z_0$ and the diameter of the vessel or particle is d. Figure 3.3a illustrates the schematic of a microvessel with diameter d in brain tissue, with $d > 2z_0$ and $d > 2w_0$. The size of the 2PF ellipse in the y–z plane (Fig. 3.3b) is the spatial convolution of the G beam and the emitting particle with diameter d. Since the refractive indices of tissue and solvent are close, the reflection coefficient with index $n = 1.375$ [3] at boundaries matches the slight change in intensity due to reflection of the vessel wall. Figure 3.4a demonstrates the convolution of a G beam and the emission of a particle at size d, where $2z_0 > d$ and $2w_0 > d$. The 2PF image pattern is an ellipse with the major axis $2z_0$ (DOF) [4] and the minor axis $2w_0$. Figure 3.4b [5] demonstrates a 2PF image of green beads (200 nm) on brain tissue layer; the image was distorted since the size of the particle was smaller than $2w_0$ and $2z_0$, and corrections for the wavefront distortion were made. Evidently, a small z_0 is necessary to capture the actual images of a vessel in the z direction without any image correction, which is typically 2 μm when using a 40× lens.

Figure 3.5 illustrates an example of the light intensity profile (I) of a cerebral microvessel in rat brain in vivo. Cerebral microvessels (diameter 20–40 μm) are observed using a two-photon microscope (40× lens, NA = 0.8, water immersion, Olympus) through the thinned part of the skull. FITC-dextran-70k (FD70s, Sigma; emits at 520 nm) solution, is injected into the cerebral circulation at concentration 1 mg/ml. The light intensity profile is digitized to demonstrate variations of the 2P emission of fluorescein dye in the y and z directions, respectively. Track lines 1, 2, and 3 represent different cross sections of the microvessel, respectively. Obviously, the peak intensity is higher and the spectral peak width is wider in cross sections close to the center of the lumen. The diameter of the cerebral microvessel imaged (20~40 μm) is much bigger than $2w_0$ and $2z_0$, no distortion occurs and no correction is needed for vessel structure reconstruction. In addition, the light intensity profile can also be used to determine the microvessel permeability and solute diffusion coefficient in the brain tissue [6].

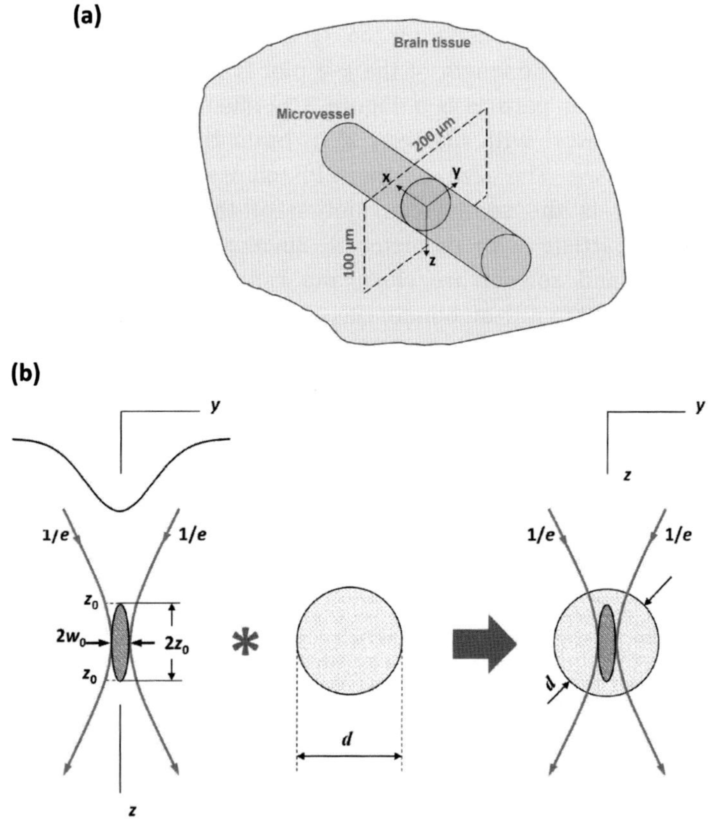

Figure 3.3 (a) Schematic of a microvessel in the brain. Laser beam scans in the z direction. (b) Convolution of Gaussian beam profile with a vessel of diameter d ($d > 2w_0$ and $d > 2z_0$) generates the 2PF pattern of scanning in the y–z plane.

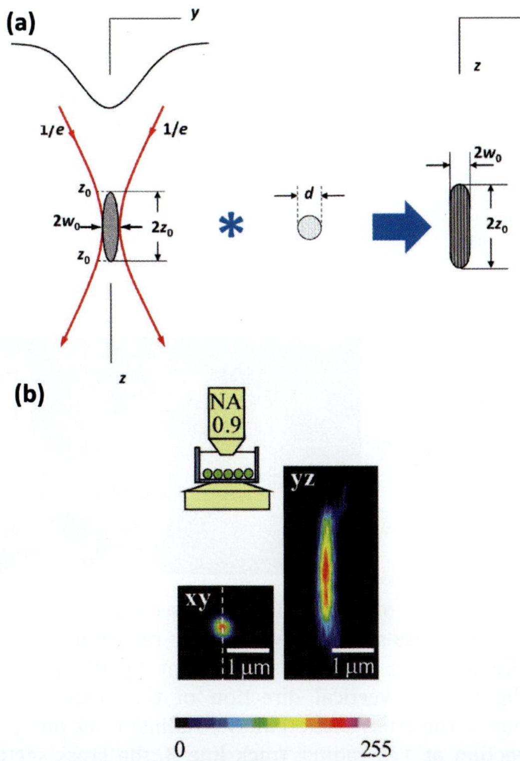

Figure 3.4 (a) Convolution of Gaussian beam profile with a vessel of diameter d ($d < 2w_0$ and $d < 2z_0$) yields the 2PF pattern of scanning in y–z plane. (b) Image of 2PF of emitter particles (beads, $d = 200$ nm $< 2z_0$) (adapted from Fig. 2a, ref. [5]).

The spatial resolution can be further improved if the G beam has a smaller beam waist w_0 or Rayleigh length z_0, which leads to the application of three-photon fluorescence (3PF) or four-photon fluorescence (4PF). For example, substituting Eq. 3.2 into Eq. 3.1 and applying Eq. 3.9 yields the intensity of 2PF

$$I_{2P} \propto \eta\sigma_2 I_G^2 = \eta\sigma_2 \left[\frac{I_0}{1+(z/z_0)^2}\right]^2 \exp\left(-\frac{4r^2}{w^2(z)}\right) \qquad (3.10)$$

The beam intensity falls off to approximately $1/z^4$ at $z >> z_0$ and the beam waist of 2P intensity profile is

$$w_0^{2P} = \frac{1}{\sqrt{2}} w_0 \qquad (3.11)$$

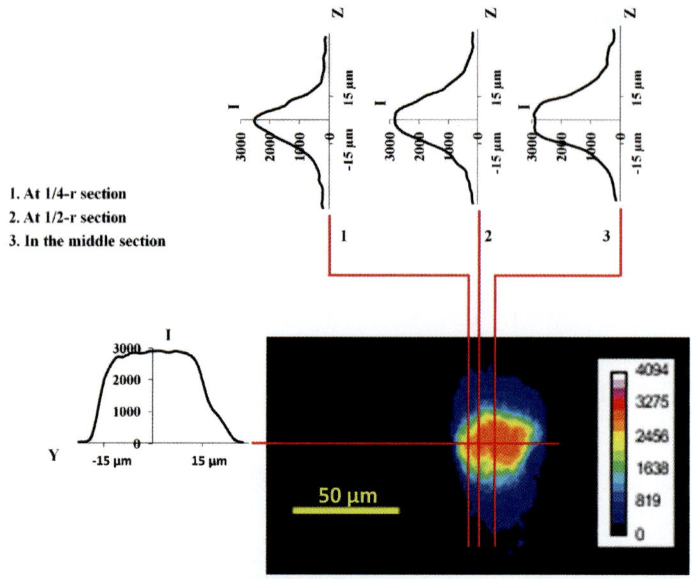

Figure 3.5 Distribution of two-photon fluorescent intensity 2PF profiles in the cross-sectional view of a rat brain microvessel in the *y* and *z* directions. Different spatial positions were digitized at vertical direction of the vessel lumen. Track line 1, the cross section at 1/4 radius; track line 2, the cross section at 1/2 radius; track line 3, the cross section in the middle. The diameter of the vessel $d > 2z_0$ and $2w_0$, which yields full resolution.

The intensity (I_{3P}) of 3PF can be obtained similarly as

$$I_{3P} \propto \eta\sigma_3 I_G^3 = \sigma_3\left[\frac{I_0}{1+(z/z_0)^2}\right]^3 \exp\left(-\frac{6r^2}{w^2(z)}\right), \tag{3.12}$$

where σ_3 is the particle 3P cross section with dye, and the unit is 10^{-80} cm^{6}s^2/photon^2 [7]. The beam intensity falls off to approximately $1/z^6$ at $z >> z_0$. The beam waist of 3P intensity profile is

$$w_0^{3P} = \frac{1}{\sqrt{3}} w_0 \tag{3.13}$$

Similarly, the intensity (I_{4P}) of 4PF is

$$I_{4P} \propto \eta\sigma_4 I_G^4 = \sigma_4\left[\frac{I_0}{1+(z/z_0)^2}\right]^4 \exp\left(-\frac{8r^2}{w^2(z)}\right) \tag{3.14}$$

where σ_4 is the particle 4P cross section with dye, and the unit is 10^{-109} $cm^8s^3/photon^3$ [7]. The beam intensity falls off to approximately $1/z^8$ at $z >> z_0$. The beam waist of 4P intensity profile is

$$w_0^{4P} = \frac{1}{2} w_0 \tag{3.15}$$

The salient parameters of 2P, 3P, and 4P imaging are summarized in Table 3.1. Reduction of beam waist w_0 from 2PF to 4PF leads to higher spatial resolutions in x–y plane. Equation 3.3 shows that a longer wavelength λ will lead to a shorter Rayleigh length z_0, hence, wavelength increase from 2PF to 4PF will generate higher spatial resolution in the z axis. Application of 3PF or 4PF other than 2PF will theoretically improve imaging resolution, however, experiments are still needed to validate the improvement both qualitatively and quantitatively.

Table 3.1 Spatial parameters for multiphoton fluorescence

Beam type	1P	2P	3P	4P
Beam waist	w_0	$w_0/\sqrt{2}$	$w_0/\sqrt{3}$	$w_0/2$
Fall off distance	$\propto \frac{1}{z^2}$	$\propto \frac{1}{z^4}$	$\propto \frac{1}{z^6}$	$\propto \frac{1}{z^8}$
DOF	$2z_0$	$1.28z_0$	z_0	$0.82z_0$
Intensity at beam waist	e^{-2}	e^{-4}	e^{-6}	e^{-8}

Gaussian beam surpasses Bessel beam for backward and forward multiphoton 3D imaging. Bessel beam is diffraction free and well collimated over long distance (>1 mm) with surrounding rings. It works well in multiphoton excitation in perpendicular viewing and scanning, and can be used in light sheet microscopy to generate 2D and, to some extent, 3D images of thick media [8–10]. However, in multiphoton excitation only ~5% of the Bessel beam is located in the central spot and a large amount is located in the rings, which are not important in the formation of images. Therefore, the long collimated Bessel beam is not ideal to locate nanometer-scaled structures in the z-axis. Gaussian beam overcomes the weaknesses of Bessel beam since its focal spot is well localized, which enables Gaussian beam to locate nanometer structures in z-stacks acquired with the backward transmission microscope. Hence, Gaussian beam is commonly employed in optical slicing and 3D reconstruction.

In summary, the spatial forms of Gaussian beam (w_0 and z_0), with resolution at micrometer scale, provide the information needed for determining microvessel permeability and solute diffusion coefficients in tissue. These properties can be used to investigate neuronal or vascular structure in physiological or biological process in vivo or ex vivo.

References

1. Yariv, A. (1989). *Quantum Electronics,* 3rd ed. (John Wiley & Sons, USA).
2. Drobizhev, M., Makarov, N. S., Tillo, S. E., Hughes, T. E., and Rebane, A. (2011). Two-photon absorption properties of fluorescent proteins. *Nat. Methods,* **8**, pp. 393–399.
3. Lister, T., Wright, P. A., and Chappell, P. H. (2012). Optical properties of human skin. *J. Biomed. Opt.*, **17**, p. 90901.
4. Ying, J. P., Liu, F., and Alfano, R. R. (1999). Spatial distribution of two-photon-excited fluorescence in scattering media. *Appl. Opt.*, **38**, pp. 224–229.
5. Chaigneau, E., Wright, A. J., Poland, S. P., Girkin, J. M., and Silber, R. A. (2011). Impact of wavefront distortion and scattering on 2-photon microscopy in mammalian brain tissue. *Opt. Express,* **19**, pp. 22755–22774.
6. Shi, L., Zeng, M., Sun, Y., and Fu, B. M. (2014). Quantification of blood-brain barrier solute permeability and brain transport by multiphoton microscopy. *J. Biomech. Eng.*, **136**, p. 031005.
7. Corrêa, D. S., Leonardo, De Boni, L., Misoguti, L., Cohanoschi, I., Hernandez, F., and Mendonca, C. R. (2007). Z-scan theoretical analysis for three-, four- and five-photon absorption. *Opt. Commun.*, 277, pp. 440–445.
8. Fahrbach, F. O., Gurchenkov, V., Alessandri, K., Nasoy, P., and Rohrbach, A. (2013). Light-sheet microscopy in thick media using scanned Bessel beams and two-photon fluorescence excitation. *Opt. Express,* **21**, pp. 13824–13839.
9. Lin, Y., Seka, W., Eberly, J. H., Huang, H., and Brown, D. L. (1992). Experimental investigation of Bessel beam characteristics, *Appl. Opt.*, **31**, pp. 2708–2713.
10. Dumin, J., Miceli, J., and Eberly, J. H. (1988). Comparison of Bessel and Gaussian beams. *Opt. Lett.*, **13**, pp. 79–80.

Chapter 4

The Optics of Deep Optical Imaging in Tissues Using Long Wavelengths

Steven L. Jacques

Departments of Biomedical Engineering and Dermatology, Oregon Health & Science University, Portland, Oregon, USA

jacquess@ohsu.edu

The ability of long wavelength (>1150 nm) photons to penetrate into a tissue or through a thin tissue section is modeled using Monte Carlo simulations based on the optical properties of breast and brain and is compared with the experiments of Shi et al. [1]. Conclusions are as follows: (1) Longer wavelengths are disadvantageous when imaging with multiply scattered photons due to increased water absorption. (2) Longer wavelengths are advantageous when using a subset of near-ballistic photons with shorter pathlengths to reach a target depth or detector. (2a) This subset increases with a lower scattering coefficient μ_s, which coincides with a lower anisotropy of scatter, g, when the experimental value for $\mu_s' = \mu_s(1 - g)$ is considered to be constant. Longer wavelengths (e.g., 1700 nm) that exceed the sub-μm size distribution of scatterers within a soft tissue yield lower g and μ_s. (2b) The narrow versus broad balance of the scattering function (g_2) can vary even though the forward versus backward balance (g or g_1) is held constant. An increasingly

Deep Imaging in Tissue and Biomedical Materials: Using Linear and Nonlinear Optical Methods
Edited by Lingyan Shi and Robert R. Alfano

ISBN 978-981-4745-88-8 (Hardcover), 978-1-315-20655-4 (eBook)
www.panstanford.com

narrow scatter function increases photon penetration. (2c) At long wavelengths, the absorption coefficient matches or exceeds the reduced scattering coefficient in magnitude, causing multiply scattered photons to be selectively absorbed relative to near-ballistic photons with the shortest path to a target depth, which improves imaging as reported by Yoo et al. [3].

4.1 Introduction

There have been tremendous strides in a variety of optical imaging modalities, some of which, e.g., absorption, reflectance, Raman scattering, IR spectroscopy, photoacoustics, might benefit by using longer wavelengths for deep imaging [1]. Deep imaging involves imaging that does not use diffuse multiply scattered photons, but instead uses near-ballistic photons whose trajectories within a tissue are only partially deflected by scattering. Deep imaging should be sensitive to the optical absorption by tissue water and the optical scattering by tissue structure. At long wavelengths above ~1150 nm, the absorption coefficient of water is comparable to or greater than the reduced scattering coefficient of soft tissues ($\mu_a \geq \mu_s'$). This chapter considers the mechanisms involved in deep imaging using such long wavelengths.

Figure 4.1 shows the absorption by water, expressed as the absorption coefficient μ_a [cm^{-1}] versus wavelength. The figure emphasizes five transmissive windows of relatively low water absorption centered at 800, 1060, 1263, 1700 and 2200 nm wavelengths. The figure also shows the reduced scattering coefficients for the five windows, based on the mean value of μ_s' [cm^{-1}] for brain and breast [2]. The μ_s' spectrum is expressed as

$$\mu_s'(\lambda) = a\left(f_{\text{ray}}\left(\frac{\lambda}{500\text{ nm}}\right)^{-4} + f_{\text{mie}}\left(\frac{\lambda}{500\text{ nm}}\right)^{-b_{\text{mie}}}\right), \tag{4.1}$$

where a equals μ_γ^f (500 nm), f_{ray} is the fraction of light scattered at 500 nm due to Rayleigh scattering by sub-μm structures much smaller than the wavelength of light, $f_{\text{mie}} = 1 - f_{\text{ray}}$ is the fraction scattered at 500 nm due to Mie scattering by larger structures comparable to or larger than the wavelength of light, b_{mie} is the scatter power for the Mie component of scatter. The mean values of breast and brain are a = 23 cm^{-1}, f_{ray} = 0.30, and b_{mie} = 0.89, from a review of the still too limited set of published tissue optical properties [2]. Equation 5.1 is approximately equivalent to the commonly cited function $\mu_s' = a(\lambda/\lambda_{ref})^{-b}$, where λ_{ref} is a chosen reference wavelength, e.g., λ_{ref} = 500 nm, for wavelengths in the range of 400–1000 nm, which are commonly measured. However, this author suspects that Eq. 4.1 is better for extrapolating to the longer wavelengths above 1150 nm where experimental measurements are lacking. Table 4.1 summarizes the optical properties of $\mu_{a,water}$ and $\mu'_{s\text{tissue}}$ (the mean of breast and brain properties) at the center wavelengths of these windows.

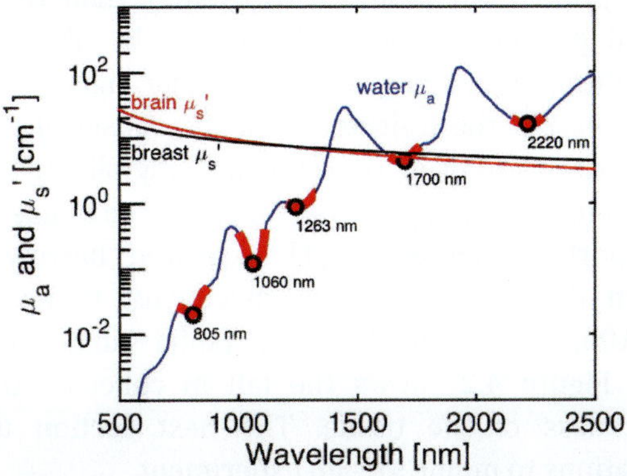

Figure 4.1 Absorption coefficient, μ_a [cm^{-1}] (blue line), and expected reduced scattering coefficient, μ_s' [cm^{-1}], versus wavelength (red line for brain, black line for breast). The μ_s' values are extrapolated from data in the 600–1000 nm range to λ > 1000 nm using Eq. 4.1 [2].

Table 4.1 Optical absorption of water and reduced scattering of soft tissues ($\mu'_{s\text{tissue}}$ equals mean μ_s' of brain and breast spectra [2])

Wavelength (nm)	$\mu_{a.water}$ (cm^{-1})	$\mu_s'_{tissue}$ (cm^{-1})
800	0.020	11.5
1060	0.120	8.5
1263	0.935	7.1
1700	4.440	5.4
2220	16.130	4.3

The regions where water absorption matches or exceeds tissue scattering include the two windows at 1700 and 2200 nm.

Multiply scattered photons on average have long pathlengths as they attempt to penetrate into a tissue to some particular target depth or detector, and hence they are strongly absorbed. In contrast, a subpopulation of near-ballistic photons that either avoid scatter or are scattered by only a slight deflection angle can penetrate directly to the target depth with a shorter pathlength. Hence, they are less strongly absorbed. Such selective absorption improves the contrast of imaging that uses near-ballistic photons. This strategy for deep imaging was first reported by Yoo et al. [3], who improved image contrast by using a wavelength more strongly absorbed by water.

Figure 4.1 shows that the 1700 and 2200 nm windows have water absorption comparable to or greater than the scattering. Such optical properties cause photons to be absorbed quickly before significant scattering can occur. The signal drops due to the absorption, but the trajectories of the subset of photons that still reach a deep target or detector are only slightly deviated by scattering. Hence, there is potential for improved image contrast.

The report by Shi et al. [1] measured narrow-solid-angle transmission through thin tissue specimens, tissue thicknesses L = 50, 100, 150, and 200 μm (solid angle of collection $\Omega \approx 0.01$). Figure 4.2 shows the fall in collected transmission versus thickness of the tissue. The next section uses Monte Carlo simulations to mimic the Shi experiment.

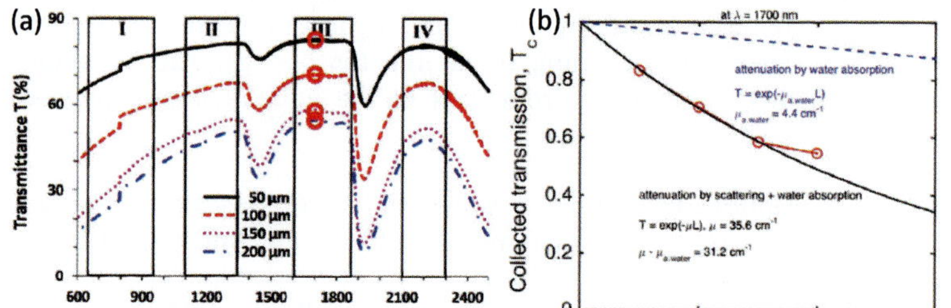

Figure 4.2 (a) The Shi experiment at 1700 nm wavelength (from [1] with permission). Red circles added to indicate the collected transmission T_c at 1700 nm. (b) The T_c (red circles) fell exponentially versus tissue thickness L with an attenuation coefficient μ = 35.6 cm^{-1} (black line). The attenuation by pure water ($\mu_{a.water}$) is shown (blue dashed line), and the difference $\mu - \mu_{a.water}$ is attributed to scattering.

4.2 Monte Carlo Simulations

This section uses Monte Carlo simulations to illustrate the penetration of light into a thick tissue or through a thin tissue specimen. First, the simulations consider the light penetration into a thick tissue at the center wavelengths of the five windows of lower water absorption. Second, the simulations recreate the Shi experiment [1] of narrow collection of transmission, T_c, through thin tissue sections.

4.2.1 Modeling Light Penetration into a Thick Tissue

When absorption is dominant over scattering, as is the case for the 1700 and 2200 nm windows, diffusion theory is not useful and Monte Carlo simulations are a computational option that properly predicts light distributions within the tissue and transmission of light through a thin tissue thickness. This study used the MCML simulation [4] to generate the relative fluence rate, $\varphi(z, r)$ [W/cm^2 per W delivered] or [cm^{-2}], which equals the relative energy concentration, C [J/cm^3 per J delivered] or [cm^{-3}], times the speed of light in the medium, c [cm/s]: $\varphi = cC$. So fluence rate essentially means photon energy concentration. The Monte Carlo simulation uses the absorption coefficient, μ_a [cm^{-1}], the scattering coefficient μ_s [cm^{-1}], and the anisotropy of scattering, g [dimensionless] = $\langle\cos\theta\rangle$ where θ is the angle of deflection by a single scattering event. The g was assumed to be 0.90 (see Discussion), and $\mu_s = \mu_s'/(1 - g)$. The average optical properties for brain and breast in Table 4.1 were used in the simulations.

Figure 4.3 shows the iso-φ contours for the center wavelengths of the 5 windows of relatively lower water absorption. Note that the contours are very similar for all wavelengths but are slightly less penetrating for the longer wavelengths due to increased water absorption. The lower scattering at longer wavelengths is not sufficient to overcome the increase in water absorption at longer wavelengths. There is no advantage to using long wavelengths with high water absorption if one is only measuring diffusely scattered photons. Shorter wavelengths that avoid water absorption are better.

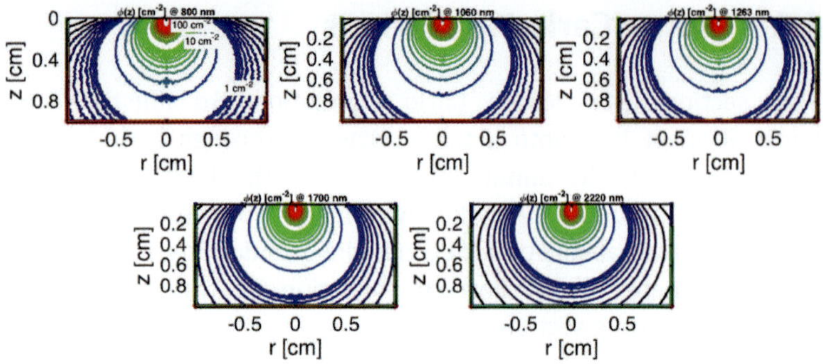

Figure 4.3 Monte Carlo simulations of relative fluence rate distributions in tissue, $\varphi(z, r)$ [W/cm^2]. Iso-φ contours for φ = 0.1 to 1 to 10 to 100 cm^{-2}. These distributions are for the total population of photons, which is dominated by multiply scattered photons. Photons diffuse down concentration gradients.

Figure 4.4 shows the attenuation of the fluence rate φ along the central z-axis, $\varphi_{\text{on-axis}}$ at $r = 0$, as a function of depth (z) within the thick tissue simulation. There is an initial fast attenuation, $K \exp(-\mu z)$, where scattering deflects photons from the central z-axis, then a transition to a slow attenuation versus depth in the diffusive regime at deeper depths, where multiply scattered light diffuses slowly into the tissue. This transition occurs at a depth $z = 1/\mu_s'$ called the transport mean free path, which equals 0.183 cm for the 1700 nm wavelength. The slope of the initial fast attenuation varies with wavelength, and the values of μ are listed in Fig. 4.4. The attenuation μ for 800 nm is ~51 cm^{-1}, while the attenuation for 1700 nm is ~44 cm^{-1}. So there is an approximately 2-fold advantage to using the 1700 nm window for penetration to a focus at a depth of 1 mm in a tissue ($e^{-44(0.1\text{ cm})}/e^{-51(0.1\text{ cm})} \approx 2$). Beyond ~2 mm, however, the diffusely scattered photons dominate, and longer wavelengths are less penetrating than shorter wavelengths due to increased water absorption.

Also shown in Fig. 4.4 is the attenuation at 1700 nm reported by Shi et al. [1], ~36 cm^{-1}, which is only 18% lower than the Monte Carlo simulation of μ at 1700 nm, ~44 cm^{-1}, indicating that the Shi experiment is consistent with the simulated behavior for on-axis transmission through a thin soft tissue like breast or brain.

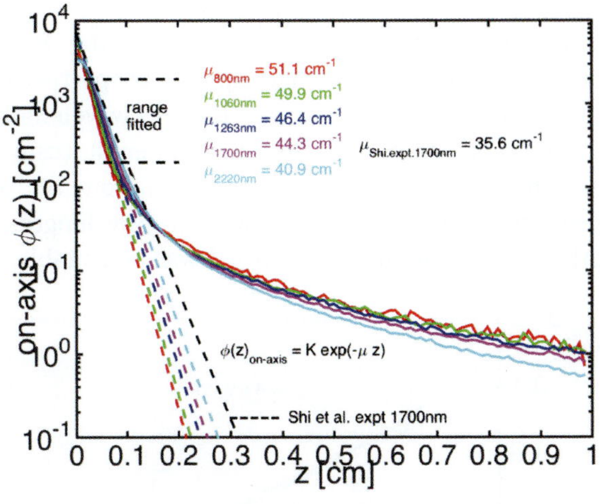

Figure 4.4 The on-axis fluence rate, $\varphi(z)_{\text{on-axis}}$ versus depth z, from Monte Carlo simulations of an average breast/brain tissue (Table 4.1). The initial rapid attenuation, K exp($-\mu z$), is primarily due to scattering (K is a scaling factor [cm^{-2}]). The slow attenuation for z beyond 0.2 cm characterizes the diffusion of multiply scattered photons. There is a modest wavelength dependence to the $\varphi_{\text{on-axis}}$. The attenuation at 1700 nm reported by Shi et al. [1] is shown as a black dashed line, scaled by the mean value of K for the Monte Carlo simulations.

4.2.2 Modeling the Shi Experiment of Narrow Transmission Using Thin Tissues

The Monte Carlo simulation mimicked the experiment of Shi et al. [1], which measured transmission into a narrow solid angle of collection, $\Omega \approx 0.01$ steradians, using a commercial spectrometer. The simulations were for the experiment using 1700 nm light and for the experimental tissue thicknesses, L = 50, 100, 150, and 200 µm. The simulated transmission was collected in 9 bins of angular deflection, from 0 to 90°. The first 10° bin corresponded to an incremental solid angle of collection of $d\Omega$ = 0.0076 sr $\approx$ 0.01 sr. Therefore, the transmission collected in the first bin (0–10°) was used to simulate the Shi experiment.

Figure 4.5 shows that T_c at 1700 nm decreased as the thickness L increased, $T_c \approx$ K$e^{-\mu L}$, with $\mu \approx$ 42 cm^{-1}, similar to the ~36 cm^{-1} reported by Shi et al. [1] (Fig. 4.2) and similar to the ~44 cm^{-1} indicated in the initial fast attenuation of $\varphi(z)_{\text{on-axis}}$ within the tissue (Fig. 4.4). Again, the Monte Carlo simulations, based on the average optical properties of breast and brain tissues extrapolated to longer wavelengths using Eq. 4.2, were consistent with the experimental results of Shi et al., which suggests that the extrapolation of published optical properties at visible/near-infrared wavelengths to long wavelengths above 1150 nm using Eq. 4.1 was reasonable.

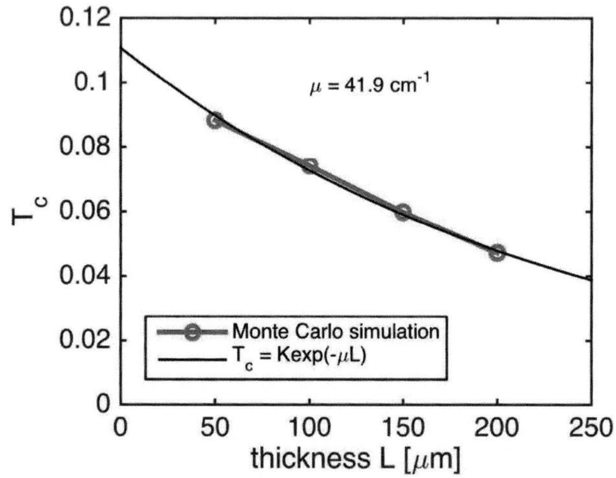

Figure 4.5 Narrow collected transmission T_c within the first 10° off central z-axis along r = 0, as a function of tissue thickness, which mimics the Shi experiment at 1700 nm. The value of μ is ~42 cm^{-1}.

4.2.3 Summary of Results

So what have we learned?

(1) Using diffusely scattered photons

Longer wavelengths, such as the 1700 window of relatively lower water absorption, are of no advantage when using diffuse multiply scattered photons, which have rather long pathlengths before reaching a target depth within a tissue or transmitting through a thin tissue thickness to a detector. Shorter wavelengths that avoid water absorption penetrate better despite higher scattering since the wavelength-dependent fall in scattering, $(\lambda/500\ \text{nm})^{-\text{bmie}}$, is relatively slow.

(2) Using near-ballistic photons

There is a subset in the population of photons penetrating a tissue, in which photons are only slightly deflected. Although small, this subset of photons is capable of better image contrast than diffusely scattered photons. For example, using a narrow solid angle of collection of transmission through a thin tissue ($<1/\mu_s'$ thickness), as in the experiment of Shi et al. [1], allows preferential collection of these near-ballistic photons relative to multiply scattered photons. As the tissue thickness increases above 2 mm, the signal from partially scattered photons drops below the tsunami of diffusely scattered photons transmitting through the tissue (see Fig. 4.4).

This model suggests that imaging beyond 2 mm will always rely on diffuse multiply scattered photons. Is deep imaging using wavelengths above 1050 nm a myth? Does the improvement in image contrast only hold for tissues less than 2 mm thickness? Not necessarily so. The modeling used in Figs. 4.3–4.5 relied on some assumptions. The following Discussion addresses these assumptions and presents the rationale for why deep imaging using longer wavelengths of light may be possible.

4.3 Discussion

The possibility of deep imaging, where imaging uses near-ballistic photons and avoids diffuse multiply scattered photons, is not dismissed by the previous Monte Carlo model, which was based on some assumptions. This Discussion addresses two key assumptions and shows how deep imaging using longer wavelengths of light may be possible.

Assumption 1: The anisotropy g equals 0.90.

The experimental measurements reviewed in [2] and summarized in Table 4.1 specify an experimental value of μ_s' at each wavelength, and these values are accepted as fixed values. By assuming a value of g, the value of μ_s' at each wavelength equals $\mu_s'/(1 - g)$. The previous model assumed $g = 0.90$. However, if the assumed value of g is 0.50, the value of μ_s decreases 5-fold, which means that there are 5-fold fewer scattering events to deplete the on-axis fluence, $\varphi(z)_{\text{on-axis}}$.

Figure 4.6 compares the pattern of light penetration for a tissue with optical properties at 1700 nm (μ_a = 4.440, μ_s' = 5.4, from Table 4.1), when assuming g = 0.50 versus g = 0.90. With g = 0.90, there is a high μ_s (54 cm^{-1}) and a scattering event occurs on average every 1/54 cm^{-1} or 185 μm. In contrast, with g = 0.50, the μ_s is lower (10.8 cm^{-1}) and scattering occurs on average every 926 μm. With g = 0.50, a subset of photons can penetrate deeply, seen in Fig. 4.6b as a tongue of more deeply penetrating photons along the central z-axis. The iso-φ contour for φ = 0.1 cm^{-2} shows the subset of on-axis photons penetrating to ~8 mm depth before being overcome by diffuse light propagation.

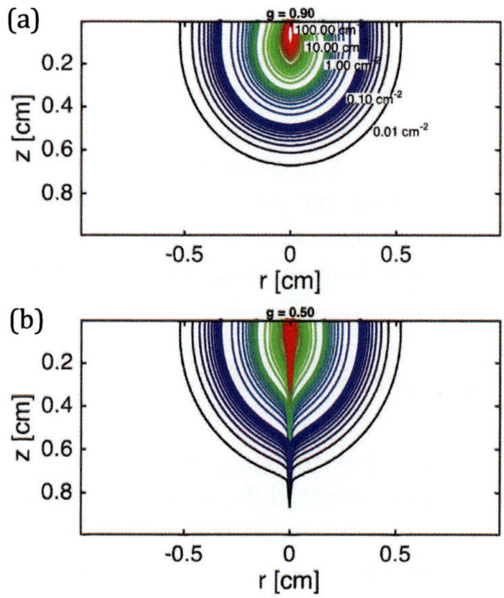

Figure 4.6 Monte Carlo simulations of fluence rate distribution for constant μ_s' = 5.4 cm^{-1} and μ_a = 4.4 cm^{-1} (Table 4.1), while assuming g equals either 0.90 (a) or 0.50 (b). More on-axis photons survive when g = 0.50 than g = 0.90, due to the corresponding drop in μ_s.

Figure 4.7a shows the on-axis fluence, $\varphi(z)_{\text{on-axis}}$ for the two cases of g = 0.90 and g = 0.50 in Fig. 4.6. The $\varphi(z)_{\text{on-axis}}$ is stronger versus depth for the case of g = 0.50. Figure 4.7b shows the ratio $\varphi(z)_{\text{on-axis.g = 0.50}}/\varphi(z)_{\text{on-axis.g = 0.90}}$, which has a maximum value of 27 at z = 0.38 mm and 5 at z = 1 cm. Hence, the assumed value of g has a big effect on the prediction of on-axis light penetration.

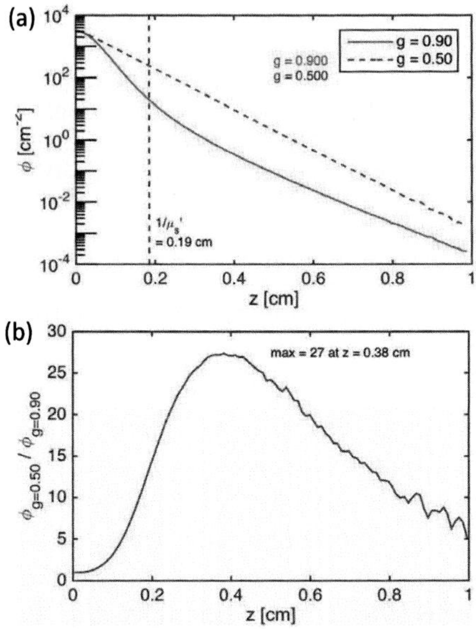

Figure 4.7 On-axis fluence rate, $\varphi(z)_{\text{on-axis}}$, for g = 0.90 and g = 0.50, while μ_s' is constant at 5.4 cm^{-1} and μ_a = 4.4 cm^{-1} (using Monte Carlo simulations of Fig. 4.6). (a) $\varphi(z)_{\text{on-axis}}$ [cm^{-2}]. (b) Ratio $\varphi(z)_{\text{on-axis.g = 0.50}}/\varphi(z)_{\text{on-axis.g=0.90}}$.

This result may seem counter-intuitive. Why should forward-directed scatter (g = 0.90) be less penetrating than broad scatter (g = 0.50)? The answer is that μ_s' is constant at the experimentally measured value, so decreasing g requires a corresponding decrease in μ_s, which increases the number of ballistic photons that penetrate without scatter.

Why would g be lower for λ = 1700 nm than for λ = 800 nm? Tissues present a range of sub-μm structure that scatters light, which can be approximately modeled using Mie scattering by spheres in the size range of 200–1000 nm. The ratio of wavelength/diameter governs the value of g, which falls as the ratio increases. So a lower g in soft tissue is expected for 1700 nm light than for 800 nm light, as indicated in Table 4.2. A more detailed analysis would consider the size distribution and close packing of spheres in developing this approximate model, but Table 4.2 illustrates the concept that g should be lower for

1700 nm light than for 800 nm light, and consequently μ_s is also lower, which allows better penetration of ballistic and near-ballistic photons.

Table 4.2 Mie theory predictions of the g values for various size scattering spheres ($n_{sphere} = 1.38$, $n_{medium} = 1.35$). For a given diameter, g is smaller for the longer wavelength.

Diameter (μm)	0.200	0.400	0.600	0.800	1.000
$\lambda = 800$ nm	0.1892	0.6590	0.8078	0.8860	0.9184
$\lambda = 1700$ nm	0.0406	0.1670	0.3785	0.6127	0.7478

Assumption 2: The scattering function is a Henyey–Greenstein function.

The Monte Carlo modeling for Figs. 4.3–4.7 used the Henyey–Greenstein scattering function to model the probability of deflection by an angle θ during a single scattering event. While this assumption works well for the total population of multiply scattered light, it is not necessarily accurate for the subset of photons that propagate on-axis in a forward direction. The g value equals $\langle\cos\theta\rangle$, which describes the balance between forward and backward scatter and governs diffusion of multiply scattered photons. But the scattering can be either narrow or broad, even if the forward/backward balance is constant.

The narrow/broad balance can be described by the average second Legendre polynomial, $g_2 = \langle(3\cos\theta^2 - 1)/2\rangle$. The anisotropy g is equal to the average first Legendre polynomial, $g = g_1 = \langle\cos\theta\rangle$. This consideration of g_1 versus g_2 has been described [5, 6].

To illustrate this effect, a Monte Carlo program that allows for an arbitrary scattering function $p(\theta)$ was prepared, based on MCML. The user selects a choice of g_1 and g_2, then the program uses least-squares fitting to find a $p(\theta)$ using the first 10 Legendre polynomials. Keeping μ_a, μ_s and g_1 constant at 4.44 cm^{-1}, 54 cm^{-1} and 0.90, respectively, three values of g_2 (0.7, 0.8, 0.9) were simulated. Figure 4.8a shows the results in terms of the $\varphi(z)_{\text{on-axis}}$. As g_2 increased, the scattering became narrower and the initial rapid attenuation μ decreased. The differences in μ for various g_2 values were similar to the differences between the Shi experiment and the Monte Carlo simulations Figs. 4.4 and 4.5). Figure 4.8b shows how the $\varphi(z)_{\text{on-axis}}$ at 0.5 cm depth varied with g_2, roughly doubling as g_2 increased from

0.7 to 0.9. Therefore, the details of the narrow/broad balance of the scattering function should modestly influence the light penetration.

In summary, assumption #1 that the anisotropy g equals 0.90 affects the deeper penetration of light. For a constant μ_s', a lower g coincides with a lower μ_s and hence more photons can penetrate without scatter. Assumption #2 that the scattering function is a Henyey–Greenstein function affects the early attenuation within ~2 mm. Although the attenuation beyond ~2 mm (i.e., the slope in Fig. 4.8b) is the same for different g_2, the advantage of higher φ due to high g_2 achieved during the initial penetration is not lost, but continues to propagate into the tissue. Narrow scatter (high g_2) yields higher φ at deeper depths than broad scatter (low g_2).

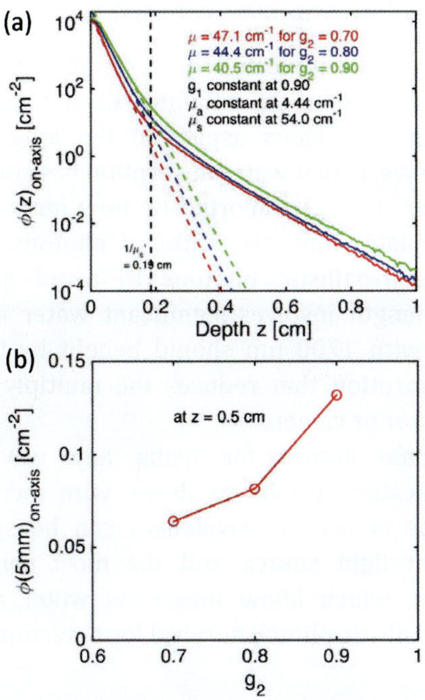

Figure 4.8 How the narrow/broad balance (g_2) in the scattering function affects photon penetration into a soft tissue. Monte Carlo simulations varied g_2 (0.7, 0.8, 0.9) while other optical properties were kept constant: g_1 = 0.90, μ_s' = 5.4 cm^{-1}, μ_a = 4.4 cm^{-1} (optical properties for 1700 nm in Table 4.1; simulation of $\varphi(z)_{\text{on-axis}}$ in a thick tissue) increasing g_2 (more narrow scatter) decreases attenuation μ[cm^{-1}] of $\varphi(z)_{\text{on-axis}}$. (a) $\varphi(z)_{\text{on-axis}}$. (b) $\varphi(z = 0.5\text{ cm})_{\text{on-axis}}$.

Keep in mind, however, that beyond 2 mm diffuse multiply scattered photons dominate and image contrast is poor. Therefore, the criterion that g is low (e.g., closer to 0.5) so that μ_s is low and a central tongue of near-ballistic photons penetrates deeply (Figs. 4.6 and 4.7) is the more important key to deep imaging.

This chapter has extrapolated published optical scattering properties to longer wavelengths, emphasizing the 1700 nm wavelength. However, there is little real data in this wavelength range. The work of Shi et al. [1] is a good beginning, and the simulations using extrapolated data are close to the Shi experiments. Nevertheless, more experimental work is needed at long wavelengths in thicker tissues greater than 200 μm.

Increased absorption improves contrast

The above discussion indicated how changes in the scattering properties of a tissue could influence deep imaging of tissues. However, the most obvious aspect of the longer wavelengths, 1700 nm and above, is that water absorption is strong.

In 1991, Yoo et al. [3] reported on how increased absorption selectively attenuates multiply scattered photons, while allowing ballistic and near-ballistic photons to avoid absorption. The 1700 nm wavelength involves significant water absorption, and hence imaging with 1700 nm should benefit by this mechanism of selective absorption that reduces the multiply scattered light reaching a detector or camera.

This rationale argues for using any wavelength where $\mu_s \geq \mu_s'$. Since scattering falls so slowly with increasingly longer wavelengths, the choice of wavelength can be specified by the most convenient light source and the most sensitive detector above 1150 cm, which allow maximum water absorption and sufficient near-ballistic photon survival for detection.

4.4 Conclusions

Following are the two major conclusions:

(1) Longer wavelengths are disadvantageous when imaging with multiply scattered photons due to increased water absorption.

(2) Longer wavelengths are advantageous when using a subset of near-ballistic photons with shorter pathlengths to reach a target depth or detector. There are three sub-conclusions:

(2a) This subset increases with lower scattering (μ_s) and anisotropy of scatter (g). Given an experimental value for $\mu_s' = \mu_s(1 - g)$, a drop in g requires a corresponding drop in μ_s, which allows photons to suffer fewer scattering events and penetrate more deeply on axis. Longer wavelengths (e.g., 1700 nm) that exceed the sub-μm size distribution of scatterers within a soft tissue yield lower g. This mechanism for deep imaging showed a 27-fold increase in the $\varphi_{\text{on-axis}}$ at a depth of 0.4 cm when using $g = 0.5$ versus $g = 0.9$ while μ_a and μ_s' were constant (see Table 4.1).

(2b) The narrow versus broad balance of the scattering function can vary even though the forward versus backward balance (g) is held constant. An increasingly narrow scatter function (higher g_2) increases photon penetration. This mechanism shows a 2-fold increase in $\varphi(z)_{\text{on-axis}}$ at a depth of 0.5 cm when $g_2 = 0.9$ rather than $g_2 = 0.7$ while μ_a and μ_s' were constant (see Table 4.1). This effect counters the effect 2a above, but is a weaker effect.

(2c) At long wavelengths ($\lambda > 1150$ nm), the absorption coefficient matches or exceeds the reduced scattering coefficient in magnitude, causing multiply scattered photons to be selectively absorbed relative to near-ballistic photons with the shortest path to a target depth or detector. This selective absorption improves imaging, as reported by Yoo et al. [3].

References

1. Shi, L., Sordillo, L. A., Rodríguez-Contreras, A., Alfano, R. R. (2016). Transmission in near-infrared optical windows for deep brain imaging. *J. Biophoton.*, **9**(1–2), pp. 38–44.

2. Jacques, S. L. (2013). Optical properties of biological tissues: A review. *Phys. Med. Biol.*, **58**, pp. R37–R61.

3. Yoo, K. M., Liu, F., Alfano, R. R. (1991). Imaging through a scattering wall using absorption. *Opt. Lett.*, **16**, pp. 1068–1070.

4. Wang, L. H., Jacques, S. L., Zheng, L. Q. (1995). MCML–Monte Carlo modeling of photon transport in multi-layered tissues. *Comput. Methods Programs Biomed.*, **48**, pp. 131–146.

5. Bevilacqua, F., Depeursinge, C. (1999). Monte Carlo study of diffuse reflectance at source–detector separations close to one transport mean free path. *J. Opt. Soc. Am. A*, **16**, p. 2935.

6. Kanick, S. C., Gamm, U. A., Schouten, M., Sterenborg, H. J. M. C., Robinson, D. J., Amelink, A. (2011). Measurement of the reduced scattering coefficient of turbid media using single fiber reflectance spectroscopy: Fiber diameter and phase function dependence. *Biomed. Opt. Express*, **2**, p. 1687.

Chapter 5

Light Propagation and Interaction in Highly Scattering Media for Deep Tissue Imaging

W. B. Wang,[a,b] Lingyan Shi,[a,c] Luyao Lu,[a,d] Laura A. Sordillo,[a] L. Wang,[a] S. K. Gayen,[b,e,f] and R. R. Alfano[a,b,e]

[a]*Institute for Ultrafast Spectroscopy and Lasers, City College of New York, New York, USA*
[b]*Physics Department, City College of New York, New York, USA*
[c]*Department of Biology, City College of New York, New York, USA*
[d]*Department of Biomedical Engineering, Wenzhou Medical University, Wenzhou, China*
[e]*Physics Program, Graduate Center of the City University of New York, New York, USA*
[f]*Center for Discovery and Innovation, City College of New York, New York, USA*

ralfano@ccny.cuny.edu

The salient properties of laser light are directionality, coherence, polarization, and wavelength, as well as ultrashort pulse width of ultrafast lasers, and center wavelength and bandwidth for continuous-wave lasers. These properties can change significantly as light propagates through a highly scattering medium such as clouds, fog, smoke, murky water, and biological tissues. This chapter provides a brief overview of key concepts needed for understanding light propagation deep into a turbid medium, and

Deep Imaging in Tissue and Biomedical Materials: Using Linear and Nonlinear Optical Methods
Edited by Lingyan Shi and Robert R. Alfano

ISBN 978-981-4745-88-8 (Hardcover), 978-1-315-20655-4 (eBook)
www.panstanford.com

parameters that are exploited to extract photons that remember initial properties and retain information for generating direct images of targets embedded deep inside the medium. More recent developments, such as identification of new near-infrared wavelength ranges and vortex light beams with higher angular momentum, provide additional advantages for deep tissue imaging. These developments along with the use of nonlinear optical processes, such as second harmonic generation (SHG) and multiphoton excitation (MPE) for deep tissue imaging, are reviewed.

5.1 Introduction

Photonics can provide new armamentarium to complement the current medical imaging modalities, such as X-rays, computer-aided tomography (CAT), magnetic resonance imaging (MRI), ultrasound, positron emission tomography (PET), etc., for imaging body parts. The interest in light-based modalities arises from a need to have smart tools, which are safe, compact, affordable, and enable noninvasive or minimally invasive, preferably in vivo monitoring of tissue biochemical components. Light offers the opportunity to overcome the limitations of many current methods.

Several scientific and technological developments make realization of the above-mentioned opportunity more probable than ever. First, development of broadly tunable ultrafast lasers, based on Ti: sapphire, Cr: Forsterite, doped fibers as active materials, as well as ultrafast supercontinuum, and application-specific semiconductor diode lasers, provides a wide range of wavelengths for spectroscopy and imaging. Concurrent developments in charge coupled device (CCD) cameras, near-infrared cameras, complementary metal oxide semiconductor (CMOS) cameras, as well as fast and compact personal computers make signal detection, data acquisition, processing, and analysis faster and efficient than ever before. Second, recent advent of vortex light beams with higher orbital angular momentum (OAM) may be advantageous for imaging and spectroscopy depending on how such beams scatter and interact with materials systems. Third, identification of 1100–1350 nm and 1600–1870 nm NIR spectral windows for biological tissue imaging, in addition to the

650–950 nm range that has commonly been used may improve penetration depth. Finally, imaging based on nonlinear optical interactions between light and tissue constituents has been proved to have better spatial resolution and ability to provide sectional images because of the higher power law dependence of nonlinear optical signal on light intensity.

This chapter provides an overview of the physics of light propagation through highly scattering turbid media, such as microspherical beads suspended in water, Intralipid-10% or similar suspensions, ex vivo tissues from brain, breast, prostate, and other body organs. Some of the techniques used to sort out unscattered or minimally scattered light for generating direct images are then presented. Linear and nonlinear propagation of optical vortex beams in model turbid media and biological tissues (brain, breast, prostate) are then explored to assess their application in deep imaging application of human body. The interaction of light with a scattering medium, characteristics of the transmitted light, key parameters that characterize the light emergent from the medium are presented. The absorption in the NIR windows from 700 to 2500 nm with reduced scattering is shown to improve imaging inside brain. Imaging based on nonlinear optical interactions is briefly mentioned, since a detailed review appears in Chapter 1, by Sangeeta Murugkar and Robert Boyd. The major application of knowledge presented in this chapter will be helpful to develop multiphoton imaging and optical tomography diagnostic systems for diagnosing diseases of breast, brain, prostate and other body organs and flow of blood through blood brain barrier (BBB). The material presented in this chapter is mainly derived from some of the seminal articles published by the researchers at the Institute for Ultrafast Spectroscopy and Lasers (IUSL) of the City College of New York over the past 25 years, and reviewed elsewhere (Alfano, R. R. Wang, W. B. Wang, L., and Gayen, S. K. 2015. "Light Propagation in Highly-Scattering Turbid Media: Concepts, Techniques, and Biomedical Applications," in *Photonics: Scientific Foundations, Technology and Applications, Volume IV*, 1st edition, edited by David L. Andrews. Published by John Wiley & Sons, Inc. pp. 367–412). This is not a review of the entire field, and of contributions made by many researchers active in the field.

5.2 Physics of Light Propagation for Imaging through a Highly Scattering Medium

Linear and nonlinear optical imaging techniques are being developed to identify, locate, and diagnose an "object" embedded in a scattering medium, such as a tumor inside a human breast or prostate, blood vessels and/or neurons in brain, vulnerable plaques in arteries and so on. Biological tissues are light scattering media composed of cells and vessel structures, of different sizes from nanometer to millimeter scales. Chemical constituents of biological tissues include water, oxy-hemoglobin (HbO_2), hemoglobin (Hb), tryptophan, tyrosine, collagen, NADH, elastin, flavin, and other proteins. The interaction of light with biological tissues and their constituents may be proportional to the light intensity (*linear* interaction) or some power of the intensity (*nonlinear* interaction). Principal linear light–tissue interactions include *elastic scattering* that does not involve any change in the wavelength of the incident light fluorescence, and *absorption*. Nonlinear interactions include *multi-photon absorption, second and higher harmonic generations, four-wave mixing, phase conjugation*, and *Raman scattering*. The signal that results from interaction of light with biological tissue can be different for normal and cancerous tissues, and may vary even for cancerous tissues depending on the stage of cancer. These differences provide the basis for optical biomedical imaging and distinguishing between normal and cancerous tissues.

The basis of transillumination imaging is the difference in the transmission of light through a turbid medium (such as normal tissue) and an "object" (such as a tumor inside tissue). This differential transmission, in principle, should lead to the formation of a "shadow" of the object (tumor) in transmitted light. Depending on wavelength of light, it is both scattered and absorbed in varying degrees by biological tissues resulting in diffusive glow [1]. However, excess absorption and scattering by the medium may be detrimental to formation and observation of a transillumination image. Absorption reduces the intensity of transmitted light and if it is too strong there may not be enough transmitted photons to form a well-defined image. Scattering also reduces intensity of light propagating in the forward direction, and in addition, blurs the shadow image as scattered light encroaches in the shadow region. The key to successful development of optical imaging techniques is twofold: (a) to choose wavelengths of light judiciously to minimize absorption; and (b) to deal with the problem of light scattering to reduce image blur. Initially, light in the 650–950 nm range used to be employed by most researchers because light in this wavelength range was less absorbed by tissue constituents than in the visible or near ultraviolet. More recently, the 1100–1350 nm range and 1600–1870 nm range have been identified as two other useful windows for imaging through biological tissues [3, 4].

Absorption results from electronic and vibrational transitions within the energy levels of the atoms and molecules in the tissue. In the NIR, the absorption bands are due mainly to vibrational transitions in water and other tissue components, their overtones, and combination of fundamental and overtone vibration modes. The 1064 nm peaks in tissue absorption spectra arises from all three of principal vibration modes of water. Scattering arises from variations of index of refraction in the microscopic and macroscopic structures of the tissue; it increases strongly as the wavelength of light becomes comparable to the size of the scattering elements. In addition to absorption and scattering of light, fluorescence emitted by the objects upon absorption of incident light may also provide the necessary contrast for imaging, and has led to fluorescence imaging approaches [4–7]. We first consider direct imaging approaches based on scattering and absorption contrasts.

5.2.1 Components of Transmitted Light from Scattering Media

Laser light characterized by high degree of directionality, spectral brightness, mono-chromaticity, coherence, and pulse duration (in the case of ultrashort pulses of light) is natural choice for imaging through highly scattering media. The forward transmitted photons of an ultrashort pulse of light are composed of three components: *ballistic*, *snake*, and *diffusive* [10], as shown schematically in Fig. 5.1. The ballistic or coherently forward scattered photons propagate in the direction of the incoming beam, traverse the shortest path, retain most of the characteristics of the incident light, and carry the maximum information for formation of a direct image. The multiple-scattered photons travel long distances within the medium, lose many of their initial characteristics, and emerge later in all directions. They constitute what is known as the diffusive component of the transmitted light. The photons that scatter slightly in the forward direction retain significant initial properties and information on structures hidden in the scattering medium. These photons are called snake photons because their trajectories resemble a wriggling snake.

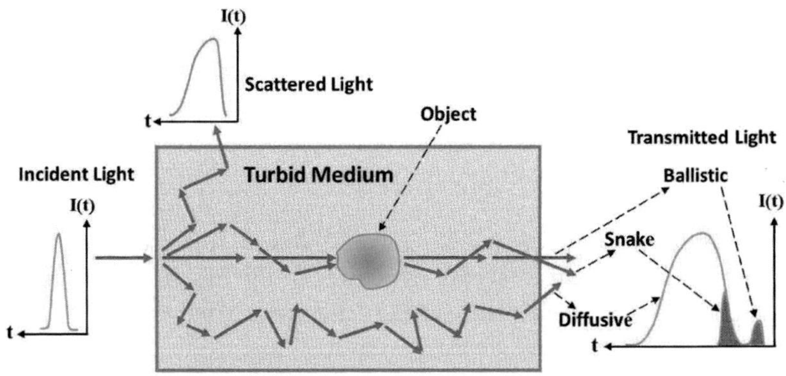

Figure 5.1 Schematic diagram of light pulse propagating in a turbid medium showing the ballistic, snake, and diffuse photons. Adapted from Refs. 8 and 9.

Ballistic photons are most effective in forming a shadow image. Snake photons also generate transillumination images whose resolution depends on the position and width of the temporal slide used in imaging. The snake photons that form a well-resolved shadow image arrive much earlier than most of the diffuse photons. The relative intensities of these three components in the transmitted beam depend on the color of light and characteristics of the sample. In a highly scattering optically thick medium, the ballistic component is extremely weak, while the diffuse component is the most intense. To form images of the structures inside the turbid medium using ballistic and snake photons, the most numerous diffusive photons need be strongly discriminated against.

For *direct imaging* of targets embedded deep inside the scattering medium, the image-bearing photons need to be extracted out from the background of multiple-scattered image-blurring photons using clever experimental techniques [8, 9, 11–37]. The second approach, sometimes referred as diffuse imaging that solves the inverse problem [38–42], is based on detection of all of the photons including multiple-scattered photons at various positions around the object. The aim is to construct the image from the measured intensities, known experimental parameters, and tissue characteristics using model of light propagation and sophisticated computer algorithms. The impetus for solving the inverse problem stems from the realization that for tissues that are more than a few centimeters thick, the only transmitted light is the scattered light.

Both frequency-domain and time-domain data may be used in the second approach. The thrust is to develop appropriate algorithms to generate a map of the interior structure of a turbid medium in a reasonable time. The Monte Carlo method focuses on the particle nature of light and promises higher accuracy at the cost of higher computational complexity and time [40]. Alternate methods based on the diffusion approximation of the radiative transfer equation are being pursued by various groups [38–40, 42–44].

Near-infrared light spanning the 700–1300 nm spectral range in the first optical transmission window appears to be a judicious choice. Light in this wavelength range is not as strongly absorbed by tissue as the visible light and so will have higher transmission and less likelihood of causing burns. The availability of broadly wavelength-tunable ultrafast solid state lasers, such as Ti:sapphire [45] and Cr:forsterite [46], to cover this spectral range is an advantage. Fiber lasers (doped with Er/Yb, or Tm) operate in the NIR at 1620 and 1960 nm, respectively, and supercontinuum covers 700 to 2500 nm. The wavelength tunability offers the possibility of monitoring different chemical species by tuning to different wavelengths. Near-infrared light is also less scattered than visible light and is not known to cause tissue damage even with prolonged exposure at intensity levels that may be needed for imaging. The balance of small absorption and scattering can improve images in far NIR at about 1600 to 1800 nm.

5.2.2 Key Optical Parameters for Describing Light Propagation in Highly Scattering Media

Knowledge of the key optical properties of scattering media (Intralipid suspension in water, milk, polystyrene beads of different sizes suspended in water, and tissues of breast, brain, and prostate) is needed for developing photonic techniques for optical imaging and tomography approaches. These optical parameters are absorption length l_a, scattering mean free path l_s, transport mean free path l_{tr}, and anisotropy factor g.

The ballistic light is considered to be the fraction of the incident light that is not affected by scattering events within the medium and retains the characteristics (coherence, directionality, polarization) of the incident beam. The attenuation of the ballistic photon intensity in traversing a length L of a turbid medium may be estimated by:

$$I_b = I_i \exp[-(L/l_a + L/l_s)], \tag{5.1}$$

where I_i is the intensity of the incident beam; l_a is the absorption length, defined as the mean distance a photon travels before being absorbed by the turbid medium; and l_s is the scattering mean free path, defined as a mean distance between consecutive scattering events for a photon. The parameters l_a and l_s are related to the absorption coefficient μ_a and scattering coefficient μ_s of the medium, respectively, as

$$l_a = \mu_a^{-1}; \text{ and } l_s = \mu_s^{-1}. \tag{5.2}$$

The total attenuation coefficient is, $\mu_T = 1/l_a + 1/l_s = 1/l_T$, where l_T is the total attenuation length.

The spatial anisotropy of scattering is given by the *anisotropy factor* g, defined as the mean cosine of the scattering angles, that is, $g = \langle\cos\theta\rangle$. The value of g is 1 for propagation in the incident direction and 0 for isotropic scattering.

Another important parameter is the transport mean free path l_{tr} defined as the distance in which a photon is fully randomized (forgets its original direction of motion) after numerous scattering events, as illustrated schematically in Fig. 5.2. The relationship between l_{tr} and l_s is given by

$$\langle l_{tr}\rangle = \Sigma\, l_s\hat{n}, \tag{5.3}$$

where $\hat{n}$ represents the vector displacement of a photon in turbid media. Written explicitly,

$$<l_{tr}> = <l_s + l_s\cos\theta + l_s\cos^2\theta + l_s\cos^3\theta + \cdots + l_s\cos^n\theta + \cdots>$$
$$<l_{tr}> = l_s/(1 - <\cos\theta>) = l_s/(1-g), \tag{5.4}$$

One also defines the *reduced scattering coefficient*,

$$\mu_s' = \mu_s(1-g) = (l_{tr})^{-1}. \tag{5.5}$$

The parameters l_a, l_s, μ_a, and μ_s are intrinsic properties of the material medium and are given by

$$l_s = \mu_s^{-1} = (N\sigma_s)^{-1}, \text{ and } l_a = \mu_a^{-1}, \tag{5.6}$$

where N is the volume concentration of particles, and σ_s and σ_a are the scattering cross section and absorption cross section, respectively.

The intensity of snake light is found, from experiments, to follow the equation [47]

$$I_s(\Delta t) = A\exp[-bz/l_{tr}], \tag{5.7}$$

in time interval Δt, where b is a parameter that depends on Δt and has an average value of 0.8. The snake light is portion of the photons that arrive before multiple-scattered diffusive photons and after the ballistic component.

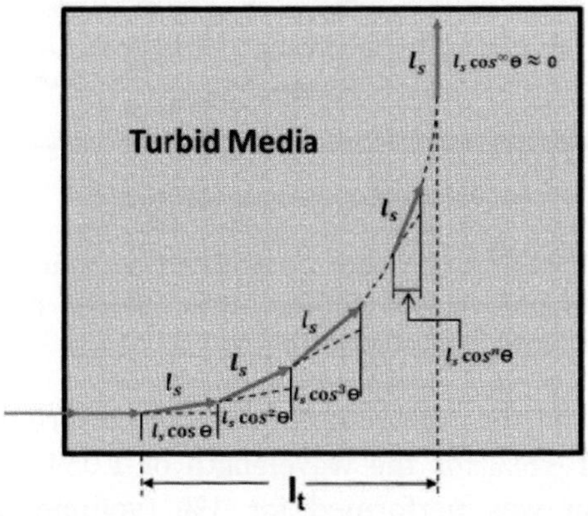

Figure 5.2 Mean free scattering distance l_s, mean free transport distance, l_{tr}, (l_t) and anisotropy factor, g.

The values of g, l_s and l_t depend on particle size and are calculated using Mie scattering theory [48] and shown in Figs. 5.3 and 5.4, respectively. In Fig. 5.3 it can be seen that the g factor greatly depends on wavelength, especially when particle size is less than 1 µm, which is close to the wavelengths of 0.527 and 1.054 µm. For the Intralipid-10% suspension with an average particle diameter of ~0.5 µm, the values of g will be ~0.9 and ~0.6 for 0.527 µm and 1.054 µm, respectively.

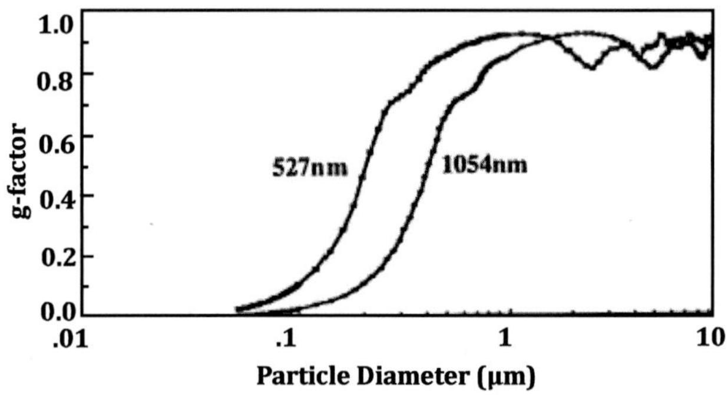

Figure 5.3 Calculated values of g-factor as a function of particle diameter.

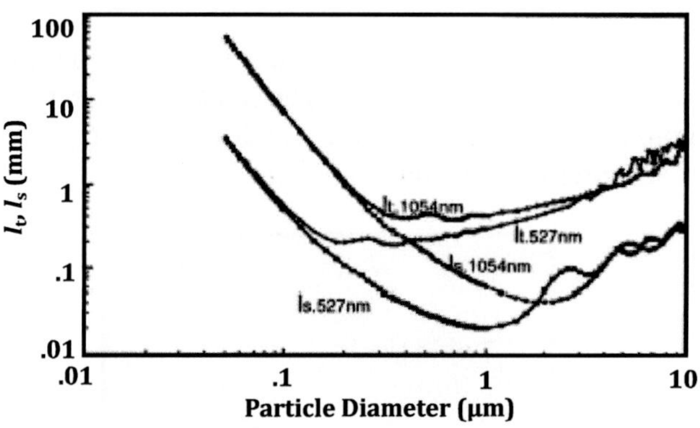

Figure 5.4 Values of l_{tr} (l_t) and l_s calculated as a function of particle diameter in 1% of the 10% diluted polystyrene bead suspension in water [49].

At larger particle diameters, g oscillates around 0.85 with a deviation of ~5% for the wavelength of 1.054 µm. In Fig. 5.4 the calculation was performed for 1% (volume density) bead solutions (10% in stock). Due to the wavelength dependence, there is smaller difference between l_{tr} and l_s for particles with smaller diameter. When the diameter increases, the difference increases. When diameter is more than 3 μm, the difference in values of both l_t and l_s increases for the two wavelengths.

5.2.3 Values of Key Optical Parameters for Human Tissues and Some Model Media

Tissue scatters light strongly in the forward direction, the g value is ~0.9. A value of g = 0.9 makes l_{tr} = $10l_s$. Intralipid-10% suspension with g ~ 0.8 and 0.6 for 530 nm and 1054 nm wavelengths, respectively, and lower absorption coefficient has been widely used as a model medium to simulate homogenized breast tissue [50, 51]. The particle size distribution in intralipid-10% solution is shown in Fig. 5.5.

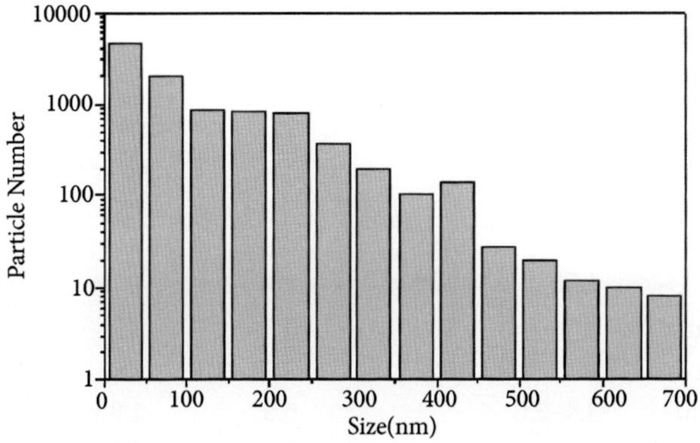

Figure 5.5 The particle size distribution in Intralipid-10% suspension. After Refs. 49 and 51.

Associated optical parameters of 2% diluted Intralipid-10% suspension and human breast tissues at visible and IR wavelengths are shown in Table 5.1 [50–54]. Intralipid 10% suspension contained 50 gm soya bean oil, 6 gm phospholipid, 11.25 gm glycerol, and 2% concentration solution was made by diluting Intralipid-10% with distilled water [51].

It can be seen from the data in Table 5.1 that the transport mean free path and absorption length of 2% diluted Intralipid-10% suspension is ~3.4–5.0, 5, and 6.6 mm at 630, 800, and 1064 nm wavelengths, respectively. In the early time detection $l_{KF.s}$ = 2.3 mm and l_t = 4.6 mm, if g ~ 0.5 at 1054 nm.

Table 5.1 Optical properties of human breast tissues and 2% Intralipid solution

Tissue type and 2% Intralipid	Transport mean free path l_t (mm)	Absorption length l_a	Anisotropy factor, g	Scattering length l_s, (mm)	Wave-length (nm)
Glandular	1.25	17	0.91	0.067	1064
	0.8				700
Adipose	0.7	155	0.95	0.0040 ±	620
(fatty breast)	1.2	10		0.0006	1064
	0.77	100			700
	1.0	100			633
	1.3	130			800
Carcinoma	1.21	230	0.88		1064
	0.5	155			700
	0.6	24			620
	0.54	62.5			633
	0.96				800
Benign	2.6	30			620
Fibroadenoma	2.08	13			1064
	1.55	100			620
2% diluted	3.0	500	0.5	1.0	620
of stock	10	500	0.71–0.8	3.3	1064
Intralipid-	3.4–5.0		0.5	1.83	633
10%	6.6		0.64	2.3	1064
	5.0			1.23	800
					1054
					527

Source: After references 50, 52–54.

For the 2% diluted Intralipid-10% suspension, the l_a = ~500 nm from 620 to 1064 nm. For breast tissues, l_{tr} = ~0.5–2.6 mm, at 620 to ~1.2–2.1 mm and l_a = ~100 mm. For the mean cosine of scattering angle g, there is a difference between human breast tissue and Intralipid suspension. The g_{tissue} is from 0.88–0.95, while $g_{intralipid}$ is from 0.5–0.83, which was calculated from Mie's theory. Table 5.2 (Ref. 22) and Table 5.3 (Ref. 26) show the calculated results.

Table 5.2 Transport mean free path and absorption length with standard deviations for various random media at 620 nm and 1064 nm measured using time-resolved transmission and diffusion approximation

	Transport mean free path, l_{tr} (mm)		Absorption length (nm)	
Random media	**1064 nm**	**620 nm**	**1064 nm**	**620 nm**
Benign human breast	~5.1*	2.6 ± 0.2	—	30 ± 5
Fatty human breast	~1.4*	0.7 ± 0.1	200	155 ± 50
Chicken breast	4.9 ± 0.3	2.5 ± 0.5	70 ± 10	60 ± 30
Chicken fat	0.74 ± 0.1	0.6	230 ± 70	155
Pork breast	3.5 ± 0.5	1.2 ± 0.2	70 ± 10	70 ± 8
Pork fat	0.58 ± 0.05	0.35 ± 0.05	400 ± 60	300 ± 50
Whole milk (25%)	—	1.0 ± 0.1	—	253 ± 50
2% solution of intralipid-10%	10 ± 1, 4.4**	3.0 ± 0.1	500 ± 50	500 ± 50

*Estimated by $\mu_{t\text{-IR}} \approx 0.5\mu_{t\text{-Visible}}$.

**Estimated by $\mu_{t\text{-}1054\,\text{nm}}$ from data.

For 2% diluted Intralipid solution, the scattering mean free path at 1054 nm, $l_s \approx 2.2$ mm, and $l_{tr} \approx l_s/(1 - 0.5) = 4.4$ mm has been measured using a K-F imaging system [51].

Breast tissue consists of fat, fiber, glandular, and blood. The mean free transport path and the absorption coefficients of four different breast tissues at 1060, 800, and 625 nm obtained by Peters et al. [52] are displayed in Table 5.3.

Table 5.3 Transport mean free path and absorption length with standard deviations for various random media at 620 nm and 1064 nm

	1060 nm		800 nm		625 nm	
Tissue types	l_{tr} (mm)	l_a (mm)	l_{tr} (mm)	l_a (mm)	l_{tr} (mm)	l_a (mm)
Normal glandular	1.25	17	0.84	25	0.60	20
Normal adipose	1.2	10	1.4	14	1.28	13
Fibroadinoma	2.08	13	1.55	100	1.04	17
Ductal carcinoma	1.21	7	1.14	33	0.79	8

Source: After Ref. [52].

The l_{tr} values for adipose (fat) tissue listed in Table 5.3 are different from values in Table 5.2 [53, 54]. The main reason of the discrepancy is that samples used by Peters et al. were compressed homogenized tissues, which no longer possessed any natural structure [52]. The values in Table 5.2 were obtained from in vitro samples. It can be seen from Table 5.3 that the transport mean free path of the normal glandular tissue strongly depends on wavelength and decreases with the increase of the wavelength.

The optical parameters for brain tissue are different from other tissues due to its unique composition that contains twice as much lipid and less than half of the protein found in muscle tissue [55]. Table 5.4 shows the fitted values of l_{tr} and reduced scattering coefficient μ_s' in each optical window, μ_a values were estimated as from the absorption coefficient of water by assuming water content of the brain is about 75% and therefore $\mu_a = 0.75\mu_{a\text{-water}}$, and corresponding values of g and μ_s that were calculated from these fitted parameters l_{tr}, μ_a and μ_s'. The value of g obtained from fitting slightly increases from 0.92 to 0.98 with the increase of wavelength because of the forward scattering generated by the large cell particles structure in tissue size from 2 to 20 μm. Note that the four windows represent the first (I, 650–950 nm), second (II, 1,100–1,350 nm), third (III, 1,600–1,870 nm), and fourth (IV, centered at 2,200 nm) NIR optical tissue windows for deep brain imaging [56].

Table 5.4 Optical parameters for brain tissue in different spectral windows

	Exp	Model				
Window	avg. peak l_{t_obs} (μm)	l_t (μm)	μ_a (mm^{-1})	μ_s (mm^{-1})	g	μ_s' (mm^{-1})
I	192.4	191.2	0.0037	5.226	0.92	0.4
II	259.4	263.9	0.0743	3.715	0.95	0.2
III	289.7	296.4	0.5448	2.829	0.97	0.08
IV	248.1	248.7	1.4376	2.829	0.98	0.04

5.2.4 Optical Absorption Spectra of Key Chromophores in Tissues

The absorption in NIR of different key chromophores in tissue, such as water (H_2O), hemoglobin (Hb), and oxy-hemoglobin (HbO_2) as a function of wavelength is shown in Fig. 5.6. It suggests that the wavelength range of 700–1100 nm is most suitable for optical mammography since there is a weaker absorption by H_2O, the main component of tissue. Attenuation spectra of distilled water and the 2% diluted intralipid-10% stock solution in 1 mm-thick cell are shown in Fig. 5.7. This curve shows that scattering dominates from 200 to 800 nm, while absorption by water dominates the attenuation for wavelength greater than 1300 nm. The absorption of water appears at ~900 nm, ~1200 nm, ~1400 nm and longer. There is another optical window in 1600 to 1800 nm region where scattering is low. This region is called Golden Window for deepest penetration in brain.

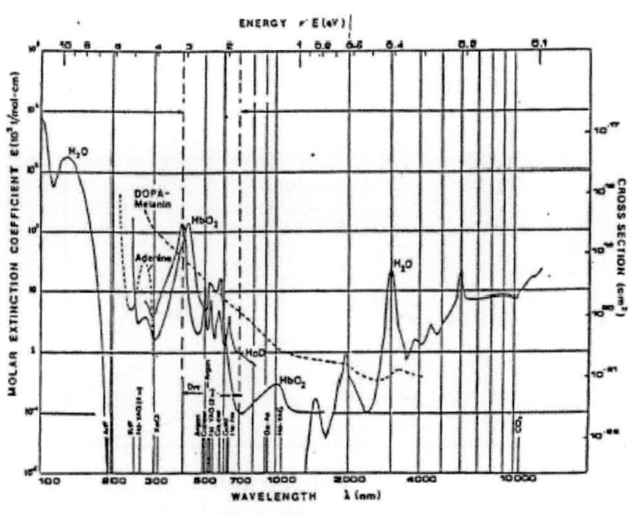

Figure 5.6 Absorption spectra of tissue constituents [57].

The absorption in 700 to 1100 nm range is shown in Fig. 5.8 for water (H_2O), oxyhemoglobin (HbO_2) and deoxyhemoglobin (Hb). In this wavelength range, Hb and HbO_2 can be used to measure hypoxia and may play an important role to separate cancer from normal breast tissues. Hypoxia can be an important marker of breast cancer. Further water absorption appears in the 1000 to 1400 nm region (see Fig. 5.7) and offers another biomarker for cancer detection.

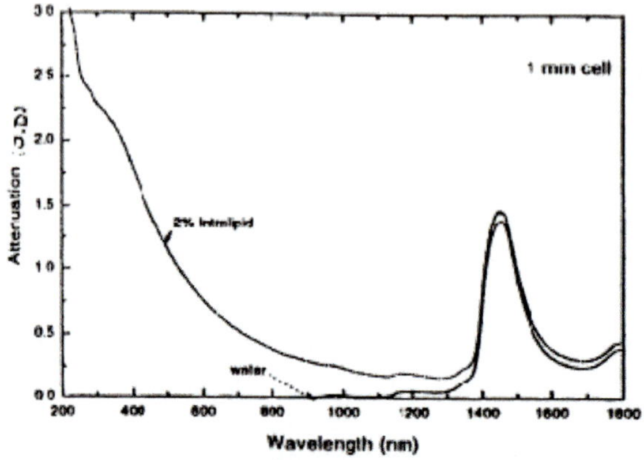

Figure 5.7 Attenuation of light by water and 2% diluted Intralipid in a 1 mm cell.

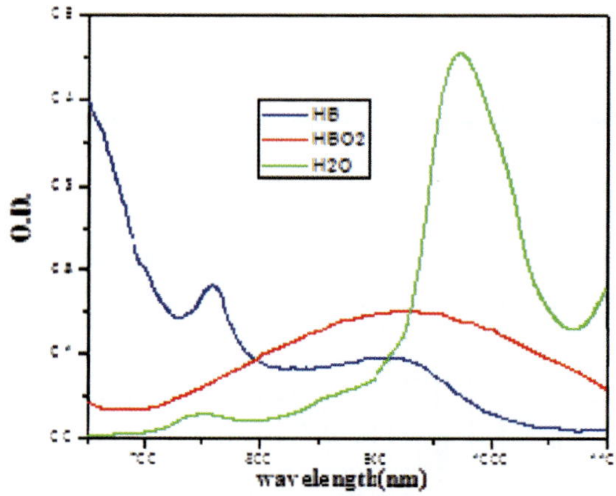

Figure 5.8 Absorption spectra of Hb, HbO_2, and H_2O.

Figure 5.9 shows the transmittance T (%) measured for the brain tissues in each optical window, and Table 5.5 summarizes the peak transmittances of brain tissues in these windows. Maximum peak transmittances are all located in window III for tissues with thicknesses of 50, 100, 150, and 200 μm. Peak transmittances in windows II and IV are close to each other and are much higher than those in window I, showing their better potential for deep imaging than window I [56].

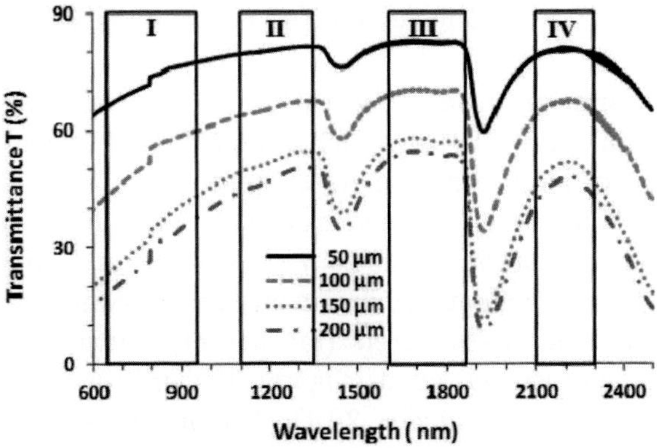

Figure 5.9 Transmittance T (%) in optical tissue windows I, II, III and IV. Brain tissue thicknesses were 50, 100, 150, and 200 μm, respectively.

Table 5.5 Peak transmittance T (%) of brain tissues measured in each optical window [56]

	Tissue thickness			
Window	**50 μm**	**100 μm**	**150 μm**	**200 μm**
I: 650–950 nm	77.3	59.8	43.0	37.4
II: 1100–1350 nm	81.5	67.5	54.5	50.4
III: 1600–1870 nm	82.9	70.5	58.1	54.6
IV: 2100–2300 nm	81.0	67.7	51.7	47.8

5.3 Study of Ballistic and Diffuse Light Components

It was believed prior to 1989 that when a light pulse enters a highly scattering medium it broadens and expands. However, Yoo and Alfano [10] observed a startling effect in the ultrashort time domain. In an experiment involving transmission of 80 fs, 620 nm pulses through a 10 mm-thick cell containing a suspension of 0.33 μm latex beads, they found a weak forward-propagating pulse that appeared earlier than the multiple-scattered components of light. A Hamamatsu streak camera of 5 ps time resolution was used in the experiment. The temporal and angular information was obtained simultaneously indicating that the light component was coherent. This coherent light pulse was designated as the "ballistic pulse" which obeys Eq. (5.1) and is the weak remnant of the input laser pulse.

The transition from coherent to incoherent scattering is presented in Fig. 5.10 for a slab sample, which consisted of a suspension of latex beads in water held in a 10 mm-thick cell. The bead diameter was varied from 0.33 to 15.8 μm. At a low scatterer concentrations of up to n = 7.0 × 10^{16} m^3 ($n\sigma_s z$ = 17.9), the coherent component was dominant in the forward-transmitted signal. The multiple-scattered (incoherent) component started becoming appreciable at a slightly higher concentration, of n = 8.0 × 10^{16} m^3 ($n\sigma_s z$ = 20.4). The incoherent component rapidly gains in intensity relative to the coherent counterpart as the concentration increases. The corresponding $n\sigma_s z$ values are 22.8, 24.3, and 26.8, respectively. When $n\sigma_s z$ is above 26, only the incoherent component is visible.

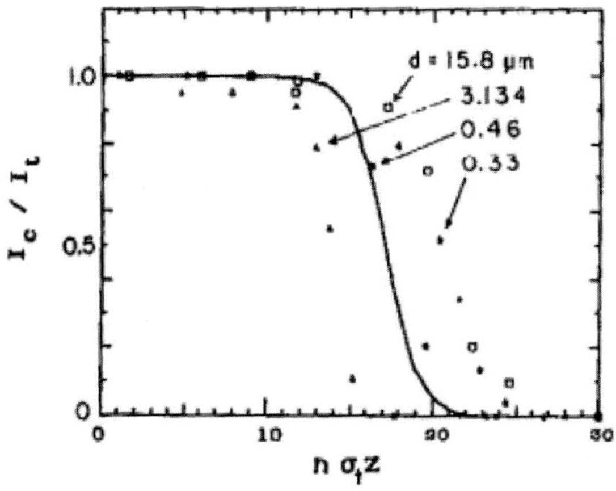

Figure 5.10 Plots of ratio of coherent ballistic (I_c) to total (I_t) intensity of scattered light in the forward direction versus ($n\sigma_s z$) for four different scatterer diameters. The solid curve is plotted from Eq. (5.1) in Ref. 10.

The coherent and incoherent components of ultrafast laser pulses passing through a slab of random media may be separated from each other using time-resolved measurement techniques. The ability to time-resolve the coherent from incoherent components is important in medical imaging and diagnostics. The diffusion approximations of radiative transfer equation (RTE) that is commonly used to describe light propagation through optically thick media (sample thickness $z >> l_{tr}$) fails [57, 58] to explain the experimental data as the value of z/l_{tr} decreases below 10.

There is a regime of scattering where the diffusion approximation fails, in particular, when $z/l_t \leq 10$ for inverse imaging. Diffusion theory is used to explain occurrences where the transport of particles and waves in random media occurs. The analytical solution for the diffusion equation are generally available and are found to be valid [59, 60] when $z/l_t >> 1$. For many physical media the wave transport occurs in the region where z/l_t is small. The diffusion approximation has been extensively used in many of the pioneering light-scattering studies [60–66], in applications of light scattering to study physical and biological media [67–69], and inverse imaging [38–40].

A direct method to probe the transport of photon pulses in a random medium is to measure the distribution of photon arrivals in the time domain. When a plane front ultrafast pulse is incident normally onto a slab of random medium, the temporal distribution of photon arrivals at a point on the opposite side of the slab is predicted by the diffusion theory as [10]

$$I_z(t) = \frac{D}{\pi z^2} \sum_{m=1}^{\infty} m \left(\frac{\pi z}{d}\right)^2 \sin\left(\frac{m\pi z}{d}\right) \times \exp\left[-Dt\left(\frac{m\pi}{d}\right)^2\right], \tag{5.8}$$

where $D = \nu l_t/3$ is the diffusion coefficient, d is transport distance, $d = z + 2z_0$, $z_0 = 0.71 l_{tr}$, ν is the speed of photons, and z and l_{tr} were defined earlier.

The experimental arrangement shown schematically in Fig. 5.11 closely imitates the geometry pertinent to Eq. (5.8). Ultrafast laser pulses of 100 fs duration were generated at the repetition rate of 82 MHz by a colliding pulse mode-locked (CPM) dye-laser system. The laser power was 10 mW at a wavelength centered on 620 nm. The laser beam of 4 mm diameter was split into two beams: a reference beam to mark the zero time of the signal beam, which was expanded to 35 mm diameter. The central portion of the expanded beam was selected by a diaphragm and passed through the random medium consisting of latex beads suspended in water contained in a cylindrical 50 mm diameter and 10 mm thickness glass cell. Photons scattered out of the cell will be lost, i.e., both sides of the cell satisfied an absorbing boundary condition. A black pinhole of 2 mm diameter was placed at the center on the opposite side of the cell. The temporal distribution of photons scattered out from this pinhole in the forward direction at an angle of 10 mrad was measured by a synchro-scan streak camera. The diameter of the incident beam was set to 20 mm; any further increase in the diameter will not change temporal profile of the scattered photons. The measured scattered pulse profile can be described by Eq. (5.8) if the diffusion approximation is valid.

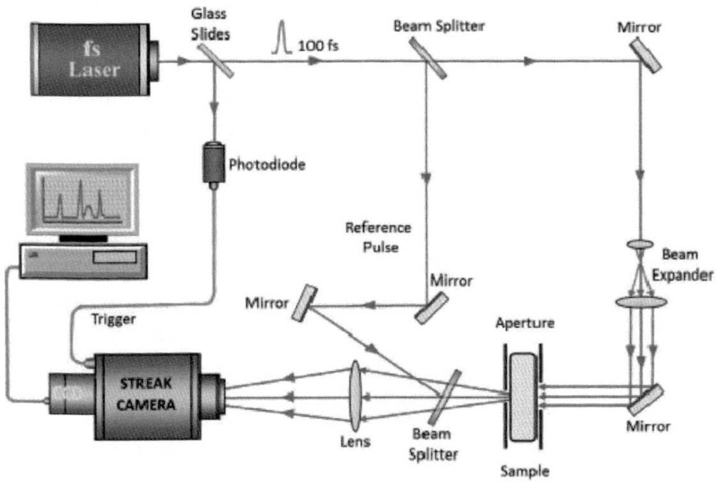

Figure 5.11 A schematic diagram of the experimental setup.

Experimental studies were carried out for different bead diameters, each at a series of different concentrations. For highly concentrated random medium, where $z/l_t >> 1$, the diffusion equation holds, which was not surprising. In this case, the transport mean free path was obtained by fitting the scattered pulse profile using Eq. (5.8). Figure 5.12 illustrates an example where the diffusion theory fits the experimental data. The photon transport mean free path l_t was found to be 0.309 mm which is close to the value of 0.25 mm computed from Mie theory using $l_t^m = 1/n\sigma_m$, where n is the number density of scattering particles and σ_m is the momentum-transfer scattering cross section.

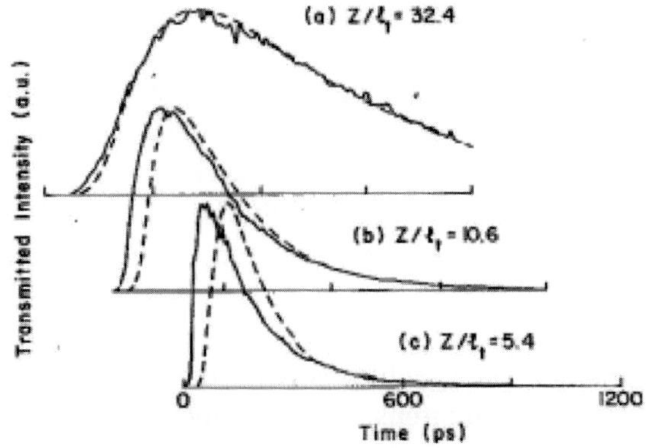

Figure 5.12 Transmitted pulse profiles of 10 mm-thick random media consisting of latex beads of diameter 0.296 μm. The bead concentration n, l_{tr}, and z/l_{tr} are, respectively, (a) 6.65 × 10^8 mm^{-3}, 0.309 mm, and 32.4; (b) 2.18 × 10^8 mm^{-3}, 0.941 mm, and 10.6; and (c) 1.10 × 10^8 mm^{-3}, 1.853 mm, and 5.4. Solid curves are the experimental results and dashed curves are computed from the diffusion theory where their peak intensity is normalized [58].

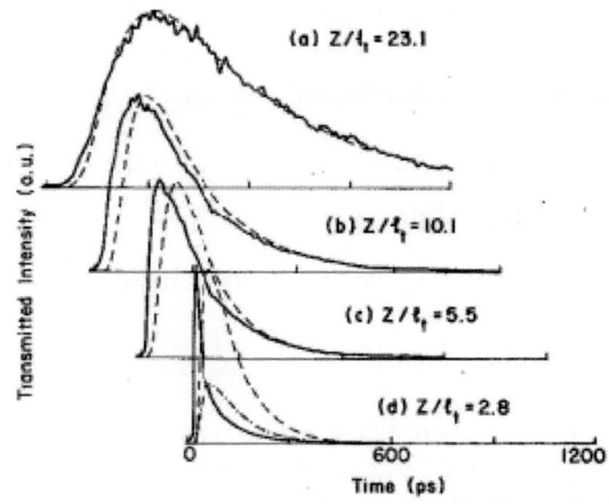

Figure 5.13 Transmitted pulse profiles of a 10 mm-thick random medium consisting of latex beads of diameter 3.134 μm. The bead concentration n, l_t, and z/l_t- are, respectively, (a) 9.62 × 10^5 mm^{-3}, 0.433 mm, and 23.1; (b) 4.20 × 10^5 mm^{-3}, 0.989 mm, and 10.1; and (c) 2.28 × 10^5 mm^{-3}, 1.823 mm, and 5.5. Solid curves trace the experimental results. Dashed and dash-dotted curves are computed from the diffusion theory where their peak and total intensity are normalized to the experimental curves, respectively.

Figures 5.12 and 5.13 display a series of transmitted pulse profiles for media with decreasing bead concentration for bead diameters of 0.296 and 3.134 μm, respectively. Figures 5.12a and 5.13a show that the experimental (solid) curves can be fitted by diffusion theory (dashed curves), where z/l_t is large as indicated in the figures. Figures 5.12b and 5.13b, with $z/l_{tr} \approx 10$, clearly show deviations. The measured photons arrive earlier than predicted by the diffusion approximation. The peaks of the theoretical (dashed) curves are normalized to the peak of the experimental data, whereas the dash-dotted curves are normalized to the total intensity of the experimental results. None of the theoretical plots from the diffusion approximation agrees with the experimental results at early time when $z/l_t < 10$.

A quantitative indication where the diffusion approximation breaks down is the average time of photon arrival from the pulse. The average arrival time is computed by

$$\bar{t} = \int_0^\infty dt' t' I(t') / \int_0^\infty dt' I(t'), \tag{5.9}$$

where $I(t')$ is the temporal distribution of the scattered pulse measured experimentally. The theoretical prediction of the average arrival time can be obtained exactly from Eq. (5.8) as

$$\bar{t} = \frac{1}{2}\frac{z(z+4z_0)}{vl_t} \sim \frac{z^2}{2vlt}. \tag{5.10a}$$

The average length $<z>$ traveled by diffusive light is

$$<z> = z^2/2l_t. \tag{5.10b}$$

The average time of photon arrival $\bar{t}$ as a function of z/l_t predicted by Eq. (5.10) is plotted by a solid curve in Fig. 5.14. The average time of arrival of the experimentally measured photons are computed from Eq. (5.9). The results from random media consisting of suspension in water of beads with diameters d = 0.296, 0.46, 1.09, 3.134, and 11.9 μm in Fig. 5.14 are represented by plots of stars, triangles, circles, squares, and pluses, respectively. The dotted curve is obtained by multiplying Eq. (5.10) by 0.9 to indicate where the experimental data deviate by more than 10%. The data displayed in Fig. 5.14 clearly demonstrate that the average time of arrival deviates by 10% when $z/l_t \approx 10$. The deviation increases as z/l_t is reduced further below 10 and many more photons arrive earlier than predicted by diffusion theory. Beads of larger diameter will increase the amount of light scattered in the forward direction. This anisotropic scattering will result in reducing the average arrival time of the photons, which is apparent in Fig. 5.14 when z/l_t. is less than 5. For a given value of z/l_t (< 5), the largest bead diameter (11.9 μm) medium (plusses) shows the smallest average arrival time, while the smallest bead diameter (1.09 μm) medium (circles) shows the largest average arrival time, and the intermediate bead diameter (3.134 μm) medium (squares) shows that its average arrival time falls between that for the 11.9 and 1.09 μm beads. So, diffusion approximation breaks down for sample thickness z when z/l_t < 5. A better theory is needed for describing light transport in scattering media in the regime where the diffusion approximation breaks down. Cai et al. have developed cumulated solutions of the radiative transport equation, which are significant improvements [70, 71], and Ntziachristos et al. have demonstrated an approach namely early photon tomography to use the ballistic photons for imaging objects embedded inside turbid media such as tissue [72].

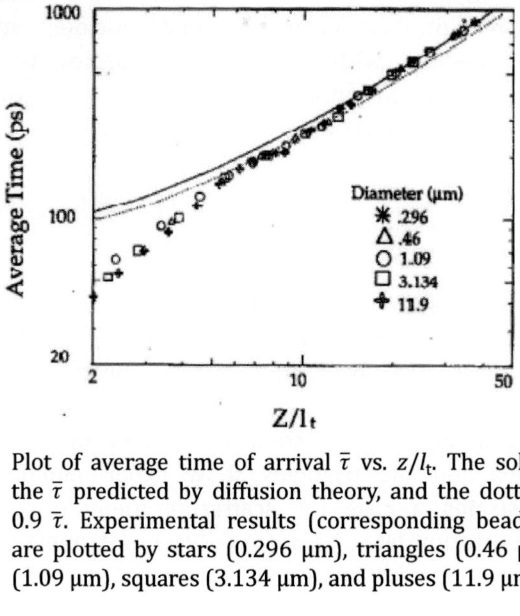

Figure 5.14 Plot of average time of arrival $\bar{\tau}$ vs. z/l_t. The solid curve is the $\bar{\tau}$ predicted by diffusion theory, and the dotted curve is 0.9 $\bar{\tau}$. Experimental results (corresponding bead diameter) are plotted by stars (0.296 μm), triangles (0.46 μm), circles (1.09 μm), squares (3.134 μm), and pluses (11.9 μm).

5.4 NIR Absorption

Scattering and absorption processes remove forward-propagating photons that are needed for forming direct images, and are generally considered detrimental to direct imaging. However, if photon scarcity is not a problem, the presence of some absorption may be useful for enhancing image contrast. Since image-blurring multiple-scattered light travels longer distances within the scattering/absorptive medium than the image-bearing ballistic and snake light, the multiple scattered light gets significantly more absorbed than the image-bearing light. Presence of moderate absorption may thus enhance the ratio of image-bearing photons to image-blurring photons and hence improve image contrast compared to when there is no absorption.

The average distance $\overline{z}$ that multiple-scattered light travels in the medium is z^2/l_t (see Eq. 5.10b), which is longer than z, the distance that the ballistic light traverses through a slab of thickness z. Light traveling over a longer path length has a higher probability of being absorbed. Thus, the introduction of absorbing dye into a random medium will cause it to absorb the multiple scattered light more than the ballistic light. The reduction of noise over signal with the use of absorption, was measured by using time- and angle-resolved ultrafast laser pulse measurements. When a plane ultrafast pulse is normally incident upon a slab of random medium with absorbing boundaries, the temporal profile of the scattered pulse exiting from a point on the opposite side of the slab predicted by the diffusion theory is

$$I_z(t)=\frac{D}{\pi z^2}\sum_{m=1}^{\infty} m\left(\frac{\pi z}{d}\right)^2 \sin\left(\frac{m\pi z}{d}\right)\times \exp\left[-Dt\left(\frac{m\pi}{d}\right)^2\right]\exp\left(-\frac{vt}{l_a}\right), \quad (5.11)$$

where $D = v\ l_t/3$ is the diffusion coefficient, $d = z + 2z_0$, $z_0 = 0.71l_{tr}$, v is the speed of the photon, z is the thickness of the slab, l_t is the transport mean free path, and l_a is the absorption length.

The intensities of the diffuse and ballistic light are computed and plotted in Fig. 5.15 by solid and dashed lines, respectively. The squares and the triangles show experimental data. This figure illustrates a case for scattering medium consisting of 0.3% concentration latex beads of 0.296 μm diameter suspended in water l_t = 1.0 mm is found experimentally and $l_s = 1/n\sigma_s$ = 0.3 mm, where $n = 2.2 \times 10^{17}$ m^{-3} is the scatterer density and $\sigma_s = 1.54 \times 10^{-14}$ m^2 is the scattering cross section computed numerically from the Mie theory. The thickness of the slab of random medium is z = 10 mm, and the absorption length is varied from 1000 mm to 1 mm. The intensity of diffuse light is much higher than the intensity of ballistic light when the medium is less absorbing, e.g., when l_a = 1000 mm, the diffuse intensity is 130 times stronger than that of the ballistic signal. The solid angle of collection Ω for the diffuse light is taken to be 1×10^{-7}. As the absorption in the medium is increased, the diffuse intensity decreases faster than the ballistic intensity. The diffuse intensity is decreased below the ballistic intensity when l_a is less than 2 mm. This reduction of diffuse light with respect to ballistic signal will substantially improve the quality of the image of an object hidden in a highly scattering random medium.

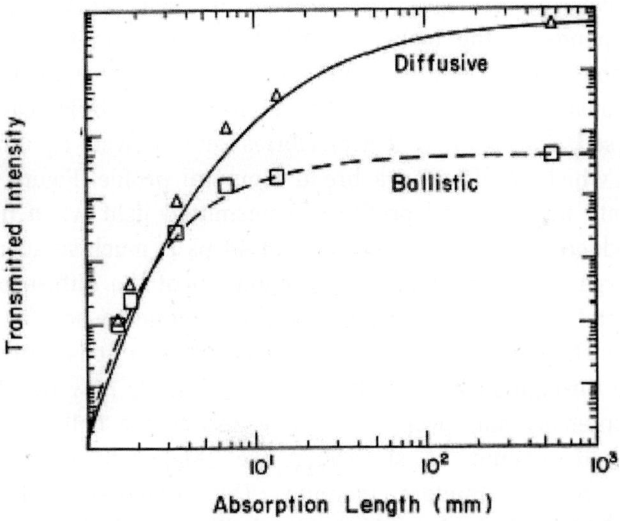

Figure 5.15 Theoretical prediction and experimental results of the total diffuse light and the signal intensity (in arbitrary units) for light transmitted through a random medium with different absorption lengths, using z = 10 mm, l_s = 0.3 mm, and l_t = 1.0 mm. The dashed and solid curves correspond to the ballistic and diffuse light intensity, respectively. The squares and triangles correspond to the measured ballistic and diffuse light intensity, respectively.

The intensities of the diffuse and ballistic signal presented in Fig. 5.15 were measured by using an angle- and time-resolved technique that was described in Ref. 58. In this method a beam of ultrafast laser pulses of 100 fs duration (repetition rate 82 MHz, average power 15 mW, wavelength centered at 620 nm, beam diameter 10 mm) from a colliding pulse mode-locked dye laser system was incident upon a slab of random medium. This random medium consisted of a 0.3% concentration of latex beads of 0.296 μm diameter suspended in water in a 10 mm-thick cylindrical glass cell with a diameter of 50 mm. A black pinhole of 2 mm diameter was placed at the center of the opposite side of the cell. The temporal distribution of the photons from this pinhole within 3 mrad in the forward direction was measured by a synchro-scan streak camera. The temporal profiles of ballistic signals and diffuse light pulses in the forward direction for different amounts of absorbing dye added in the random medium are displayed in Fig. 5.16. The narrow peak at time 44 ps corresponds to the ballistic signal that travels in the incident direction and arrives first. In contrast, the diffuse light pulse that consists of photons that had undergone multiple scattering traversed many different trajectories and arrived at different times which resulting in a broad temporal profile. Figure 5.16a presents the temporal profile of transmitted light when no dye was added. The tiny ballistic peak at 44 ps is much smaller than the large and broad temporal distribution of the diffuse pulses. The total intensity of the diffuse light is found to be 130 times greater than the intensity of the ballistic signal. It is impossible to see through such a cloudy medium. To study how the diffuse light intensity may decrease with respect to the ballistic signal, increased amount of absorbing dye (Malachite Green) was added to the random medium. The temporal profiles in Figs. 5.16b–5.16d show both the ballistic and diffuse signal intensities decreased when the absorptive dye concentration increased, but the intensity of the diffuse light decreased more, enhancing the relative intensity of the ballistic signal to the diffuse signal.

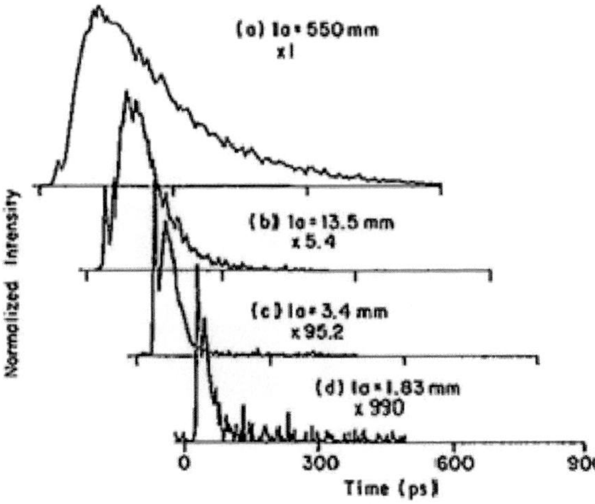

Figure 5.16 Transmitted pulse profiles through a slab of random medium 10 mm-thick with a 0.3% concentration of 0.296 μm diameter latex beads at different absorbing dye concentrations. The absorption lengths of these media are indicated on the curves.

A demonstration that an object hidden behind a scattering wall can in fact be imaged with higher clarity using the aid of absorption [73] is shown in Fig. 5.17. A cross of 3 and 2 mm-wide bars serves as an object and the scattering medium is placed between the object and the imaging video camera. The image of the object with the interference pattern which arises from light diffraction at the edge of the bar is shown in Fig. 5.17a when there were no scatterers in the sample cell. This image disappears when the medium behind the cross is made scattering (Fig. 5.17b), reappears when the scattering medium is made absorbing (Fig. 5.17c), and disappears again when the laser wavelength is tuned away from the absorbing spectral range of the medium (Fig. 5.17d). So, absorption can help to reduce the effects of scattering. It can be used for coding information as light passes through fog or clouds by selecting the light wavelength in the water absorption range from 800 to 1100 nm (Fig. 5.8) to 2500 nm and improve images of tumor growing in breast. Water absorption in breast tissue can enhance image contrast [74].

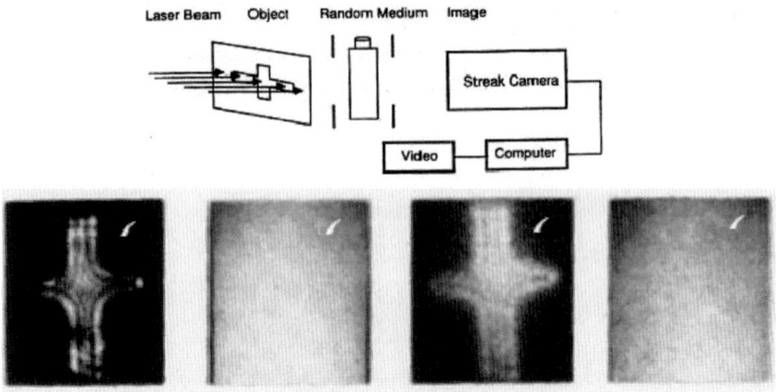

Figure 5.17 Imaging setup (top) and photographic images of a cross obtained with: (a) no scatterer in the cell; (b) scatterers are added into the cell where the thickness is 20 scattering mean free path, the laser wavelength is 620 nm; (c) Malachite Green dye is added to the scattering medium in the cell with a resulting absorption length of 4.6 mm; (d) the laser is tuned to 515 nm where the absorption length of the scattering medium increased to 75 mm.

Figure 5.18 shows the attenuation spectra of brain tissues of different thicknesses. Evidently, a trough of the attenuation spectra falls within window III (1600–1870 nm), indicating the minimal absorbance in this window [56].

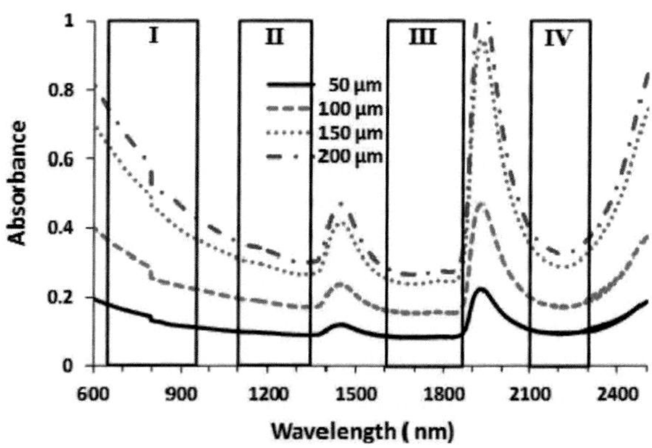

Figure 5.18 Attenuation spectrum of brain tissues spanning the 4 NIR spectral windows. Brain tissue thicknesses were 50, 100, 150, and 200 μm, respectively.

5.5 Transition from Ballistic to Diffuse in Model Scattering Media and Brain

There are three components called ballistic, snake and diffusive in optical pulse propagation in a highly scattering turbid medium. Three key length parameters –scattering mean free path, l_s; transport mean free path, l_t; and absorption length l_a (alternatively, the corresponding coefficients, μ_s, μ_s' and μ_a, respectively) are used to describe light propagation characteristics through the medium. Another key parameter is the anisotropy factor, g defined as the $<\cos\theta>$, where θ is the scattering angle. The relationship between these parameters and the relations governing the attenuation of the intensity of the ballistic and snake light as they propagate through the medium have been presented in Section 5.2.2. Ballistic and snake light may provide direct images of targets in turbid media, provided the deleterious effect of diffusive light (noise for direct imaging) could be sufficiently reduced. Section 5.4 reviewed some of the techniques for reducing the diffusive light noise. However, for sufficiently thick samples such as human breast or prostate the ballistic and snake light intensities are reduced so much and the diffusive noise becomes so dominant that direct imaging is not feasible. Diffusive light, that is multiple scattered within the medium, carries a wealth of information about the interior structures of the medium. Different approaches, referred to as inverse image reconstructions, that use knowledge of incident light properties, measurement of light emergent from the medium at the boundaries, average optical properties of the medium, an appropriate model for light propagation through the medium, and numerical algorithm to estimate optical properties of every voxel of the sample, are needed to extract the relevant information about the embedded objects, such as tumors in a breast [6, 38–40, 44].

The diffuse light intensity I_D described in diffusion approximation is:

$$I_D = I_0 \exp(-\mu_{eff} z), \tag{5.12}$$

where I_0 is the incident intensity, z is the sample thickness, and μ_{eff} is the effective scattering coefficient:

$$\mu_{\mathrm{eff}} = \sqrt{3\mu_{\mathrm{a}}(\mu_{\mathrm{s}}' + \mu_{\mathrm{a}})} \tag{5.13}$$

that is related to the diffusion coefficient $D = c/3\mu_{\mathrm{a}}(\mu_{\mathrm{s}}' + \mu_{\mathrm{a}})$.

Another area of interest is the transition regime between ballistic and diffuse light transport because it carries valuable information about characteristics of scattering [75, 76]. Assuming quasi-one-dimensional propagation and a small detection solid angle $\delta\Omega$ the detected light intensity was shown to be a sum of ballistic light intensity and a fraction of diffuse light intensity [75]:

$$I/I_0 = \exp(-\mu_{\mathrm{T}} z) + (\delta\Omega/4\pi)\exp(-\mu_{\mathrm{eff}} z). \tag{5.14}$$

Equation (5.14) does not include a term for snake (quasi-ballistic) photons and makes the assumption that absorption is negligible, $\mu_{\mathrm{a}} << \mu_{\mathrm{s}}$, that means the medium is diffusive (not absorptive). A transition depth, z_c is defined as

$$z_c = (\mu_{\mathrm{eff}} - \mu_{\mathrm{T}})^{-1}\ln(d\Omega/4\pi), \tag{5.15}$$

such that for $z << z_c$ the photons detected are mainly ballistic, and for $z >> z_c$ most photons are diffusive in nature.

Figure 5.19 shows the confocal signal as a function of the thickness (L) of the scattering medium (0.48 μm latex spheres suspensions in water with l_s = 74 μm) obtained by Kempe et al. [76]. The ballistic regime characterized by an exponential decay extends to $L_{\max}$ (~1 mm) at which ballistic and diffuse light intensities are equal, followed by the regime with much slower decay of intensity where diffuse light dominates.

In thin media the light intensity is governed by ballistic transport while for thick media by diffusive transport. The dynamic range of detection system must be $>10^5$. The use of diffusion equations for inverse problem requires the sample thickness to be $>5l_t$.

Light propagation through a tissue medium undergoes multiple scattering for NIR wavelengths. Multiple light scattering in medium splits the incident light into two components: a coherent (ballistic) and an incoherent (diffusive) component. The ballistic signal propagates undeviated in the forward direction and carries object information, while the diffusive light undergoes random walk in all directions and undergoes more optical paths and absorption than the ballistic light [56]. The ballistic photons in scattering media are governed by the Beer-Lambert law:

$$T = I/I_0 = \exp(-\mu_t z) = \exp(-z/l_t) \tag{5.16}$$

Incorporating the diffusive mode, Eq. (5.16) becomes

$$T = \frac{I}{I_0} = \exp\left(-\frac{z}{l_t}\right) + \frac{\Omega}{4p}\frac{\left[\sinh\left(\sqrt{\frac{3d^2}{l_{tr}l_a}}\left(1-\frac{z}{d}\right)\right)\right]}{\left[2\sinh\left(\sqrt{\frac{3d^2}{l_{tr}l_a}}\right)\right]}, \tag{5.17}$$

where the first exponential term is the ballistic mode, and the second term is the diffusive mode; l_a is absorption length, and l_{tr} is transport length; Ω is the solid angle, for thin samples we use a small Ω, measured as $\sim 10^{-2}$; $d = z + 2z_0$, where $z_0 = 0.71\ l_{tr}$. Eq. (5.17) can also be written as

$$T = \frac{I}{I_0} = \exp\left(\frac{-z}{l_t}\right) + \frac{\Omega}{4\pi}\exp(-\mu_{eff}z), \tag{5.18}$$

where $\mu_{eff} = \sqrt{3\mu_a(\mu_s' + \mu_a)}$.

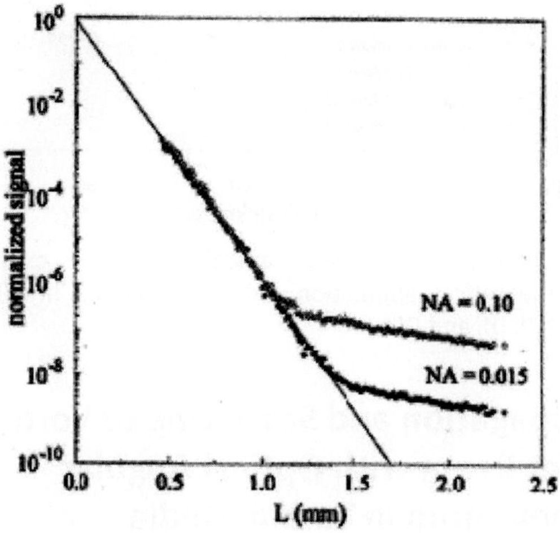

Figure 5.19 Semilogarithmic plot of the signal as a function of sample thickness for a suspension with l_s = 74 μm. Shown is the signal obtained with a NA of 0.10 for a pinhole with v_p = 1.24 and for a NA of 0.015 with v_p = 1.16. The solid line is the fit to the part of the curves that exhibits an exponential falloff.

To evaluate the role that the ballistic and diffusive modes each plays in light transmission through thin brain tissues, simulations were performed using Eq. (5.18) for tissue thickness from 50 to 200 μm in the four optical windows and compared with experimental results of the peak transmittance [56]. Model simulation results match well with the experiments in all four windows (Fig. 5.20). The absorption coefficient μ_a chosen is in the range of 0.005–0.04 mm^{-1}. The reduced scattering coefficient μ_s' is between 0.5–10 mm^{-1}. In addition, μ_s' is known to decrease with wavelength and by using fitting equations and parameters of rat brain tissue given by Jacques [77], μ_s' was chosen within the range of 1–10 mm^{-1}, 0.2–4 mm^{-1}, 0.08–2 mm^{-1}, and 0.04–1 mm^{-1} for windows I, II, III, and IV, respectively.

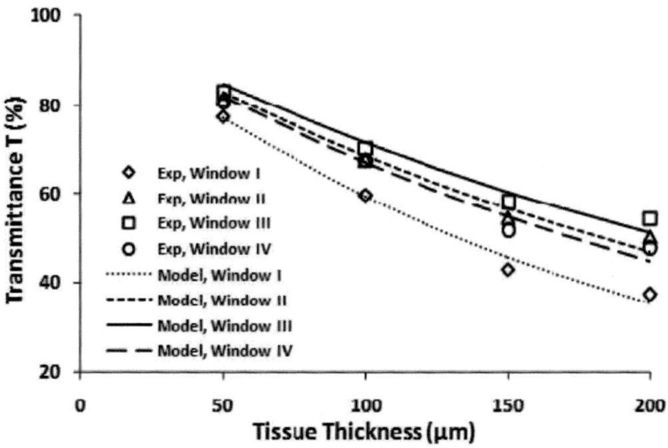

Figure 5.20 Comparison of peak transmittance from experiment and theoretical simulation for brain tissues in NIR windows I, II, III, and IV.

5.6 Propagation and Scattering of Vortex Light Beam with Optical Angular Momentum in Turbid Media

Optical Laguerre Gaussian (LG) vortex beams are relatively recent developments that pertain to generation of light beams with higher orbital angular momentum than that of a circularly polarized beam. We have explored use of LG beams for propagation and imaging through highly scattering media, and will briefly review the results in this section [78]. In particular, we investigated scattering and transmission of LG vortex beams with different OAM states (L) in turbid media and compared its performance with that of a Gaussian (G) beam using suspensions of large and small scattering particles in water as the scattering media. The transmission of LG and G beams were measured with different ratios of sample thickness (z) to scattering mean free path (l_s) of the turbid media, z/l_s. In the ballistic domain, LG and G beams exhibited comparable transmission, while in the diffusive region, LG beams show higher transmission than G beam, and the LG beams with higher L values show higher transmission than that with lower L values. The transition points from ballistic to diffusive regions for different scattering media were studied and determined. This newly observed transmission difference for different LG beams may be used for deep target detection in turbid media through LG beam imaging.

A schematic diagram of the experimental setup is shown in Fig. 5.21. A 5 mW, 633 nm beam from a He-Ne laser was used as a light source. The laser beam was focused into a single-mode optical fiber (SMOF). The output beam from the SMOF was used to illuminate a spatial light modulator (SLM, HoloEye LC-R 720) working in the reflection mode. The LG beams with different OAM states were generated using the SLM with different forked diffraction patterns. The generated LG beams were then divided by a beam splitter into two separate beams used as sample and reference beams, respectively, as shown in Fig. 5.21. A sample cell containing scattering medium was placed in the path of the sample beam. A reference beam that does not pass through the scattering medium was used, as in conventional transmittance measurements, to monitor and account for changes and fluctuations in the incident light power. The images of the sample beam output from the turbid media and the reference beam were recorded simultaneously using the same 16-bit CCD camera (Photometrics Model CH250 L). The transmittances of light in the different scattering media were calculated based on the ratio of intensity of the sample beam image over the intensity of the reference beam image.

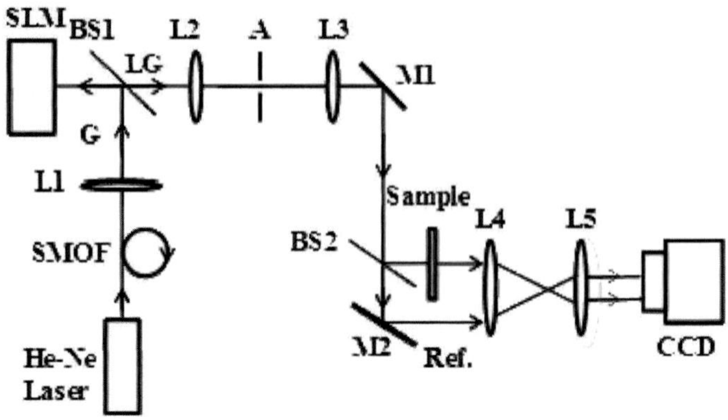

Figure 5.21 Experimental setup where SMOF-single mode optical fiber, L-lenses, G-Gaussian beam, BS-beam splitter, SLM-spatial light modulator, LG-Laguerre Gaussian beam, A-aperture, M-mirror, CCD-charge coupled device camera.

The SLM diffraction patterns and the generated corresponding LG beam shapes used in this study are shown in Fig. 5.22. The forked diffraction patterns were created using a MatLab-based software (HoloEye SLM application software). The forked diffraction patterns correspond to interference patterns of a TEM_{00} beam and a LG_0^L beam. Therefore, when a TEM_{00} Gaussian beam is used to illuminate the different forked diffraction patterns in SLM, the LG beams with different values of *L* will be generated, which have helical and twisted wave fronts.

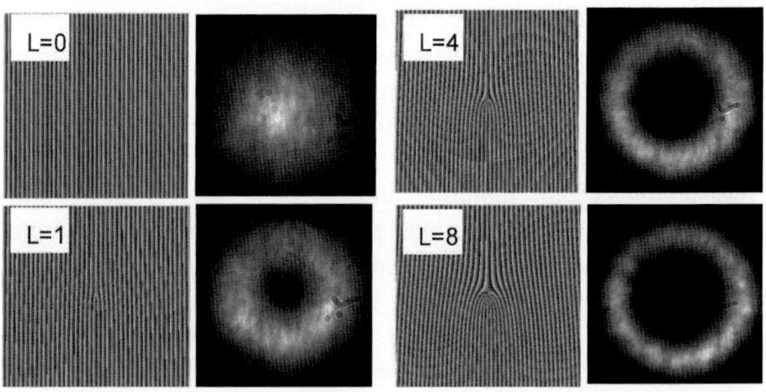

Figure 5.22 HoloEye SLM diffraction patterns and corresponding generated LG beams.

The scattering media used in the experiments were suspensions in water of polystyrene latex microspheres (beads) of various sizes and concentrations. The thicknesses (z) of the turbid media were varied to study transmission of LG beams as a function of ratio of sample thickness to the scattering mean free path (l_s) of the media, i.e., T versus z/l_s. Two types of beads with diameters of 0.107 and 3.12 μm were used as typical small and large scattering particles compared with the wavelength of light, respectively, to study the effect of particle size to the scattering and transmission of LG beams. The concentration of the scattering media is 1% diluted from the commercial bead solution (10% in stock). The scattering mean free path, l_s, the transport mean free path (l_t), and the anisotropy factor g as a function of particle size and wavelength for 1% of the polystyrene lax bead solutions (10% in stock) were investigated elsewhere [79], and discussed in Section 5.2.2 of this chapter. Using Fig. 5.4 in Section 5.2.2, the values of l_s, l_t and g at 633 nm for the 1% concentration turbid medium consisting of d = 0.107 μm small bead particles were estimated as l_s = 500 μm, l_t = 500 μm and g = 0, respectively. In contrast, the values of l_s, l_{tr} and g at 633 nm for the 1% concentration turbid medium consisting of d = 3.12 μm large bead particles were estimated as l_s = 100 μm, l_t = 800 μm and g = 0.875, respectively.

The transmission images of the G and LG beams with L = 0, 1, 4, and 8 were recorded for both small and large beads solutions with fourteen (14) thicknesses of the turbid media varied in a wide range from 0 to 40 mm (i.e., z = 0, 0.1, 0.2, 0.5, 1, 2, 3, 4, 5, 6, 8, 10, 20, and 40 mm). Based on the recorded transmission images data, the transmittances T of the LG and G beams in scattering media with different particle sizes, sample thicknesses and OAM values L were calculated.

As an example, Fig. 5.23 shows log T as a function of ratio z/l_s and topological charge L of LG beams for the scattering media consisting of small particles (d = 0.107 μm) in 1% concentration solution (volume density) of the commercial bead solution (10% in stock). The following four salient features can be seen from Fig. 5.23: (1) within the ballistic region, where z/l_s is small and ballistic transition is the dominant process, the LG and G beams show no significant differences on transmission; (2) within the diffusive region, where z/l_s is large and the multiple scattering is the dominant process, the LG beams show higher transmission than the G beam; (3) within the diffusive region, the LG beams with larger L-values show higher transmission than the LG beams with small L values; and (4) the transition point from the ballistic region to the diffusive region [80] for this small particle solution was observed at $z/l_s = 10$.

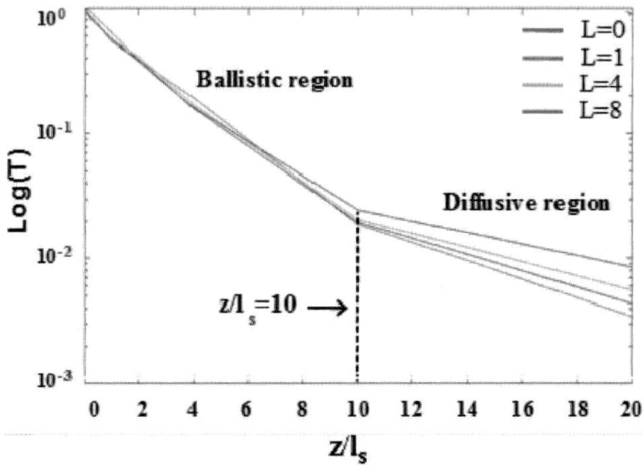

Figure 5.23 Log-T as a function of z/l_s where z is the thickness of turbid media and l_s is the scattering mean free path of the turbid medium (beads in water, d = 0.107 μm, l_s = 500 μm).

Similar log T versus z/l_s features shown in Fig. 5.23 were observed for the scattering media consisting of large bead particles (d = 3.12 μm): (1) LG and G beams show no significant differences on transmission in the ballistic region; (2) LG beams show higher T than G beam in the diffusive region; and (3) the LG beams with larger L values show higher T than the LG beams with small L values in the diffusive region [78].

It is an interesting topic to establish a physical model to explain why transmission of LG beams in turbid media is higher than that of G beams in the diffusive region. Palacios et al. [81]. used the dark core of an optical vortex to detect on-axis, forward-scattering light from a colloidal solution in the single and multiple scattering regions. Their results show that since the propagation phase fronts of LG beams are helical and unless acted upon by a scattering body, a vortex in coherent beam will propagate to infinity. However, scattering may superimpose other modes into dark vortex core region, shifting or brightening the core if the light is, respectively, coherent or incoherent [81]. It is believed that this brightening may contribute to the higher transmission of LG beams in comparison with G beams because the beam intensity recorded in the CCD camera includes contributions from both ballistic and forward multiple scattering photons. The observed higher transmissions of LG beams show different forward multiple scattering for LG and G beams. Since the spatial anisotropy of scattering is mainly given by the anisotropy factor g, the different forward scattering feature of LG and G beams in a turbid medium may indicate their different g values, which may be related to the vortex nature of LG beams.

This newly observed higher transmission of LG beams in comparison with G beams may be used for deep target detection in turbid media through LG beam imaging. We expect these observations to stimulate further theoretical and experimental studies to assess the efficacy of LG beams for signal transmission and imaging applications.

5.7 Nonlinear Optical Subsurface Imaging of Tissues

The nonlinear optical processes of second harmonic generation (SHG) and two-photon excitation fluorescence (2PEF) from biological tissues have been reported by several authors [82–89]. These processes enable imaging techniques that provide maps of local symmetry and molecular content of subsurface structures of tissues. Figure 5.24 shows a typical SHG and 2PEF spectrum of native chicken tissue. A combination of the confocal optical approach and 2PEF has been used to visualize the structure of biological tissues [85]. Scanning SHG microscopy in transmission geometry was developed for transparent, non-scattering media [83]. More recently, we have adapted and extended SHG and 2PEF-based techniques for imaging subsurface structures in highly scattering media using back-reflection geometry [82]. This novel technique shows promise for generating two-dimensional histopathological maps of tissue layers at different depths within a range of a few millimeters from the surface. A salient feature of the SHG-based techniques is its potential to yield information about the local symmetry that may help pathologists in their diagnosis.

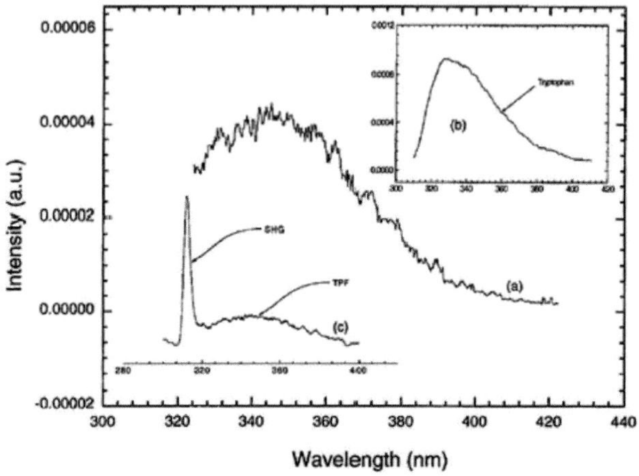

Figure 5.24 TPF fluorescence spectrum excited at 600 nm from (a) chicken muscle tissue, (b) tryptophan powder, and (c) the SHG and TPF spectrum of chicken muscle tissue excited at 625 nm [82].

A schematic diagram of the experimental arrangement for SHG and 2PEF tomography is shown in Fig. 5.25. A beam of 100 MHz repetition-rate, 625 nm, 100 fs duration pulses from an amplified colliding pulse mode-locked dye laser were delivered to a sample of chicken tissue with skin using a 27× microscope objective. The signal was collected in a back-reflection geometry by the same objective and phase sensitively monitored by a photomultiplier tube and lock-in amplifier. The objective could be translated in the *x*-*y* plane to generate area scans and in the *z*-direction to obtain profiles as a function of depth.

Figure 5.26a shows a typical axial (*y*-*z* plane) scan in logarithmic pseudo-color scale. Figure 5.26b shows an axial intensity profile along the *z*-direction (into the tissue). The first peak in the intensity profile corresponds to the interface between the skin and muscle, and the subsequent peaks are associated with the sub-subsurface layered structures in the tissue [86].

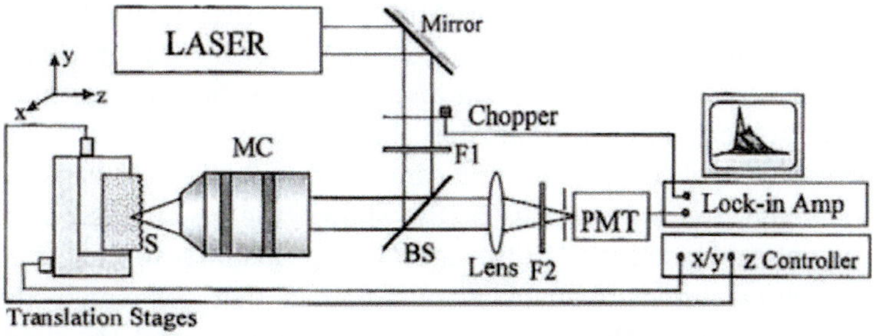

Figure 5.25 Schematic of the experimental setup: MC, microscope objective; BS, beam splitter; F1, F2, long and narrow bandpass filters, respectively; PMT, photomultiplier tube [86].

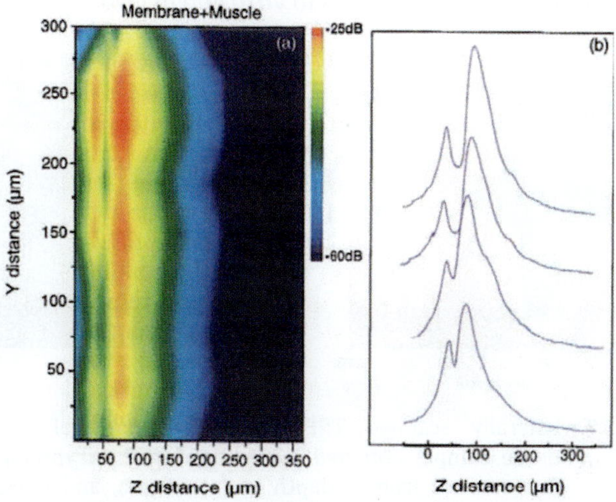

Figure 5.26 (a) 2D SHG depth image of a fascia membrane attached to chicken muscle tissue, log I_{SHG} versus *y*, *z*. (b) Analog axially scanned SHG profiles in linear scale [86].

Typical 2PEF images of chicken tissue samples obtained by monitoring the blue fluorescence of tryptophan centered on 400 nm and excited by 625 nm, 100 fs pulses are displayed in Figs. 5.27a and 5.27b. The 625 nm laser beam was scanned in a plane perpendicular to the orientation of the fibers in the tissue. Colored regions in the resulting map display the spatial distribution of tryptophan in the tissue. These 2PEF maps clearly demonstrate the potential of the technique for imaging the histology of tissue surface and layers at different depths within a few millimeters below the surface [87].

Furthermore, the use of SHG noninvasive technology to evaluate subsurface tumor progression stages without cutting into the tissue has been demonstrated [88]. Second harmonic tomographic imaging in hamster cheek pouch tissues, including normal control, dysplasia, and cancer after 8 and 14 weeks of carcinogen treatment (Fig. 5.28), shows the usefulness of in situ optical histology. With lateral and axial scanning, two-dimensional intensity maps were obtained that correlated SHG signal intensity with histological findings and showed the depth of tumor invasion at different stages of progression. This technology provides a noninvasive histological and medical diagnostic modality, which may help physicians to evaluate noninvasively the state of tissues in vitro and in vivo.

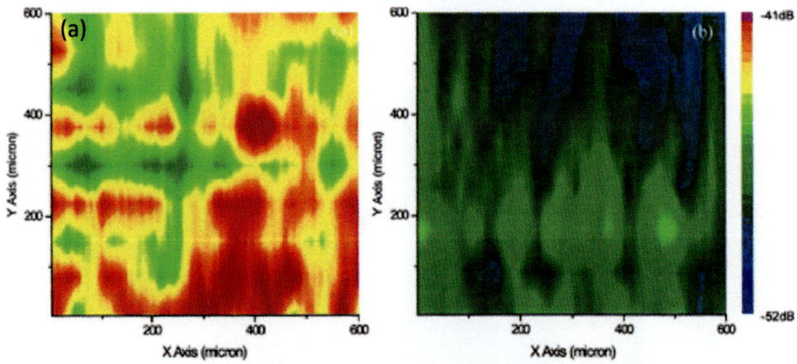

Figure 5.27 Laterally scanned TPF histological maps of a chicken tissue sample obtained by monitoring the tryptophan blue fluorescence from a depth of (a) 50 μm, and (b) 250 μm from the surface [87].

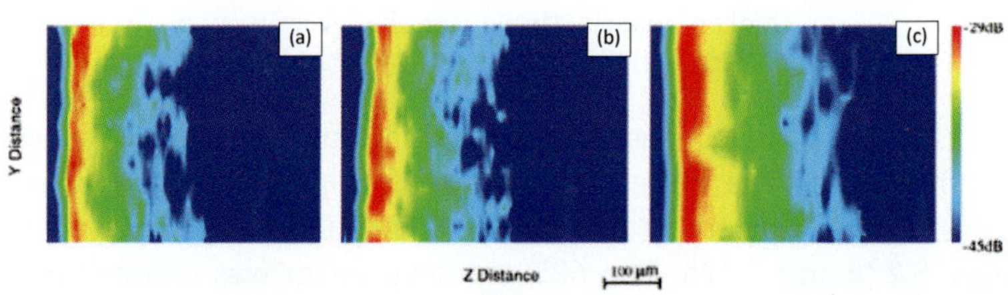

Figure 5.28 Two-dimensional tomographic second harmonic images obtained by 120 fs laser pulse excitation at 800 nm. (a) Normal hamster pouch tissue. (b) Hamster pouch tissue after 8 weeks of carcinogen (DMBA) treatment. (c) Hamster pouch tissue after 14 weeks of carcinogen (DMBA) treatment [88].

5.8 Summary

The research on use of light for probing and imaging biomedical media derived from the promise for development of safe, noninvasive, and inexpensive clinical imaging modalities with diagnostic ability. This chapter presents an overview of pulsed and continuous-wave, Gaussian and Laguerre Gaussian, low and high orbital-angular-momentumlight propagation through highly-scattering turbid media. It reviews the early-light transillumination imaging techniques as well as nonlinear optical tomography approaches that have been developed for imaging targets in turbid media with potential applications for imaging maladies deep inside body organs by a pioneering group over the years, and provides glimpses of the historical development of the field.

Development of detailed understanding of linear and nonlinear interactions of light as it propagates through a highly-scattering medium, and how those change the salient properties of light have led to the development of imaging and biopsy techniques.

Until recently the 650–950 nm wavelength range was considered to be "optical biomedical imaging window" as light in this spectral range is less absorbed than the visible and near-ultraviolet light. More recent work has identified three more spectral ranges: 1100–1350 nm, 1500–1860 nm, and 2100–2300 nm to be advantageous for the spectroscopic study and optical imaging of brain, breast and prostate tissues. Development in laser technology now provides high power laser outputs at 1500 nm, 1620 nm, 1700 nm; and 1996 nm from compact fiber lasers; while supercontinuum lases provide tunable output from 450–2250 nm. Advent of frequency modulation methods and devices has greatly enhanced frequency-domain methods. Beam shaping techniques now enable generation of Gaussian and Laguerre-Gaussian beams with very low to high orbital angular momenta. Beams with higher orbital angular momenta may provide further advantages in probing biomedical media. Nonlinear optical techniques enable sensitive probing of sub-surface maladies.

Concomitant developments in detector technology using bulk or quantum structures of InSb, InGaAs, PbS and other semi-conductors made highly sensitive area cameras available. These developments, along with advances in optical fiber beam delivery and signal acquisition systems, numerical algorithms, and personal computers with high memory and computational prowess are making development of frequency-agile and highly sensitive imaging systems with robust image reconstruction ability. The progress made so far indicates thatphotonic approaches will play an important role in deep imaging of brain, breast, prostate and other body organs, and helpcombat diseases through prompt detection and diagnosis.

Acknowledgment

We acknowledge IUSL researchers, G. E. Andersen, Feng Liu, K. M. Yoo, X. Ni, W. Cai, J. Dolne, X. Liang, P. P. Ho, S. Demos, B. B. Das, Yici Guo, M. Alrubaiee, Yang Pu, Q. Z. Wang, and G. Zhang for their contribution to the work reviewed here, and acknowledge Y. Budansky for technical help. The reviewed research was supported in part by US Army Medical Research and Materiel Command, NASA, and NYSTAR.

References

1. Cutler, M. (1929). Transillumination as an aid in the diagnosis of breast lesion, *Surg. Gynecol. Obstet.* **48**, pp. 721–730.
2. Gannot, I., Garashi, A., Chernomordik, V., and Gandjbachkhe, A., (2004). Quantitative optical imaging of the pharmacokinetics of fluorescent-specific antibodies to tumor markers through tissue-like turbid media, *Opt. Lett.* **29**, pp. 742–744.
3. Sordillo, L. A., Pu, Y., Pratavieira, S., Budansky, Y., and Alfano, R. R. (2014). Deep optical imaging of tissue using the second and third near-infrared spectral windows, *J. Biomed. Opt.* **19**, p. 056004.
4. Sordillo, L. A., Lindwasser, L., Budansky, Y., Leproux, P., and Alfano, R. R. (2015). Near-infrared supercontinuum laser beam source in the second and third near-infrared optical windows used to image more deeply through thick tissue as compared with images from a lamp source, *J. Biomed. Opt.* **20**, p. 030501.
5. Godavarty, A., Eppstein, M. J., Zhang, C., Theru, S., Thompson, A. B., Gurfinkel, M., and Sevick-Muraca, E. M. (2003). Fluorescence-enhanced optical imaging in large tissue volumes using a gain-modulated ICCD camera, *Phys. Med. Biol.* **48**, pp. 1701–1720.

6. Corlu, A., Choe, R., Durduran, T., Rosen, M. A., Schweiger, M., Arridge, S. R., Schnall, M. D., and Yodh, A. G. (2007). Three-dimensional in vivo fluorescence diffuse optical tomography of breast cancer in humans, *Opt. Express* **15**, pp. 6696–6716.
7. Alrubaiee, M., Xu, M., Gayen, S. K., and Alfano, R. R. (2006). Localization and cross section reconstruction of fluorescent targets in ex vivo breast tissue using independent component analysis, *Appl. Phys. Lett.* **89**, p. 133902.
8. Gayen, S. K., and Alfano, R. R. (1996). Emerging optical biomedical imaging techniques, *Opt. Photon. News* **7**(3), pp. 17–22 and references therein.
9. Wang, L., Ho, P. P., Liu, C., Zhang, G., and Alfano, R. R. (1991). Ballistic 2-D imaging through scattering wall using an ultrafast Kerr gate, *Science* **253**, pp. 769–771.
10. Yoo, K. M., and Alfano, R. R. (1990). Time-resolved coherent and incoherent components of forward light scattering in random media, *Opt. Lett.* **15**, pp. 320–322.
11. Huang, D., Swanson, E. A., Lin, C. P., Shuman, J. S., Stinson, W. G., Chang, W., Hee, M. R., Flotte, T., Gregory, K., Puliafito, C. A., and Fujimoto, J. G. (1991). Optical coherence tomography, *Science* **254**, pp. 1178–1181.
12. Drexler, W., and Fujimoto, J. G. (2008). *Optical Coherence Tomography: Technology and Applications* (Springer).
13. Dilworth, D. S., Leith, E. N., and Lopez, J. L. (1991). Three-dimensional confocal imaging of objects embedded within thick diffusing media, *Appl. Opt.* **30**, pp. 1796–1803.
14. Horinaka, H., Hashimoto, K., Wada, K., and Cho, Y. (1995). Extraction of quasi-straight-forward-propagating-photons from diffused light transmitting through a scattering medium by polarization modulation, *Opt. Lett.* **20**, pp. 1501–1503.
15. Demos, S. G., and Alfano, R. R. (1996). Temporal gating in highly scattering media by the degree of optical polarization, *Opt. Lett.* **21**, pp. 161–163.
16. Demos, S. G., and Alfano, R. R. (1997). Optical polarization imaging, *Appl. Opt.* **36**, pp. 150–163.
17. Demos, S. G., Savage, H., Heerdt, A. S., Schantz, S., and Alfano, R. R. (1996). Time-resolved degree of polarization for human breast tissue, *Opt. Commun.* **124**, pp. 439–442.

18. Wang, L., Ho, P. P., Liang, X. Dai, H., and Alfano, R. R. (1993). Fourier-Kerr imaging in thick turbid media, *Opt. Lett.* **18**, pp. 241–243.

19. Yoo, K. M., Xing, Q. R., and Alfano, R. R. (1991). Imaging objects hidden in highly scattering media using femtosecond second-harmonic generation cross-correlation time gating, *Opt. Lett.* **16**, pp. 1019–1021.

20. Faris, G. W., and Banks, M. (1994). Imaging through highly scattering media with a novel upconverting time gate, in *Proceeding on Advances in Optical Imaging and Photon Migration 21* (Alfano, R. R., ed.), Optical Society of America, Washington, D C., 1994, pp. 139–142.

21. Reintjes, J., Bashkansky, M., Duncan, M. D., Mahon, R., Tankersley, L. L., Moon, J. A., Adler, C. L., and Prewitt, J. M. S. (1993). Time-gated imaging with nonlinear optical Raman interactions, *Opt. Phot. News* **4**(10), pp. 28–32.

22. Jarlman, V., Berg, R., and Svanberg, S. (1992). Time-resolved tansillumination of the breast, *Acta Radiol.* **33**, pp. 277–2779.

23. Yoo, K. M., Das, B. B., and Alfano, R. R. (1992). Imaging of a translucent object hidden in a highly scattering medium from early portion of the diffuse component of a transmitted ultrafast laser pulse, *Opt. Lett.* **17**, pp. 958–960.

24. Spears, K. G., Serafin, J., Abramson, N. H., Zhu, X. M., and Bjelkhagen, H. (1989). Chronocoherent imaging for medicine, *IEEE Trans. Biomed. Eng.* **36**, pp. 1210–1221.

25. Leith, E., Arons, E., Chen, H., Chen, Y., Dilworth, D., Lopez, J., Shih, M., Sun, P. C., and Vossler, G. (1993). Electronic holography for imaging through tissue, *Opt. Phot. News* **4**(10), pp. 19–23.

26. Arons, E., Chen, H., Clay, K., Dilworth, D., Draper, R., Lopez, J., Leith, E., Shih, M., and Sun, P. C. (1994). New holographic methods for improved imagery through scattering media, in *Proceedings on Advances in Optical imaging and Photon Migration 21* (Alfano, R. R., ed.), Optical Society of America, Washington, D.C., 1994, pp. 239–243.

27. Sappey, A. D. (1994). Optical imaging through turbid media with a degenerate four wave mixing correlation time gate. *Appl. Opt.* **33**, pp. 8346–8353.

28. O'Leary, M. A., Boas, D. A., Chance, B., and Yodh, A. G. (1995). Experimental images of heterogeneous turbid media by frequency-domain diffusing-photon tomography, *Opt. Lett.* **20**, pp. 426–428.

29. Sevick, E. M., Hutchinson, C. L., Frisoli, J. K., Johnson, M. L., Nowaczyk, K., Szmacinski, H., and Lakowicz, J. R. (1993). The physical basis of

biomedical optical imaging using time-dependent measurements of photon migration in the frequency domain, in *Medical Optical Tomography: Functional Imaging and Monitoring* (Muller, G. J., Alfano, R. R., Arridge, S. R., Beuthan, J., Gratton, E., Kaschke, M., Masters, B. R., Svanberg, S., and van der Zeect, P., eds.), vol. IS 11, SPIE Institute Series, SPIE, Bellingham, Washington, 1993, pp. 485–512.

30. Gratton, E., Mantulin, W. W., vandeVen, M. J., Fishkin, J. B., Maris, M. B., and Chance, B. (1993). A novel approach to laser tomography, *Bioimaging* **1**, pp. 40–46.

31. Yamashita, Y., and Kaneko, M. (1993). Visible and infrared diaphanoscopy for medical diagnosis, in *Medical Optical Tomography: Functional Imaging and Monitoring* (Muller, G. J., Alfano, R. R., Arridge, S. R., Beuthan, J., Gratton, E., Kaschke, M., Masters, B. R., Svanberg, S., and van der Zeect, P., eds.), vol. IS 11, SPIE Institute Series, SPIE, Bellingham, Washington, 1993, pp. 283–316.

32. Beuthan, J. (1993). IR-diaphanoscopy in medicine, in *Medical Optical Tomography: Functional Imaging and Monitoring* (Muller, G. J., Alfano, R. R., Arridge, S. R., Beuthan, J., Gratton, E., Kaschke, M., Masters, B. R., Svanberg, S., and van der Zeect, P., eds.), vol. IS 11, SPIE Institute Series, SPIE, Bellingham, Washington, 1993, pp. 263–282.

33. Chan, K. P., Yamada, M., Devaraj, B., and Inaba, H. (1995). Optical imaging through highly scattering media by use of heterodyne detection in the 1.3 μm wavelength region, *Opt. Lett.* **20**, pp. 492–494.

34. Inaba, H. (1993). Coherent detection imaging for medical laser tomography, in *Medical Optical Tomography: Functional Imaging and Monitoring* (Muller, G. J., Alfano, R. R., Arridge, S. R., Beuthan, J., Gratton, E., Kaschke, M., Masters, B. R., Svanberg, S., and van der Zeect, P., eds.), vol. IS 11, SPIE Institute Series, SPIE, Bellingham, Washington, 1993, pp. 317–347.

35. Schmidt, A., Corey, R., and Saulnier, P. (1995). Imaging through random media by use of low-coherence optical heterodyning, *Opt. Lett.* **20**, pp. 404–406.

36. Frecher, A. F. Drexler, W., Hitzenberger, C. K., and Lasser, T. (1993). In vivo optical coherence tomography in ophthalmology, in *Medical Optical Tomography: Functional Imaging and Monitoring* (Muller, G. J., Alfano, R. R., Arridge, S. R., Beuthan, J., Gratton, E., Kaschke, M., Masters, B. R., Svanberg, S., and van der Zeect, P., eds.), vol. IS 11, SPIE Institute Series, SPIE, Bellingham, Washington, 1993, pp. 355–370.

37. Kalpaxis, L., Wang, L. M., Galland, P., Liang, X., Ho, P. P., and Alfano, R. R. (1993). Three-dimensional temporal image reconstruction

of an object hidden in highly scattering media by time-gated optical tomography, *Opt. Lett.* **18**, pp. 1691–1693.

38. Hebden, J. C., Arridge, S. R., and Deply, D. T. (1997). Optical imaging in medicine: I. Experimental Techniques, *Phys. Med. Biol.* **42**, pp. 825–840.
39. Arridge, S. R., and Hebden, J. C. (1999). Optical imaging in medicine: II. Modeling and reconstruction, *Phys. Med. Biol.* **42**, pp. 841–853.
40. Arridge, S. R. (1999). Optical tomography in medical imaging, *Inverse Probl.* **15**, pp. R41–R93.
41. Hiraoka, M., Firbank, M., Essenpreis, M. E., Cope, M., Arridge, S. R., Zee, P., and Delpy, D. (1993). Monte Carlo simulation of light transport through inhomogeneous tissue, in *Proc. SPIE*—The International Society for Optical Engineering, 1888, pp. 149–159.
42. Arrige, S. R. (1993). The forward and inverse problems in time-resolved infrared imaging, in *Medical Optical Tomography: Functional Imaging and Monitoring* (Muller, G. J., Alfano, R. R., Arridge, S. R., Beuthan, J., Gratton, E., Kaschke, M., Masters, B. R., Svanberg, S., and van der Zeect, P., eds.), vol. IS 11, SPIE Institute Series, SPIE, Bellingham, Washington, 1993, pp. 35–64.
43. Cai, W., Gayen, S. K., Xu, M., Zevallos, M., Alrubaiee, M., Lax, M., and Alfano, R. R. (1999). Optical three-dimensional inverse image reconstruction of objects in turbid media from ultrafast time-sliced optical transmission measurements, *Appl. Opt.* **38**, pp. 4237–4246.
44. Xu, M., Alrubaiee, M., Gayen, S. K., and Alfano, R. R. (2008). Optical diffuse imaging of an ex vivo model cancerous human breast using independent component analysis, *IEEE J. Select. Top. Quantum Electron.* **14**, p. 43.
45. Moulton, P. F. (1986). Spectroscopic and laser properties of $Ti^{3+}Al_2O_3$, *J. Opt. Soc. Am. B* **3**, pp. 125–133.
46. Petricevic, V., Gayen, S. K., and Alfano, R. R. (1988). Laser action in chromium-activated forsterite for near-infrared excitation: Is Cr^{4+} the lasing ion? *Appl. Phys. Lett.* **53**, pp. 2590–2593.
47. Liu, F., Yoo, K. M., and Alfano, R. R. (1994). Transmitted photon intensity through biological tissues within various time windows, *Opt. Lett.* **10**, pp. 740–742.
48. Ishimaru, A. (1978). *Wave Propagation in Random Media*, vols. 1 and 2, (Academic, New York).
49. Wang, L. (1995). Picosecond Kerr gated imaging of phantoms in turbid media, Ph. D. Thesis, The City University of New York.

50. Moes, C. J. M., van Gemert, M. C., Star, W. M., Marijnissen, J. P. A., and Prahl, S. A. (1989). Measurements and calculations of energy fluence rate in a scattering and absorbing phantom, *Appl. Opt.* **28**, pp. 2292–2296.

51. van Staveren, H. J., Moes, C. J. M., van Marle, J., Prahl, S. A., and van Gemert, M. J. C. (1991). Light scattering in Intralipid-10% in the wavelength range of 400–1100 nm, *Appl. Opt.* **31**, pp. 4507–4514.

52. Peters, V. G., Wyman, D. R., Patterson, M. S., and Frank, G. L. (1990). Optical properties of normal and diseased human breast tissues in the visible and near infrared, *Phys. Med. Biol.* **9**, pp. 1317–1334.

53. Key, H., Davies, E. R., Jackson, P. C., and Wells, P. N. (1991). Optical attenuation characteristics of breast tissues at visible and near-infrared wavelengths, *Phys. Med. Biol.* **5**, pp. 579–590.

54. Gandjbakhche, A. H., Nossal, R. J., Dadmarz, R., and Schwartzentruber, D. (1995). Expected resolution and detectability of adenocarcinoma tumors within human breast in time-resolved images, in *Proc. SPIE*, the International Society for Optical Engineering, **2387**, Paper 2387-2318, p. 111.

55. McIlwain, H., and Bachelard, H. S. (1995). *Biochemistry and the Central Nervous System* (Churchill Livingstone, Edinburgh; New York).

56. Shi, L., Sordillo, L. A., Rodríguez-Contreras, A., and Alfano, R. R. (2016). Transmission in near-infrared optical windows for deep brain imaging, *J. Biophoton.*, **9**(1–2), pp. 38–43.

57. Svanberg, S., and Metcalf, H. (1992). *Atomic and Molecular Spectroscopy: Basic Aspects and Practical Applications*, Springer-Verlag.

58. Yoo, K. M., Liu, F., and Alfano, R. R. (1990). When does the diffusion approximation fail to describe photon transport in random media? *Phys. Rev. Lett.* **64,** pp. 2647–2650; and *Phys. Rev. Lett.* **65,** pp. 2210–2211.

59. Lax, M., Narayanamurti, V., and Fulton, R. C. (1987). Classical diffusive photon transport in a slab, in *Proceedings of the Symposium on Laser Optics of Condensed Matter* (Birman, J. L., Cummins, H. Z., and Kaplyanskij, A. A., eds.), Leningrad, June 1987, Plenum, New York.

60. Watson, G. H., Fleury, P. A., and McCall, S. L. (1987). Searching for photon localization in the time domain, *Phys. Rev. Lett.* **58**, pp. 945–948.

61. Akkermans, E., Wolf, P. E., and Maynard, R. (1986). Coherent backscattering of light by disordered media: Analysis of the peak line shape, *Phys Rev. Lett.* **56**, pp. 1471–1474.

62. Stephen, M. J., and Cwilich, G. (1986). Rayleigh scattering and weak localization: Effects of polarization, *Phys Rev. B* **34**, pp. 7564–7572.

63. van der Mark, M. B., van Albada, M. P., and Lagendijk, A. (1988). Light scattering in strongly scattering media: Multiple scattering and weak localization, *Phys. Rev. B* **37**, pp. 3575–3592.
64. Rosenbluh, M., Edrei, I., Kaveh, M., and Freund, I. (1987). Precision determination of the line shape for coherently backscattered light from disordered solids: Comparison of vector and scalar theories, *Phys. Rev. A* **35**, pp. 4458–4460.
65. Feng, S., Kane, C., Lee, P. A., and Stone, A. D. (1988). Correlation and fluctuations of coherent wave transmission through disordered media, *Phys. Rev. Lett.* **61**, pp. 834–837.
66. Stephen, M. J. (1988). Temporal fluctuations in wave propagation in random media, *Phys. Rev. B* **37**, pp. 1–5.
67. Pine, D. J., Weitz, D. A., Chaikin, P. M., and Herbolzheimer, E. (1988). Diffusing wave spectroscopy, *Phys. Rev. Lett.* **60**, pp. 1134–1137.
68. Weitz, D. A., Pine, D. J., Pusey, P. N., and Tough, R. J. A. (1989). Nondiffusive Brownian motion studied by diffusing-wave spectroscopy, *Phys. Rev. Lett.* **63**, pp. 1747–1750.
69. Freund, I., Kaveh, M., and Rosenbluh, M. (1988). Dynamic multiple scattering: Ballistic photons and the breakdown of the photon-diffusion approximation, *Phys. Rev. Lett.* **60**, pp. 1130–1133.
70. Cai, W., Lax, M., and Alfano, R. R. (2000). Analytical solution of the elastic Boltzmann transport equation in an infinite uniform medium using cumulant expansion, *J. Phys. Chem. B* **104**, pp. 3996–4000.
71. Cai, W., Lax, M., and Alfano, R. R. (2000). Cumulant solution of the elastic Boltzmann transport equation in an infinite uniform medium, *Phys. Rev. E* **61**, pp. 3871–3876.
72. Niedre, M. J., de Kleine, R. H., Aikawa, E., Kirsch, D. G., Weissleder, R., and Ntziachristos, V. (2008). Early photon tomography allows fluorescence detection of lung carcinomas and disease progression in mice in vivo, *PNAS*, **105**, pp. 19126–19131.
73. Yoo, K. M., Liu, F., and Alfano, R. R. (1991). Imaging through a scattering wall using absorption, *Opt. Lett.* **16**, pp. 1068–1070.
74. Gayen, S. K., Zevallos, M. E., Alrubaiee, M., Evans, J. M., and Alfano, R. R. (1998). Two-dimensional near-infrared transillumination imaging of biomedical media with a chromium-doped forsterite laser, *Appl. Opt.* **37**, 5327.
75. Yaroshevsky, A., Glasser, Z., Granot, E., and Sternklar, S. (2011). Transition from the ballistic to diffusive regime in a turbid medium, *Opt. Lett.* **36**, pp. 1395–1397.

76. Kempe, M., Genack, A. Z., Rudolph, W., and Dorn, P. (1997). Ballistic and diffuse light detection in confocal and heterodyne imaging systems, *J. Opt. Soc. Am. A* **14**, pp. 216–223.

77. Jacques, S. L. (2013). Optical properties of biological tissues: A review. *Phys. Med. Biol.*, **58**(11), pp. R37–R61.

78. Wang, W. B., Gozali, R., Nguyen, T. A., and Alfano, R. R., Propagation and transmission of optical vortex beams through turbid scattering wall with orbital angular momentums, *Proc. SPIE*, **9318**, p. 931805.

79. Alfano, R. R., Wang, W. B., Wang, L., and Gayen, S. K. (2015). Chapter 9: Light propagation in highly scattering turbid media: Concept, techniques, and biomedical applications, in *Photonics, Volume IV: Biomedical Photonics, Spectroscopy, and Microscopy* (Andrews, D., ed.), John Wiley & Sons, Inc., pp. 367–412.

80. Kempe, M., Genack, A. Z., Rudolph, W., and Dorn, P. (1997). Ballistic and diffuse light detection in confocal and heterodyne imaging systems, *J. Opt. Soc. Am. A* **14**, pp. 216–223.

81. Palacios, D., Rozas, D., and Swartzlander, Jr., G. A. (2002). Observed scattering into a dark optical vortex core, *Phys. Rev. Lett.* **88**, pp. 103902-1-4.

82. Guo, Y., Wang, Q. Z., Zhadin, N., Liu, F., Demos, S., Calistru, D., Tirksliunas, A., Katz, A., Budansky, Y., Ho, P. P., and Alfano, R. R. (1997). Two-photon excitation of fluorescence from chicken tissue, *Appl. Opt.* **36**, pp. 968–970.

83. Freund, I., Deutsch, M., and Sprecher. A. (1986). Connective tissue polarity. Optical second-harmonic microscopy, crossed-beam summation, and small-angle scattering in rat-tail tendon, *Biophys. J.* **50**, pp. 693–712.

84. Fine, S., and Hansen, W. P. (1971). Optical second harmonic generation in biological systems, *Appl. Opt.*, **10**, pp. 2350–2353.

85. Denk, W. (1996). Two-photon excitation in functional biological imaging, *J. Biomed. Opt.* **1**, pp. 296–304.

86. Guo, Y., Ho, P. P., Savage, H., Harris, D., Sacks, P., Schantz, S., Liu, F. Zhadin, N., and Alfano, R. R. (1997). Second-harmonic tomography of tissues, *Opt. Lett.* **22,** pp. 1323–1325.

87. Guo, Y., Ho, P. P., Liu, F., Wang, Q. Z., and Alfano, R. R. (1998). Noninvasive two-photon-excitation imaging of tryptophan distribution in highly scattering biological tissues. *Opt. Commun.* **154**, pp. 383–389.

88. Guo, Y., Savage, H., Liu, F., Schantz, S., Ho, P. P., and Alfano. R. R. (1999). Subsurface tumor progression investigated by noninvasive

optical second harmonic tomography. *Proc. Nat. Acad. Sci.* **96**, pp. 10854–10856.

89. Alfano, R. R., Demos, S. G., Galland, P., Gayen, S. K., Guo, Y., Ho, P. P., Liang, X., Liu, F., Wang, L., Wang, Q. Z., and Wang, W. B. (1998). Time-resolved and nonlinear optical imaging for medical applications, in Proceedings of the NYAS conference on "Advances in Optical Biopsy and Imaging," *Ann. N. Y. Acad. Sci.*, **838**, pp. 14–28.

Chapter 6

Application of Nonlinear Microscopy in Life Sciences

Zdenek Svindrych and Ammasi Periasamy

W.M. Keck Center for Cellular Imaging, University of Virginia, Department of Biology, Gilmer Hall, Charlottesville, VA 22904, USA

ap3t@virginia.edu

As was the case with the introduction of confocal fluorescence microscopy in life sciences, the application of second harmonic and multiphoton imaging was another milestone in biological imaging, since it enabled researchers to move from cultured cells deep into tissues and whole organs, even in live animal models. The lower scattering, absorption, and phototoxicity of near infrared light together with the inherent optical sectioning capability (that is, no need for confocal pinhole in the detection path) are the main factors that allow deep tissue imaging without the need for tissue clearing and other steps incompatible with live samples. Because immunofluorescent staining is not feasible in native samples, other methods of selective fluorescent labeling must be used, such as fluorescent fusion proteins. Alternative approaches take advantage of endogenous fluorescent molecules, such as tryptophan, NADH, NADPH, flavins; and the effects of relevant biochemical processes on their fluorescent properties.

Deep Imaging in Tissue and Biomedical Materials: Using Linear and Nonlinear Optical Methods
Edited by Lingyan Shi and Robert R. Alfano

ISBN 978-981-4745-88-8 (Hardcover), 978-1-315-20655-4 (eBook)
www.panstanford.com

6.1 Introduction

While the most widely used microscopy modality in clinical settings and histological laboratories is simple widefield microscopy (the whole field of view is illuminated and observed at once), either in transmission arrangement—hematoxylin and eosin stained sections, or in reflection—detailed observation of tissue surfaces, endoscopy [1]; primary research in life sciences relies heavily on fluorescence microscopy and its further technological improvements, such as spinning disc confocal microscopy, confocal laser scanning microscopy (CLSM) [2], laser scanning multiphoton microscopy (MPM) [3], second harmonic generation (SHG) imaging [4], selective plane illumination microscopy (SPIM) with both single photon and multiphoton excitation [5], and various superresolution microscopy techniques based either on widefield mode (structured illumination methods (SIM) [6], single molecule localization [7, 8] or fluctuations [9]) or laser scanning mode (stimulated emission depletion—STED [10], image scanning methods [11], such as AiryScan [12] and MSIM [13, 14]). Some of these methods can also readily be enhanced using time-resolved detection (fluorescence lifetime imaging—FLIM [15]), multispectral detection [16], or extended using fiber optic probes for endoscopic imaging [17] or remote sensing in behaving animals [18].

The main advantage of fluorescence microscopy is its ability to conveniently visualize the distribution of minute quantities of fluorescent molecules with sub-micrometer lateral spatial resolution and millisecond temporal resolution; also several fluorescent labels can be distinguished based on their photophysical properties (excitation and emission wavelengths, fluorescence lifetime, etc. [19]). Then, the wealth of available methods to target the fluorescent molecules to structures and/or events of interest turn fluorescence microscopy into the most powerful tool in cell biology.

On a single cell level, fluorescent antibodies and fusion proteins are available that can label individual proteins inside the cell, delineate subcellular structures, mark the phase of a cell cycle and cell differentiation. Fluorescent sensors (based on engineered fluorescent molecules [20] and fusion proteins [21]) are available to measure membrane potentials, ion concentrations and other local physico-chemical properties, enabling, for example, studies of membrane function [22] and fast Ca^{2+} transients in neurons [23]. Longer timescale studies can use, e.g., artificial and fluorescent nucleotides to study the details of replication and transcription [24]. On larger spatial scale, synaptic uptake of injected fluorescent agents can be used to trace neuronal circuits [25], or even the whole bloodstream can be made visible by injecting fluorescently conjugated dextrans or lectins [26].

Quite different approach is to focus on endogenous fluorescent molecules, that is, fluorescent species already naturally present in organelles, cells, and tissues. NADH (a coenzyme crucial in energy metabolism), amino acid tryptophan (and the closely chemically related neurotransmitter serotonin) and flavins are common examples of such autofluorescent markers [27]. Clearly, these molecules cannot be manipulated by the researcher to explore the behavior of thousands of cellular proteins independently, as is the case of "exogenously" labeled samples; one can only study those processes that "use" the endogenous markers in one way or the other. The complication that arises from the non-trivial relation between the endogenous marker photophysical properties and the studied biological processes is overweighed by gross simplification of sample preparation (no transfection or other labeling required), moreover, all labeling-related artifacts are naturally avoided.

The goal here is not to enumerate all the possible combinations of imaging modalities and sample labeling strategies; it's mere to show the breadth of the field of fluorescence microscopy and to help identify areas where nonlinear microscopy with its unique attributes may enable new opportunities in improving and advancing the life sciences research.

6.2 Basic Principles of Multiphoton Microscopy

6.2.1 The Missing Cone Problem

The missing cone problem is a limitation common to many imaging methods, where a 3D object is projected on a 2D detector and geometrical constraints forbid observing the sample from some angles. This includes tilt-series transmission electron tomography, some forms of x-ray (computer) tomography, and also widefield fluorescence microscopy. It can be fully appreciated when the image formation process is formulated in spatial frequency domain [28], where the extent of the optical transfer function (OTF, a Fourier transform of the point spread function—PSF) drops to zero for angles near the optical axis. A real-life demonstration of this limitation is when one attempts to image a perfectly homogeneous fluorescent horizontal layer of infinite extent (much larger than the field of view, with perfectly even illumination): one cannot determine whether the layer is in focus, as the intensity of the (otherwise perfectly uniform) image of the layer does not change appreciably with defocus.

There are a number of ways to "fill" the missing cone, including methods based solely on image processing (deconvolution). The culprit of applying deconvolution on widefield fluorescence image stacks is the fact that the missing information is recovered using additional constraints, such as sparsity of the fluorescent structures [29], and may thus lead to noticeable artifacts (most algorithms would not cope well with the uniform layer example mentioned above).

6.2.2 Confocal Detection

It is now well understood that the missing cone problem can be solved using patterned excitation illumination [30] (with the example above, when a pattern with sharp edges is projected onto the fluorescent layer, we can immediately tell when the layer is in focus based on the observed contrast of the pattern); the finer details are present in the pattern, the better axial resolution is achieved. While some patterns, such as grid lines in the case of SIM, may require some image post-processing; the simplest "pattern," that is, a single spot or single line scanning across the sample, may achieve optical sectioning with all-optical means. When the excitation is provided by single diffraction-limited laser spot scanned in a raster fashion and the detector only "sees" the fluorescence originating from the very same spot, we arrive at CLSM. The first commercially available confocal systems (before the introduction of lasers, then called "optical masers") were using an array of moving spots created (and then filtered on the detection side) by a spinning Nipkow disk [31].

The introduction of confocal microscopy into life sciences was a major breakthrough in imaging of cells and tissues. Unlike widefield fluorescence microscopy, CLSM provides optically sectioned images without the out-of-focus fluorescence; the lateral resolution is the same as can be achieved in widefield mode (the Abbe's diffraction limit):

$$d_{xy} = 0.61\frac{\lambda}{\text{NA}} \tag{6.1}$$

where λ is the wavelength of light (excitation for confocal, emission for widefield) and NA the numerical aperture of the objective. The axial resolution of a confocal microscope with a pinhole set to 1 Airy diameter is [32]

$$d_z = \frac{\lambda}{n - \sqrt{n^2 - \text{NA}^2}}. \tag{6.2}$$

A well-aligned confocal microscope provides faithful representation of the 3D distribution of the fluorophores without image-processing artifacts. Five or more different labels can be imaged simultaneously and distinguished based on their excitation and emission spectra, chromatically corrected objective lenses offer precise colocalization over the whole visible range. Those indispensable properties made CLSM the golden standard in cellular and tissue imaging in biological research.

6.2.3 Multiphoton Microscopy

In the framework of nonlinear microscopy, traditional fluorescence microscopy (regardless of widefield or confocal modality) is termed "linear," that is, the intensity of fluorescence emitted from each part of the sample is directly proportional to the local intensity of excitation, $I_{\text{em}} \propto I_{\text{exc}}$. While any deviation from linearity (such as saturation of fluorescence) can be exploited to achieve optical sectioning, or even increased resolution [33], some sorts of nonlinearities (such as multiphoton excitation and SHG) are particularly suitable, as they can be used for 3D microscopy with relatively simple instrumentation and without additional image processing steps.

In a standard (one-photon, linear) confocal microscope the detection pinhole is responsible for rejecting the out-of-focus light. If we increase the pinhole size to infinity, so that all fluorescence originating from the sample (and entering the objective) is collected, the optical-sectioning ability is lost and such system behaves just like widefield (except that the resolution is governed by the excitation wavelength, as opposed to the emission wavelength in widefield mode). This can be understood intuitively (in the context of the fluorescent layer example mentioned above) on the basis of energy conservation: each lateral section of the focused excitation beam profile contains the same total light flux, no matter whether it's high intensity in a tight focused spot or larger region of weaker intensity further from focus. Then, in the linear case, the total amount of fluorescence from the fluorescent layer is constant, whether the layer is in focus or not; a hallmark of the missing cone problem.

If, on the other hand, the emission intensity behavior can be written as $I_{em} \propto I_{exc}^m$, where $m > 1$, then the emission will originate predominantly from the focus of the excitation beam where I_{exc} is highest; and the optical sectioning ability is restored even without the detection pinhole. In fact, it is highly advantageous to detect all photons escaping from the sample, even the scattered ones, as all photons originate predominantly from the focus. Open pinhole, or even detectors placed close to the sample (before the scanner, non-descanned detectors, NDD) are preferred. It can be shown that for $m = 2$ the behavior of such nonlinear scanning microscope is equivalent to that of a CLSM [34].

Two-photon (2P) absorption is one of processes that yield the $I_{em} \propto I_{exc}^2$ behavior (see Fig. 6.1). It is a quantum process where two photons are absorbed simultaneously to excite the fluorescent molecule. The molecule then emits a photon during the transition to the ground state in the same fashion as in single-photon excited fluorescence (that is, the fluorescence lifetime and the emission spectrum are largely unchanged). Because the 2P absorption cross-section is extremely low, two conditions must be met: (1) the instantaneous excitation intensity is sufficiently high, and (2) single-photon excitation is avoided. The second condition can be easily fulfilled by selecting long enough excitation wavelength (often close to twice the optimal 1P absorption wavelength, 800–1400 nm for common dyes). But the first condition would require excitation powers of several tens of Watts per μm^2, a value hard to achieve in a point scanning configuration, and prohibitive in terms of sample damage. The solution is to concentrate the excitation power into short pulses with a duty cycle of 10^{-5} and lower (100 fs pulses at 80 MHz repetition rate as an example), so one can achieve the same absorption rates (and fluorescence intensities) with less than 10 mW of average laser power (and kilowatt peak power). Note that the average intensity of 1P fluorescence is the same for pulsed and continuous excitation, as long as the average excitation intensity is constant, while for $I_{em} \propto I_{exc}^2$ the average fluorescence intensity would be proportional to the peak excitation intensity (inversely proportional to the duty cycle).

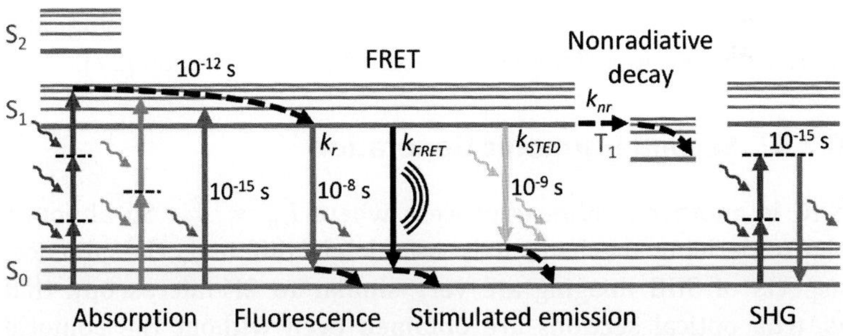

Figure 6.1 Schematic Jablonski diagram of multiphoton (three- and two-photon) and linear absorption, bringing molecule into the first excited singlet state. Various radiative decay pathways (fluorescence, stimulated emission) and non-radiative decay pathways (e.g., Forster resonance energy transfer, crossing to triplet state, chemical binding or decomposition, etc.) are depicted. For comparison, Second Harmonic Generation (SHG) does not involve any electronic transitions. Note, that multiphoton excitation into higher states (e.g., S_2) is also possible, and may lead to increased phototoxicity and rapid photobleaching. Characteristic timescales for important processes are mentioned.

As with concentrating the excitation in time domain, focusing the excitation spatially is crucial for efficient 2P excitation. Single point laser scanning configuration with high NA objective is the imaging modality of choice, whereas line-scanning and spinning-disc configurations are significantly less feasible [35] and efficient 2P excitation is possible only over limited field of view.

Multiphoton fluorescence is generally exited by absorbing two or more photons simultaneously (see Fig. 6.1). While the probability of such process decreases rapidly with increasing order of the transition, the strength of the nonlinearity increases (e.g., $I_{em} \propto I_{exc}^3$ in the case of three-photon excited fluorescence), so short-pulsed excitation may also effectively excite multiphoton processes. Of course, one has to use progressively longer excitation wavelengths to excite given fluorescent molecule with higher order processes. One of the more common examples is three-photon excitation of tryptophan [27] and other molecules excited in the UV. The longer wavelengths of commercial tunable pulsed lasers that recently became available (up to 1300 nm) enabled three-photon excitation of some common fluorescent dyes and 2P excitation even with red dyes [36]. Higher order multiphoton absorption may also be responsible for rapid bleaching and increased phototoxicity when very short pulses are used [37].

6.2.4 Second Harmonic Generation

SHG is another nonlinear process where $I_{em} \propto I_{exc}^2$. So although the underlying mechanism is completely different, the practical aspects of SHG imaging are very similar to 2P microscopy, that is, true optical sections are obtained even without the confocal pinhole and concentrating the illumination into short pulses helps to improve the yield of this process considerably. Indeed, SHG and 2P imaging are often performed simultaneously, with the same excitation source and similar imaging conditions.

SHG is a coherent process that stems from the nonlinear dependence of the polarization of a molecule on the local electric field, namely it comes from the (nonzero) second-order dielectric susceptibility [38] (it can also be viewed quantum-mechanically, see Fig. 6.1). So there are two crucial requirements for the sample to be an SHG source: (1) it has to be crystalline (to allow coherent addition of the second harmonic electric field) over a distance that corresponds to the size of the "focus" of the incident light, that is about 1 μm, and (2) a nonzero second-order dielectric susceptibility that generally requires low crystal symmetry (the crystal structure needs to lack a center of symmetry). Some surfaces and interfaces may also generate second harmonic due to local symmetry breaking [39].

The most common application of SHG in biological imaging is imaging ordered structures of collagen [4], but other ordered protein structures (myosin, tubulin) may also be imaged.

The main practical differences between SHG and 2P microscopy are summarized below: the wavelength of the second harmonic is exactly one half of the excitation wavelength, and thus narrow-band, but it is longer in the case of 2P fluorescence due to Stokes shift and the emission spectrum is broad (as with normal fluorescence); the SHG is directed away from the illuminating objective, and some additional scattering is needed to direct the generated photons back to the objective (alternatively, photons may be detected through the microscope condenser); SHG is an instantaneous process (it has zero lifetime), in contrast with fluorescence lifetime of the order of nanoseconds.

6.2.5 Absorption, Scattering and Wavefront Distortion in Tissues

Deep tissue imaging, however important in life sciences, is very difficult and faces many limitations and challenges. Methods based on x-ray absorption (CT), nuclear magnetic resonance (MRI) or emission from radioactive markers (PET), as mere examples, do allow whole body imaging, but lack the resolution and specificity of light microscopy. When single-cell or subcellular resolution in intact tissue is required, together with specific labeling of structures, proteins or cellular events, fluorescence microscopy is the first choice.

The limits of the depth from which fluorescence can be collected can be purely geometrical (the size of the objective lens, available working distance and numerical aperture) or more fundamental: absorption, scattering and wavefront distortion of the illumination and/or the emitted light. Background fluorescence is often the limiting factor when widefield approach is used.

Unwanted absorption of light leads to decrease of the excitation intensity and of the collected fluorescence according to Beer's law [40]. Most of the energy of the absorbed photons is converted to heat; it may also contribute to breakdown of some light-sensitive molecules, generation of free radicals and other phototoxic effects. The absorption coefficient generally decreases with increasing wavelength, blue light being more readily absorbed in tissue than red or near infrared, with minimum absorption typically around 1000–1500 nm, depending on the content of water, hemoglobin, melanin and other pigments [41]. Longer wavelength light is also significantly less phototoxic, because electronic transitions responsible for those effects are of higher energy.

Light scattering is caused by abrupt changes of refractive index on small length scale, such as sub-micron particles, lipid droplets, protein assemblies, or individual proteins. The scattered photons appear with random direction and they no longer can contribute to the focused excitation spot or to confocal detection, but they can still be detected with widefield NDD detectors in a multiphoton microscope. Due to the λ^4 dependence of Rayleigh scattering (scattering from sub-wavelength features), longer excitation and emission wavelengths tend to be much less affected.

Variations of local refractive index on much longer length scale than the light wavelength causes weak refraction of light rays and distortion of the spherical wavefronts of the focused excitation light or the spherical wavefronts of the generated fluorescence (later to be focused on the detection pinhole in the case of CLSM). Such distortion, much like other optical aberrations, reduce the peak intensity in the focus, both on excitation, in the sample, as well as on detection, on the pinhole; and reduce the optical resolution. With two photon excitation and widefield detection the wavefront distortion of the emitted fluorescence does not play any role, but the quadratic dependence of the fluorescence intensity on the excitation intensity any distortion of the diffraction-limited excitation spot has more severe consequences in terms of signal intensity, and even more so in higher order excitation processes (three-photon, etc.). On the other hand, because of the slow variation of these aberrations, these can be pre-compensated using adaptive optics, so the state-of-the art imaging modality for deep tissue imaging in live tissue is multiphoton microscopy combined with adaptive optics [42].

To summarize, it is the synergy of several factors why nonlinear microscopy is advantageous in non-invasive deep live-tissue imaging: the longer wavelengths of the light used for multiphoton excitation and SHG are less sensitive to scattering and absorption in common tissues, longer wavelengths cause less photodamage, and the widefield (non-descanned) detection of emitted photons nullifies the effect of scattering and wavefront distortions on the detection part. Also, since the nonlinear processes take place only in the focal plane being imaged, multiphoton microscopy causes much less photodamage than single-photon CLSM, especially when large stacks of planes are being collected.

It is worth mentioning that for fixed samples there is a wealth of clearing methods that dramatically reduce light absorption and scattering [43]. These often employ removal of lipids and pigments from permeabilized tissue, either with water-based detergents (such as SDS), or organic solvents. In the subsequent step, all low-refractive-index solvents are replaced with index-matching medium (generally n = 1.4–1.5) to reduce light scattering and optical aberrations. Penetration depths of 10 mm are achievable in cleared tissue with confocal or SPIM techniques [44].

6.3 Instrumentation for Nonlinear Microscopy

6.3.1 Light Sources in Nonlinear Microscopy

The high power of incident light required for appreciable 2P absorption in fluorescently labeled samples necessitates the use of lasers as excitation sources. All biological applications use pulsed lasers for multiphoton excitation (with duty cycle of 10^{-4} and lower), as this helps to lower the time-averaged illumination power by several orders of magnitude. Historically, many techniques (Q-switching, mode-locking with subsequent amplification or optical parametric conversion, harmonic generation, Raman conversion, etc.) have been used to obtain short pulses in a wide range of wavelengths from a wide range of gain media (mostly optically pumped solid state media or dye solutions). But the true boom in nonlinear microscopy in life sciences was closely related to the introduction of turn-key, reliable, widely tunable femtosecond lasers with high repetition rates [45].

The most prevalent are tunable mode-locked Ti:sapphire oscillators, delivering sub-100 fs pulses at around 80 MHz repetition rate, covering the range of 680–1080 nm with up to 3 W of average output power. With 2P excitation these cover the range of 350–550 nm. To cover longer excitation wavelengths (to efficiently excite red fluorophores and to take advantage of the lower scattering and absorption in tissues), pulsed lasers covering 680–1300 nm, based on Yb-doped fibers, became available recently. These longer wavelengths also allow for three-photon excitation of some fluorescent proteins.

While wide tunability is a key benefit in a research environment, the complexity and price of these systems make fixed wavelength pulsed lasers a viable alternative in biological research. Fiber lasers (the fiber core, doped with fluorescent ions, is the gain medium, so high pump power can be maintained over extended length) are especially attractive for their ruggedness, reliability, and low price. As examples, Er-doped fiber lasers produce 1560 nm (780 nm after frequency doubling), 100 fs pulses; Yb-doped fiber lasers emit at 1040 nm. With these lasers, multiphoton imaging of some fluorescent proteins and SHG imaging are readily achievable [46].

6.3.2 Point Scanning Nonlinear Microscopy

The most common two-photon microscopes used in biology research are, in fact, point-scanning CLSMs adapted for two-photon excitation by adding a femtosecond near-infrared (often tunable) laser and proper dichroic (or rather polychroic) beam splitters and combiners [3]. While internal confocal detectors (behind the scanner and the pinhole, often also called "descanned") may be used to detect the fluorescence, the "field of view" of these detectors is quite small even with the pinhole fully open. To unveil the full potential of 2P microscopy, NDD need to be added somewhere between the objective and the scanning mirrors, often as close to the sample as possible to eliminate unnecessary optical elements in the detection path. Properly designed NDDs can receive every photon generated inside the sample and entering the objective (within its FOV and NA limits), which is crucial for imaging deep in scattering tissue, as explained above.

The same instrumentation can be used for SHG imaging as well, tissue scattering even helps to direct the second harmonic photons back towards the objective. Alternatively, imaging through the microscope condenser and the transmission detector (also common on many confocal microscopes) may be used with weakly scattering samples. High NA immersion condenser (quite uncommon on modern fluorescence microscopes) or even immersion objective lens mounted instead of the condenser lens can boost the signal considerably, due to the specific angular distribution of the second harmonic signal [38].

Due to its relative simplicity, point-scanning multiphoton microscope is an ideal platform for implementing more advanced aspects of nonlinear microscopy, such as dispersion compensation, adaptive optics and pulse shaping. Dispersion compensation (also called "prechirp"), a feature often seamlessly integrated into femtosecond laser sources, helps to achieve shortest possible pulse duration inside the sample by compensating for the group velocity dispersion (GVD, a manifestation of wavelength-dependence of refractive index) caused by various elements along the excitation beam path.

Adaptive optics (AO) helps in achieving high intensity in the laser focus inside the sample by compensating the wavefront distortions the laser beam experiences in a biological sample. Typically, a deformable mirror (with tens to hundreds actuators, or degrees of freedom) is placed in a position conjugate to the back focal plane of the objective. When the optimal deformation is applied, the excitation light can be focused perfectly even hundreds of micrometers deep in native tissue [42]. There are various strategies how to estimate the correct deformation, some relying on the measurement of the aberrated wavefront with, e.g., Shack-Hartmann sensor, to close the control loop. Others may iteratively search for optimum deformation, one that would cause highest intensity of some non-linear process, 2P excited fluorescence being employed most often.

Spectral shaping of ultrafast pulses is another advanced concept, where the phase of a broadband pulse is engineered in such a way, that it selectively excites one fluorophore, and not the other [47, 48]. While simple tuning of Ti:sapphire laser to appropriate wavelengths may perform similarly, spectral shapers are much faster, at can "switch" between fluorophores in milliseconds (when based on liquid crystals) or even microsecond (when based on acousto-optic interaction); typical Ti:sapphire laser can sweep across its tuning range in 10 s.

6.3.3 Multipoint Scanning Two-Photon Microscopy

Usually only a small fraction of the available laser power is used for multiphoton excitation in single focus. To improve efficiency and imaging speed, multipoint scanning systems are available, where a number of closely spaced foci (up to 64) is scanned in parallel across the sample [49]. The fluorescence emanating from individual foci then cannot be discriminated using single NDD detector, so either de-scanned configuration with a matched multi-point detector or widefield detection with a pixelated detector (CCD or CMOS camera) is needed. While being much faster, achieving 30 fps at 4 megapixel resolution, it tends to be more affected by the scattering of the fluorescence photons than single-point scanning microscope with NDD detection.

A closely related technique, spinning-disc confocal microscopy with 2P excitation, was also demonstrated with a customized Yokogawa microlens-based spinning disk [35]. Again, the main advantage of this approach, the fast imaging using CCD or CMOS camera, is offset by the limited FOV where 2P fluorescence can be excited with the available laser power. Also, due to the crosstalk between adjacent pinholes, the performance in densely labeled or strongly scattering tissue is worse than in the case single point excitation.

6.3.4 Line Scanning Nonlinear Microscopy

As with multipoint scanning systems, line scanning approaches were developed to achieve faster (30 fps) imaging. Typical pulsed lasers are powerful enough to excite 2P fluorescence along modestly long line of focus. For wide field of view amplified lasers with lower repetition rate (to allow high peak intensity while maintaining low average laser power) may be used. There are several strategies of detection of fluorescence from the line focus. Linear detector behind a confocal slit is the traditional method, alternatively, the light passing the slit can be once again scanned with a mirror and detected with a CCD or CMOS detector [50]. Equivalently, rolling shutter of a CMOS camera, synchronized with the line of excitation can be used in simple widefield detection arrangement [51].

6.3.5 Temporal Focusing

In principle, widefield two-photon excitation is possible over limited FOV. While the peak illumination intensity is several orders of magnitude lower (compare the area of a quite small FOV, e.g., 10^3 um^2 to the area of diffraction-limited focus with high NA illumination, about 0.1 um^2), which would translate to 10^8 times decreased 2P fluorescence, this difference can be partially offset by increasing laser power (often only 1% of the available laser power is used in the point-scanning microscope) and by the much longer widefield camera exposure time (100 ms) than the pixel dwell time (microseconds) during point scanning. But the main advantage of 2P excitation, the confinement of the excitation only to the focal plane, is lost in widefield illumination.

Temporal focusing restores the optical sectioning ability (to some degree) even with widefield illumination. The idea behind this method is to spread the short-pulsed illumination spectrally using a grating, so that each spectral component takes different path and all components recombine in the focal plane to form a transform-limited pulse (see Fig. 6.2). Away from focus, the individual components do not combine with the right phase relations and do not form short pulse with high peak intensity. Practically, this is achieved by placing a grating at a position conjugate to the focal plane of the objective, with parameters adjusted such that the diffracted spectrum of the original pulse extends over the whole back focal aperture of the objective [52].

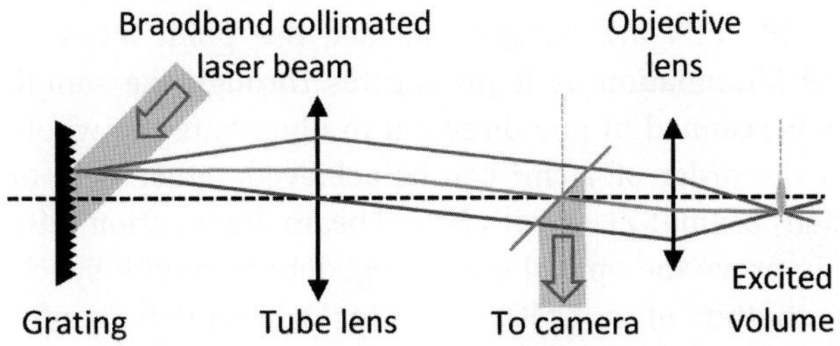

Figure 6.2 Schematic principle of widefield temporal focusing. The broadband laser pulse is spread spectrally so that the shortest pulse is restored in the focus of the objective lens.

The same approach can be combined with line-scanning illumination to achieve fast two-photon imaging with improved optical sectioning capabilities [53] due to more focused excitation of a line focus. Further improvement in rejecting out-of-focus light can be achieved using synchronized rolling shutter detection (with CMOS cameras), that emulates confocal slit [54].

Widefield temporal focusing has also been combined with adaptive optics to improve temporal-focusing two-photon imaging in live tissues [55]. Other method of improving the sectioning ability of widefield temporal focusing microscopy is based on structured illumination [56].

6.3.6 Two-Photon Selective Plane Illumination Microscopy

Even in its linear (one-photon) version, SPIM is one of the fastest imaging modalities with inherent optical sectioning and extremely low phototoxicity. These properties stem from the geometrical arrangement, where the illumination and detection objectives are placed with their axes perpendicular to each other. The (low NA) illumination objective creates a line focus with extended depth of focus, and illuminates a thin plane as it propagates through the sample. The (high NA) objective images the fluorescence from that plane onto a CMOS camera. The excitation light is used extremely efficiently, as it generates useful fluorescence as it passes throughout the whole thickness of the sample.

While it is possible to use line-focused pulsed laser for 2P excitation in SPIM, this would only allow limited FOV observation. Current 2P SPIM microscopes, in fact, use point focus (creating a line of illumination as it propagates through the sample), and the line is scanned in one direction to illuminate the whole plane. FOVs of the order of 1 cm^2 can be achieved, either with ordinary (Gaussian) beam [57], or with Bessel beam illumination [58].

To improve the optical sectioning, detection with synchronous rolling shutter of a CMOS detector to emulate confocal slit aperture [59] has been implemented (with single-photon fluorescence excitation).

6.3.7 Superresolution Nonlinear Microscopy

Superresolution techniques reach resolution that is higher than the Abbe limit (Eqs. 6.1 and 6.2). The main strategies to overcome the diffraction limit are: structured illumination (SIM) and saturated SIM, localization of single blinking molecules (PALM, STORM) or capturing intensity fluctuations of overlapping molecules (SOFI), and methods based on nonlinear behavior of fluorescence intensity (STED, RESOLFT). While all methods, that allow better than 2× improvement in resolution, rely on some nonlinear or random effect, all these methods may use multiphoton processes for excitation or activation of fluorescence. As examples, PALM with widefield temporal-focus 2P photoactivation in single plane [60] and with line-scan temporal focus 2P photoactivation [61] were demonstrated. Also, STED with 2P excitation using femtosecond pulses, and 1P depletion with stretched pulses of the same wavelength has been demonstrated [62].

6.3.8 Opto-Acoustic Intravital Imaging with Multiphoton Excitation

The difficulty of detecting fluorescence photons through thick layers of scattering tissue can be mitigated by detecting sound waves generated as excitation light gets absorbed by the tissue [129]. As acoustic waves get absorbed much less than visible photons in typical tissues, the penetration depth in opto-acoustic (OA) imaging is limited by scattering and absorption of the excitation light. In a typical (single-photon excited) OA experiment the lateral resolution is given by the diffraction-limited focal spot size, that is sub-micrometer range, whereas the axial resolution is given by the speed of sound in the tissue and the bandwidth of the receiving transducer, and is limited to several tens of micrometer even with high-bandwidth (100 MHz) transducers. This severe limitation can be overcome with multi-photon absorption, that is confined to the focus of the laser, allowing true optical resolution in all three dimensions even with low-bandwidth (1 MHz) transducers [63].

6.3.9 Multiphoton Endoscopy with GRIN Needle Lenses

An alternative approach to excite and collect fluorescence from deep within a live tissue is to use GRIN objective lenses. Instead of having direct the excitation and emission photons through thick layers of scattering tissue or to remove large parts of the tissue, a thin (down to 0.35 mm) GRIN lens can be inserted up to 10 mm into the tissue and relay the image plane deep inside the tissue to a plane just outside the protruding end of the lens [64]. A common 2P microscope (with suitable objective lens) is then used to image this relayed plane. While the observed animal still needs to be rigidly connected to the microscope, this approach is less invasive than, for example, a cranial window; and images with subcellular resolution can be obtained from very deep layers of tissue. Moreover, adaptive optics can be used to correct for any optical aberrations introduced by the GRIN lens [65].

6.3.10 Intravital 2P Microscopy with Optical Fibers

Remote fluorescence sensing using optical fibers is an attractive alternative to fixed-objective microscopes. The main applications are endoscopic imaging, tumor imaging and mapping the brain activity in behaving (and freely moving) transgenic animals. While linear (one-photon) fluorescence systems based on coherent fiber bundles [66], as well as single-core fibers with distal scanning [67] are readily available, two photon excitation through optical fibers prove to be challenging.

The technical difficulties stem from the basic requirements for efficient 2P excitation, that is, tight focus (diffraction limited, which calls for single-mode propagation of the excitation light) and short pulses with high peak power at the sample are necessary. The large GVD a laser pulse experiences in a long optical fiber can be pre-compensated with suitable pulse shaper, but the small mode area of typical SM fibers (less than 100 μm^2) is not compatible with high peak power, thanks to nonlinear pulse broadening (mainly self-phase modulation, SPM) [68]. These limitations can be partially offset using large mode area fibers (often fabricated as photonic-crystal fibers) or multimode fibers provided that only single mode is excited and the light energy does not couple into higher modes. When such fiber is coupled to a distal scanner with a miniature objective lens (often a GRIN lens), such system can be used for multiphoton microscopy. To improve the efficiency with which the fluorescence is collected, double-clad fiber (i.e., the cladding of the central single-mode core acts itself as a large core for fluorescence collection) may be used [2].

SPM can, in principle, be also suppressed by stretching the pulses before launching them into the fiber, and compressing them back at the distal end, but miniaturized spectral shapers are not available [69]. Alternatively, approaches based on nonlinear spectral broadening (in a piece of SM fiber), with subsequent phase conjugation and large GVD compensation have been demonstrated. The goal is to use the last inches of the delivery fiber to perform the spectral compression to restore the (nearly) transform-limited pulse [70].

Two-photon excitation has been demonstrated with coherent fiber bundles (up to 10^5 individual cores packed into single fiber of 1 mm diameter) terminated with a GRIN objective lens at the distal end; the scanning is performed at the proximal end in a raster-scanning fashion [71]. The main advantages of this approach are: (1) very compact and rugged distal end (only a GRIN rod lens glued to the end facet of the fiber), and (2) as the excitation light is guided through a single core at any instant, all the other cores can serve for collection of (scattered) fluorescence, this improves the detection efficiency and reduces the effects of fiber autofluorescence. The current drawbacks of this approach are following: the resolution is given by the number and spacing of cores, the individual cores are multimode, there is strong leaking of the excitation light into neighboring cores, and the nonlinear effects in the core material are not well understood. These drawbacks stem from the fact that the currently available imaging fiber bundles are not optimized for 2P excitation. As a result, the available power at the fiber end is not sufficient for practical intravital microscopy. Attempts to improve the efficiency using the methods above (namely, the spectral broadening/conjugation/compression approach) have not been reported yet.

New technologies are emerging that attempt imaging through multimode fibers or custom single-mode fiber bundles [72] without distal scanning. These approaches try to measure [73] or calculate [74] the phase scrambling a wavefront experiences on passing though the fiber. The effect of the fiber then can be inverted using adaptive optics and the image of the distal object can be recovered. In nonlinear microscopy the phase profile of a beam entering the fiber would be engineered in a way to produce a focused spot at the distal end, the spot could be scanned using different phase profiles and the nonlinear signal (2P fluorescence) generated in the focus would be guided back by the fiber to a detector. The main disadvantages are the strong dependence of the phase function on the bending of the fiber and the limited speed of scanning (currently limited by the spatial modulator update speed to about 10^3 points per second).

Approach based on double-clad fiber with a large-mode area photonic crystal fiber core combined with distal scanning was also successfully demonstrated [67].

6.3.11 Fluorescence Lifetime Imaging Microscopy (FLIM)

While not limited to multiphoton microscopy, FLIM, especially its time-correlated single photon counting (TCSPC) variant, can be easily integrated into point-scanning multiphoton microscopes. The instrumental requirements for TCSPC fluorescence lifetime measurements are as follows: (1) pulsed excitation source with pulse width less than 1 ns and repetition rate not much higher than 100 MHz—a condition fulfilled by all common pulsed lasers used for multiphoton excitation, (2) single photon counting detector with timing jitter less than 1 ns—that is all modern multialkali and GaAsP PMTs, SPADs, etc., and (3) hardware capable of measuring the arrival time of each detected photon with sub-1 ns resolution and jitter, with respect to the excitation laser pulse.

An excited fluorescence molecule will decay to the ground state with a rate constant k that is the sum of the rate constants of all mechanisms through which the molecule can leave the excited state:

$$k = k_r + k_{nr}, \tag{6.3}$$

where k_r is the rate of radiative decay (that is mainly fluorescence, or stimulated emission in the case of STED microscopy) and k_{nr} is the sum of all nonradiative decay rates (e.g., collision quenching, triplet state formation, conformational or chemical changes in the excited state, including fluorophore breakdown, and energy transfer to nearby molecules, including FRET; see also Fig. 6.1). After a short excitation pulse the number of molecules n remaining in the excited state decreases exponentially and so does the intensity of fluorescence emission I:

$$I(t) \propto n(t) \propto e^{-kt} = e^{-t/\tau}, \tag{1.4}$$

where k is the overall rate constant of the decay, t is the time elapsed since the excitation pulse, and τ is the fluorescence lifetime defined as $\tau = 1/k$.

The exponential decay is built as a histogram from the arrival times of individual photons (note, that there is typically less than one photon detected per 100 laser pulses in a TCSPC experiment) and the lifetime extracted by fitting with exponentials or by more elaborate convolution methods (see Fig. 6.3). If more different fluorescent species (or different forms of the same fluorophore) contribute to the decay, multiple-exponential fitting (the model is a linear combination of several exponential decays with different lifetimes) can be used to extract the fluorescence lifetimes of the individual components.

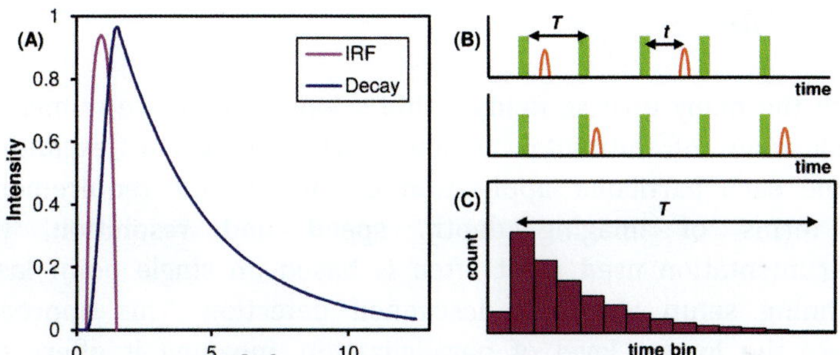

Figure 6.3 The Fluorescence lifetime and its measurement. (a) The fluorescence decay of an ensemble of identical molecules after a laser pulse. The Instrumental Response Function (IRF) combines the excitation pulse duration and the time resolution of the detector and the jitter of the associated hardware. (b) In a practical TCSPC experiment there is typically much less than one photon detected for every laser pulse. The time of photon arrival t within the laser pulse period T is measured for every photon. In the typical case of 80 MHz laser pulse repetition rate the period T = 12.5 ns. (c) A histogram of photon arrival times that is built from the single photon events approaches the theoretical ensemble decay curve of (a). This histogram can be built in each pixel of a confocal image and the lifetime information at each pixel is then extracted by fitting an exponential model, see Eq. (1.4). With permission from [128].

FLIM is also invaluable when interpreting Forster Resonance Energy Transfer (FRET) experiments, that is, energy transfer from an excited donor fluorescent molecule to an acceptor at a distance less than 10 nm. Rather than from fluorescence intensity images taken at several excitation and emission wavelengths [75] FRET can be precisely quantified based on the shortening of fluorescence lifetime of the donor, as the energy transfer introduces a new nonradiative pathway for the decay of the excited state of the donor fluorophore [76]. Also, the fraction of donor molecules participating in FRET can be determined from bi-exponential fitting of fluorescence decay histograms; the longer lifetime component corresponding to donors that do not have an acceptor within FRET range [77].

6.4 Biological Applications of Nonlinear Microscopy

With the many diverse fields of life sciences there are numerous application of intravital imaging with nonlinear excitation. While each particular application has its specific requirements in terms of imaging depth, speed and resolution, the instrumentation used most often is based on single point laser scanning setup with non-descanned detection. This approach offers the lowest level of parallelization, meaning it offers the best optical sectioning and detection efficiency in scattering tissue, but the acquisition itself is rather slow. It also allows straightforward implementation of advanced concepts, like adaptive optics and FLIM. On the other hand, methods offering higher level of parallelization of detection, namely line-scanning and widefield multiphoton excitation with temporal focusing, and multiphoton SPIM, can reach very high imaging speeds, but are less efficient in scattering tissue, because scattered photons do not contribute to the in-focus image.

6.4.1 Application of Nonlinear Microscopy in Neuroscience

Biological imaging of nervous system is closely linked to the development in the field of deep-tissue microscopy and the introduction of fluorescent labeling techniques, mainly genetically encoded neural activity indicators [78] and dyes for dendritic and axonal tracing [79, 80].

Traditionally, genetically encoded calcium indicators of the GCaMP family (derived from GFP) were used for neuronal activity imaging in wide range of animal models [81]. Imaging depths up to 800 μm in live mouse brain were reported with 2P excitation (amplified pulses at rather low repetition rate of 200 kHz) and sparse labeling [82]. To allow deep 2P imaging of densely labeled zebrafish brain, an elaborate adaptive optics scheme was developed [42]. Alternatively, imaging depths over 1 mm were achieved using three-photon excitation (at 1675 nm) of RFP in pyramidal neurons of transgenic mice [83]. Injectable calcium-sensitive dyes with emission in the red part of the spectrum have also been used for deep two-photon brain imaging [84]. Endogenous fluorophores, such as NADH, can also be used to study the metabolic state of brain tissue, see Fig. 6.4.

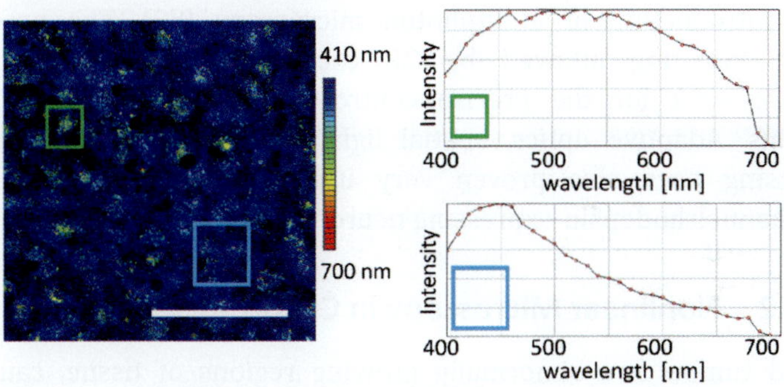

Figure 6.4 Two-photon imaging of endogenous molecules in live brain tissue. Left: spectral image of live mouse brain tissue section, imaged cca 50 μm below surface. NADH (blue) and flavin autofluorescence (green) are visible. Right: representative emission spectra of NADH and flavins. 740 nm excitation, 40×/1.1NA W, scalebar 50 μm.

Microendoscopic probes (those of sub-1 mm diameter) are promising tools in studying of activity in deep structures live brain. But due to difficulties associated with short-pulse light delivery with fiber bundles, and the relative bulkiness of distal scanners, approaches based on GRIN endoscopic lenses are likely to provide the first deep-brain activity images [85].

Neuronal circuit tracing can be performed using transgenic models, where the fluorescent protein is controlled by a specific promoter, such as Thy1-YFP-H mouse, where YFP is expressed in the axons of motor and sensory neurons [25]; other methods of transfection can also be used [86]. Alternatively, injecting fluorescent dyes or nanoparticles that can be taken up by the synapses, or using trans-synaptic tracing strategies, based, e.g., on recombinant rabies virus [80] are useful tools to gather information about neural circuits. With the light penetration depth limited to or below 1 mm in point-scanning microscopy setups, even with three-photon excitation and red-shifted fluorophores, other approaches, generally incompatible with sample viability, are required to map whole-brain circuits. One approach is thick mechanical sectioning combined with multiphoton microscopy [87], or whole-brain clearing and SPIM imaging [44].

Optogenetics, a method of neuromodulation using light, acting on light-sensitive ion channels (opsins), also benefits from the introduction of multiphoton microscopy [88]. The precise light targeting allowed by 2P excitation relaxes the strict requirement for the precise control of the expression of the opsins. Adaptive optics, spatial light modulators and temporal focusing have also proven very useful when precise control of channelrhodopsin-expressing neurons is required [89, 90].

6.4.2 Nonlinear Microscopy in Cancer Research

Solid tumors are abnormally growing regions of tissue, caused by mutation in one of the cell types comprising the parent tissue. High rate of cell division, increased metabolic rate supported by local changes in vasculature and loss of original tissue function are associated with tumor growth. All these manifestations are exploited in the study, diagnosis and treatment of tumors.

Multiphoton microscopy helps in cancer research to visualize immunolabeled tissues [91], transgenic models expressing fluorescent proteins [92] or tumor cell lines GFP-transfected with suitable promoters [93], changes in vasculature, oxygenation and blood flow [94], as well as general changes in metabolic state of neoplastic cells using endogenous NADH [95, 96], see Fig. 6.5.

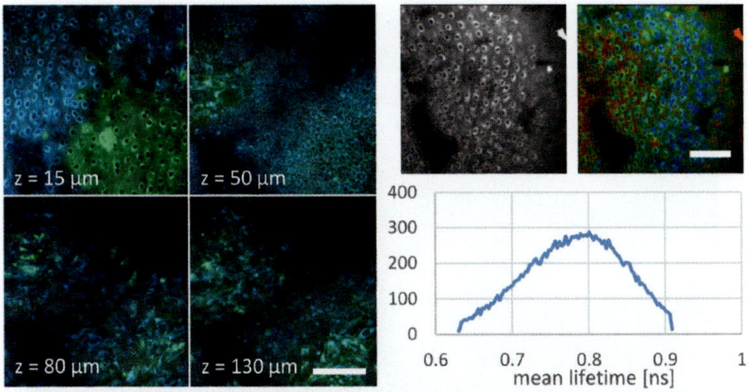

Figure 6.5 Two-photon imaging of endogenous molecules in malignant mouse ear tissue. Left: montage of selected planes of a *z*-stack through the tissue (blue: NADH, green: FAD). Right: intensity and lifetime image of NADH in the tumor tissue, mean lifetime calculated from bi-exponential fit. The histogram shows distribution of mean lifetime in the image. 740 nm excitation, 40×/1.1 NA W, scalebar 100 µm.

6.4.3 Multiphoton Microscopy in Developmental Biology

Due to the lower photodamage and low scattering and absorption in tissue, multiphoton microscopy is an ideal tool in early embryonal development in wide range of model animals, 40× NA 1.1 W. Although less efficient in scattering samples, SPIM approach is preferred for its high speed and extremely low photodamage. Although not yet commercially available with multiphoton excitation, 2P-SPIM was successfully tested with drosophila [97] and zebrafish embryos [98].

Laser-scanning two-photon microscopy has been used to study number of organs and systems in developing embryos, such as early mouse development [99] (see also Fig. 6.6), motoneurons in mouse spinal cord [100], retina [101], mouse brain vasculature [83], mouse kidney [102], etc., Due to the wide availability of 2P-excitation enabled scanning microscopes, multiphoton microscopy becomes standard tool in embryonic imaging.

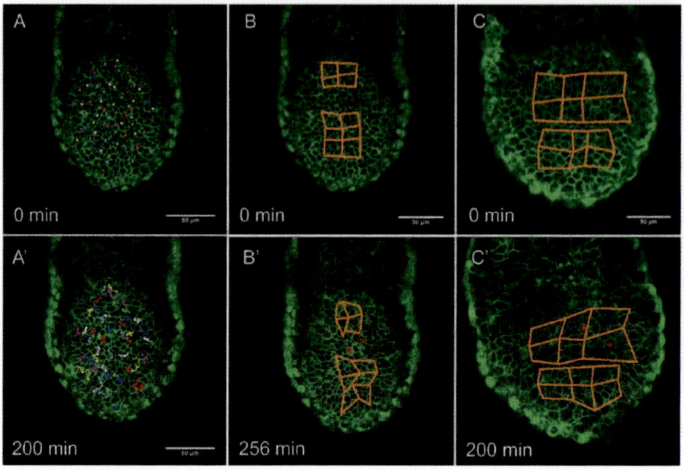

Figure 6.6 Two-photon snapshots of EGFP-labeled mT/mG prestreak embryo development. A, A′: Cell tracking in the posterior epiblast over a time course of 200 min. B–C′: Distortion diagrams in the posterior epiblast (red asterisks—nascent mesoderm in the later time points), distal tip is down. 880 nm excitation, 40×/1.4NA Oil objective. Adapted from [99].

6.4.4 Nonlinear Microscopy in Tissue Engineering

Highly keratinous tissues, such as skin, are notoriously difficult to image due to strong scattering and absorption of light. Yet deep tissue imaging of both skin in vivo and various tissue engineering model systems is of importance [103]. Multiphoton microscopy allows deeper penetration into tissues, but equally important application of two-photon absorption is the microfabrication of biomaterial scaffolds by means of two-photon photopolymerization [104]. Even STED-like approach was reported for fabrication with resolution surpassing the Abbe optical limit [105].

Due to the high abundance of collagen both in connective tissues and in tissue engineering scaffolds, SHG is an indispensable technique allowing high contrast label-free imaging of the local density and orientation of collagen fibrils [106], as illustrated in Fig. 6.7. SHG imaging has been used for the studies of collagen in skin [107], tendon [108], fascia [109], liver damage [110], artificial scaffolds [111] as well as aging and disease of various connective tissues [112]. Since type I collagen is capable of producing SHG only in its fully mature state, the extracellular collagen modification and maturation can be studied in in vitro cell cultures by comparing SHG image to immunofluorescence of collagen [113].

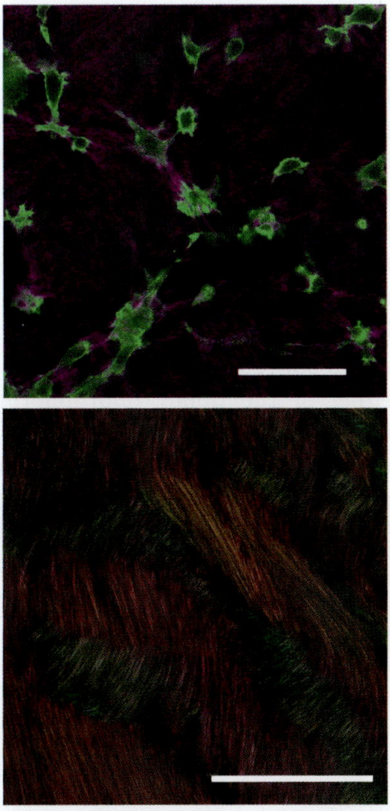

Figure 6.7 Up: 2P image of Alexa 488—Phalloidin stained fibroblasts in collagen scaffold (SHG). 900 nm excitation, 40×/1.1NA W objective, scalebar 100 um. Dwone: SHG of rabbit tendon, color-coded three angles of polarization of the excitation light (red—0°, green—60°, blue—120°). 860 nm excitation, 63×/1.2NA W objective, scalebar 50 um.

With the availability of long wavelength pulsed lasers, the third harmonic signal (THG) originating from live tissue can also be detected and used to form an image from deeper within scattering tissue, without the need for labeling [114]. THG imaging offers new contrasting method in tissue engineering, as the third order susceptibility is nonzero in a wide range of biological materials.

6.4.5 Deep Tissue FRET Imaging

Forster resonance energy transfer is a process of nonradiative energy transfer between an excited donor molecule and an acceptor molecule in ground state within close proximity (several nanometers). Figure 6.8 depicts one of requirements of FRET—appreciable spectral overlap between donor emission and acceptor absorption. FRET is often used to study protein-protein interactions on the nanometer scale. The donor quenching and the increased emission of the acceptor due to FRET [75] can be measured in cultured cells, but is unreliable in tissues due to high levels of autofluorescence and spectrally dependent and variable light absorption in tissue. The shortening of the fluorescence lifetime of the donor fluorophores, on the other hand, can be measured precisely even deep in scattering and absorbing tissue. Point scanning multiphoton microscope with TCSPC detection is the prevalent instrumentation for FLIM-FRET measurements [76].

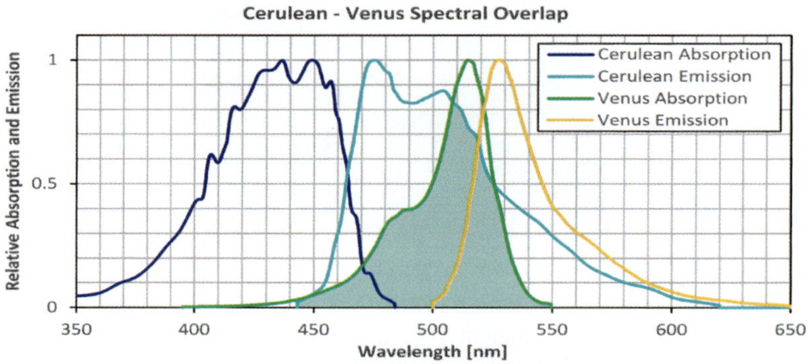

Figure 6.8 Absorption and emission spectra of a typical FRET pair: Cerulean—Venus. The shaded area is the spectral overlap between donor (Cerulean FP) emission and acceptor (Venus FP) absorption.

FLIM-FRET with two-photon excitation is common even in in vitro cell culture studies [115, 116] due to the inherent pulsed nature of the two-photon excitation. Multiphoton excitation is even more beneficial in deep tissue imaging, enabling studies of protein-protein interactions deep in tissues [117, 118], (see Fig. 6.9). Nevertheless, while both the instrumentation and the biological models or FRET-based biosensors are widely available, deep-tissue FLIM-FRET studies are scarce.

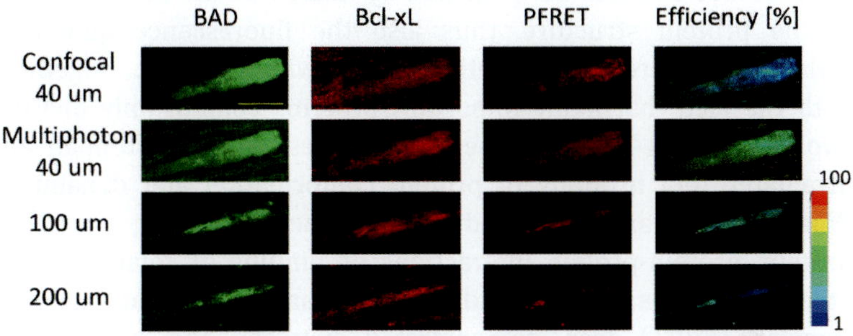

Figure 6.9 Tissue FRET in Traumatic Axonal Injury. Six hours post injury; tissue labeled with BAD/Alexa 488 (donor) and Bcl-xL/Alexa 555 (acceptor) demonstrates energy transfer consistent with BAD-Bcl-xL heterodimerization. The same tissue was used for Confocal FRET and two-photon FRET imaging using Biorad Radiance2100 confocal/multiphoton microscopy. Confocal PFRET image was obtained at 40-microns depth and was unable to get decent signal beyond that depth. On the other hand, we obtained two-photon PFRET images up to 200 microns depth. As shown in the figure the signal appears to be less when we moved from 40 to 200 microns. This can be attributed to the concentration of the fluorophore deep inside the tissue and also to the visible FRET signal lost in the tissue before reaching the detector. Bar = 10 μm; 20× MIMM NA 0.75; Confocal-ExD 488 nm, EmD 528/30 nm, ExA 543 nm, EmA 590/70 nm; Two-photon-ExD 790 nm, ExA 730 nm. D—Donor; A—Acceptor. Adapted from [117].

6.4.6 Multiphoton Imaging of Endogenous Molecules

Fluorescent molecules already naturally present and functioning in cells and tissues are very attractive for label-free biological imaging, as minimal sample preparation (and associated risk of introducing artifacts) is needed. While many cellular metabolites and other small molecules are fluorescent to some extent (especially with UV excitation), the most important endogenous markers in mammalian tissues and cell cultures are tryptophan, NAD(P)H and flavins.

Tryptophan is an essential amino acid, that absorbs in the UV (around 280 nm) and its emission spectrum depends strongly on the local environment of the molecule, with emission maximum ranging from 300 to 370 nm [119]. The tryptophan fluorescence can also be quenched by nearby amino acid residues in the protein structure, thus also the fluorescence quantum yield and fluorescence lifetime vary widely [120]. Together with the fact, that many important proteins contain only one or two tryptophan residues, these properties make tryptophan an invaluable tool in study of protein conformation and dynamics. When tryptophan is exposed near the active binding site of a protein, protein-protein interactions or binding of cofactors and other substrates can be studied. For example, resonant energy transfer (FRET) was demonstrated between the tryptophan residue of estradiol dehydrogenase and NADPH [121].

NADH, a reduced form of nicotinamide adenine dinucleotide (NAD) is a coenzyme important in cellular metabolism. Its fluorescence can be excited around 350 nm and the emission range is 450–550 nm. Because NADH takes part in oxidative phosphorylation, beta oxidation and the citric acid cycle, it is a vital part of the cellular metabolism and can be used to study metabolic state of native and cancerous tissues [122]. Free NADH experiences significant quenching [19] and displays short fluorescence lifetime (around 0.4 ns) and proportionally low quantum yield, when compared to protein-bound NADH, whose lifetime also depends on the particular protein, leading to wide range of lifetimes (1.5–4.5 ns). With these unique properties, the NADH fluorescence intensity (or lifetime, for more quantitative results) can be used to monitor cellular redox state [123].

NADPH, the phosphate analog of NADH, is a cofactor and energy carrier in biosynthetic reactions. It is also involved in redox reactions that protect the cell against reactive oxygen species, and in other important biological pathways. The emission and excitation spectra are essentially identical to that of NADH, but since these two coenzymes interact with different host proteins, the degrees to which they are quenched, and thus their fluorescence lifetimes, differ. It is possible to measure relative concentration of NADH and NADPH [124] based on empirical lifetime measurements, with results applicable to a wide range of tissues.

Flavins, fluorescent molecules derived from riboflavin, are important sources of autofluorescence in the green spectral range. Biologically most important are flavin adenine dinucleotide (FAD) and flavin mononucleotide (FMN). These are cofactors of many enzymatic reactions, often connected to cellular metabolism (FMN takes part in the electron transport chain, FAD is involved in citric acid cycle and in oxidative phosphorylation). FAD, together with NADH, was used as a basis for calculating redox ratio [123] to enhance the sensitivity with which precancerous cells can be discriminated. The fluorescence lifetime of FAD is also strongly dependent on its binding state, but unlike that of NADH, the lifetime of bound FAD is much shorter than that of the free molecule [19].

All these endogenous fluorophores are strongly connected to cellular metabolism, and it is highly advantageous to image them simultaneously to be able to extract the information about the cellular and tissue metabolic state with robustness and accuracy. Figure 6.10 shows the fluorescence of endogenous molecules (NADH, FAD) in mouse ear and brain tissue. In our metabolic studies of brain tissue and cancer cell cultures we take advantage of the simultaneous 3P excitation of tryptophan and 2P excitation of NADH with a train of femtosecond pulses centered at 740 nm [27]. The emission windows are 340–380 nm for tryptophan and 460–500 nm for NADH. Flavins have significant 2P absorption in a broad range of wavelengths (700–900 nm). Because their emission, too, overlaps spectrally with NADH emission [125], we have found it necessary to isolate the FAD contribution alone using 2P excitation at 890 nm and 520–570 nm emission window.

Because intensity measurements may be strongly affected by scattering and absorption, all three detectors also measure the fluorescence lifetime (in time-correlated single photon counting mode [15]) for more robust quantitative analysis. The two-component decays of NAD(P)H and FAD allow us to separate the contributions from bound and free dinucleotides. The longer component of the NAD(P)H can be correlated to the NADH/NADPH ratio in cells and tissues [124]. The tryptophan fluorescence can be used as a general proxy [126], or together with lifetime information can be used for more detailed analysis of cellular metabolic state, e.g., through possible tryptophan—NADH energy transfer [27]. FAD fluorescence intensity is used for calculating redox ratio, an important metabolic parameter [123]. With lifetime information available, the effect of fluorophore quenching on the intensity can be corrected for and more precise concentration measurements can be performed [127].

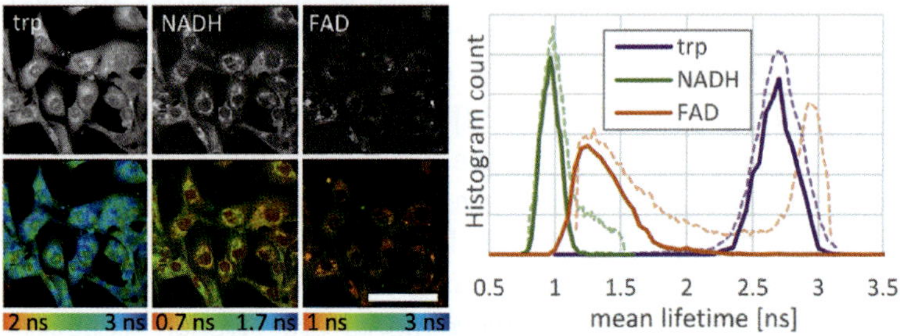

Figure 6.10 Left: intensity and lifetime images of African-American prostate cancer cell culture, tryptophan (three-photon excitation at 740 nm, 340–380 nm detection), NADH (two-photon excitation at 740 nm, 460–500 nm detection), and FAD (two-photon excitation at 890 nm, 520–570 nm detection). Note that NADH fluorescence originates predominantly from mitochondria. Right: histograms of the mean lifetimes (calculated from bi-exponential fits) of tryptophan (around 2.7 ns), NADH (around 0.9 ns) and FAD (around 1.3 ns) based on the whole field of view (dashed lines) and thresholded for mitochondria (solid lines). 40×/1.3NA Oil objective, scalebar 50 μm.

6.5 Conclusion

Nonlinear microscopy (multiphoton-absorption excited fluorescence and harmonic generation) is well established in biological imaging, both on the subcellular level and in live-tissue studies. It was demonstrated on wide range of platforms, ranging from single spot laser scanning configuration (offering the least degree of parallelization, but most versatile and most immune to light scattering), to line scanning configurations, widefield temporal focusing and multiphoton SPIM (fast, highly parallel detection, but unable to exploit scattered fluorescence photons for image formation).

By far the most common multiphoton instrumentation employed for biological imaging is single spot laser scanning configuration based on a confocal laser scanning microscope. It is readily available, and further advances in femtosecond laser sources will make it even more common imaging modality. It is being used extensively in many areas of biological imaging, ranging from subcellular studies to whole organ or whole body imaging in neuroscience, developmental biology and cancer research, to name just a few. While adding fluorescence lifetime measurement capabilities is straightforward and not uncommon, there are only few studies using lifetime-based sensors or FLIM-FRET in the context of deep tissue imaging. On the other hand, FLIM-based quantitative studies of endogenous fluorophores (NADH, FAD) are quickly gaining popularity.

During the last decade or two, a number of new imaging techniques were devised, mainly to overcome the speed limitation of point-scanning microscopy. These often employ array detectors (CCD or CMOS) to increase the speed and quantum efficiency and often they rely on specifically tailored laser sources (various amplification schemes, OPOs) to excite larger volume of the sample simultaneously. Because only a handful of imaging facilities can offer such instruments, truly biological studies based on this class of advanced microscopy yet have to appear.

Vivid development can be seen in the area of fiber-based nonlinear microendoscopy. While several vital concepts of multiphoton intravital imaging using optical fibers have been demonstrated, some technological obstacles still have to be overcome to render multiphoton excitation advantageous. The most challenging part is the delivery of short, intense, and tightly focused illumination through the optical fiber. New custom fibers (photonic crystals, large mode area fibers, coherent fiber bundles) and new laser sources will eventually enable efficient multiphoton excitation. Another area of technological development are adaptive/diffractive optics and advanced spectral shaping of laser pulses to further increase the efficiency of multiphoton absorption and fluorescence. Methods that do not rely on the detection of the fluorescence photons at all, such as opto-acoustic imaging, are also quickly gaining popularity and will allow new probes to be used to gather more information from deeper within living tissues and animal.

References

1. Kiernan, J. (1981). *Histological and Histochemical Methods: Theory and Practice*. Oxford: Pergamon Press.
2. Pawley, J. B. (2006). *Handbook of Biological Confocal Microscopy* (Pawley, J. B., ed.), 3rd ed., New York: Springer Science & Business Media.
3. Denk, W., Strickler, J. H., and Webb, W. W. (1990). Two-photon laser scanning fluorescence microscopy. *Science*, **248**(4951), pp. 73–76.
4. Campagnola, P. J., Lewis, A., Loew, L. M., Clark, H. A., and Mohler, W. A. (2001). Second-harmonic imaging microscopy of living cells. *J. Biomed. Opt.*, **6**(3), pp. 277–286.
5. Palero, J., Santos, S. I. C. O., Artigas, D., and Loza-Alvarez, P. (2010). A simple scanless two-photon fluorescence microscope using selective plane illumination. *Opt. Express*, **18**(8), pp. 8491–8498.
6. Shao, L., Kner, P., Rego, E. H., and Gustafsson, M. G. L. (2011). Super-resolution 3D microscopy of live whole cells using structured illumination. *Nat. Methods*, **8**(12), pp. 1044–1046.
7. Křížek, P., Raška, I., and Hagen, G. M. (2012). Flexible structured illumination microscope with a programmable illumination array. *Opt. Express*, **20**(22), pp. 24585–24599.
8. Baddeley, D., Crossman, D., Rossberger, S., Cheyne, J. E., Montgomery, J. M., Jayasinghe, I. D., Cremer, C., Cannell, M. B., and Soeller, C. (2011). 4D super-resolution microscopy with conventional fluorophores and single wavelength excitation in optically thick cells and tissues. *PLoS One*, **6**(5), p. e20645.
9. Dertinger, T., Colyer, R., Vogel, R., Heilemann, M., Sauer, M., Enderlein, J., and Weiss, S. (2012). Superresolution optical fluctuation imaging (SOFI). *Adv. Exp. Med. Biol.*, **733**, pp. 17–21.
10. Kempf, C., Staudt, T., Bingen, P., Horstmann, H., Engelhardt, J., Hell, S. W., and Kuner, T. (2013). Tissue multicolor STED nanoscopy of presynaptic proteins in the calyx of Held. *PLoS One*, **8**(4), p. e62893.
11. Müller, C. B., and Enderlein, J. (2010). Image scanning microscopy. *Phys. Rev. Lett.*, **104**(19), p. 198101.
12. Huff, J. (2015). The Airyscan detector from ZEISS: Confocal imaging with improved signal-to-noise ratio and super-resolution. *Nat. Methods*, **12**(12), pp. i–ii.
13. York, A. G., Chandris, P., Nogare, D. D., Head, J., Wawrzusin, P., Fischer, R. S., Chitnis, A., and Shroff, H. (2013). Instant super-

resolution imaging in live cells and embryos via analog image processing. *Nat. Methods*, **10**(11), pp. 1122–1126.

14. Ingaramo, M., York, A. G., Wawrzusin, P., Milberg, O., Hong, A., Weigert, R., Shroff, H., and Patterson, G. H. (2014). Two-photon excitation improves multifocal structured illumination microscopy in thick scattering tissue. *Proc. Natl. Acad. Sci. U. S. A.*, **111**(14), pp. 5254–5259.

15. Becker, W. (2012). Fluorescence lifetime imaging--techniques and applications. *J. Microsc.*, **247**(2), pp. 119–136.

16. Zimmermann, T. (2005). Spectral imaging and linear unmixing in light microscopy. *Adv. Biochem. Eng. Biotechnol.*, **95**, pp. 245–265.

17. Lee, C. M., Engelbrecht, C. J., Soper, T. D., Helmchen, F., and Seibel, E. J. (2010). Scanning fiber endoscopy with highly flexible, 1 mm catheterscopes for wide-field, full-color imaging. *J. Biophotonics*, **3**(5–6), pp. 385-407.

18. Cui, G., Jun, S. B., Jin, X., Luo, G., Pham, M. D., Lovinger, D. M., Vogel, S. S., and Costa, R. M. (2014). Deep brain optical measurements of cell type–specific neural activity in behaving mice. *Nat. Protoc.*, **9**(6), pp. 1213–1228.

19. Lakowicz, J. R. (2006). *Principles of Fluorescence Spectroscopy*. 3rd ed., New York: Springer US.

20. Domaille, D. W., Que, E. L., and Chang, C. J. (2008). Synthetic fluorescent sensors for studying the cell biology of metals. *Nat. Chem. Biol.*, **4**(3), pp. 168–175.

21. Cabantous, S., Nguyen, H. B., Pedelacq, J.-D., Koraïchi, F., Chaudhary, A., Ganguly, K., Lockard, M. A., Favre, G., Terwilliger, T. C., and Waldo, G. S. (2013). A new protein-protein interaction sensor based on tripartite split-GFP association. *Sci. Rep.*, **3**, p. 2854.

22. Baker, B. J., Mutoh, H., Dimitrov, D., Akemann, W., Perron, A., Iwamoto, Y., Jin, L., Cohen, L. B., Isacoff, E. Y., Pieribone, V. A., Hughes, T., and Knopfel, T. (2008). Genetically encoded fluorescent sensors of membrane potential. *Brain Cell Biol.*, **36**, pp. 53–67.

23. Stosiek, C., Garaschuk, O., Holthoff, K., and Konnerth, A. (2003). In vivo two-photon calcium imaging of neuronal networks. *Proc. Natl. Acad. Sci. U. S. A.*, **100**(12), pp. 7319–7324.

24. Smirnov, E., Borkovec, J., Kováčik, L., Svidenská, S., Schröfel, A., Skalníková, M., Svindrych, Z., Křížek, P., Ovesný, M., Hagen, G. M., Juda, P., Michalová, K., Cardoso, M. V., Cmarko, D., and Raška, I. (2014). Separation of replication and transcription domains in nucleoli. *J. Struct. Biol.*, **188**(3), pp. 259–266.

25. Tang, P., Zhang, Y., Chen, C., Ji, X., Ju, F., Liu, X., Gan, W.-B., He, Z., Zhang, S., Li, W., and Zhang, L. (2015). In vivo two-photon imaging of axonal dieback, blood flow, and calcium influx with methylprednisolone therapy after spinal cord injury. *Sci. Rep.*, **5**, p. 9691.

26. Yoder, E. J. (2002). In vivo Microscopy of the mouse brain using multiphoton laser scanning techniques. *Proc. Soc. Photo. Opt. Instrum. Eng.*, **4620**, pp. 14–29.

27. Jyothikumar, V., Sun, Y., and Periasamy, A. (2013). Investigation of tryptophan-NADH interactions in live human cells using three-photon fluorescence lifetime imaging and Förster resonance energy transfer microscopy. *J. Biomed. Opt.*, **18**(6), p. 060501.

28. Macias-Garza, F., Bovik, A. C., Diller, K. R., Aggarwal, S. J., and Aggarwal, J. K. (1988). The missing cone problem and low-pass distortion in optical serial sectioning microscopy. In *ICASSP-88., IEEE International Conference on Acoustics, Speech, and Signal*, pp. 890–893.

29. Pnevmatikakis, E. A., and Paninski, L. (2013). Sparse nonnegative deconvolution for compressive calcium imaging: algorithms and phase transitions. In C. J. C. Burges, L. Bottou, M. Welling, Z. Ghahramani, and K. Q. Weinberger (Eds.), *Adv. Neural Inf. Process. Syst.*, **26** (pp. 1250–1258). La Jolla, CA: Neural Information Processing Systems Foundation, Inc.

30. Wilson, T. (2011). Optical sectioning in fluorescence microscopy. *J. Microsc.*, **242**(2), pp. 111–116.

31. Boyde, A. (1992). Three-dimensional images of Ramón y Cajal's original preparations, as viewed by confocal microscopy. *Trends Neurosci.*, **15**(7), pp. 246–248.

32. Stelzer, E. H. K. (2006). The intermediate optical system of laser-scanning confocal microscopes. In (Pawley, J. B., ed.), *Handbook of Biological Confocal Microscopy*, 3rd ed., New York: Springer Science & Business Media, pp. 207–220.

33. Gustafsson, M. G. L. (2005). Nonlinear structured-illumination microscopy: Wide-field fluorescence imaging with theoretically unlimited resolution. *Proc. Natl. Acad. Sci. U. S. A.*, **102**(37), pp. 13081–13086.

34. Denk, W., Piston, D. W., and Webb, W. W. (2006). Multi-photon molecular excitation in laser-scanning microscopy. In (Pawley, J. B., ed.), *Handbook of Biological Confocal Microscopy*, 3rd ed., New York: Springer Science & Business Media, pp. 535–549.

35. Shimozawa, T., Yamagata, K., Kondo, T., Hayashi, S., Shitamukai, A., Konno, D., Matsuzaki, F., Takayama, J., Onami, S., Nakayama, H.,

Kosugi, Y., Watanabe, T. M., Fujita, K., and Mimori-Kiyosue, Y. (2013). Improving spinning disk confocal microscopy by preventing pinhole cross-talk for intravital imaging. *Proc. Natl. Acad. Sci.*, **110**(9), pp. 3399–3404.

36. Kobat, D., Durst, M. E., Nishimura, N., Wong, A. W., Schaffer, C. B., and Xu, C. (2009). Deep tissue multiphoton microscopy using longer wavelength excitation. *Opt. Express*, **17**(16), p. 13354.

37. Patterson, G. H., and Piston, D. W. (2000). Photobleaching in two-photon excitation microscopy. *Biophys. J.*, **78**(April), pp. 2159–2162.

38. Bianchini, P., and Diaspro, A. (2008). Three-dimensional (3D) backward and forward second harmonic generation (SHG) microscopy of biological tissues. *J. Biophotonics*, **1**(6), pp. 443–450.

39. Bloembergen, N., and Pershan, P. S. (1962). Light waves at the boundary of nonlinear media. *Phys. Rev.*, **128**(2), pp. 606–622.

40. Sauer, M., Hofkens, J., and Enderlein, J. (2011). Basic principles of fluorescence spectroscopy. In *Handbook of Fluorescence Spectroscopy and Imaging*, Winheim, Germany: Wiley-VCH Verlag GmbH & Co. KGaA, pp. 1–30.

41. Jacques, S. L. (2013). Optical properties of biological tissues: A review. *Phys. Med. Biol.*, **58**(11), pp. R37–R61.

42. Wang, K., Milkie, D. E., Saxena, A., Engerer, P., Misgeld, T., Bronner, M. E., Mumm, J., and Betzig, E. (2014). Rapid adaptive optical recovery of optimal resolution over large volumes. *Nat. Meth.*, **11**(6), pp. 625–628.

43. Richardson, D. S., and Lichtman, J. W. (2015). Clarifying tissue clearing. *Cell*, **162**(2), pp. 246–257.

44. Susaki, E. A., Tainaka, K., Perrin, D., Yukinaga, H., Kuno, A., and Ueda, H. R. (2015). Advanced CUBIC protocols for whole-brain and whole-body clearing and imaging. *Nat. Protoc.*, **10**(11), pp. 1709–1727.

45. Girkin, J. M., and McConnell, G. (2005). Advances in laser sources for confocal and multiphoton microscopy. *Microsc. Res. Tech.*, **67**(1), pp. 8–14.

46. Yildirim, M., Durr, N., and Ben-Yakar, A. (2015). Tripling the maximum imaging depth with third-harmonic generation microscopy. *J. Biomed. Opt.*, **20**(9), p. 096013.

47. Pastirk, I., Dela Cruz, J., Walowicz, K., Lozovoy, V., and Dantus, M. (2003). Selective two-photon microscopy with shaped femtosecond pulses. *Opt. Express*, **11**(14), pp. 1695–1701.

48. Ogilvie, J. P., Débarre, D., Solinas, X., Martin, J.-L., Beaurepaire, E., and Joffre, M. (2006). Use of coherent control for selective two-photon fluorescence microscopy in live organisms. *Opt. Express*, **14**(2), pp. 759–766.

49. Kumar, S., Dunsby, C., De Beule, P., Owen, D. M., Anand, U., Lanigan, P. M. P., Benninger, R. K. P., Davis, D. M., Neil, M. A. A., Anand, P., Benham, C., Naylor, A., and French, P. M. W. (2007). Multifocal multiphoton excitation and time correlated single photon counting detection for 3-D fluorescence lifetime imaging. *Opt. Express*, **15**(20), pp. 12548–12561.

50. Brakenhoff, G. J., Squier, J., Norris, T., Bliton, A. C., Wade, M. H., and Athey, B. (1996). Real-time two-photon confocal microscopy using a femtosecond, amplified Ti: Sapphire system. *J. Microsc.*, **181**(3), pp. 253–259.

51. Mei, E., Fomitchov, P. A., Graves, R., and Campion, M. (2012). A line scanning confocal fluorescent microscope using a CMOS rolling shutter as an adjustable aperture. *J. Microsc.*, **247**(3), pp. 269–276.

52. Yew, E. Y. S., Sheppard, C. J. R., and So, P. T. C. (2013). Temporally focused wide-field two-photon microscopy: Paraxial to vectorial. *Opt. Express*, **21**(10), p. 12951.

53. Tal, E., Oron, D., and Silberberg, Y. (2005). Improved depth resolution in video-rate line-scanning multiphoton microscopy using temporal focusing. *Opt. Lett.*, **30**(13), pp. 1686–1688.

54. Rupprecht, P., Prevedel, R., Groessl, F., Haubensak, W. E., and Vaziri, A. (2015). Optimizing and extending light-sculpting microscopy for fast functional imaging in neuroscience. *Biomed. Opt. Express*, **6**(2), p. 353.

55. Chang, C.-Y., Cheng, L.-C., Su, H.-W., Hu, Y. Y., Cho, K.-C., Yen, W.-C., Xu, C., Dong, C. Y., and Chen, S.-J. (2014). Wavefront sensorless adaptive optics temporal focusing-based multiphoton microscopy. *Biomed. Opt. Express*, **5**(6), p. 1768.

56. Choi, H., Yew, E. Y. S., Hallacoglu, B., Fantini, S., Sheppard, C. J. R., and So, P. T. C. (2013). Improvement of axial resolution and contrast in temporally focused widefield two-photon microscopy with structured light illumination. *Biomed. Opt. Express*, **4**(7), pp. 995–1005.

57. Lavagnino, Z., Zanacchi, F. C., Ronzitti, E., and Diaspro, A. (2013). Two-photon excitation selective plane illumination microscopy (2PE-SPIM) of highly scattering samples: Characterization and application. *Opt. Express*, **21**(5), pp. 5998–6008.

58. Planchon, T. A., Gao, L., Milkie, D. E., Davidson, M. W., Galbraith, J. A., Galbraith, C. G., and Betzig, E. (2011). Rapid three-dimensional isotropic imaging of living cells using Bessel beam plane illumination. *Nat. Methods*, **8**(5), pp. 417–23.

59. de Medeiros, G., Norlin, N., Gunther, S., Albert, M., Panavaite, L., Fiuza, U.-M., Peri, F., Hiiragi, T., Krzic, U., and Hufnagel, L. (2015). Confocal multiview light-sheet microscopy. *Nat. Commun.*, **6**, p. 8881.

60. Vaziri, A., Tang, J., Shroff, H., and Shank, C. V. (2008). Multilayer three-dimensional super resolution imaging of thick biological samples. *Proc. Natl. Acad. Sci. U. S. A.*, **105**(51), pp. 20221–20226.

61. York, A. G., Ghitani, A., Vaziri, A., Davidson, M. W., and Shroff, H. (2011). Confined activation and subdiffractive localization enables whole-cell PALM with genetically expressed probes. *Nat. Methods*, **8**(4), pp. 327–333.

62. Bianchini, P., Harke, B., Galiani, S., Vicidomini, G., and Diaspro, A. (2012). Single-wavelength two-photon excitation-stimulated emission depletion (SW2PE-STED) superresolution imaging. *Proc. Natl. Acad. Sci. U. S. A.*, **109**(17), pp. 6390–6393.

63. Langer, G., Bouchal, K.-D., Grün, H., Burgholzer, P., and Berer, T. (2013). Two-photon absorption-induced photoacoustic imaging of Rhodamine B dyed polyethylene spheres using a femtosecond laser. *Opt. Express*, **21**(19), p. 22410.

64. Jung, J. C., and Schnitzer, M. J. (2003). Multiphoton endoscopy. *Opt. Lett.*, **28**(11), pp. 902–904.

65. Bortoletto, F., Bonoli, C., Panizzolo, P., Ciubotaru, C. D., and Mammano, F. (2011). Multiphoton fluorescence microscopy with grin objective aberration correction by low order adaptive optics. *PLoS One*, **6**(7), p. e22321.

66. Santos, S., Chu, K. K., Lim, D., Bozinovic, N., Ford, T. N., Hourtoule, C., Bartoo, A. C., Singh, S. K., and Mertz, J. (2009). Optically sectioned fluorescence endomicroscopy with hybrid-illumination imaging through a flexible fiber bundle. *J. Biomed. Opt.*, **14**(3), p. 030502.

67. Rivera, D. R., Brown, C. M., Ouzounov, D. G., Pavlova, I., Kobat, D., Webb, W. W., and Xu, C. (2011). Compact and flexible raster scanning multiphoton endoscope capable of imaging unstained tissue. *Proc. Natl. Acad. Sci.*, **108**(43), pp. 17598–17603.

68. Braun, A., Kane, S., and Norris, T. (1997). Compensation of self-phase modulation in chirped-pulse amplification laser systems. *Opt. Lett.*, **22**(9), pp. 615–617.

69. Lee, S. H., Cavalieri, A. L., Fritz, D. M., Myaing, M., and Reis, D. A. (2004). Adaptive dispersion compensation for remote fiber delivery of near-infrared femtosecond pulses. *Opt. Lett.*, **29**(22), pp. 2602–2604.

70. Clark, S. W., Ilday, F. O., and Wise, F. W. (2001). Fiber delivery of femtosecond pulses from a Ti: Sapphire laser. *Opt. Lett.*, **26**(17), pp. 1320–1322.

71. Göbel, W., Kerr, J. N. D., Nimmerjahn, A., and Helmchen, F. (2004). Miniaturized two-photon microscope based on a flexible coherent fiber bundle and a gradient-index lens objective. *Opt. Lett.*, **29**(21), pp. 2521–2523.

72. Andresen, E. R., Bouwmans, G., Monneret, S., and Rigneault, H. (2013). Toward endoscopes with no distal optics: Video-rate scanning microscopy through a fiber bundle. *Opt. Lett.*, **38**(5), pp. 609–611.

73. Papadopoulos, I. N., Farahi, S., Moser, C., and Psaltis, D. (2013). High-resolution, lensless endoscope based on digital scanning through a multimode optical fiber. *Biomed. Opt. Express*, **4**(2), pp. 260–270.

74. Plöschner, M., Tyc, T., and Čižmár, T. (2015). Seeing through chaos in multimode fibres. *Nat. Photonics*, **9**(8), pp. 529–535.

75. Sun, Y., and Periasamy, A. (2010). Additional correction for energy transfer efficiency calculation in filter-based Forster resonance energy transfer microscopy for more accurate results. *J. Biomed. Opt.*, **15**(2), p. 020513.

76. Sun, Y., Day, R. N., and Periasamy, A. (2011). Investigating protein-protein interactions in living cells using fluorescence lifetime imaging microscopy. *Nat. Protoc.*, **6**(9), pp. 1324–1340.

77. Becker, W., Bergmann, A., Hink, M. A., König, K., Benndorf, K., and Biskup, C. (2004). Fluorescence lifetime imaging by time-correlated single-photon counting. *Microsc. Res. Tech.*, **63**(1), pp. 58–66.

78. Looger, L. L., and Griesbeck, O. (2012). Genetically encoded neural activity indicators. *Curr. Opin. Neurobiol.*, **22**(1), pp. 18–23.

79. Lichtman, J. W., and Denk, W. (2011). The big and the small: Challenges of imaging the Brain's circuits. *Science*, **334**, pp. 618–623.

80. Ginger, M., Haberl, M., Conzelmann, K.-K., Schwarz, M. K., and Frick, A. (2013). Revealing the secrets of neuronal circuits with recombinant rabies virus technology. *Front. Neural Circ.*, **7**, p. 2.

81. Tian, L., Hires, S. A., Mao, T., Huber, D., Chiappe, M. E., Chalasani, S. H., Petreanu, L., Akerboom, J., McKinney, S. A., Schreiter, E. R., Bargmann, C. I., Jayaraman, V., Svoboda, K., and Looger, L. L. (2009).

Imaging neural activity in worms, flies and mice with improved GCaMP calcium indicators. *Nat. Methods*, **6**(12), pp. 875–881.

82. Mittmann, W., Wallace, D. J., Czubayko, U., Herb, J. T., Schaefer, A. T., Looger, L. L., Denk, W., and Kerr, J. N. D. (2011). Two-photon calcium imaging of evoked activity from L5 somatosensory neurons in vivo. *Nat. Neurosci.*, **14**(8), pp. 1089–1093.

83. Horton, N. G., Wang, K., Kobat, D., Clark, C. G., Wise, F. W., Schaffer, C. B., and Xu, C. (2013). In vivo three-photon microscopy of subcortical structures within an intact mouse brain. *Nat. Photonics*, **7**(3), pp. 205–209.

84. Tischbirek, C., Birkner, A., Jia, H., Sakmann, B., and Konnerth, A. (2015). Deep two-photon brain imaging with a red-shifted fluorometric Ca^{2+} indicator. *Proc. Natl. Acad. Sci.*, **112**(36), pp. 11377–11382.

85. Methodology, I., Jung, J. C., Mehta, A. D., Aksay, E., Stepnoski, R., and Schnitzer, M. J. (2004). In vivo mammalian brain imaging using one- and two-photon fluorescence microendoscopy. *J. Neurophysiol.*, **92**(5), pp. 3121–3133.

86. Mahou, P., Zimmerley, M., Loulier, K., Matho, K. S., Labroille, G., Morin, X., Supatto, W., Livet, J., Débarre, D., and Beaurepaire, E. (2012). Multicolor two-photon tissue imaging by wavelength mixing. *Nat. Methods*, **9**(8), pp. 815–818.

87. Ragan, T., Kadiri, L. R., Venkataraju, K. U., Bahlmann, K., Sutin, J., Taranda, J., Arganda-Carreras, I., Kim, Y., Seung, H. S., and Osten, P. (2012). Serial two-photon tomography for automated ex vivo mouse brain imaging. *Nat. Methods*, **9**(3), pp. 255–258.

88. Prakash, R., Yizhar, O., Grewe, B., Ramakrishnan, C., Wang, N., Goshen, I., Packer, A. M., Peterka, D. S., Yuste, R., Schnitzer, M. J., and Deisseroth, K. (2012). Two-photon optogenetic toolbox for fast inhibition, excitation and bistable modulation. *Nat. Methods*, **9**(12), pp. 1171–1179.

89. Andrasfalvy, B. K., Zemelman, B. V., Tang, J., and Vaziri, A. (2010). Two-photon single-cell optogenetic control of neuronal activity by sculpted light. *Proc. Natl. Acad. Sci.*, **107**(26), pp. 11981–11986.

90. Papagiakoumou, E., Anselmi, F., Bègue, A., de Sars, V., Glückstad, J., Isacoff, E. Y., and Emiliani, V. (2010). Scanless two-photon excitation of channelrhodopsin-2. *Nat. Methods*, **7**(10), pp. 848–854.

91. Koga, S., Oshima, Y., Honkura, N., Iimura, T., Kameda, K., Sato, K., Yoshida, M., Yamamoto, Y., Watanabe, Y., Hikita, A., and Imamura, T. (2014). In vivo subcellular imaging of tumors in mouse models using a fluorophore-conjugated anti-carcinoembryonic antigen

antibody in two-photon excitation microscopy. *Cancer Sci.*, **105**(10), pp. 1299–1306.

92. Hoffman, R. M. (2015). Application of GFP imaging in cancer. *Lab. Invest.*, **95**, pp. 432–452.

93. Le Dévédec, S. E., Lalai, R., Pont, C., de Bont, H., and van de Water, B. (2011). Two-photon intravital multicolor imaging combined with inducible gene expression to distinguish metastatic behavior of breast cancer cells in vivo. *Mol. Imaging Biol.*, **13**(1), pp. 67–77.

94. Tanaka, K., Morimoto, Y., Toiyama, Y., Matsushita, K., Kawamura, M., Koike, Y., Okugawa, Y., Inoue, Y., Uchida, K., Araki, T., Mizoguchi, A., and Kusunoki, M. (2012). In vivo time-course imaging of tumor angiogenesis in colorectal liver metastases in the same living mice using two-photon laser scanning microscopy. *J. Oncol.*, **2012**, p. 265487.

95. Varone, A., Xylas, J., Quinn, K. P., Pouli, D., Sridharan, G., McLaughlin-Drubin, M. E., Alonzo, C., Lee, K., Münger, K., and Georgakoudi, I. (2014). Endogenous two-photon fluorescence imaging elucidates metabolic changes related to enhanced glycolysis and glutamine consumption in precancerous epithelial tissues. *Cancer Res.*, **74**(11), pp. 3067–3075.

96. Skala, M., and Ramanujam, N. (2010). Multiphoton redox ratio imaging for metabolic monitoring in vivo. *Methods Mol. Biol.*, **594**, pp. 155–162.

97. Truong, T. V., Supatto, W., Koos, D. S., Choi, J. M., and Fraser, S. E. (2011). Deep and fast live imaging with two-photon scanned light-sheet microscopy. *Nat. Methods*, **8**(9), pp. 757–760.

98. Keller, P. J., and Stelzer, E. H. (2008). Quantitative in vivo imaging of entire embryos with digital scanned laser light sheet fluorescence microscopy. *Curr. Opin. Neurobiol.*, **18**(6), pp. 624–632.

99. Williams, M., Burdsal, C., Periasamy, A., Lewandoski, M., and Sutherland, A. (2012). Mouse primitive streak forms in situ by initiation of epithelial to mesenchymal transition without migration of a cell population. *Dev. Dyn.*, **241**(2), pp. 270–283.

100. Hinckley, C. A., Alaynick, W. A., Gallarda, B. W., Hayashi, M., Hilde, K. L., Driscoll, S. P., Dekker, J. D., Tucker, H. O., Sharpee, T. O., and Pfaff, S. L. (2015). Spinal locomotor circuits develop using hierarchical rules based on motorneuron position and identity. *Neuron*, **87**(5), pp. 1008–1021.

101. Duan, X., Krishnaswamy, A., De la Huerta, I., and Sanes, J. R. (2014). Type II cadherins guide assembly of a direction-selective retinal circuit. *Cell*, **158**(4), pp. 793–807.

102. Phillips, C. L., Arend, L. J., Filson, A. J., Kojetin, D. J., Clendenon, J. L., Fang, S., and Dunn, K. W. (2001). Three-dimensional imaging of embryonic mouse kidney by two-photon microscopy. *Am. J. Pathol.*, **158**(1), pp. 49–55.

103. Schade, R., Weiß, T., Berg, A., Schnabelrauch, M., and Liefeith, K. (2010). Two-photon techniques in tissue engineering. *Int. J. Artif. Organs*, **33**(4), pp. 219–227.

104. Raimondi, M. T., Eaton, S. M., Nava, M. M., Laganà, M., Cerullo, G., and Osellame, R. (2012). Two-photon laser polymerization: From fundamentals to biomedical application in tissue engineering and regenerative medicine. *J. Appl. Biomater. Biomech.*, **10**(1), pp. 55–65.

105. Fischer, J., and Wegener, M. (2011). Three-dimensional direct laser writing inspired by stimulated-emission-depletion microscopy. *Opt. Mater. Express*, **1**(4), pp. 2363–2366.

106. Chen, X., Nadiarynkh, O., Plotnikov, S., and Campagnola, P. J. (2012). Second harmonic generation microscopy for quantitative analysis of collagen fibrillar structure. *Nat. Protoc.*, **7**(4), pp. 654–669.

107. Chernyavskiy, O., Vannucci, L., Bianchini, P., Difato, F., Saieh, M., and Kubínová, L. (2009). Imaging of mouse experimental melanoma in vivo and ex vivo by combination of confocal and nonlinear microscopy. *Microsc. Res. Tech.*, **72**(6), pp. 411–423.

108. Williams, R. M., Zipfel, W. R., and Webb, W. W. (2005). Interpreting second-harmonic generation images of collagen I fibrils. *Biophys. J.*, **88**(2), pp. 1377–1386.

109. Légaré, F., Pfeffer, C., and Olsen, B. R. (2007). The role of backscattering in SHG tissue imaging. *Biophys. J.*, **93**(4), pp. 1312–1320.

110. Cox, G. C., Xu, P., Sheppard, C. J. R., and Ramshaww, J. A. (2003). Characterization of the second harmonic signal from collagen. In (Periasamy, A., and So, P. T. C., eds.), *Multiphoton Microscopy in the Biomedical Sciences III* (*Proceedings of SPIE*), **4963**, pp. 32–40.

111. Filová, E., Burdíková, Z., Rampichová, M., Bianchini, P., Čapek, M., Košťáková, E., Amler, E., and Kubínová, L. (2010). Analysis and three-dimensional visualization of collagen in artificial scaffolds using nonlinear microscopy techniques. *J. Biomed. Opt.*, **15**(6), pp. 66011–66017.

112. Brockbank, K. G. M., MacLellan, W. R., Xie, J., Hamm-Alvarez, S. F., Chen, Z. Z., and Schenke-Layland, K. (2008). Quantitative second harmonic generation imaging of cartilage damage. *Cell Tissue Bank.*, **9**(4), pp. 299–307.

113. Liskova, J., Babchenko, O., Varga, M., Kromka, A., Hadraba, D., Svindrych, Z., Burdikova, Z., and Bacakova, L. (2015). Osteogenic cell differentiation on H-terminated and O-terminated nanocrystalline diamond films. *Int. J. Nanomed.*, **10**, pp. 869–884.

114. Yildirim, M., Durr, N., and Ben-Yakar, A. (2015). Tripling the maximum imaging depth with third-harmonic generation microscopy. *J. Biomed. Opt.*, **20**(9), p. 096013.

115. Kim, J., Li, X., Kang, M. S., Im, K. Bin, Genovesio, A., and Grailhe, R. (2012). Quantification of protein interaction in living cells by two-photon spectral imaging with fluorescent protein fluorescence resonance energy transfer pair devoid of acceptor bleed-through. *Cytom. Part A*, **81**A(2), pp. 112–119.

116. Peter, M., Ameer-Beg, S. M., Hughes, M. K. Y., Keppler, M. D., Prag, S., Marsh, M., Vojnovic, B., and Ng, T. (2005). Multiphoton-FLIM quantification of the EGFP-mRFP1 FRET pair for localization of membrane receptor-kinase interactions. *Biophys. J.*, **88**(2), pp. 1224–1237.

117. Chen, Y., Mills, J. D., and Periasamy, A. (2003). Protein localization in living cells and tissues using FRET and FLIM. *Differentiation*, **71**(9–10), pp. 528–541.

118. Cahalan, M. D., Parker, I., Wei, S. H., and Miller, M. J. (2002). Two-photon tissue imaging: Seeing the immune system in a fresh light. *Nat. Rev. Immunol.*, **2**(11), pp. 872–880.

119. Vivian, J. T., and Callis, P. R. (2001). Mechanisms of tryptophan fluorescence shifts in proteins. *Biophys. J.*, **80**(5), pp. 2093–2109.

120. Eftink, M. R. (2000). Intrinsic fluorescence of proteins. In (Lakowicz, J. R., ed.), *Topics in Fluorescence Spectroscopy Vol. 6: Protein Fluorescence*, Boston, MA: Springer US, pp. 1–15.

121. Li, B., and Lin, S. X. (1996). Fluorescence-energy transfer in human estradiol 17 beta-dehydrogenase-NADPH complex and studies on the coenzyme binding. *Eur. J. Biochem.*, **235**(1–2), pp. 180–186.

122. Bird, D. K., Yan, L., Vrotsos, K. M., Eliceiri, K. W., Vaughan, E. M., Keely, P. J., White, J. G., and Ramanujam, N. (2005). Metabolic mapping of MCF10A human breast cells via multiphoton fluorescence lifetime imaging of the coenzyme NADH. *Cancer Res.*, **65**(19), pp. 8766–8773.

123. Skala, M. C., Riching, K. M., Gendron-Fitzpatrick, A., Eickhoff, J., Eliceiri, K. W., White, J. G., and Ramanujam, N. (2007). In vivo multiphoton microscopy of NADH and FAD redox states, fluorescence lifetimes, and cellular morphology in precancerous epithelia. *Proc. Natl. Acad. Sci.*, **104**(49), pp. 19494–19499.

124. Blacker, T. S., Mann, Z. F., Gale, J. E., Ziegler, M., Bain, A. J., Szabadkai, G., and Duchen, M. R. (2014). Separating NADH and NADPH fluorescence in live cells and tissues using FLIM. *Nat. Commun.*, **5**, p. 3936.

125. Islam, M. S., Honma, M., Nakabayashi, T., Kinjo, M., and Ohta, N. (2013). pH dependence of the fluorescence lifetime of FAD in solution and in cells. *Int. J. Mol. Sci.*, **14**(1), p. 1952.

126. Guo, Y., Ho, P. P., Liu, F., Wang, Q. Z., and Alfano, R. R. (1998). Noninvasive two-photon-excitation imaging of tryptophan distribution in highly scattering biological tissues. *Opt. Commun.*, **154**(5–6), pp. 383–389.

127. Palo, K., Brand, L., Eggeling, C., Jäger, S., Kask, P., and Gall, K. (2002). Fluorescence intensity and lifetime distribution analysis: Toward higher accuracy in fluorescence fluctuation spectroscopy. *Biophys. J.*, **83**(2), pp. 605–618.

128. Burdikova, Z., Svindrych, Z., Pala, J., Hickey, C., Wilkinson, M. G., Panek, J., Auty, M. A. E., Periasamy, A., and Sheehan, J. J. (2015). Measurement of pH micro-heterogeneity in natural cheese matrices by fluorescence lifetime imaging. *Front. Microbiol.*, **6**, p. 183.

129. Wang, L. V., and Hu, S. (2012). Photoacoustic tomography: In vivo imaging from organelles to organs. *Science*, **335**(6075), pp. 1458–1462.

Chapter 7

Smart Biomarker-Coated PbS Quantum Dots for Deeper Near-Infrared Fluorescence Imaging in the Second Optical Window

Takashi Jin,[a] Akira Sasaki,[a,b] and Yukio Imamura[a]

[a]*Quantitative Biology Center (QBiC), Riken, Suita, Osaka 565-0874, Japan,*
[b]*Biomedical Research Institute, National Institute of Advanced Industrial Science and Technology (AIST), Tsukuba, Ibaraki 305-8566, Japan*

tjin@riken.jp

7.1 Introduction

In the conventional fluorescence imaging, visible-emitting fluorescent probes are usually used to visualize biomolecules and organelles in cells [1, 2]. In the case of deep-tissue imaging, visible-emitting fluorescent probes cannot be applied because of the strong absorption and scattering of visible light by intrinsic chromophores (e.g., hemoglobin, melanin, and flavin) and organelles (e.g., mitochondria and cytoskeleton) in tissues. Compared with visible light (400–700 nm), near-infrared (NIR)

Deep Imaging in Tissue and Biomedical Materials: Using Linear and Nonlinear Optical Methods
Edited by Lingyan Shi and Robert R. Alfano

ISBN 978-981-4745-88-8 (Hardcover), 978-1-315-20655-4 (eBook)
www.panstanford.com

light (700–2500 nm) is highly permeable in living tissues. In addition, tissue label-free fluorescence (autofluorescence, native fluorescence) and scattering induced by NIR light excitation are much lower than those by visible light excitation in tissue imaging [3]. Owing to these properties, NIR light is widely used for the non-invasive visualization of deep tissues at the whole body level for one and multiphoton excitation. In most of the commercially available in vivo imaging systems, the NIR region ranging from 700 to 900 nm called the therapeutic window (1st-NIR optical window) has been used for deep tissue imaging. This is because the conventional NIR photodetectors (silicon CCD camera) are sensitive in the 1st-NIR region, and 1st-NIR emitting probes (e.g., Indocyanine green, Cy 7, and CdSeTe quantum dots (QDs)) are commercially available. Although 1st-NIR fluorescence imaging is useful for the non-invasive visualization of organs and tissues at the whole body level, its spatial resolution is not enough to observe cellular dynamics in vivo. As the label-free native fluorophores (autofluorescence) and light scattering by tissues significantly decrease in the NIR region ranging from 1000 to 1400 nm (2nd-NIR optical window), NIR fluorescence imaging in the 2nd-NIR region is expected to offer better spatiotemporal resolution in deep tissue imaging [4, 5]. In the NIR region, there is less scattering from Rayleigh and Mie scattering, which reduces the effect of blurring. There are several optical transmission zones in the NIR region for deepest penetration in brain [6].

Recently, 2nd-NIR fluorescence imaging has been applied to non-invasive visualization of deep-tissues because of the development of a 2nd-NIR photodetector (InGaAs CCD camera) and 2nd-NIR emitting fluorescent probes [7–44]. Compared with the 1st-NIR emitting probes, a limited number of fluorescent materials that emit in the 2nd-NIR region have been reported. Only several groups have developed single-walled carbon nanotubes (SWNTs) [7–23], Ag_2S QDs [24–35], Cr^{3+} and Cr^{4+} ions in crystals [36] and rare-earth-doped nanocomposites [37] as 2nd-NIR fluorescent probes. Using these probes, the capability of 2nd-NIR fluorescence imaging has been examined for non-invasive visualization of organs, tumors, and blood vessels in mice [7–37]. However, the fluorescence brightness of these 2nd-NIR emitting probes is not enough (quantum yield (QY) <6%

[7, 27, 45, 46]) for the deep-tissue imaging with high spatiotemporal resolution. Moreover, the emission wavelengths of these probes cannot be tuned in the 2nd-NIR region. For instance, SWNTs have a broad emission spectrum over the wavelength range from 1000 to 1500 nm, and its emission peak cannot be tuned. In the case of Ag_2S QDs with their particle size from 5.4 to 10 nm, their emission peaks do not show size-tunability [27]. In Fig. 7.1, 2nd-NIR emitting probes are summarized together with representative visible and 1st-NIR emitting probes. There are no known native molecules that emit in the NIR from 700 to 1600 nm.

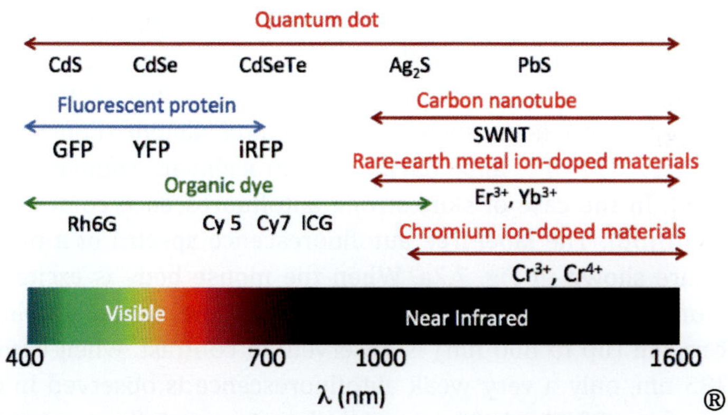

Figure 7.1 Representative fluorescent probes that can be used for fluorescence imaging at the wavelengths of visible (400–700 nm) and near-infrared (700–1600 nm) region.

Biological and medical studies often need to simultaneously analyze multiple components in biological system; therefore, emission-tunable fluorescent probes are needed. PbS QDs are alternative candidates as 2nd-NIR fluorescent probes which have size-dependent emissions. PbS QDs emit at the 2nd-NIR region, because their band gap energy (0.41 eV) corresponds to the energy gap in the 2nd-NIR window [47–55]. In addition, the emission wavelengths (1000–1400 nm) of PbS QDs are easily controlled by changing by their QD size. Thus, multi-colored fluorescence imaging at the 2nd-NIR region can be achieved by using different sizes of PbS QDs.

In this chapter, we focus on the applications of 2nd-NIR emitting PbS QDs for non-invasive fluorescence imaging of deep-tissues [39–42]. To clarify the optical properties of tissues, we first describe the label-free autofluorescence, absorption, and scattering of mouse skin, heart, and liver in vitro. Next we summarize our recent work regarding the synthesis of 2nd-NIR emitting PbS QDs, and their applications to non-invasive imaging of lymph nodes, cerebral blood vessels, breast tumors, and phagocytic cell migration in mice.

7.2 Optical Properties of Tissues

In non-invasive fluorescence imaging of deep tissues, autofluorescence, absorption, and scattering by tissues significantly affect the signal to back ground (S/B) ratios of fluorescence images. Autofluorescence in the visible region results mainly from the intrinsic chromophores such as nicotinamide adenine dinucleotide phosphate (NADP) and flavin in intracellular compartments [56–58]. In the case of skin, strong autofluorescence comes from melanin [58]. The label-free autofluorescence spectra of a mouse body are shown in Fig. 7.2a. When the mouse body is excited at 482 or 670 nm, strong native autofluorescene in the visible to NIR region (up to 800 nm) is observed. In contrast, when excited at 785 nm, only a very weak autofluorescence is observed in the region from 800 to 1000 nm, and almost no autofluorescence is observed over 1000 nm. These findings suggest that the background fluorescence resulting from tissue label-free autofluorescence in the 2nd-NIR and 3rd-NIR optical window [59] are much lower than that in the 1st-NIR optical window.

In addition to the label-free autofluorescence, light absorption and scattering by tissues also affect the S/B ratios of fluorescence images. The absorption spectra of skin, heart, and brain slices of a mouse are shown in Fig. 7.2b. There is no significant difference in tissue absorptions between the 1st-NIR and 2nd-NIR optical windows. However, it should be noted that the NIR-light absorption by tissues significantly increases around 1500 nm, where water molecules strongly absorb NIR-light. Figure 7.3 shows the scattering effects on the fluorescence images of QD-doped silica beads (15 μm in diameter). Fluorescence images of the silica beads are taken across the tissue slices of skin, brain, and heart. Slice thickness of the tissues is 120 μm for skin, 100 μm for brain, and 200 μm for heart. In the case of fluorescence images observed at 520 and 720 nm, the images are significantly blurred by the strong light scattering by the tissues. In contrast, the effect of blurring on the fluorescence images over 1000 nm is remarkably decreased, and the loss of fluorescence intensity of silica beads by light scattering is suppressed. Thus, the S/B ratios of fluorescence images of deep tissues in the 2nd-NIR optical window are much better than that in the 1st-NIR optical window.

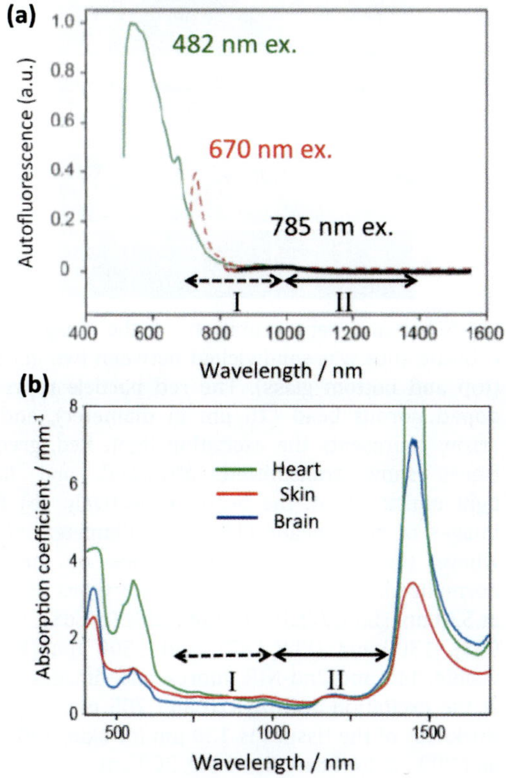

Figure 7.2 (a) Native label free, autofluorescence spectra of the dorsal side of a mouse body. The autofluorescence spectra were taken by excitation of 482, 670, and 785 nm. The dotted and solid arrows show the wavelength rage of 1st-NIR optical window (I) and 2nd-NIR optical window (II), respectively. (b) Absorption spectra of tissue slices of mouse skin, brain, and heart. Slice thickness of the skin, brain, and heart is 120, 100, and 200 μm, respectively.

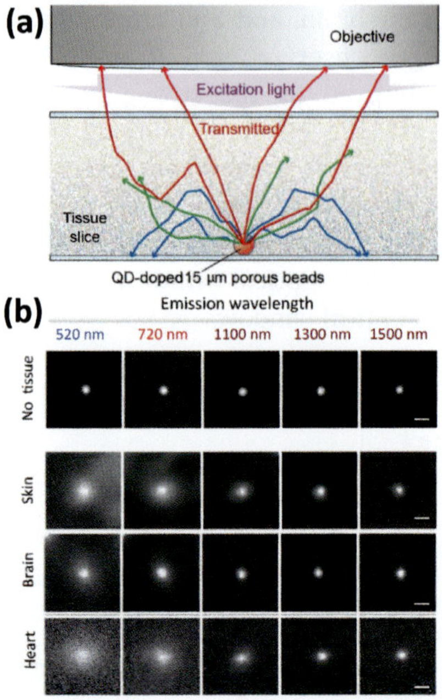

Figure 7.3 (a) Schematic representation of the imaging experiment. A tissue slice was sandwiched between two glass cover slips (top and bottom glass). The red particle represents a QD-doped porous bead (15 μm in diameter), and the purple arrow represents the excitation light. Red, green, and blue traces show transmitted, absorbed, and backscattered light emitted from the bead, respectively. (b) Fluorescence images of silica beads (15 μm in diameter) across tissues, where the fluorescence intensities of the images are normalized. Silica beads contain five types of QDs that emit at 520 nm (CdSe/ZnS) 720 nm (CdSeTe/CdS), 1100 nm (PbS/CdS), 1300 nm (PbS/CdS), and 1500 nm (PbS/CdS). The visible, 1st- and 2nd-NIR fluorescence images were obtained at the excitation at 482, 670 and 785 nm, respectively. Slice thickness of the tissues is 120 μm for skin, 100 μm for brain, and 200 μm for heart. Scale bar, 200 μm.

7.3 NIR Probes in First and Second Optical Window

Currently, there are no commercially available fluorescent probes that can be used for the 2nd-NIR deep-tissue imaging. Several groups have reported 2nd-NIR emitting fluorescent probes such as SWNTs, Ag_2S QDs, and rare-earth-doped nanocomposites [7–35]. SWNTs, with rod-like shapes several hundred of a nanometer, have broad emission spectra in the wavelength from 1000 to 1600 nm. The QYs of fluorescent SWNTs are less than 1% [10]. Dai et al. have developed highly fluorescent SWNTs and their capabilities for deep-tissue imaging of organs and cerebral blood vessels [12, 14, 16–18, 20–21, 23]. They also developed organic-dye (IR-1061) based 2nd-NIR fluorescent probes for deep-tissue imaging [43]. Ag_2S QDs are relatively small probes (<10 nm in diameter) compared to SWNTs, and their QY is 5.8% [27]. Rare-earth doped nanocomposites also act as 2nd-NIR emitting probes (1185, 1310, 1475, and 1525 nm emission) by incorporation of the different rare-earth metal ions to $NaYF_4$ nanoparticles [37]. The QYs of rare-earth-doped (Er, Yb) solid-state materials are less than 2%.

Our group has synthesized PbS QDs as 2nd-NIR fluorescent probes for deep-tissue imaging [39–42]. We developed a novel synthetic method for preparing highly fluorescent PbS/CdS core/shell QDs (QY >40%) in organic phase [40]. The surface of the PbS/CdS QDs is easily modified to be water-soluble and the resulting QDs are used for deep-tissue imaging. We also developed aqueous synthetic methods for preparing water-soluble PbS QDs which are coated with glutathione and recombinant proteins [39, 42]. The QYs of these PbS QDs are over 10% in aqueous solution. Figure 7.4a shows the fluorescence spectra of aqueous

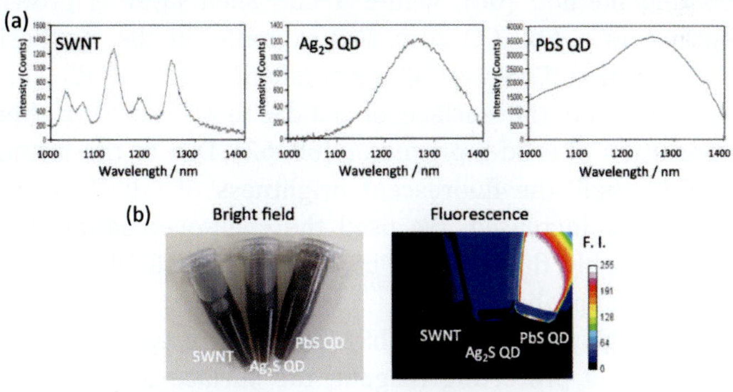

Figure 7.4 (a) Fluorescence spectra of nanomaterials that emit in the 2nd-NIR window: SWNT, Ag_2S QD, and PbS QD. (b) Bright field and fluorescence images (>1000 nm) of SWNT, Ag_2S QD, and PbS QD. To compare the fluorescence brightness, absorbance at the excitation wavelength (720 nm) was adjusted to be the same value of 0.5 for SWNT, Ag_2S QD, and PbS QD.

solutions of SWNTs, Ag_2S QDs, and PbS QDs. The surface of SWNTs (HIPCO SWNT, NanoIntegris) is coated with sodium cholate to be soluble to water. The Ag_2S and PbS QDs are prepared by aqueous synthetic methods, and their surface is coated with glutathione [39]. In the case of Ag_2S QDs, Ag_2S nanoparticles were grown at 80°C. To compare fluorescence brightness, absorbance of the NIR probes was set to a same value (Ab = 0.5) at the excitation wavelength of 760 nm. As shown in Fig. 7.4b, the fluorescence of PbS QDs is at least ten-times brighter than that of SWNT and Ag_2S QDs in aqueous solution.

7.4 Synthesis of Quantum Dots

7.4.1 Highly Fluorescent PbS/CdS QDs

To date, PbS/CdS core/shell QDs have been synthesized by using a cation-exchange method [48, 49], where Pb^{2+} cations on the surface of PbS QDs are gradually exchanged with Cd^{2+} cations. In this case, the size of a PbS core decreases with increasing the thickness of a CdS shell. Thus, the emission peak of the resulting PbS/CdS QDs shifts to shorter wavelengths compared with the initial PbS QDs. A disadvantage of the cation-exchange method is the long reaction time (1–48 h) [51] to form a CdS shell in PbS QDs. An alternative for preparing PbS/CdS QDs is a shell-overcoating method [60], where a CdS shell layer is grown on the surface of a PbS QD core. For example, in the synthesis of 1st-NIR emitting CdSeTe/CdS (core/shell) QDs, a CdS shell is easily formed on the surface of a CdSeTe QD core by thermal decomposition of a Cd-S precursor [60–62]. Due to the formation of the CdS shell, the fluorescent brightness of CdSeTe/CdS QDs is significantly increased. We used the shell-overcoating method to prepare highly fluorescent PbS/CdS core/shell QDs as a 2nd-NIR fluorescent probe (Fig. 7.5).

As the highly fluorescent PbS/CdS QDs are synthesized using oleylamine as a protecting reagent, the surface of PbS/CdS QDs is hydrophobic and they are insoluble in water. To use the PbS/CdS QDs in aqueous solution, the QDs have to be modified by surface coating. For deep-tissue imaging, small-sized fluorescent QD probes (<ca. 10 nm) are desirable to promote the circulation of probes in tissues and their renal clearance [63]. So far, two

approaches have been used for the chemical modification of QD surface to be hydrophilic. One method is the encapsulation of QDs with amphiphilic polymers [64]. In principle, this method maintains the fluorescence brightness of the initial QDs, because the QD surface does not change after the encapsulation. However, the hydrodynamic size of the resulting QDs increases depending on the molecular weights of the amphiphilic polymers. Another method is the ligand-exchange using small molecules such as hydrophilic thiol compounds [65]. In this case, the size of QDs is not significantly changed when using the thiol compounds with small molecular weights. However, the ligand-exchange method often results in significant decrease in the fluorescence brightness and colloidal stability of QDs [65].

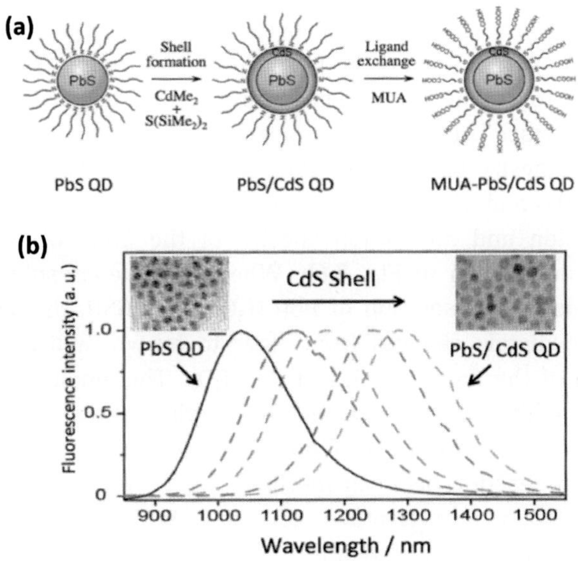

Figure 7.5 (a) Schematic representation of the synthesis of MUA-coated PbS/CdS QDs. A CdS shell onto a PbS core was formed using a Cd-S precursor of dimethylcadmium ($CdMe_2$) and hexamethyldisilathiane ($S(SiMe_2)_2$). Ligand-exchange with mercaptoundecanoic acid (MUA) is performed to obtain water-soluble PbS/CdS QDs. (b) Spectral changes of PbS QDs upon formation of the CdS shell. Black line shows the fluorescence spectrum of the PbS QDs before the shell formation. Inset shows the transmission electron microscopy images of PbS QDs and PbS/CdS QDs before and after a CdS shell formation. Scale bar, 10 nm.

7.4.2 Water-Soluble PbS QDs

Water-soluble PbS QDs can be prepared in aqueous phase using protecting agents such as 1-thioglycerol/dithioglycerol, dihydrolipotic acid, L-Cys, apoferritin, and luciferase [66–70]. Among the QDs coated with the above compounds, only the QDs coated with 1-thioglycerol/dithioglycerol are emission tunable in the 2nd-NIR optical window [66]. However, PbS QDs coated with thiol compounds such as 1-thioglycerol and dithioglycerol are cytotoxic and non-biocompatible. We therefore used GSH as a coating agent for the preparation of biocompatible 2nd-NIR fluorescent PbS QDs. GSH is a natural mono-thiol compound that exists in most organs at the mM level and has no cytotoxicity [60]. In addition, surface modifications of GSH-coated QDs with biomolecules are relatively easy, because GSH has two functional groups, two carboxyl and one primary amino groups, that make it possible to conjugate biological compounds such as antibody to the QD surface [71, 72].

We synthesize GSH-coated water-soluble PbS QDs by reacting $Pb(CH_3COO)_2$ and Na_2S in the presence of GSH (Fig. 7.6a) [39]. The emission and absorption spectra of the 2nd-NIR emitting PbS QDs are shown in Fig. 7.6b. When an aqueous solution of Na_2S is added to a solution of $Pb(CH_3COO)_2$ + GSH, the solution color immediately changes from clear to brown, indicating the formation of the nanoparticles of PbS QDs. The amount of Na_2S significantly affects the emission wavelength of the resulting PbS QDs because the formation of the QD nanoparticles follows the bimolecular reaction between the Pb^{2+} and S^{2-} ions. Additionally, the pH and/or temperature of the solution of reactants ($Pb(CH_3COO)_2$ + GSH) also affects the emission wavelength. A higher pH may increase the concentration of the complex between Pb^{2+} and GSH, because the pKa of the sulfhydryl group of GSH is 9.65 [73]. At the same time, a higher temperature can also increase the formation rate of the PbS QDs. Thus, by controlling the amounts of Na_2S and the pH and temperature of the solution of reactants, the emission wavelength of GSH-coated PbS QDs can be tuned in the 2nd-NIR optical window.

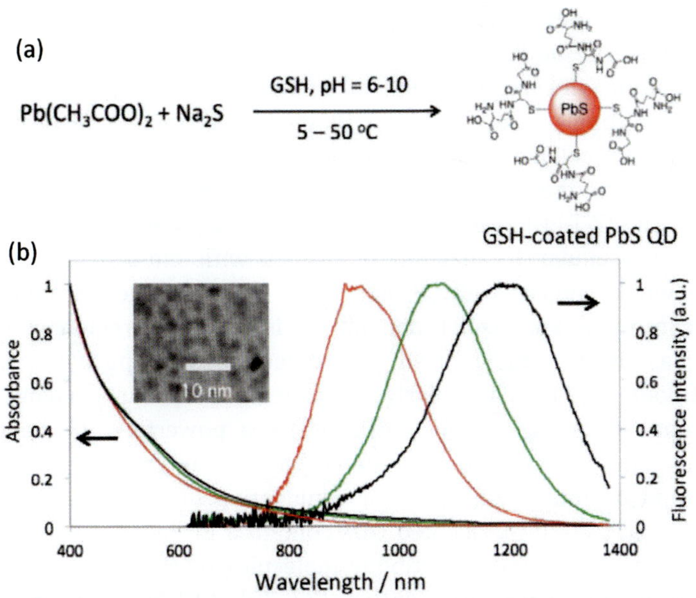

Figure 7.6 (a) Schematic of the aqueous synthesis of GSH-coated PbS QDs. (b) Fluorescence and absorption spectra of GSH-coated PbS QDs with emission peaks at 925 nm (red), 1050 nm (green), and 1200 nm (black). The inset picture shows a transmission electron microscopy image of GSH-coated PbS QDs with an emission peak of 1000 nm.

7.4.3 Dual Emitting PbS QDs

We developed a facile synthetic method for the preparation of recombinant protein coated PbS QDs with dual emission (visible and 2nd-NIR fluorescence), which can bind immunoglobulin G (IgG) in aqueous solution [42]. We used an enhanced green fluorescent protein (EGFP)-Protein G B1 domain [74] tagged with glutathione S-transferase (GST) as a template [70, 75] (GST-EGFP-GB1) to grow PbS QDs in aqueous phase (Fig. 7.7a). The recombinant protein, GST-EGFP-GB1 was genetically designed and prepared by a conventional method for recombinant protein synthesis in *Escherichia coli*. The fluorescent protein EGFP was chosen to add a function of visible fluorescence emission to the 2nd-NIR emitting PbS QDs. 2nd-NIR fluorescence has an advantage in its high permeability for deep tissue imaging; however, the technique is saddled with difficulties arising from its low spatial resolution and requirement of special photo detectors. This dual-emitting property of the QDs enables a conduction of high-resolution cellular confocal imaging by EGFP fluorescence in addition to the deep-tissue imaging by 2nd-NIR fluorescence. Thus the GST-EGFP-GB1-coated QDs make it possible to correlate the intracellular localization of probes with bio-distribution in the whole mouse body. Protein G is a cell surface protein of *Streptococcus* and its B1 domain binds to the Fc region of IgG. It has been widely used for detecting and purifying IgG antibodies [74]. Particularly in cancer molecular imaging, a receptor-targeting antibody can act as a powerful tool via the conjugation of antibodies to the QD surface. In this case, a large amount of antibody is used for conjugation to QDs [39, 41, 76]. As the GST-EGFP-GB1-coated QDs contain a Protein G B1 domain, they can easily bind to antibody molecules by the simple mixing of antibodies with QD solution [77–79]. GST-EGFP-GB1-coated PbS QDs can be prepared by using a protein-mediated QD synthetic method. The synthesis is carried out by reacting GST-EGFP-GB1 with lead acetate ($Pb(Ac)_2$) at ambient conditions to promote the binding of Pb^{2+} to GST-EGFP-GB1. The addition of an aqueous solution of sodium sulfide (Na_2S) immediately forms PbS QDs coated with the GST-EGFP-GB1 protein. By this self-assembly synthetic approach, we can easily prepare protein-coated PbS QDs within 10 min. To characterize the GST-EGFP-GB1-coated PbS QDs, we have measured their transmission electron microscopic (TEM) images, dynamic light scattering (DLS), and fluorescence/absorption spectra. Figure 7.7b shows a TEM image of GST-EGFP-GB1-coated PbS QDs. The PbS QDs have spherical shapes (ca. 5 nm) with a selected area diffraction (SAED) pattern and lattice fringes, confirming the crystalline nature of the PbS QDs with a cubic structure. Hydrodynamic size of the PbS QDs was determined to be ca. 30 nm in diameter by dynamic light scattering. This value is ca. 20 nm larger than that determined by TEM analysis. This result indicates that the GST-EGFP-GB1 proteins (10 nm in diameter) surround the surface of the PbS QDs. Fluorescence spectrum of the GST-EGFP-GB1-coated PbS QDs showed dual emission, with visible fluorescence at 515 nm and a 2nd-NIR fluorescence at 1150 nm, under excitation at

450 nm (Fig. 7.7c). The visible and NIR fluorescence emissions are from EGFP proteins and PbS QDs, respectively. The QY of the 2nd-NIR emitting PbS QDs (1150 nm emission peak) was measured by a relative method with CdSeTe/CdS QDs (840 nm emission, QY = 50%) as a standard. The QY value was determined to be 10% in water.

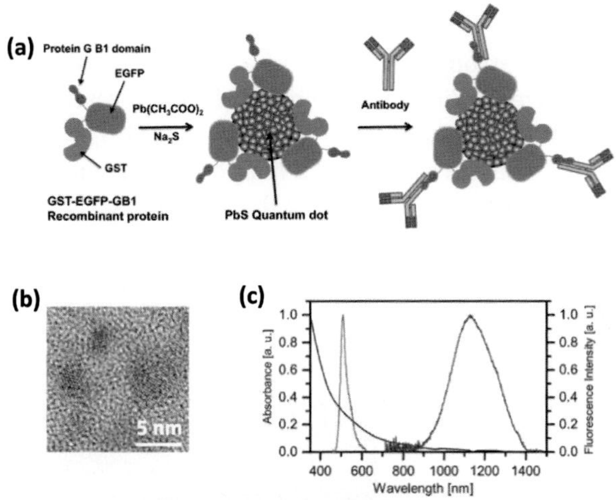

Figure 7.7 (a) A schematic representation of the synthesis of recombinant protein (GST-EGFP-GB1)-coated PbS QDs. GST-EGFP-GB1 protein is used as a template for growing PbS QDs in aqueous phase. Antibody molecules bind to the Protein G B1 domain on the QD by simply mixing the antibody with the QD solution. (b) A high-resolution TEM image of protein (GST-EGFP-GB1)-coated PbS QDs. (c) Fluorescence and absorption spectra of the protein-coated QDs with emission peaks at 515 nm (EGFP) and 1150 nm (QD).

7.5 Non-Invasive Fluorescence Imaging

7.5.1 Lymph System

Fluorescence imaging of a mouse lymph system [80] was performed using visible, 1st-NIR and 2nd-NIR fluorescent QDs (CdSe/ZnS, CdSeTe/CdS, and PbS/CdS) labeled with bovine serum albumin (BSA) to reduce non-specific binding of the QDs to tissues (Fig. 7.8a) [40]. A mixture of each QD solution with normalized fluorescence intensities was directly injected into the mouse footpad and fluorescence images of the lymph system were taken. Label-free fluorescence (autofluorescence) of the mouse body dramatically decreases in the lymphangiography done by 2nd-NIR fluorescence imaging compared with that by conventional visible and NIR imaging (Fig. 7.8a). Lymph node images by 2nd-NIR fluorescence, especially at 1100 and 1300 nm, are much clearer than those by visible and 1st-NIR imaging due to the higher penetration of the 2nd-NIR light in tissues. The weak fluorescence signal in the lymph node observed at 1500 nm results from the strong absorption of NIR light by water molecules (the right image in Fig. 7.8a). The 2nd-NIR fluorescence imaging at 1300 nm with excitation of 785 nm gives a brightest image to visualize both lymph vessels and a lymph node. Figure 7.8b shows fluorescence profiles of the QDs for the lymph vessels. The spatial-resolution of the lymph vessel in 2nd-NIR fluorescent imaging is improved by increasing the imaging wavelength, and the signal to background (S/B) ratio can be made 2.5~8 times better than that of the NIR imaging at 720 nm (Fig. 7.8b).

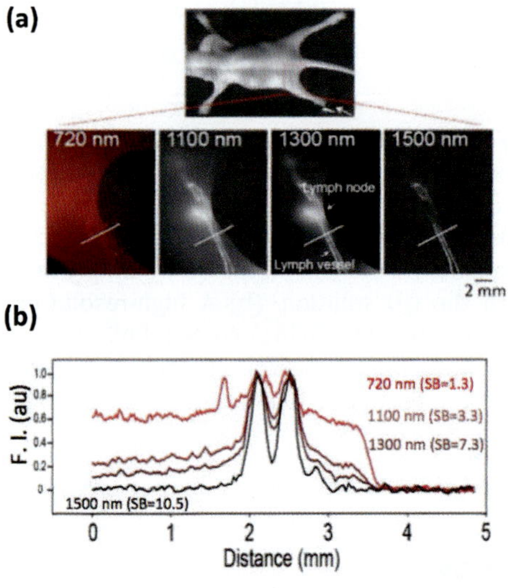

Figure 7.8 (a) An image of a whole body mouse, and fluorescence images of a mouse lymph system. The mixture of QDs with 720, 1100, 1300, and 1500 nm emission was injected into a footpad of the mouse. The 1st and 2nd-NIR fluorescence images were obtained at the excitation at 670 and 785 nm, respectively. The white lines in the fluorescence images are used for the fluorescence profile analysis in (b). (b) Line profile analysis of fluorescence intensities of the QDs in the lymph system.

7.5.2 Cerebral Blood Vessels

In the cerebrovascular fluorescence imaging, the fluorescence from the probes should penetrate into brain thorough the scalp and skull (ca. 2 mm thickness). To visualize the cerebral blood vessels, fluorescence imaging in the 2nd-NIR window is valid, since the 2nd-NIR fluorescence shows less absorption and scattering by tissues. The fluorescence imaging of cerebral blood vessels of a mouse is conducted using GSH-coated 1st-NIR CdSeTe/CdS QDs (800 nm emission) and MUA-coated 2nd-NIR PbS/CdS QDs (1300 nm emission) as shown in Fig. 7.9a. Each QD solution was injected into the tail vein of a mouse, and cerebrovascular fluorescence images were taken in the 1st-NIR and 2nd-NIR optical window. The cerebral vessels showed a clearer image in the 2nd-NIR window due to the higher penetration of the 2nd-NIR light with lower autofluorescence and scattering in the tissue. It should be noted that the spatial-resolution of the fluorescence image of the cerebral blood vessels is improved by increasing the imaging wavelength (Fig. 7.9b), which also increased the signal to background ratio of the 2nd-NIR imaging compared with visible or 1st-NIR imaging. Time-lapse images of cerebral blood vessels provide the changes in the blood flow in the cerebral vessels are shown in Fig. 7.9c.

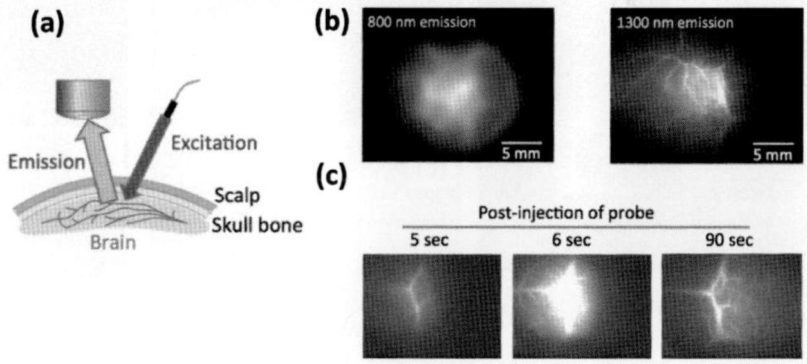

Figure 7.9 (a) Schematic of fluorescence imaging of cerebral blood vessels in a living mouse. An aqueous solution of QDs (CdSeTe/CdS QDs with a 800 nm emission and PbS QDs with a 1300 nm emission) was injected into a tail vein, and the fluorescence images of the blood vessels were taken. (b) Comparison of fluorescence images of the blood vessels taken at 800 and 1300 nm. The 1st- and 2nd-NIR fluorescence images were obtained at the excitation at 670 and 785 nm, respectively. (c) Time-lapse of fluorescence images of the blood vessels after the injection of PbS QDs (1300 nm emission).

7.5.3 Breast Tumor

For the fluorescence imaging of breast tumor, we used a nude mouse xenograft (Fig. 7.10a) implanted with human breast tumor cells (KPL-4) [81] which are over-expressing human epidermal growth receptor 2 (HER2) [42, 82]. 2nd-NIR fluorescence imaging of the breast tumor is shown in Figs. 7.10b and 7.10c. An aqueous solution (200 μL) of GST-EGFP-GB1-coated QDs (1 μM), which has dual emitting property and binding ability with antibody, mixed with anti-HER2 antibody was injected into the tail vein, and second-NIR fluorescence images (excitation: 670 nm, emission: 1300 nm) of a live mouse were taken after 48 h after the QD injection. As shown in Fig. 7.9c, intense 2nd-NIR fluorescence from the breast tumor is clearly observed, compared to the control image without QD injection (not shown). This indicates the accumulation of anti-HER2 antibody bound QDs in the tumor. In contrast, no significant 2nd-NIR fluorescence signal is observed in the area without tumors (Fig. 7.10b). To confirm actual accumulation of the QDs in the breast tumor cells, we carried out ex vivo fluorescence imaging of an isolated tumor at a higher magnification by a confocal microscope with visible light and a

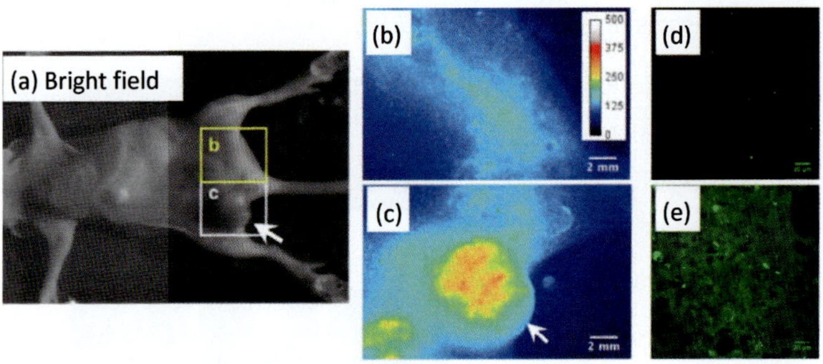

Figure 7.10 (a) Bright field image of a breast tumor xenograft in a living mouse. Fluorescence images (1300 nm) were taken in the region (b) as a negative control and (c) as a breast tumor. The location of the breast tumor is indicated by the arrows in the bright field image and fluorescence image. (d) Ex vivo fluorescence images (515 nm) of the breast tumor without QD injection. (e) Ex vivo fluorescence images (515 nm) of the breast tumor 48 h after QD injection. The fluorescence images at 515 and 1300 nm were taken by the excitation of 482 and 785 nm, respectively.

2nd-NIR microscope. Compared with the negative control without QDs injection (Fig. 7.10d), intense fluorescence of EGFP emitting from the anti-HER2 antibody-conjugated QDs is observed in the breast tumor (Fig. 7.10e). These findings indicate that the conjugates of anti-HER2 antibody and GST-EGFP-GB1-coated QDs pass through the blood vessels of the breast tumor and bind to the HER2 receptors on the tumor cells.

7.5.4 Phagocytic Cell Migration

Phagocytic cells such as dendritic cells and macrophages are the body's gate keepers for guarding against foreign substances such as antigens and pathogens [83]. Non-invasive imaging of phagocytic cell migration in vivo can lead to a better understanding of pathophysiological situations including foreign substance induced contact dermatitis, allergy and viral infection. Currently, non-invasive imaging modalities such as MRI, PET, and bioluminescence have been employed for visualizing phagocytosis dynamics in vivo [84–88]. However, these modalities cannot be applied for the study of non-invasive visualization of cellular dynamics with a high temporal resolution (<1 s). To overcome the drawbacks of the conventional modalities, we have developed multimodal (2nd-NIR fluorescence and magnetic resonance) imaging techniques able to visualize phagocyte cell dynamics in a mouse lymph system. It has been shown that the size of particles influences both the efficiency of phagocytosis and permeability across the lymphatic wall [89]. Although micrometer-sized particles cannot pass through the lymphatic wall, they can be actively transported by phagocyte cells [90, 91]. For 2nd-NIR fluorescence imaging of the active transport, we used two types of probes: 2nd-NIR fluorescent multimodal probe containing Fe_3O_4 (1.5 μm diameter), and BSA-coated PbS/CdS QDs (10–20 nm diameter). Both probes were injected into a mouse lymphatic system via the footpad. BSA-coated PbS/CdS QDs were quickly and passively transported from the footpad to a nearest lymph node with high permeability within 30 s. In contrast, the multimodal probe engulfed by the phagocyte cells required up to 24 h to complete the active transport, which was confirmed by an inhibitor of cell migration, PTX [92] (middle images in Fig. 7.11a). Alternatively we confirmed the migration of the phagocyte cells to a lymph node by MRI. As shown in Fig. 7.11b, the multimodal probes are detected inside the swelling lymph node at a depth of 1–2 mm. These results show that the 2nd-NIR fluorescent multimodal-probe makes it possible to visualize the dynamics and localization of phagocytic cells in a mouse lymph system.

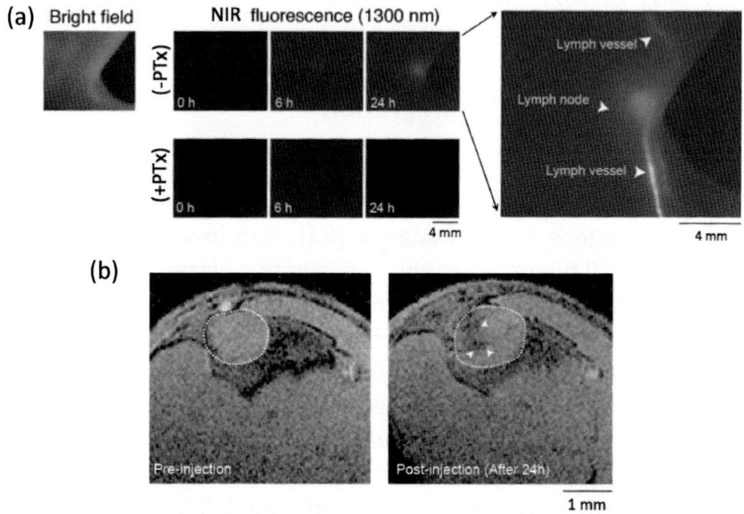

Figure 7.11 (a) Bright field image of a mouse foot (left). Time-lapse of fluorescence image (1300 nm) of the mouse foot after the injection of silica beads containing PbS QDs and Fe_3O_4 nanoparticles in the presence and absence of a cell migration inhibitor, PTX (middle). A magnified image of a lymph system shows the migration of phagocytic cells to a lymph node (blue). The lymph vessels (yellow) were visualized by injection of BSA-coated PbS QDs with a 1100 nm emission (left). (b) Magnetic resonance images of a lymph node before and after the injection of silica beads. The white arrowheads indicate phagocytic cells that exist in a lymph node.

7.6 Future Prospects

During the past few years, significant progress has been made in the synthesis of PbS QDs for non-invasive deep-tissue imaging in the 2nd-NIR window [39–42]. Although PbS QDs contain heavy metal ions, these QDs coated with MUA, GSH and recombinant proteins do not show significant cytotoxicity [39, 40, 42]. These

PbS QDs act as highly fluorescent, biocompatible probes for deep-tissue imaging. Since the surface of PbS QDs is easily modified with biomolecules such as antibodies, PbS QDs can be used as molecular probes for in vivo cellular imaging regarding inflammatory and cancer metastasis. With the aid of the light-sheet fluorescence microscopy [93] or two-photon excitation microscopy [94], the spatial resolution of the 2nd-NIR fluorescence imaging would be remarkably improved. We have demonstrated that the 2nd-NIR imaging can be applied for non-invasive tissue imaging of a lymph system, breast tumors, and brain. This imaging technique may have potential to use in pre-clinical studies of possible molecular diagnosis after septic encephalopathy (i.e., brain inflammatory syndrome) [95].

Currently, the novel non-invasive imaging technologies are required to uncover microvascular mechanisms of diseases. So far, the validity for microvascular imaging including stroke [96], cardiac dysfunction [97], pulmonary neovascularization [98] has been reported. However, conventional imaging technologies (i.e., MRI, PET, CT) show limitations for such microvascular imaging, because of their high costs, and low flexibility using various pathological biomarkers. We have developed 2nd-NIR emitting PbS QD-probes [41] with binding activity to G-protein coupled receptor [99] which associates with lipid metabolisms of pathological conditions [100]. Such functional 2nd-NIR probes can be applied to the visualization for microvascular malfunctions.

Non-invasive NIR fluorescence imaging could be extend to the longer wavelength regions including 3rd-NIR (1600–1870 nm) [59] and the 4th optical window (around 2200 nm) [59] for deeper tissue imaging. The non-invasive NIR fluorescence imaging over 1000 nm will be an important modality for the visualization of biomolecular and cellular dynamics in living tissues. In the near future, combined with other imaging modality such as MRI, NIR fluorescence imaging over 1000 nm may lead to a breakthrough in non-invasive deep tissue imaging and future molecular diagnosis.

References

1. Terai, T., and Nagano, T. (2006). Fluorescent probes for bioimaging applications, *Curr. Opin. Chem. Biol.*, 12, pp. 515–521.

2. Joo, C., Balci, H., Ishitsuka, Y., Buranachai, C., and Ha, T. (2008). Advances in single-molecule fluorescence methods for molecular biology, *Ann. Rev. Biochem.*, 77, pp. 51–76.
3. Weissleder, R. (2001). A clearer vision for in vivo imaging, *Nat. Biotechnol.*, 19, pp. 316–317.
4. Smith, A. M., Mancini, M. C., and Nie, S. (2009). Second window for in vivo imaging, *Nat. Nanotechnol.*, 4, pp. 710–711.
5. Lim, Y. T., Kim, S., Nakayama, A., Stott, N. E., Bawendi, M. G., and Frangioni, J. V. (2003). Selection of quantum dot wavelengths for biomedical assays and imaging, *Mol. Imaging*, 2, pp. 50–64.
6. Shi, L., Sordillo, L. A., Rodoríguez-Contrera, A., and Alfano, R. (2016). Transmission in near-infrared optical windows for deep brain imaging, *J. Biophotonics*, 9, pp. 38–43.
7. O'Connell, M. J., Bachilo, S. M., Huffman, C. B., Moore, V. C., Strano, M. S., Haroz, E. H., Rialon, K. L., Boul, P. J., Noon, W. H., Kittrell, C., Ma, J. P., Hauge, R. H., Weisman, R. B., and Smalley, R. E. (2002). Band gap fluorescence from individual single-walled carbon nanotubes, *Science*, 297, pp. 593–596.
8. Cherukuri, P., Bachilo, S. M., Litovsky, S. H., and Weisman, R. B. (2004). Near-infrared fluorescence microscopy of single-walled carbon nanotubes in phagocytic cells, *J. Am. Chem. Soc.*, 126, pp.15638–15639.
9. Lefebvre, J., Austing. D. G., Bond, J., and Finnie, P. (2006). Photoluminescence imaging of suspended single-walled carbon nanotubes, *Nano Lett.*, 6, pp. 1603–1608.
10. Crochet, J., Clemens, M., and Hertel, T. (2007). Quantum yield heterogeneities of aqueous single-wall carbon nanotube suspensions, *J. Am. Chem. Soc.*, 129, pp. 8058–8059.
11. Leeuw, T. K., Reith, R. M., Simonette, R. A., Harden, M. E., Cherukuri, P., Tsybouiski, D. A., Beckingham, K. M., and Weisman, R. B. (2007). Single-walled carbon nanotubes in the intact organism: near-IR imaging and biocompatibility studies in drosophila, *Nano Lett.*, 9, pp. 2650–2654.
12. Liu, Z., Cai, W., He, L., Nakayama, N., Chen, K., Sun, X., Chen, X., and Dai, H. (2007). In vivo biodistribution and highly efficient tumour targeting of carbon nanotubes in mice, *Nat. Biotechnol.*, 2, p. 52.
13. Jin, H., Heller, D. A., and Strano, M. S. (2008). Single-particle tracking of endocytosis and exocytosis of single-walled carbon nanotubes in NIH-3T3 cells, *Nano Lett.*, 8, pp. 1577–1585.

14. Welsher, K., Liu, Z., Daranciang, D., and Dai, H. (2008). Selective probing and imaging of cells with single walled carbon nanotubes as near-infrared fluorescent molecules, *Nano Lett.*, 8, pp. 586–590.

15. Jin, H., Heller, D. A., Sharma, R., and Strano, M. S. (2009). Size-dependent cellular uptake and expulsion of single-walled carbon nanotubes: Single particle tracking and a generic uptake model for nanoparticles, *ACS Nano*, 3, pp. 149–158.

16. Welsher, K., Liu, Z., Sherlock, S. P., Robinson, J. T., Chen, Z., Daranciang, D., and Dai, H. (2009). A route to brightly fluorescent carbon nanotubes for near-infrared imaging in mice, *Nat. Nanotechnol.*, 4, pp. 773–780.

17. Weisher, K., Sherlocka, S. P., and Dai, H. (2011). Deep-tissue anatomical imaging of mice using carbon nanotube fluorophores in the second near-infrared window, *Proc. Natl. Acad. Sci. U. S. A.*, 108, pp. 8943–8948.

18. Hong, G., Lee, J. C., Robinson, J. T., Raaz, U., Xie, L., Huang, N. F., Cooke, J. P., and Dai, H. (2012). Multifunctional in vivo vascular imaging using near-infrared II fluorescence, *Nat. Med.*, 18, pp. 1841–1846.

19. Yi, H., Ghosh, D., Ham, M. H., Qi, J., Barone, P. W., Strano, M. S., and Belcher, A. M. (2012). M13 phage-functionalized single-walled carbon nanotubes as nanoprobes for second near-infrared window fluorescence imaging of targeted tumors, *Nano Lett.*, 12, pp. 1176–1183.

20. Robinson, J. T., Hong, G., Liang, Y., Zhang, B., Yaghi, O. K., and Dai, H. (2012). In vivo fluorescence imaging in the second near-infrared window with long circulating carbon nanotubes capable of ultrahigh tumor uptake, *J. Am. Chem. Soc.*, 134, pp. 10664–10669.

21. Diao, S., Hong, G., Robinson, J. T., Jiao, L., Antaris, A. L., Wu, J. Z., Choi, C. L., and Dai, H. (2012). Chirality enriched (12,1) and (11,3) single-walled carbon nanotubes for biological imaging, *J. Am. Chem. Soc.*, 134, 16971–16971.

22. Ghosh, D., Bagley, A. F., Na, Y. J., Birrer, M. J., Bhatia, S. N., and Belcher, A. M. (2014). Deep, non-invasive imaging and surgical guidance of submillimeter tumors using targeted M13-stabilized single-walled carbon nanotubes, *Proc. Natl. Acad. Sci. U. S. A.*, 111, pp. 13948–13953.

23. Hong, G., Diano, S., Chang, J., Antaris, A. L., Chen, C., Zhang, B., Zhao, S., Atochin, D. N., Huang, P. L., Andreasson, K. L., Cuo, C. J., and Dai, H. (2014). Through-skull fluorescence imaging of the brain in a new near-infrared window, *Nat. Photonics*, 8, pp. 723–730.

24. Du, Y., Xu, B., Fu, T., Cai, M., Li, F., Zhang,Y., and Wang, Q. (2010). Near-infrared photoluminescent Ag_2S quantum dots from a single source precursor, *J. Am. Chem. Soc.*, 132, p. 1470.
25. Hong, G., Robinson, J. T., Zhang, Y., Diao, S., Antaris, A. L., Wang, Q., and Dai, H. (2012). In vivo fluorescence imaging with Ag_2S quantum dots in the second near-infrared region, *Angew. Chem. Int. Ed.*, 51, 9818–9821.
26. Jiang, P., Tian, Z. Q., Zhu, C. N., Zhang, Z. L., and Pang, D. W. (2012). Emission-tunable near-infrared Ag_2S QDs, *Chem. Mater.*, 24, pp. 3–4.
27. Zhang, Y., Hong, G., Zhang, Y., Chen, G., Li, F. Dai, H., and Wang, Q. (2012). Ag_2S Quantum dot: A bright and biocompatible fluorescent nanoprobe in the second near-infrared window, *ACS Nano*, 5, pp. 3695–3702.
28. Jiang, P., Zhu, C. N., Zhang, Z. L., Tian, Z. Q., and Pang, D. W. (2012). Water-soluble Ag_2S quantum dots for near-infrared fluorescence imaging in vivo, *Biomaterials*, 33, pp. 5130–5135.
29. Yang, H. Y., Zhao, Y. W., Zhang, Z. Y., Xiong, H. M., and Yu, S. N. (2013). One-pot synthesis of water-dispersible Ag_2S quantum dots with bright fluorescent emission in the second near-infrared window, *Nanotechnology*, 24, p. 055706.
30. Zhang, Y., Zhang, Y., Hong, G., Hei, W., Zhou, K., Yang, K., Li, F., Chen, G., Liu, Z., Dai, H., and Wang, Q. (2013). Biodistribution, pharmacokinetics and toxicology of Ag_2S near-infrared quantum dots in mice, *Biomaterials*, 34, pp. 3639–3646.
31. Zhang, Y., Liu, Y., Li, C., Chen, X., and Wang, Q. (2014). Controlled synthesis of Ag_2S quantum dots and experimental determination of the exciton Bohr radius, *J. Phys. Chem. C*, 118, pp. 4918–4923.
32. Gui, R., Wan, A., Liu, X., Yuan, W., and Jin, H. (2014). Water-soluble multidentate polymers compactly coating Ag_2S quantum dots with minimized hydrodynamic size and bright emission tunable from red to second near-infrared region, *Nanoscale*, 6, pp. 5467–5473.
33. Li, C., Zhang, Y., Wang, M., Zhang, Y., Chen, G., Li, L., Wu, D., and Wang, Q. (2014). In vivo real-time visualization of tissue blood flow and angiogenesis using Ag_2S quantum dots in the NIR-II window, *Biomaterials*, 35, pp. 393–400.
34. Chen, G., Tian, F. Zhang, Y., Zhang, Y., Li, C. and Wang, Q. (2014). Tracking of transplanted human mesenchymal stem cells in living mice using near-infrared Ag_2S quantum dots, *Adv. Funct. Mater.*, 24, pp. 2481–2488.

35. Chen, G., Tian, F., Li, C., Zhang, Y., Weng, Z., Zhang, Y., Peng, R., and Wang, Q. (2015). In vivo real-time visualization of mesenchymal stem cells tropism for cutaneous regeneration using NIR-II fluorescence imaging, *Biomaterials*, 53, pp. 265–273.

36. Sharonov, F., Petricevic, V., Bykov, A., and Alfano, R. R. (2015). Near-infrared laser operation of Cr^{3+} centers in chromium-doped $LiInGeO_4$ and $LiScGeO_4$ crystals, *Opt. Lett.*, 8, pp. 851–853.

37. Naczynski, D. J., Tan, M. C., Zevon, M., Wall, B., Kohl, J., Kulesa, A., Chen, S., Roth, C. M., Riman, R. E., and Moghe, P. V. (2013). Rare-earth-doped biological composites as in vivo shortwave infrared reporters, *Nat. Commun.*, 4, p. 2199.

38. Hemmer, E., Venkatachalam, N., Hyodo, H., Hattori, A., Ebina, Y., Kishimoto, H., and Soga, K. (2013). Upconverting and NIR emitting rare earth based nanostructures for NIR-bioimaging, *Nanoscale*, 5, pp. 11339–11361.

39. Nakane, Y., Tsukasaki, Y., Sakata, T., Yasuda, H., and Jin, T. (2013). Aqueous synthesis of glutathione-coated PbS quantum dots with tunable emission for non-invasive fluorescence imaging in the second near-infrared biological window (1000–1400 nm), *Chem. Commun.*, 49, pp. 7584–7586.

40. Tsukasaki, Y., Morimatsu, M., Nishimura, G., Sakata, T., Yasuda, H., Komaytsuzaki, A., Watanabe, T. M., and Jin, T. (2014). Synthesis and optical properties of emission-tunable PbS/CdS core–shell quantum dots for in vivo fluorescence imaging in the second near-infrared window, *RSC Adv.*, 4, pp. 41164–41171.

41. Tsukasaki, Y., Komatsuzaki, A., Mori, Y., Ma, Q., Yoshioka, Y., and Jin, T. (2014). A short-wavelength infrared emitting multimodal probe for non-invasive visualization of phagocyte cell migration in living mice, *Chem. Commun.*, 50, pp. 14356–14359.

42. Sasaki, A., Tsukasaki, Y., Komatsuzaki, A., Sakata, T., Yasuda, H., and Jin, T. (2015). Recombinant protein (EGFP-Protein G)-coated PbS quantum dots for in vitro and in vivo dual fluorescence (visible and second-NIR) imaging of breast tumors, *Nanoscale*, 7, pp. 5115–5119.

43. Tao, Z., Hong, G., Shingi, C., Chen, C., Diao, S., Antaris, A. L., Zhang, B., Zou, Y., and Dai, H. (2013). Biological imaging using nanoparticles of small organic molecules with fluorescence emission at wavelengths longer than 1000 nm, *Angew. Chem. Int. Ed.*, 52, pp. 13002–13006.

44. Hong, G., Zou, Y., Antaris, A. L., Diao, S., Wu, D., Cheng, K., Zhang, X., Chen, C., Liu, B., He, Y., Wu, J. Z., Yuan, J., Zhang, B., Tao, Z., Fukunaga, C., and Dai, H. (2014). Ultrafast fluorescence imaging in vivo with

conjugated polymer fluorophores in the second near-infrared window, *Nat. Commun.*, 5, p. 4206.

45. Bünzli, J. C. G. (2010). Lanthanide luminescence for biomedical analyses and imaging, *Chem. Rev.*, 110, pp. 2729–2755.
46. Eliseeva, S. V., and Bünzli, J. C. G. (2010). Lanthanide luminescence for functional materials and bio-sciences, *Chem. Soc. Rev.*, 39, pp. 189–227.
47. Sargent, E. H. (2005), Infrared QDs, *Adv. Mater.*, 17, pp. 515–522.
48. Pietryga, J. M., Werder, D. J., Williams, D. J., Casson, J. L., Schaller, R. D., Kilimov, V. I., and Hollingsworth, J. A. (2008). Utilizing the lability of lead selenide to produce heterostructured nanocrystals with bright, stable infrared emission, *J. Am. Chem. Soc.*, 130, pp. 4879–4885.
49. Zhao, H., Wang, D., Zhang, T., Chaker, M., and Ma, D. (2010). Two-step synthesis of high-quality water-soluble near-infrared emitting quantum dots via amphiphilic polymers, *Chem. Commun.*, 46, pp. 5301–5303.
50. Neo, M. S., Venkatram, N. G., Li, S., Chin, W. S., and Ji, W. (2010). Synthesis of PbS/CdS core-shell QDs and their nonlinear optical properties, *J. Phys. Chem. C*, 114, pp. 18037–18044.
51. Zhao, H., Chaker, M., Wu, N., and Ma, D. J. (2011). Towards controlled synthesis and better understanding of highly luminescent PbS/CdS core/shell quantum dots, *J. Mater. Chem.*, 21, pp. 8898–8904.
52. Zhao, H., Chaker, M., and Ma, D. (2011). Effect of CdS shell thickness on the optical properties of water-soluble, amphiphilic polymer-encapsulated PbS/CdS core/shell quantum dots, *J. Mater. Chem.*, 21, pp. 17483–17491.
53. Kovalenko, M. V., Schaller, R. D., Jarzab, D., Loi, M. A., and Talapin, D. V. (2012). Inorganically functionalized PbS-CdS colloidal nanocrystals: Integration into amorphous chalcogenide glass and luminescent properties, *J. Am. Chem. Soc.*, 134, pp. 2457–2460.
54. Justo, Y., Geiregat, P. K., Hoecke, K. V., Vanhaecke, F., Donega, C. D. M., and Hens, Z. (2008). Optical Properties of PbS/CdS core/shell quantum dots, *J. Phys. Chem. C*, 117, 20171–20177.
55. Ren, F., Zhao, H., Vetrone, F., and Ma, D. (2013). Microwave-assisted cation exchange toward synthesis of near-infrared emitting PbS/CdS core/shell quantum dots with significantly improved quantum yields through a uniform growth path, *Nanoscale*, 5, pp. 7800–7804.
56. Georgakoudi, I., Jacobson, B. C., Müller, M. G., Sheets, E. E., Badizadegan, K., Carr-Locke, D. L., Crum, C. P., Boone, C. W., Dasari, R. R., Van

Bam, J., and Feld, M. S. (2002). NAD(P)H and collagen as in vivo quantitative fluorescent biomarkers of epithelial precancerous changes, *Cancer Res.,* 62, pp. 682–687.

57. Zipfel, W. R., Williams, R. M., Christie, R., Nikitin, A. Y., Hyman, B. T., and Webb, W. W. (2003). Live tissue intrinsic emission microscopy using multiphoton-excited native fluorescence and second harmonic generation, *Proc. Natl. Acad. Sci. U. S. A.,* 100, pp. 7075–7080.

58. Gallas, J. M., and Eisner, M. (1987). Fluorescence of melanin-dependence upon excitation wavelength and concentration, *Photochem. Photobiol.,* 45, pp. 595–600.

59. Sordillo, L. A., Pu, Y., Pratavieira, S., Budansky, Y., and Alfano, R. R. (2014). Deep optical imaging of tissue using the second and third near-infrared spectral windows, *J. Biomed. Opt.*, 19, p. 056004.

60. Jin, T., Fujii, F., Komai, Y., Seki, J., Seiyama, A., and Yoshioka, Y. (2008). Preparation and characterization of highly fluorescent, glutathione-coated near infrared quantum dots for in vivo fluorescence imaging, *Int. J. Mol. Sci.*, 9, p. 2044.

61. Ma, Q., Nakane, Y., Mori, Y., Hasegawa, M., Yoshioka, Y., Watanabe, T. M., Gonda, K., Ohuchi, N., and Jin, T. (2012). Multilayered, core/shell nanoprobes based on magnetic ferric oxide particles and quantum dots for multimodality imaging of breast cancer tumors, *Biomaterials,* 33, pp. 8486–8494.

62. Hasegawa, M., Tsukasaki, Y., Ohyanagi, T., and Jin, T. (2013). Bioluminescence resonance energy transfer coupled near-infrared quantum dots using GST-tagged luciferase for in vivo imaging, *Chem. Commun.*, 49, pp. 228–230.

63. Choi, H. S., Liu, W., Misra, P., Tanaka, E., Zimmer, J. P., Ipe, B. I., Bawendi, M. G., and Franginoi, J. V. (2007). Renal clearance of quantum dots, *Nat. Biotechnol.*, 25, pp. 1165–1170.

64. Michalet, X., Bentolila, F. F., Tsay, J. M., Doose, S., Li, J. J., Sundaresan, G., Wu, A. M., Gambhir, S. S., and Wiss, S. (2005). Quantum dots for live cells, in vivo imaging, and diagnostics, *Science,* 307, pp. 538–544.

65. Jin, T., Fujii, F., Yamada, E., Nodasak, Y., and Kinjo, M. (2006). Control of the optical properties of quantum dots by surface coating with calix[n]arene carboxylic acids, *J. Am. Chem. Soc.*, 128, pp. 9288–9289.

66. Zhao, X., Gorelikov, I., Musikhin, S., Cauchi, S., Sukhovatkin, V., Sargent, E. H., and Kumacheva, E. (2005). Synthesis and optical properties of thiol-stabilized PbS nanocrystals, *Langmuir,* 21, pp. 1086–1090.

67. Deng, D., Zhang, W., Chen, X., Liu, F., Zhang, J., Gu, Y., and Hong, J. (2009). Facile synthesis of high-quality, water-soluble, near-infrared-emitting PbS quantum dots, *Eur. J. Inorg. Chem.*, 2009, pp. 3440–3446.

68. Yu, Y., Zhang, K., and Sun, S. (2012). One-pot aqueous synthesis of near infrared emitting PbS quantum dots, *Appl. Surf. Sci.*, 258, pp. 7181–7187.

69. Hennequin, B., Turyanska, L., Ben, T., Betrán, A. M., Molina, S. I., Li, M., Mann, S., Patané, A., and Thomas, N. R. (2008). Aqueous near-infrared fluorescent composites based on apoferritin-encapsulated PbS quantum dots, *Adv. Mater.*, 20, pp. 3592–3596.

70. Ma, N., Marshall, A. F., and Rao, J. (2010). Near-infrared emitting luciferase via biomineralization, *J. Am. Chem. Soc.*, 132, pp. 6884–6885.

71. Nakane, Y., Sasaki, A., Kinjo, M., and Jin, T. (2012). Bovine serum albumin-coated quantum dots as a cytoplasmic viscosity probe in a single living cell, *Anal. Methods*, 4, pp. 1903–1905.

72. Tiwari, D. K., Tanaka, S.-I., Inouye, Y., Yoshizawa, K., Watanabe, T. M., and Jin, T. (2009). Synthesis and characterization of anti-HER2 antibody conjugated CdSe/CdZnS quantum dots for fluorescence imaging of breast cancer cells, *Sensors*, 9, pp. 9332–9354.

73. Dawson, R. M. C. (1987). *Data for Biochemical Research*, p16, 3rd ed., Oxford University Press, New York.

74. Gronenborn, A. M., Filpula, D. R., Essig, N. Z., Achari, A., Whitlow, M., Wingfield, P. T., and Clore, G. M. (1991). A novel, highly stable fold of the immunoglobulin binding domain of streptococcal protein G, *Science*, 253, pp. 657–611.

75. He, X., Gao, L., and Ma, N. (2013). One-step instant synthesis of protein-conjugated quantum dots at room temperature, *Sci. Rep.*, 3, 2825.

76. He, X., Gao, J., Gambhir, S., and Cheng, Z. (2010). Near-infrared fluorescent nanoprobes for cancer molecular imaging: Status and challenges, *Trends Mol. Med.*, 16, pp. 574–583.

77. Elshafey, R., Tavares, A. C., Siaj, M., and Zourob, M. (2013). Electrochemical impedance immunosensor based on gold nanoparticles-protein G for the detection of cancer marker epidermal growth factor receptor in human plasma and brain tissue, *Biosens. Bioelectron.*, 50, pp. 143–149.

78. Niide, T., Shimojo, K., Wakabayashi, R., Goto, M., and Kamiya, N. (2013). Enzymatic fabrication of protein-decorated gold nanoparticles by the aid of artificial peptides with gold-binding affinity, *Langmuir*, 29, pp. 15596–15605.

79. Sun, X., Zhang, G., Keynton, R. S., O'Toole, M. G., Patel, D., and Gobin, A. M. (2013). Enhanced drug delivery via hyperthermal membrane disruption using targeted gold nanoparticles with PEGylated Protein-G as a cofactor, *Nanomedicine,* 9, pp. 1214–1222.

80. Kobayashi, H., Hama, Y., Koyama, Y., Barrett, T., Regino, C. A., Urano, Y. and Choyke, P. L. (2007). Simultaneous multicolor imaging of five different lymphatic basins using quantum dots, *Nano Lett.,* 7, pp. 1711–1716.

81. Kurebayashi, J., Otsuki, T., Tang, C. K., Kurosumi, M., Yamamoto, S., Tanaka, K., Mochizuki, M., Nakamura, H., and Sonoo, H. (1999). Isolation and characterization of a new human breast cancer cell line, KPL-4, expressing the Erb B family receptors and interleukin-6, *Br. J. Cancer,* 79, pp. 707–717.

82. Coussens, L., Yang-Feng, T. L., Liao, Y. C., Chen, E., Gray, A., McGrath, J., Seeburg, P. H., Lebermann, T. A., Schlessinger, J., Francke, U., Levinson, A., and Ullrich, A. (1985). Tyrosine kinase receptor with extensive homology to EGF receptor shares chromosomal location with neu oncogene, *Science,* 230, pp. 1132–1139.

83. Janeway, C. A. Jr., and Medzhitov, R. (2002). Innate immune recognition *Annu. Rev. Immunol.,* 20, pp. 197–216.

84. Baumjohann, D., and Lutz, M. B. (2006). Non-invasive imaging of dendritic cell migration in vivo, *Immnobiology,* 211, pp. 587–597.

85. Pham, W., Xie, J., and Gore, J. C. (2007). Tracking the migration of dendritic cells by in vivo optical imaging, *Neoplasia,* 9, p. 1130.

86. Sen, D., Deerinck, T. J., Ellisman, M. H., Parke, I., and Cahalan, M. D. (2008). Quantum dots for tracking dendritic cells and priming an immune response in vitro and in vivo, *PLoS One,* 3, p. e3290.

87. Noh, Y. W., Lim, Y. T., and Chung, B. H. (2008). Non-invasive imaging of dendritic cell migration into lymph nodes using near-infrared fluorescent semiconductor nanocrystals, *FASEB J.,* 22, pp. 3908–3918.

88. Noh, Y. W., Jang, Y. S., Ahn, K. J., Lim, Y. T., and Chung, B. H. (2011). Simultaneous in vivo tracking of dendritic cells and priming of an antigen-specific immune response, *Biomaterials,* 32, pp. 6254–62631.

89. Ikomi, F., Hanna, G. K., and Schmid-Schonbein, G. W. (1995). Mechanism of colloidal particle uptake into the lymphatic system: Basic study with percutaneous lymphography, *Radiology,* 196, pp. 107–113.

90. Wu, Y. L., Ye, Q., Foley, L. M., Hitchens, T. K., Sato, K., Williams, J. B., and Ho, C. (2006). In situ labeling of immune cells with iron oxide particles: An approach to detect organ rejection by cellular MRI, *Proc. Natl. Acad. Sci. U. S. A.,* 103, pp. 1852–1857.

91. Mori, Y., Umeda, M., Fukunaga, M., Ogasawara, K., and Yoshioka, Y. (2011). MR contrast in mouse lymph nodes with subcutaneous administration of iron oxide particles: Size dependency, *Magn. Reson. Med. Sci.,* 10, pp. 219–227.

92. Lee, H. K., Zamora, M., Linehan, M. M., Iijima, N., Gonzalez, D., Haberman, A., and Iwasaki, A. (2009). Differential roles of migratory and resident DCs in T cell priming after mucosal or skin HSV-1 infection. *J. Exp. Med.,* 206, pp. 359–370.

93. Huisken, J., Swoger, J., Del Bene, F., Wittbrodt, J., and Stelzer, E. H. (2004). Optical sectioning deep inside live embryos by selective plane illumination microscopy, *Science,* 305, p. 1007.

94. Denk, W., Strickler, J., and Webb, W. (1990). Two-photon laser scanning fluorescence microscopy, *Science,* 248, pp. 73–76.

95. Imamura, Y., Wang, H., Matsumoto, N., Ogura, H., Shimazu, T., Jin, T., and Seiyama, A. (2011). *Neurochemistry in the Pathophysiology of Septic Encephalopathy,* In-Tech open access publisher, p. 149.

96. Hall, C. N., Reynell, C., Gesslein, B., Hamilton, N. B., Mishra, A., Sutherland, B. A., O'Farrell, F. M., Buchan, A. M., Lauritzen, M., and Attwell, D. (2014). Capillary pericytes regulate cerebral blood flow in health and disease, *Nature,* 508, pp. 55–60.

97. Camici, P. G., d'Amati, G., and Rimoldi, O. (2015). Coronary microvascular dysfunction: Mechanisms and functional assessment, *Nat. Rev. Cardiol.,* 12, pp. 48–62.

98. Jones, R. C., Capen, D. E., Cohen, K. S., Munn, L. L., Jain, R. K., and Duda, D. G. (2008). A protocol for phenotypic detection and characterization of vascular cells of different origins in a lung neovascularization model in rodents, *Nat. Protoc.,* 3, pp. 388–397.

99. Komatsuzaki, A., Ohyanagi, T., Tsukasaki, Y., Miyanaga, Y., Ueda, M., and Jin, T. (2015). Compact halo-ligand-conjugated quantum dots for multicolored single-molecule imaging of overcrowding GPCR proteins on cell membranes, *Small,* 12, pp. 1396–1401.

100. Marrache, A. M., Gobeil, F., Zhu, T., and Chemtob, S. (2005). Intracellular signaling of lipid mediators via cognate nuclear G protein-coupled receptors, *Endothelium,* 12, pp. 63–72.

Chapter 8

Biomedical Applications in Probing Deep Tissue Using Mid-Infrared Supercontinuum Optical Biopsy

Angela B. Seddon

Mid-Infrared Photonics Group,
George Green Institute for Electromagnetics Research,
University of Nottingham, University Park,
Nottingham, NG7 2RD, UK

angela.seddon@nottingham.ac.uk

The demonstration of chalcogenide fiber-based supercontinuum (SC) lasers that reach from 1.4 μm to beyond a wavelength of 10 μm up to 13.3 μm, together with ultra-low loss chalcogenide fiber for routing, is set to have a major impact on spectroscopic molecular sensing. In medicine, this promises to transform mid-infrared tissue imaging into an in vivo clinical tool, for instance, for rapid, early cancer diagnosis to assist treatment planning, with promise of increased sensitivity and specificity compared to current Raman in vivo molecular imaging. MIR discrete fiber lasers to pump the wideband fiber SC lasers are under development, for a compact all-fiber solution to MIR spectral molecular imaging, in vivo, in real time.

Deep Imaging in Tissue and Biomedical Materials: Using Linear and Nonlinear Optical Methods
Edited by Lingyan Shi and Robert R. Alfano

ISBN 978-981-4745-88-8 (Hardcover), 978-1-315-20655-4 (eBook)
www.panstanford.com

8.1 Mid-Infrared Electromagnetic Spectral Region

Sir William Herschel, the German-born, British astronomer, working in Bath, England, in about 1800 was the first to observe the existence of infrared radiation. Passing sunlight through a prism, he projected a visible rainbow and, placing a thermometer "beyond the red" ($<4 \times 10^{14}$ Hz; >750 nm wavelength), he detected "heat rays" in sunlight [1]. In this way he demonstrated that there is continued dispersion of sunlight into the so-called infra ("Latin, below")-red region. The infrared light is transformed to internal energy ("heat") in the absorbing material.

It is useful to consider the infrared electromagnetic region as three separate spectral regions comprising: near-infrared light (NIR: 0.75–<3 μm wavelength), mid-infrared light (MIR, 3–25 μm) and far-infrared light (FIR >25–50 μm) (see Fig. 8.1). The MIR region covers the important atmospheric windows of: 3–5 μm, referred to as the MW (midwave) IR range and 8–12 μm, the LW (longwave) IR range; wavelength limits of these atmospheric windows are for transmittance through about a 1 mile sea-level path [2].

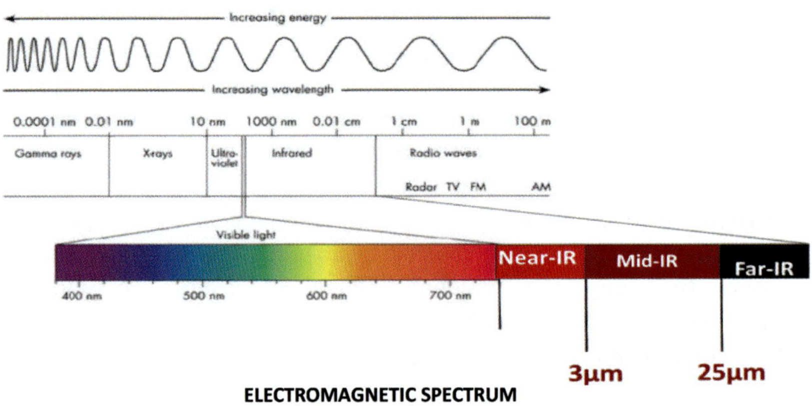

Figure 8.1 The electromagnetic spectrum showing the relative position of the MIR region. Adapted from online color at: www.pss-b.com.

The MIR spectral region encompasses the molecular fingerprints, i.e. characteristic set of MIR fundamental molecular vibrational absorptions of numerous gases, liquids and solids as diverse as: greenhouse gases; ground, water and air pollutants; pharmaceuticals; toxic agents; food and drink; oil, oil products and plastics and, as addressed in this chapter, biological tissues and cells. The aspiration here is to diagnose and monitor cancer, using MIR light for biomedical spectral imaging of tissue specifically in vivo and in real-time, as described more fully in the next few sections. Progress in developing new, bright, wideband supercontinuum MIR fiber laser sources, and MIR fiber for routing, to underpin this goal is fully described. This chapter is aimed at stimulating new MIR research amongst early-stage researchers in MIR biophotonics, MIR photonics and in MIR glass science and technology.

8.2 MIR Spectroscopy

MIR absorption spectroscopy enables interrogation of the molecular structure of matter using MIR radiation. Single molecules, dimers and more extended molecular structures in the gaseous, liquid or solid states all manifest their internal energy in part through constant molecular vibrational motion. The vibrations involve stretching or bending of *intra*-molecular bonds (i.e. the constituent molecular bonds) at particular frequencies. A simple example is the H-O-H stretching and bending bond vibrations of water molecules (see Figs. 8.2(a),(b) and (c)).

Even at absolute zero temperature, molecules possess a zero-point energy and, on raising the temperature above absolute zero, the internal energy increases and increasingly energetic types of vibrational modes are set in motion and sustained as the temperature is raised. The energy associated with each vibrational mode is quantized and the vibrational mode lies in the ground state energy level at normal room temperature.

The vibrational frequency of a molecular bond is influenced by the relative position and type of neighboring *intra*-molecular bond types, e.g. the O-H bond vibration of a metal-O-H (hydroxide) species is of slightly different frequency from that of analogous O-H, attached to H, of a water molecule. Additionally, the vibrational frequency of a particular bond is influenced by neighboring, weaker *inter*-(between)-molecular bonds, for instance hydrogen bonding between two adjacent water (H_2O) molecules, "red-shifts" (shift to lower photon energy, lower frequency) and broadens the frequency range of the water O-H stretching vibrational absorptions compared to that of free (i.e. non-hydrogen-bonded) water molecules (see Fig. 8.2(d)).

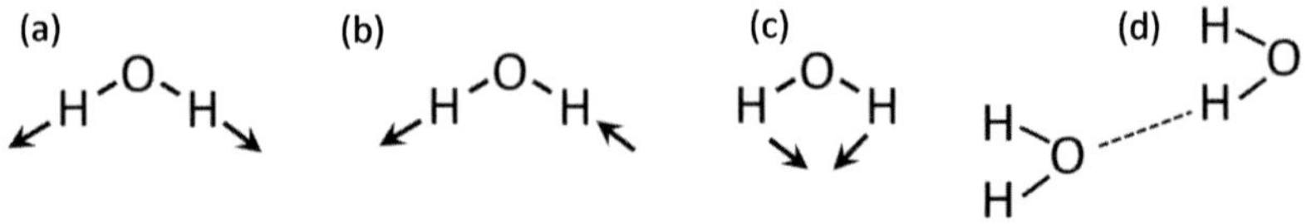

Figure 8.2 MIR-active fundamental vibrations of water: (a) symmetric stretching; (b) anti-symmetric stretching and (c) bending [3]. (d) Inter-molecular hydrogen-bonded water molecules (hydrogen bond is the dashed line) exhibit symmetric and anti-symmetric stretching vibrations which are red shifted and broadened compared to those of isolated water molecules, which are non-hydrogen bonded.

The frequency (ν) of a stretching vibrational mode of a molecular bond is approximately calculable, by assuming the behavior of a simple oscillator, as embodied in the Szigeti equation [4]:

$$\nu \propto (f/m)^{1/2}, \tag{8.1}$$

where m is the geometric mean of the two vibrating atomic masses m_1, m_2 ($m = m_1m_2/(m_1 + m_2)$) and f is the bond force constant, which is related to the strength of the molecular bond.

The photon energy (E) range of MIR electromagnetic radiation has an associated frequency (ν) range (h is Planck's constant):

$$E = h\nu, \tag{8.2}$$

which happily coincides with the frequency range of molecular vibrational motion. Energy uptake from the electrical component of incoming MIR radiation of a particular frequency is quantized and may be thought of as occurring via resonance to induce a particular molecular vibration to vibrate with greater amplitude and unchanged frequency. Any molecular shape has a calculable number of different vibrational modes (and associated vibrational frequencies) depending on its molecular symmetry. Usually, only those vibrations of bonds undergoing a net dipole change during the vibration absorb (i.e. are stimulated by) the MIR radiation.

Incoming broadband MIR radiation is absorbed at particular frequencies by a material. The frequencies of the light absorbed correspond to the innate, vibrational frequencies of the molecular entity, which in turn are dependent absolutely on the molecular structure, i.e. molecular shape (bond angles and bond lengths), thus atomic masses and bond strengths, and hence provide the means for, often unique, molecular identification. It is this phenomenon that is exploited in MIR spectroscopy. The 1971 Nobel Prize in chemistry was won by Gerhard Herzberg of Ottawa, Canada, not least for the understanding he brought to bear on the link between MIR spectroscopy and molecular structure [5(a)]. The wavelength of light absorbed (λ) is related to the vibrational frequency of the light absorbed (and of course the particular vibrational frequency of the molecular bond concerned) by the speed of light in the molecular medium (c):

$$\lambda = c/\nu \tag{8.3}$$

Henceforward here, the absorption of MIR light due to molecular vibration will be discussed in terms of the associated wavelength of the light absorbed.

In the MIR range, molecular species exhibit fundamental absorption bands with large extinction coefficients thus MIR spectroscopy potentially provides extremely sensitive chemical analysis. A molar extinction coefficient (or molar absorptivity coefficient) is defined by the Beer-Lambert law and is a measure of the relative light absorption of a particular vibrational mode of a molecular species normalized to the molar concentration of the absorbing species. Knowledge of the molar extinction coefficient allows quantification of the amount of the molecular species present. The response to MIR spectroscopy is collected as a spectrum, of the material concerned, which is a plot of intensity of MIR radiation versus wavelength and shows the uptake of radiation intensity at particular wavelengths due to vibrational absorption of the material which is characteristic of the molecular structure of the material.

These days a simple Fourier transform infrared (FT-IR) analytical instrument may be used for MIR spectroscopic characterization of molecular species in the 2.5–25 μm wavelength range. This wavelength range is a pragmatic choice to take advantage of the optimal sensitivity ranges of simple sources, detectors and beam-splitters. This is also the reason for the chemists' definition of the MIR as 2.5–25 μm. It should be noted that at shorter wavelengths than *circa* 2.5 μm, in the near-infrared, only weak overtone and combination vibrational absorption bands occur relative to the far stronger fundamental absorption bands in the MIR; while beyond 25 μm wavelength, in the far-infrared, the vibrational spectra should be collected in vacuo due to unwanted absorption, mainly of water vapor.

FT-IR spectroscopy is a potentially fast analytical technique; the entire MIR spectral information is collected at one go using an interferometer and a Fourier transform is performed to achieve a spectrum. If necessary, averaging many spectra improves the signal-to-noise ratio. A typical workhorse MIR spectrometer comprises a broadband source (e.g. Globar©: a silicon carbide rod electrically heated from 1000–1650°C which acts approximately as a thermal blackbody, with peak emission at ca. 5 μm wavelength) a broadband beamsplitter and a room temperature broadband detector (such as uncooled DTGS, deuterated triglycine sulfate). Small desktop and portable models are available with about an A4 footprint. More sophisticated (larger footprint) spectrometers have the option of more sensitive detection for instance using a cooled mercury-cadmium-telluride (MCT) detector from ~3 μm wavelength out to ~11 μm wavelength.

The biomedical spectroscopist Max Diem has been in the vanguard of in vitro and ex vivo mid-infrared biomolecular spectroscopy of human cells and tissue. His seminal paper [6] of 1999 is critical of work prior to 1996, as being too simplistic about the use of mid-infrared spectroscopy to detect malignant, as opposed to normal, cells and tissue. There is no single "cancer peak." Diem highlighted that mid-infrared spectroscopy of cells and tissue could potentially provide an immense amount of information on cellular composition, packing of cellular components, organ and cell architecture, metabolic processes and the absence/presence of disease.

Most important, Diem was the first to recognize that the MIR molecular spectroscopic information revealed by cells and tissue may revolutionize the way that pathology is performed in the 21st Century [6]. Pathology (*Greek:* pathos is suffering) is the study and diagnosis of disease through examination of organs, tissue and cells and bodily fluids.

Table 8.1 is a listing taken mainly from [7] of the infrared band assignments of some key biological molecules (bio-molecules) 3–14 μm, in general from aqueous-solution transmission measurements. This listing is not exhaustive. In Table 8.1 is included the detection of water vibrational absorption bands using Fourier transform ATR (attenuated total reflectance) spectroscopy, in vivo in human stratum corneum, which is the upper layer of external (skin) tissue [8]. Note that the water molar absorptivity in the range 4000 to 1000 cm^{-1} (2.5–10 μm) for all types of vibrational modes is several orders of magnitude weaker that the strongest bands of biomolecules [9]; nevertheless, the low relative molecular mass of water means that its molar concentration may be high.

Table 8.1 Infrared band assignments of key biological molecules (bio-molecules) 3–14 μm, compiled from Tables 11.1 and 11.2 in [7] and water bands in [8]

Absorption band/ cm^{-1}	Absorption band/μm	Chemical compound	Assignment
# 3615	**2.77**	**Water**	***(very weak)(suggested υ_3)***
# 3420	**2.92**	**Water**	***(strong), (suggested as υ_1)***
~3300	3.03	Proteins	N-H stretching vibration (amide A)
# 3220	**3.11**	**Water**	***(medium strong) (suggested $2\upsilon_2$)***
~3100	3.23	Proteins	N–H stretching vibration (amide B)
~3030	3.30	Lipids	=C–H stretching in alkenes
2955 and 2872	3.38 and 3.48	Lipids, proteins carbohydrates, nucleic acids	Asymmetric and symmetric C–H stretching vibration in CH_3
2923 and 2853	3.42 and 3.51	Lipids, proteins carbohydrates, nucleic acids	Asymmetric and symmetric C–H stretching vibration in CH_2
2125	**4.71**	**Water**	**Combination band ($\upsilon_2 + \upsilon_L$)**

(*Continued*)

Table 8.1 (*Continued*)

Absorption band/ cm^{-1}	Absorption band/µm	Chemical compound	Assignment
1741	5.74	Phospholipids, thymine, uracyl	C=O stretching vibration of esters
~1680–1695	~5.90–5.95	Anti-parallel B-sheet	
1650 to 1690	5.92–6.06	Turns and loops	
~1650–1665	~6.00–6.06	α-helix	Combine to form C=O stretch of amide group in proteins **(AMIDE I, strong band)**
~1650–1660	~6.02–6.06	Disordered	
~1625–1640	~6.10–6.15	B-sheet	
# 1640	**6.10**	**Water (bending υ_3)**	
~1615–1625	~6.15–6.19	Aggregated strands (intermolecular B-sheets)	
~1550	~6.45	Proteins	N–H bending, C–H stretching, C–O bending, C–C and N–C stretching vibrations **(AMIDE II, strong band)**
~1515	~6.60	Proteins	Tyrosine
	Below is the so-called "fingerprint" region		
~1467	~6.82	Lipids, proteins, nucleic acids	Symmetric C–H scissoring of $-CH_2$
~1454	~6.88	Lipids, proteins	Asymmetric C–H scissoring of $-CH_3$
~1400	~7.14	Fatty acids, amino acid side chains	C=O stretching vibrations of $-COO^-$
~1390	~7.19	Proteins, lipids	Symmetric C–H deformation of CH_3
~1310–1200	~7.63–8.33	Proteins	C–H/N–H deformation **(AMIDE III)**

Table 8.1 (*Continued*)

Absorption band/ cm^{-1}	Absorption band/μm	Chemical compound	Assignment
~1250–1220 (1235)	~8.000–8.196	Nucleic acids, Phospholipids	Asymmetric P=O stretching vibrations in PO_2^-
~1200–900	~8.33–11.11	Polysaccharides	C-O-P and C-O-C stretching vibrations and ring vibrations
~1173 and 1154	~8.53–8.66	Carbohydrates	Symmetric C–O stretching coupled to C–O–H bending
~1085	~9.21	Nucleic acids (DNA), Phospholipids	Symmetric P=O stretching in PO_2^-, CO–O–C symmetric stretching vibration
~1070	~9.35	Lipids	Symmetric C–O–C, CO–O–C symmetric stretching vibration
~1041 and 1055	~9.61–9.48	Carbohydrates	Symmetric C–O–C stretching vibration
~1023	~9.78	Carbohydrates	Symmetric C–O stretching vibration
900–800	11.11–12.50	Nucleotides	C=C, C=N, C–H in ring structure
700	14.29	Proteins, lipids, nucleic acids	$-CH_2$ rocking

Note: # denotes water bands found in stratum corneum in vivo in [8] (water bands in bold). Amide I, II and III bands are also in bold.

Water is present in biological cells and tissue. The isolated water molecule (effectively water vapor) has three fundamental vibrational modes (see Figs 8.2(a)–(c)). These modes are: symmetrical stretching (υ_1); asymmetrical stretching (υ_3) and bending (υ_2); the latter changes the H–O–H angle [5(c)]. MIR absorption spectra of liquid-water are more complex than that of isolated water molecules, due to hydrogen-bonding-based *inter*-molecular interactions in the liquid state. As introduced above, the molar absorptivity, ε (i.e. molar absorption coefficient), is defined as absorbance (A = ln I_o (incident light intensity)/I (transmitted light intensity)) *per* mole of absorbing species and per cm path length (traditional units, non-SI). For liquid-water [9], at 25°C, the most strongly absorbing vibrational band is centered at 3404 cm^{-1} (2.94 μm; ε = 99.9 ± 0.8 $M^{-1}cm^{-1}$) and comprises three components: (i) second overtone ($2\upsilon_2$) at ~3250 cm^{-1} (~3.08 μm) of the bending mode (i.e. of the fundamental υ_2 at 1644 cm^{-1} (6.08 μm wavelength)); (ii) enhanced by Fermi resonance with the υ_1 symmetric stretching vibration at ~3450 cm^{-1} (2.90 μm) and, finally, (iii) the third component is the υ_3 asymmetrical stretching vibrational mode at ~3600 cm^{-1} (2.78 μm). The fundamental υ_2 bending vibrational absorption at 1644 cm^{-1} (6.08 μm wavelength) has ε = 21.8 ± 0.3 $M^{-1}cm^{-1}$. A broad absorption band in the range ~500–800 cm^{-1} (~12–20 μm wavelength) is due to librations: a collective, concerted normal mode (υ_L) involving many water molecules. There is also observed in MIR absorption spectra of liquid water a small combination band of bending and libration vibrations (υ_2 + υ_L) at 2127 cm^{-1} (4.70 μm) [9]. It should be noted that there are likely to be changes incurred in both band position and intensity where water molecules are hydrogen-bonded to biomolecules, such as to the amide residue (–N–H—O(H)$_2$) instead of, or as well as, to other water molecules, and also in conditions of extreme pH and/or in the presence of polar molecules.

When water is in the presence of biomolecules, then the water bending vibration, υ_2 at 1644 cm^{-1} (6.08 μm) of ε = 21.8 ± 0.3 $M^{-1}cm^{-1}$ is practically coincident with the amide I band (at 1650 cm^{-1}, i.e. 6.06 μm). However, the molar absorptivity of different biomolecules in the 1500–1800 cm^{-1} (5.56–7.67 μm) region, even if calculated per residue, is significantly higher; for instance, the molar absorptivity of the amide I band of proteins and polypeptides is in the range ε = 300–1000 $M^{-1}cm^{-1}$. Also, liquid-water exhibits a small molar absorptivity of: ε = 6.55 ± 0.3 $M^{-1}cm^{-1}$ coincident with the amide II band, at 1550 cm^{-1}, i.e. 6.45 μm wavelength [9].

8.3 Motivation and Aspiration

Despite increasing efforts to understand cancer, its incidence is still rising. The main reason for this is that we are living longer and there is a correlation between cancer and age. It was formerly thought that this was due to a lowered-immunity in old age. It is now believed that this trend reflects the time required to accumulate a critical number of genetic defects [10].

An epidemiological study in 2012 [11] (epidemiology–study of relation of disparate factors that determine frequency and distribution of disease through the general population) predicted that cancer incidence in the UK in the over-65-years' age group will have trebled by 2040 relative to 2010, when there were 1.3 million new cases. Moreover in February 2015, Cancer Research UK [12] concluded that 1 in 2 people in the UK will be diagnosed with some form of cancer during their lifetime [12a].

Cancers are typically classified into four stages, and part of the diagnostic process is to stage the cancer. **Stage 1** usually means a cancer is relatively small and contained within the organ where it started. **Stage 2** usually means the cancer has not started to spread into surrounding tissue, but the tumor is larger than in Stage 1. **Stage 3** usually means the cancer is larger. It may have started to spread into surrounding tissues and there are cancer cells in the lymph nodes in the area. **Stage 4** means the cancer has undergone metastatic spread (metastasis is transfer of cancer from one body site to another, the secondary tumor being discontinuous with first tumor, but often of the same type).

Over 90% of cancers start in epithelial tissues (epithelial tissue—a major category of body tissue which is usually a covering tissue of internal and external body surfaces, but in a few cases is formative tissue (glands, liver)) [10]. Nowadays much effort is expended in defining the optimum treatment for each type of primary tumor yet two factors confound a successful outcome: (i) metastatic spread and (ii) the treatment is toxic to normal cells. Tumors often metastasize before diagnosis and therefore the greatest single advancement would be early detection. Effective cancer-screening is required that ideally is non-invasive and cost-effective. *Could MIR light have a role in such cancer screening (Figs. 8.3(a),(b))?*

External (skin) cancer is the most common human cancer (100,000 UK cases *p.a.* [12]) and the most common skin cancer is Basal Cell Carcinoma (BCC), which tends not to metastasize. Other skin cancers and pre-cancerous skin conditions are Squamous Cell Carcinoma (SCC), which may metastasize, and actinic (also called solar) keratoses and Bowen's disease which are both precursors to SCC; Malignant Melanoma (MM) is the most serious form of skin cancer and is now the fifth most common cancer in the UK with ~2000 cases *p.a.*, including 16% diagnosed at late Stages 3 and 4. According to statistics released in April 2014 [12]—the incidence of MM is now 5× greater than in the 1970s, ascribed in part to UK "package holidays" in southern Europe and sunbed use. Most telling of all, MM incidence rates in the UK rose by 55% from 1999–2001 to 2008–2010 [12]. Could MIR light have a role to play in fast (real-time) and effective in vivo screening for skin-cancers [13] (Fig. 8.3(a)) moreover carried out conveniently at the local medical practice, rather than at regional hospitals or centers of excellence as currently done in the UK?

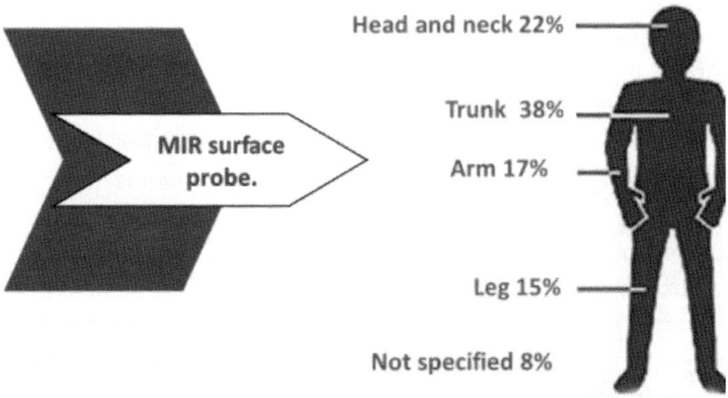

Figure 8.3(a) Could MIR light imaging of skin moles in vivo help diagnose external (skin) cancer at an early stage? (Shows % distribution of male skin cancers by location on body. Adapted from [12].)

For external (skin) cancer diagnosis, initially a visual examination of the skin is carried out by an experienced consultant dermatologist; this may be dermoscope-assisted—a dermoscope is a hand-held ~×10 magnification optical microscope. If deemed necessary, this examination is followed up by the "Gold Standard" comprising excision (removal) of a small-tissue section for histopathology (literally—study of diseased tissue). Histopathology to date refers to high-resolution optical-microscopic study of the morphology of dye-stained (hematoxylin and eosin (H&E) dyes) excised tissue, or shed cells, by a clinician (medical professional interacting directly with patients) to discover disease. The current skin cancer diagnostic process is time-consuming and dependent on human judgment. Moreover, you had to have it cut out, to know. MIR remote, in vivo, spectroscopic-imaging could offer assistance in skin cancer diagnosis that is fast, objective, non-invasive, sensitive (minimal false negatives [14]) and specific (minimal false positives [14]).

The situation is worse where disease is hidden from view in deep tissue. Cancer of the esophagus (the part of the alimentary canal which connects the throat to the stomach) is a typical example, and one whose incident rate is increasing; 73% of esophageal cancers are diagnosed in the UK at Stages 3 and 4 (2014, [15]). Conventional visible-light endoscopic diagnosis, enabling examination of the tissue topology, of early gastric cancer is not definitive. A study by Yao in 2013 [16] showed that using conventional visible endoscopy, a visualized depressed tissue region in the esophagus was not viewed as suspicious for cancer, whereas magnifying visible endoscopy yielded findings possibly consistent with cancer. Confirmation of the adenocarcinoma (a malignant tumor formed from glandular structures in epithelial tissue) was by the "Gold Standard", as described above, that is the taking of an excision biopsy followed by conventional histopathological analysis. So deep tissue must be cut out to get disease confirmation, with the associated patient discomfort and delay in diagnosis. *Could a new MIR light endoscopy have a role to play in objective, instant-feedback,* in vivo, *on early diagnosis of tumor type and margins in internal (deep) tissue [17] (Fig. 8.3(b))?*

The 2014 UK figures show overall that about 4 out of 10 patients are diagnosed with cancer at the late Stages 3 and 4 [15]. For every 1% increase in the number of patients diagnosed at an early stage (Stages 1 and 2), an additional 1,200 patients would be alive 5 years after the diagnosis. 10,000 UK deaths could be avoided each year, if 1-year survival rates were the same as the best-performing countries in Europe [15].

This late diagnosis confounds successful treatment and impacts 5-year survival rates [15]. The greatest single advancement would be early detection, when cancers are more treatable. Thus effective cancer screening is needed, ideally that is non-invasive, cost-effective and in vivo. Prior to neoplastic growth (benign or malignant tumor, consisting of newly proliferated cells), cells often undergo an abnormal pattern of growth, such as metaplasia (reversible replacement of one differentiated cell type with another mature differentiated cell type), dysplasia (proliferation of cells of an abnormal type) or hyperplasia (increase in reproduction rate of cells). Yet, these do not always progress to neoplasia. *Could MIR light provide information on pre-neoplastic cellular change for cancer prevention (Fig. 8.3(a),(b))?*

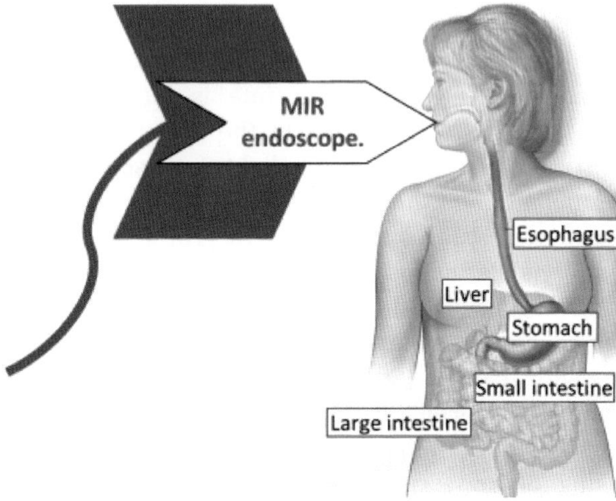

Figure 8.3(b) What about a new type of medical endoscopy based on MIR light imaging? MIR light molecular mapping and imaging of deep internal tissue in vivo—to give immediate feedback on cancer status for early diagnosis [17]?

Could MIR light molecular imaging of tissue assist the efficacy of medical surgery? Moh's micrographic section surgery is an important surgical technique for removing cancerous tumors. The Author has accompanied a clinician for observation of the clinical procedure [18] of external (skin) cancer surgery using Moh's micrographic surgery (Fig. 8.3(c) (A)–(E)) and the same technique is often applied to internal (deep tissue) tumor removal. The patient had presented with a BCC lesion on the lower-eye-rim. The patient was taken into the surgical theatre and a "banana-shaped" tissue section (Moh's section) was cut out as one piece just beyond the visually perceived cancer margins. The banana-shaped piece of tissue was carefully laid out, bisected and labeled by the surgeon. It was taken to the Hospital Pathology Laboratory where it was wax-impregnated, cryo-microtomed (*Greek*: mikros, "small," and temnein, "to cut"), and the ~10 μm-thick slices were

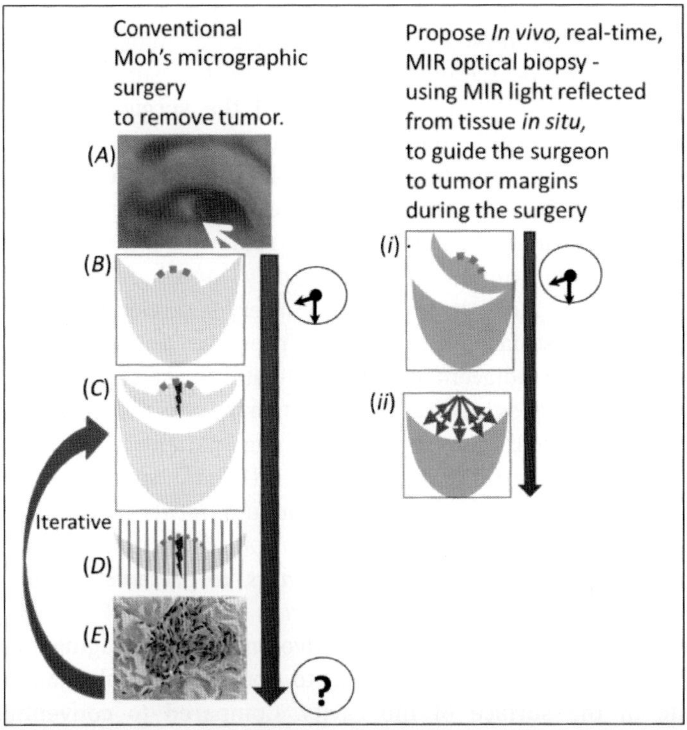

Figure 8.3(c) Location of a BCC tumor on underlid of eye is indicated in (*A*); (*B*) is a cartoon of (*A*); (*C*)–(*E*) show iterative nature of conventional Moh's micrographical section surgery for step-wise tumor removal; applies to both external (skin) tissue, as here, and in internal (deep) tissue. Tissue removed is bisected (*C*), impregnated with wax and cryo-microtomed (*D*), then H&E stained; the stained sections are examined (*E*) using histopathology (at present is examination of die-stained tissue under optical microscope to identify disease). Process (*C*) to (*E*) is repeated until the medical clinician is sure that tumor margins have been removed. Could MIR light imaging be used instead to monitor tumor margins during medical surgery? (i) to (ii) indicate MIR light in vivo molecular imaging of the cavity left after tumor removal to locate residual malignant tissue in the surface of the cavity. The aim is precise excision of the cancer tumor at its extreme borders to cellular tolerances, to preserve anatomical features as far as possible, in one go during the surgery. Compared to conventional Moh's micrographic surgery, MIR imaging could cut medical consultant time and minimize patient suffering.

mounted on silicate-glass microscope slides and stained. The surgeon next examined each stained section under the optical microscope to determine the success of the surgical removal by searching for the tumor margins. If the surgical removal is deemed unsuccessful, that is if tumorous cells are observed at the extreme margin of the excised Moh's section, then the Moh's micrographic section process is repeated iteratively until the surgeon is assured of complete tumor removal—if not completely removed, then tumor re-growth may occur from residual cancerous cells. Moh's micrographic surgery is carried out carefully and in step-wise fashion in order to preserve, as far as possible, the integrity of adjacent anatomical features. At the end of the stepwise surgery, which can stretch over hours, and occasionally days for internal tumor removal, the patient is taken for reconstructive plastic surgery. *Could MIR light spectral imaging be used in situ to guide the surgeon's hand to locate tumor margins during medical surgery for precise excision of the cancer tumor at its extreme borders to cellular tolerances, to preserve anatomical features as far as possible, in one go during the surgery?* Figure 8.3(c) (i) to (ii) indicates MIR light in vivo molecular imaging of the cavity left after tumor removal to locate residual malignant tissue in the surface of the cavity. Compared to conventional Moh's micrographic surgery, MIR spectral imaging could cut medical surgical time and minimize patient suffering.

We propose MIR light reflection from tissue for real-time, in vivo cancer diagnosis by means of MIR molecular spectral imaging using a new MIR endoscopy for internal (deep) tissue and MIR surface probe for surface-examination of external (skin) tissue (note—using MIR light not as a therapy, but as a diagnostic tool) (Fig. 8.3). In summary, we propose the in vivo, real-time MIR optical biopsy (*Greek:* "bio" is living (tissue) and "opsy" is look) (equivalent to non-stand-off, MIR hyperspectral medical imaging) to obviate the traditional type of tissue-excision for biopsy.

Practical deployment of in vivo MIR spectral imaging requires the means to generate bright MIR light and rout it to where needed. But conventional silica glass fiber-optics are opaque to MIR light and MIR light sources are weak. We propose chalcogenide-glass, low optical loss, fiber-optics to achieve in vivo molecular spectral imaging, for both passive optical fiber routing of MIR light, and as bright optical fiber MIR lasers.

The European Commission through Framework Seven (FP7) currently supports the project **MINERVA** (**MI**d- to **NE**a**R** infrared spectroscopy for impro**V**ed medical di**A**gnostics, 317803; www.minerva-project.eu) which is for the first time addressing such developments. The University of Nottingham, UK, is making MIR fiber-optic devices within MINERVA, as a first step for in vivo, real-time external tissue surface examination (i.e. non-invasive investigation of suspected skin cancer). Furthermore, chalcogenide glass fiber-optic and waveguide devices and systems could be key in new MIR communications for molecular sensing to inform decision-taking in sectors as diverse as manufacturing, energy, the environment, agriculture and security.

The term biomolecule was only coined in 1963 (see Table 8.1), and vibrational spectroscopy of biomolecules as inanimate molecules was pursued. But nowadays to understand the vibrational spectroscopy of a collection of biomolecules as a living entity is a compelling notion. All life is composed of cells in which there are biomolecules, for instance DNA involved in cell-replication; tissue is a collection of cells. Biomedical MIR absorption spectroscopy and Raman spectra of human cells and tissue may be conducted either in vitro (*Latin:* "in glass") meaning in the laboratory as opposed to in vivo (*Latin:* "in life" (in a live body)) and ex vivo (*Latin:* "outside life" meaning in the laboratory but with minimum alteration of natural conditions). Raman and MIR spectroscopies are complementary; in vivo Raman is an emerging technology and the next section explores this success and why MIR spectroscopic imaging is not yet being applied in vivo.

8.4 Raman Spectroscopy vis-à-vis MIR Spectroscopy for Medical Spectral Imaging

The BBC (8-26-2015) reported on the first Raman-assisted brain surgery in Europe carried out at Charing Cross Hospital, London, on a London Imperial College Physics PhD student who presented originally with a "golf-ball sized mid-brain tumor". The surgery was reported by the media thus: "Surgeons in London have used lasers to diagnose abnormal tissue during an operation to remove a brain tumor. The non-invasive technique measures light reflected off tissue to determine whether it is cancerous or healthy. During the operation, surgeons used a near-infrared laser probe, pointing the beam of light on to the exposed brain. This causes molecules in the cells to vibrate. Fibre optics in the probe collect the scattered light that bounces off the tissue. This is analyzed using Raman spectroscopy, which can measure the frequency of vibrations. Healthy and abnormal tissue has a slightly different 'signature'. The whole process takes a couple of seconds, and is entirely non-invasive. In all cancer surgery, the aim is to remove all abnormal tissue while sparing healthy cells." The tumor was found non-malignant and the patient was making good recovery, two months on. "Kevin O'Neill, head of neurosurgery at Imperial, said new technologies would revolutionize brain surgery. 'This is bringing the laboratory into theatre, giving a real-time molecular fingerprint of tissue—the potential is amazing, not just to differentiate between normal brain and tumor, but whether the patient is likely to respond to specific treatments'." "Neurosurgeon Babar Vaqas, the trial chief investigator, said: 'Optical technologies like this are the future. They are fast and don't destroy any tissue and could be used during many types of cancer surgery or when dealing with infection like a brain abscess.'"

This first advent in Europe, of Raman-assisted brain surgery, is the culmination of much research worldwide. It is dependent on the Raman effect being carried out at near-infrared (NIR: 0.75 to <3 μm) wavelengths of the electromagnetic spectrum to access MIR (3 to 25 μm) vibrational signatures of molecular species. The NIR operation of Raman has allowed the mature field of silica glass fiber-optics to be taken advantage of to transform Raman from "benchtop" origins into a convincing platform for in vivo, real-time medical imaging. Purified silica glass drawn to fiber is the most NIR transparent condensed medium, per unit optical pathlength, on Earth, developed in the latter part of the 20th century to underpin Internet telecommunications and photonics. (Photonics is using light to process and deliver information; biophotonics is using light as a therapy or, as here, for medical diagnosis of disease to assist clinical decision-taking for best healthcare outcomes. Light sensing, mapping and imaging of diseased human tissue and cells are at the heart of biophotonic medical diagnoses.)

The application of Raman for in vivo molecular spectral monitoring of tissue relies on the silica glass optical fiber to deliver the Raman excitation NIR laser light to the tissue and then to collect and rout the inelastically scattered NIR laser light to the detector after interaction with the tissue (see Fig. 8.4).

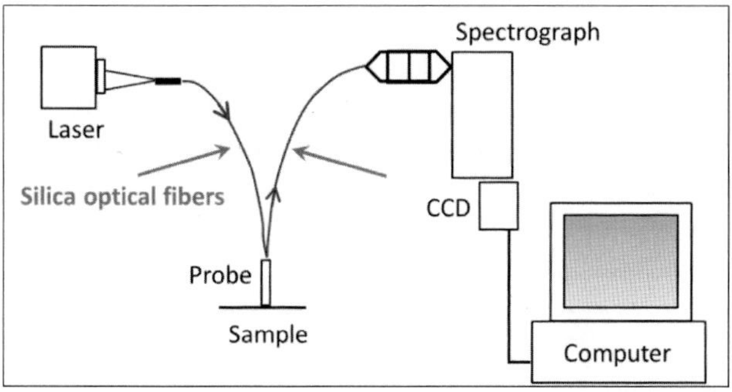

Figure 8.4 Typical Raman probe system used for in vivo tissue studies. Adapted from [19].

The Raman technique [20] is based on stimulated inelastic scattering of NIR radiation by molecules (Fig. 8.5). Spontaneous Raman scattering is a statistically rare effect, suggested to occur in 1 in 10^7 events of photon scattering by molecular species. Raman spectra are generally of weak intensity and poor contrast (see Fig. 8.6). Attempts are currently being made to boost the Raman effect by adsorbing molecular species on the surface of metal nanoparticles, to make use of plasmon enhancing effects, so-called SERS (surface enhanced Raman spectroscopy). But this has yet to be played out in an in vivo setting and the long-term health effects of using metal nanoparticles in a biologic setting are still unknown.

Early Raman studies were based on rather small sample sets yet the observed spectral trends indicated the promise of Raman spectroscopy for cancer diagnosis, as first reported by Alfano et al. 1991 [21, 22]. This initial work relied on peak height ratios for data analysis and was unable to differentiate benign and malignant lesions [23]. Multivariate statistical analysis of the entire Raman spectral region was demonstrated to enable benign and malignant tumors to be differentiated in excised tissue [24]. Now a wealth of Raman data exists from different authors to demonstrate high levels of specificity and sensitivity in distinguishing disease in this way, including cancer of the brain, breast, lung, skin, esophagus, prostate and colorectum and bone disease [25–29]. Multivariate classification models can provide an objective diagnosis of independent tissue samples obtained from new patients [26] but give almost no insight into the biochemical changes responsible for disease. Yet Raman vibrational spectra intrinsically reflect the molecular status of tissue. Teasing out the vibrational Raman-band "biomarkers" of biochemically altered cellular constituents and the associated morphological changes accompanying different types of cancer is of great interest [30, 31].

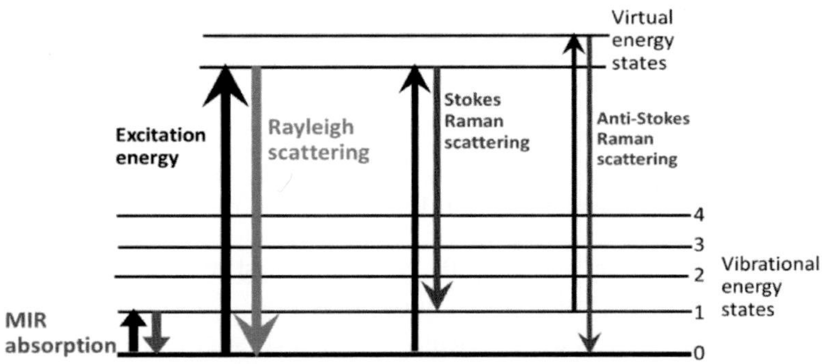

Figure 8.5 Schematic transition states during Rayleigh and Raman scattering in a material compared to MIR absorption at room temperature.

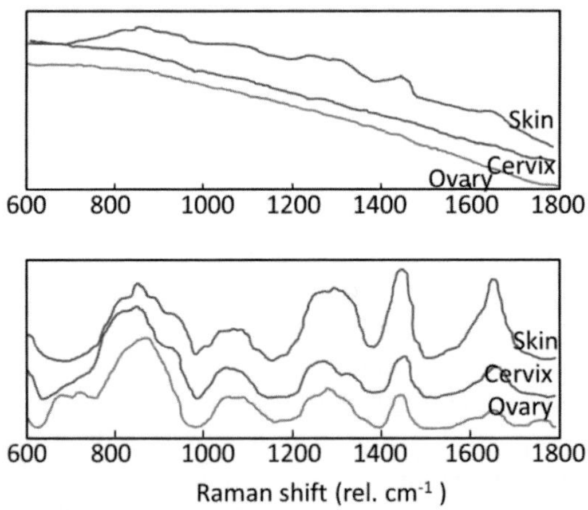

Figure 8.6 Raman spectra for three tissue types. Shown are: (a) raw and (b) processed spectra. Adapted from [19].

The relatively low speed of Raman spectral collection has impeded clinical translation to date but recent advances in multimodal imaging and non-linear microscopy demonstrate high-contrast molecular imaging on clinically relevant sample dimensions at speeds compatible with clinical use [26].

Fiber-optic and needle Raman probes are being developed for in vivo diagnostics and sub-surface deep in vivo Raman spectroscopy [26]. Initial skepticism regarding the ability of in vivo Raman spectroscopy to distinguish disease over conventional pathological approaches is being overcome. Using Raman-assisted endoscopy, Mahadevan-Jansen and her team improved on their own results for cervical cancer progression by considering potential confounding factors and stratifying spectra according to the hormonal/menopausal status of the patient before applying a separate dysplasia diagnostic algorithm. They commented in 2009 that restricting the dysplasia diagnostic to pre-menopausal women improved the overall accuracy in discriminating dysplastic tissue from normal tissue using in vivo Raman to 94%, "bringing Raman one step closer to its eventual clinical use" [31]. As recently reviewed by Nottingher, Stone et al., in vivo collection of Raman spectra has been demonstrated in cancers of the brain, breast, skin and esophagus [26 and references therein].

To summarize, inelastic Raman scattering for in vivo optical imaging is an emergent clinical tool which has been enabled by silica optical fiber to rout the light to the patient and for analysis. Rapid progress has been possible because excellent quality silica fibers are already available. However, the Raman effect is an intrinsically extremely weak phenomenon. In comparison, MIR light imaging based on direct fundamental vibrational MIR molecular absorption (see Fig. 8.5) is potentially a more sensitive optical technique, than Raman, and with very good contrast.

Both Raman spectroscopy and MIR absorption spectroscopy are vibrational spectroscopies governed by different quantum mechanical selection rules and yielding complementary spectral information.

Excellent progress has been made on in vitro and ex vivo MIR spectral imaging over the past decade (see later Section 8.5). So what is holding back the development of MIR light optical imaging in vivo? The answer is two-fold. First, common MIR blackbody light sources are too weak. Second, silica glass optical fibers do not transmit MIR light. These are the reasons why MIR light molecular spectral imaging has been restricted to date to excised biological tissue and exfoliated cells (see Section 8.5) and it has not been possible to carry out this type of imaging in vivo.

8.4.1 Active and Passive MIR Chalcogenide Glass Fibers

Hollow-core silica-glass fiber can be transparent to ~3 μm wavelength because the bandgap principle overcomes material absorption [32]. The low attenuation is possible because the majority of the transmitted power in such fiber is confined to the hollow core region and little propagates through the highly absorbing material regions of the structured fiber. The projected loss of the silica glass material itself becomes prohibitively high in the MIR reaching an estimated 10,000 dBkm^{-1} at 3 μm wavelength [4] (see Fig. 8.7(a)).

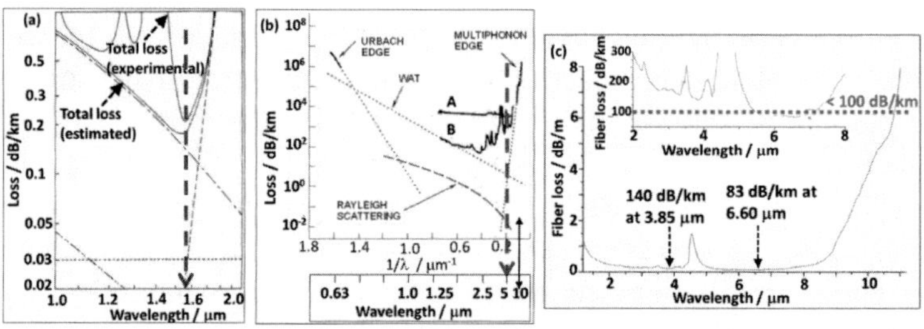

Figure 8.7 Colored arrows in (a), (b), indicate wavelength of lowest theoretical loss. (a) **vitreous silica**: UV and NIR absorption edges, Rayleigh scattering loss, experimental loss curve (*adapted from [41]*); (b) **chalcogenide glass fiber** ($As_{40}S_{55}Se_5$ atomic%) loss curve, fabricated from: (A) as-received chemicals and (B) purified chemicals. In (b) are Urbach tail, multiphonon edge, weak absorption tail, calculated Rayleigh scattering loss (*adapted from [33]*) and (c) **chalcogenide fiber record** (unclad $Ge_{10}As_{21}Se_{69}$ atomic%) optical loss spectrum. Inset shows more clearly the sub-100 dBkm^{-1} loss characteristic [35].

Chalcogenide glasses, on the other hand, provide transparency across the MIR wavelength range (Figs. 8.7(b),(c)) [33–35]. They are sufficiently chemically and physically robust materials for development [36–40] with Young's modulus ~1/5 that of silica glass and other mechanical properties scaling *pro rata*; they are chemically resistant to water. In our laboratories, MIR light transmission through 52 m of chalcogenide glass fiber has been recorded, with a usefully wide MIR window and minimum optical loss of 83 $dBkm^{-1}$ at 6.6 μm wavelength (Fig. 8.7(c); similar loss has also been recorded by several groups (see [35], and references within) in chalcogenide glass fiber of various glass compositions.

Chalcogenide glass fibers and waveguides not only exhibit MIR passive transmission but also, being optically non-linear, have the potential to generate MIR supercontinuum laser light and, when suitably doped, to be direct-emission narrowline fiber lasers, to underpin portable and compact bright MIR fiber-optic sources. MIR fiber-optic devices would facilitate remote molecular sensing and spectral imaging as well as MIR coherent imaging and MIR power delivery. These are the nuts and bolts of MIR communication. MIR spectral sensing, mapping and imaging in vivo systems could provide real-time molecular information to inform clinical decision-taking. Exploiting the MIR effectively requires the development of a raft of new, MIR fiber and waveguide based, narrow and broadband bright MIR sources and sensors and imaging systems and MIR power delivery. For instance, to date there are no MIR fiber lasers available for general use >3 μm wavelength, and none demonstrated at ≥4 μm.

To summarize, chalcogenide glass MIR fiber optics for biophotonics have potential in medical endoscopic systems, and surface probe systems, for in vivo, real-time analysis, i.e. sensing, mapping and imaging tissue, for early disease screening and diagnosis, so informing medical decisions and treatment planning, and also for fiber laser surgery at new MIR wavelengths, for a complete MIR solution. Until now MIR spectral imaging has solely been carried out on *excised* tissue and the excellent progress made in this field is highlighted next in Section 8.5.

8.5 MIR Light Molecular Spectral Imaging on Excised Tissue

MIR light molecular spectral imaging is a non-destructive and label-free technique that is becoming an increasing important analytical tool for biomedical research [42]. However, at present MIR light molecular spectral imaging is only being carried out on excised tissue due to a hitherto lack of bright, portable MIR light sources and MIR routing fiber. Below, (Sections 8.5.1 and 8.5.2) take some illustrative examples to summarize progress in MIR molecular spectral imaging, in terms of excised external (skin) tissue and of excised internal (deep) tissue. In Section 8.5.3, MIR coherent imaging is introduced. This is followed by Section 8.6 which describes the progress made towards developing the MIR devices and systems needed to bring the in vivo MIR optical biopsy into the clinic, by means of MIR light molecular spectral imaging.

8.5.1 MIR Light Spectral Imaging of Excised External (Skin) Tissue

Increasing the wavelength from visible to NIR light, enables deeper imaging due to a reduced scattering coefficient, allowing the absorption coefficient to become the main determinant of image quality [43]. There are three NIR windows: 650–750 nm, 1100–1350 nm and 1600–1870 nm and the latter is the so-called "golden-window" for brain tissue [44]; this is the NIR "sweetspot" for deepest penetration, of potentially several mm. The NIR molar absorptivity is due to characteristic, but weak, overtone and combination absorption bands of tissue, which are far weaker than the associated fundamental MIR absorption bands (see Table 8.1).

MIR light has provided the means to analyze spectroscopically and image cutaneous (skin) neoplasms. MIR light reflected from tissue tends to be far less scattered by small-scale (sub-micron) tissue inhomogeneities than visible or NIR light, as Rayleigh and Mie scattering are dependent on reciprocal wavelength $\propto \lambda^{-4}$, and $\propto \lambda^{-2}$, respectively. However, individual cells and cell features of sizes approaching the MIR wavelength may act as discrete MIR scattering centers [45]. MIR light reflected from, or transflected through (sample supported on a MIR-mirror-backed substrate; incident radiation traverses sample depth twice: i.e. before and after the back-reflection) or transmitted through tissue is rich in characteristic tissue MIR spectral signatures comprising the fundamental vibrational absorption bands of biomolecular species of which the tissue is composed. Unlike the current Gold Standard of diagnosis by histopathology of biopsied tissue (Section 8.3), MIR spectra are obtained from unstained tissue yet MIR spectral pseudo-mapping of tissue has the potential to be as, if not more, informative (see later this section and Figs. 8.9–8.14).

In an early (1999) influential study, Mantsch et al. [46] demonstrated that MIR spectra of skin BCC (see Section 8.3) are distinct from non-tumor-bearing skin components. Their study was in vitro and conducted on biopsied tissue samples. After collecting the MIR spectra from the tissue biopsies, several consecutive steps of analysis of the acquired spectra were employed. First, spectral bands were assigned using the classical group assignment based on photon frequency approach (e.g. see Fig. 8.8 [46]).

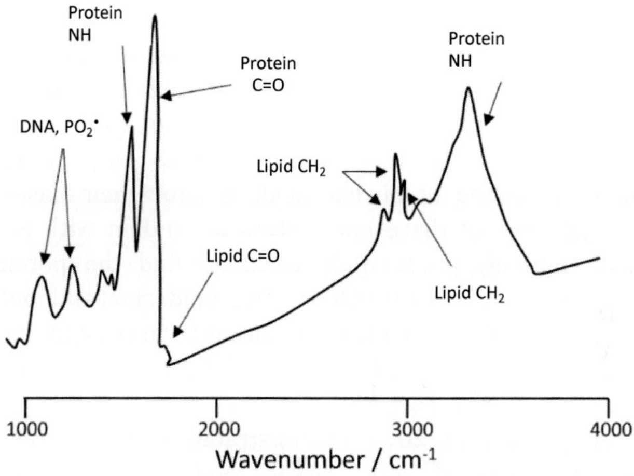

Figure 8.8 A representative MIR spectrum from cultured human melanoma cells (B16f1). The origin of the major absorption bands is indicated [46]. Reproduced by kind permission of The Society for Investigative Dermatology Inc.

Spectra were normalized to remove variability of sample thickness, etc. and then student-*t*-tests were applied to the second-derivatives of the spectra. A non-subjective classification of spectra was performed by means of LDA (linear discriminant analysis) which determined the boundaries that best separated classes of tissue type by computing the distance of each spectrum from class centroids. Data were split into "training sets" and "test sets." Thus Mantsch et al. [46] reported, for instance, that the MIR spectral discrimination of BCC from epidermis ex vivo classification was 98.7%. Even more interestingly, epidermis-overlying-BCC was distinguished from epidermis-overlying-non-tumor-skin by 98.0% (epidermis is the outmost layer of skin and is itself composed of several distinct morphological layers). The Authors remarked that these ex vivo results showed promise for future in vivo MIR spectral discrimination.

Mostaço-Guidolin et al. [47] reviewed publications prior to 2009 on the application of Fourier transform (FT) mid-infrared spectroscopy to distinguish skin cancer. Of the reviewed publications, it is worth noting the work of Hammody et al. [48] in 2008; they collected MIR spectra ex vivo within skin biopsies of both melanoma tissues and the surrounding epidermis, with the aim of identifying diagnostic parameters and suitable computational/statistical methods of analysis. It was found that malignant melanoma could be differentiated from the normal epidermis in the excised tissues of 55 patients, using parameters derived from the MIR absorbance bands originating from molecular vibrations of nucleic acids and/or their bases. Two-dimensional plots of these parameters in tandem with advanced statistical methods successfully demonstrated the potential of MIR spectroscopy to distinguish between epidermal and melanoma regions with a high classification success. They remarked that: "*infrared spectroscopy is widely perceived as a future technology for cancer detection and grading.*"

Ly et al. [49] in 2009 demonstrated MIR spectral micro-imaging in vitro of skin tissue and reported successful differentiation of skin cancer types. The skin tissue had been paraffin-embedded which enabled this retrospective spectral analysis; the paraffin contribution to spectra was removed numerically. An FT-IR spectrophotometer with MIR-microscope attachment was used in conjunction with a standard arrayed MCT (mercury cadmium telluride) MIR detector (16 pixel-line of 6.25 μm × 6.25 μm of pixels) (see Section 8.6.1); MIR spectral acquisition entailed 2-D (dimensional) rastoring spectral mapping of tissue. k-means clustering was applied to re-group spectra that showed similar characteristics (i.e. manifested similar molecular status). The group of Diem [50] had earlier demonstrated this type of unsupervised method for processing MIR spectral data of cancerous tissue. The cluster-membership information was then plotted as pseudo-color maps by assigning a color to each cluster of spectra-type. A pseudo-color map is reproduced here (Fig. 8.9 from [49]) for each of BCC, SCC and Bowen's disease, respectively, shown together with a photomicrograph in each case of adjacent tissue that had been H&E stained after the MIR-spectral acquisition.

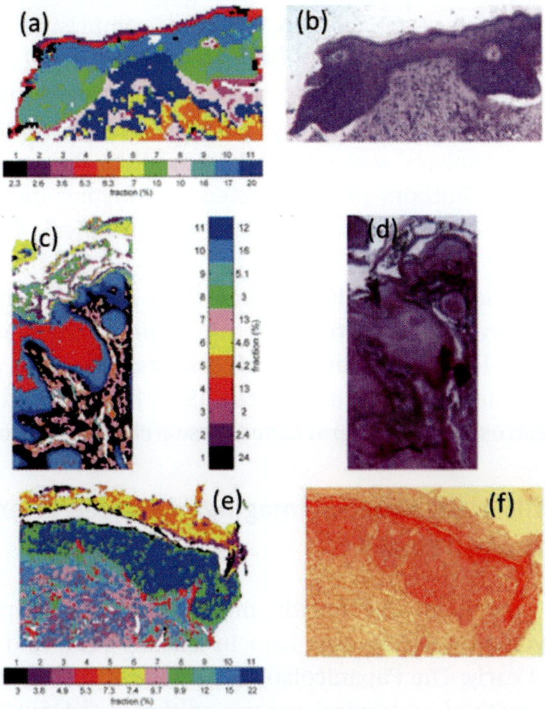

Figure 8.9 Paraffin-embedded external (skin) tissue analysis using MIR spectroscopy and k-means clustering. Panels (a), (c) and (e): pseudo-color images (maps) of tissue exhibiting BCC (a), SCC (c) and Bowen's disease (e) and photomicrographs of their corresponding (taken from adjacent tissue) H&E stained sections (panels (b), (d) and (f), respectively). Note that each image was processed independently thus the color code is arbitrary and not transferable between maps. White areas correspond to where there was paraffin and no tissue, i.e. outlier spectra. Reproduced from Ly et al. [49] by kind permission of the Royal Society of Chemistry.

From Fig. 8.9, there appears to be good agreement between, on the one hand, the MIR-spectral pseudo-color map of the non-stained-tissue and, on the other hand, the tissue morphology evident in the subsequently stained tissue-sections.

Ly et al. [49] state that the k-means clustering did not permit differentiation between multiple pathologies and maybe better suited to exploratory or descriptive investigation. They therefore chose a LDA supervised pattern recognition for the differential diagnosis of skin carcinomas (i.e. malignant tumors of epithelial origin). An 8-class model was adopted comprising BCC, SCC, Bowen's disease, normal epidermis and reactional epidermis and so on. Working directly on the full MIR "fingerprint" region (900–1800 cm^{-1} wavenumber, i.e. 5.5–11 μm wavelength) for the discriminant analysis was reported to produce the most reliable outcomes. The Authors [49] suggested that further benefit might be had from considering in tandem mathematical morphology analysis.

Stone et al. [51] and Krafft et al. [52] presented reviews in 2009 on vibrational spectroscopy as a clinical tool for cancer, including FT-IR of dermatological cancers. A more generalist review due to Bellisola and Sorio [53] has focused on MIR micro-spectroscopy in general cancer research and diagnosis.

8.5.2 MIR Light Spectral Imaging of Excised Internal (Deep) Tissue

Cervical cancer is the fourth most common malignancy in women worldwide [54], and 12th in the UK [12], and treatable if detected early. The Papanicolaou (PAP) smear test is the current screening method of cervical cancer; epithelial (i.e. surface tissue layer) cells are exfoliated from the cervix and stained with dyes for histopathology. The PAP test has reduced known mortality rates of invasive cervical cancer but, while the specificity of the PAP smear is generally very high (95%), its sensitivity can be as low as 20% to 50%, depending on the prevalence of the disease within the population [31]. This may be due to inadequate sample collection and failure to identify the telltale cell morphology, which is difficult and can amount to finding only a few suspect cells in 10,000 cells [55] by morphological examination under optical microscopy.

The group of Diem has used MIR micro-spectroscopy of excised cervical tissue in ex vivo transflection to demonstrate that variations in molecular composition of cells can be used to construct tissue spectral-maps using MIR microspectroscopy of the tissue [56].

Tissue spectra from MIR micro-spectroscopic rastoring across the sample can be treated by the mathematical statistical methods of principal component analysis or clustering techniques to obtain a visual representation of the molecular make-up of a tissue sample in terms of false color maps. The false color map is obtained by establishing similarities in spectral patterns (mathematically), then grouping spectra accordingly and finally assigning color codes to for an objective map, free from subjective assignment [56].

In early work, Diem reported the MIR spectra of major and minor biomolecular cell components and the influence on spectra of cell division [6]. Armed with understanding of the constituent MIR vibrational spectra of biomolecules, one of Diem's objectives has been to establish the MIR spectral signatures of cells and tissues in different stages of disease, as in the example that now follows.

Figure 8.10 depicts results for a patient diagnosed with a category of PAP smear of cause for concern for onset of cancer (LSIL: low-grade squamous intra-epithelial lesion) [56]. Figure 8.10(A) is an optical micrograph of excised cervical tissue stained with H&E dyes after collection of the MIR spectra. Figure 8.10 (B)–(D) illustrate the sensitivity towards chemical composition, i.e. molecular structure, afforded by cluster analysis of mid-infrared spectra in different spectral regions (note: one color per cluster and color displayed at spatial *x*–*y* coordinates at which each spectrum was collected). Figure 8.10(E) is mean extracted spectra corresponding to Fig. 8.10(C). Figure 8.11(B) is tissue H&E stained, adjacent to that shown in Fig. 8.10(A). Figure 8.11(A) is equivalent to Fig. 8.10(C) (but different colors have been used) in that both show cluster analysis of the entire spectral region (5.6–12.5 μm wavelength) but for Fig. 8.11(A) second derivatives of the spectra have been taken—a simple mathematical device to enhance subtle spectral differences such as shoulders on peaks. As Diem et al. state: the cluster maps of Fig. 8.11(A) show "*in exquisite detail*" different tissue layers including cell maturation [56].

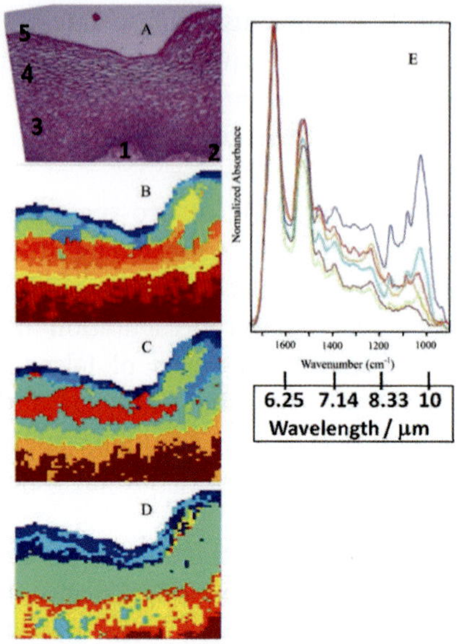

Figure 8.10 (A) Optical photomicrograph of a section of cervical epithelium from a patient diagnosed with LSIL (low-grade squamous intra-epithelial lesion). The section was stained (H&E) after FT-IR data acquisition. 1–5 in (A) are different layers in the tissue. (B) 9-cluster maps using the entire mid-infrared spectral range of 5.6–12.5 µm wavelength. (C) and (D) are each 9-cluster maps from smaller spectral regions of 5.75–6.80 µm and 8.3–10 µm, respectively. (E) Mean cluster spectra from 6 of the 7 clusters color coded to match (C). (Note individual colors of (B)-(D) are not transferable). For more detail about this figure, see [56]. Reproduced with permission from Elsevier.

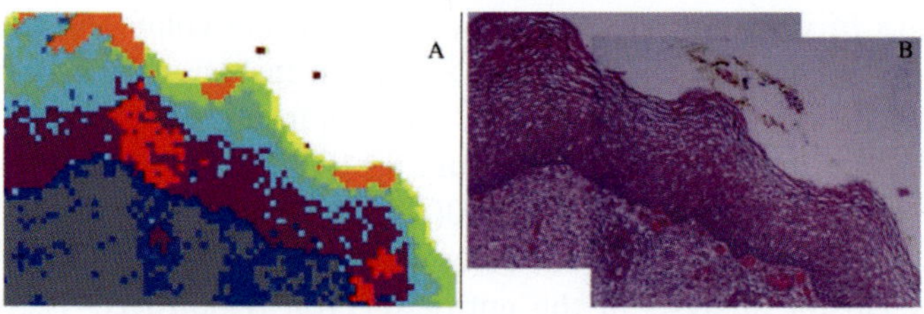

Figure 8.11 (B) Photomicrograph of a section of cervical epithelium from patient diagnosed with LSIL. This section is a continuation, to the right, of the one shown in Fig. 8.10(A), and was stained (H&E) after FTIR data acquisition. (A) 9-cluster map, constructed from the second derivative spectra and the entire mid-infrared spectral range of 5.6–12.5 µm wavelength [56]. Reproduced with permission from Elsevier.

Diem is swift to point out that the work does not seek "*cancer markers*" as specific absorption bands in the mid-infrared spectra, rather mathematical procedures are utilized (such as unsupervised principal component analysis) to analyze subtle spectral differences of the entire spectrum—and it has been demonstrated that subtle differences exist and that normal and diseased areas of tissue can be distinguished spectroscopically [56].

This type of automated pattern recognition of the families of vibrational absorption bands in MIR spectra may aid clinicians in future in diagnosis and medical-decision taking on treatment.

Using the approach described above has importantly allowed Diem and others to preserve the anatomical arrangement of tissue, allowing evaluation of the pathological features and histological architecture of tissue from patients to be analyzed ex vivo using MIR spectroscopy.

In more recent work by Nallala et al. [57], MIR spectral imaging combined with multivariate statistical analysis was applied to colonic tissue to identify spectral markers representing the biochemical changes associated with malignancy. This formed the novel concept of spectral barcodes. To implement this concept, MIR spectral imaging of non-tumoral/tumoral sample pairs from different patients was performed independently. Cluster analysis, using k-means, permitted identification of the normal and tumoral epithelial components of the colonic tissue in a label-free manner. k-means clustering was used for its rapidity and ability to handle large datasets. Spectra with similar biochemical characteristics were grouped in the same cluster in an iterative manner, where each cluster corresponded to a histological feature. The k-means-generated clusters were annotated into their corresponding histological classes by using the reference H&E stained images and pathologist guidance. Using this, the organizational levels of the colonic tissue, such as crypts, etc. were identified. These were prevalent in the non-tumoral (normal) colonic tissue but the tumoral samples were devoid of this organization, owing to loss of differentiation of the glands. The spectral difference between different clusters corresponding to the endogenous (growing within the living being) biochemical tissue signature of the histological classes was then visualized as a dendrogram (Fig. 8.12) by hierarchical clustering analysis (based on Ward's linkage algorithm).

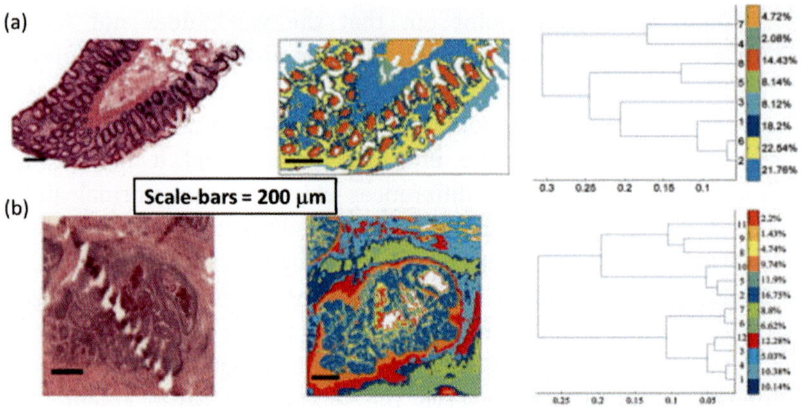

Figure 8.12 k-means clustering of non-tumoral (a)/tumoral (b) colonic MIR spectral images (mid-panel), with respective dendrograms (right-panel), compared to H&E stained sections. (a) Classified with 8 clusters representing major normal colonic tissue features (random pseudo-colors); (b) moderately differentiated adenocarcinoma in colon tissue section, classified with 12 clusters. Adapted from [57].

The MIR spectra from the tumoral and non-tumoral clusters were then extracted for further statistical analysis. The Mann–Whitney U-test was performed to identify the most discriminant MIR spectral wavelengths between the normal and malignant tissue. Only tissue pairs from a single patient were compared to avoid inter-patient variability. The most discriminant wavelengths were arranged as spectral barcodes (Fig. 8.13). The novel use of spectral barcodes was found to be a rapid method for tumor characterization: 3 min to collect spectra, 1 min for k-cluster analysis and 13 min for the spectra barcode. The spectral barcode was in essence a visualization of intra-patient biochemical variability. It did not address the inherent heterogeneity of different patients but it did show a direction forward for data-handling where data banks may be utilized in the future.

Finally, work from the Group of N. Stone (EU MINERVA) shows up-to-the-minute high resolution colonic cancer MIR micro-spectral imaging involving magnifying the image and showing a nominal resolution beyond the diffraction limit (Fig. 8.14) [58].

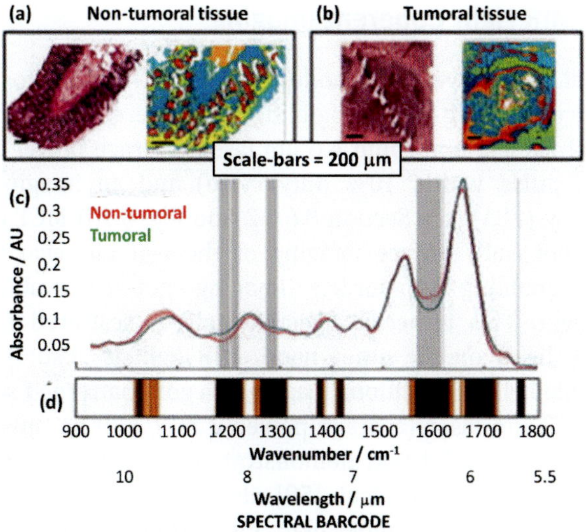

Figure 8.13 Construction of spectral barcode. MIR spectra (color coded as Fig. 8.12) corresponding to non-tumoral and tumoral epithelial components from (a) and (b) were retrieved from the k-means cluster images and compared using a statistical test with different *p*-values of significance, to find out significant discrimination wavelengths (c). For example, discriminant wavelengths at $P < 0.00016$ are shown as gray bars in (c), and spectral barcodes (black) in (d). Adapted from [57].

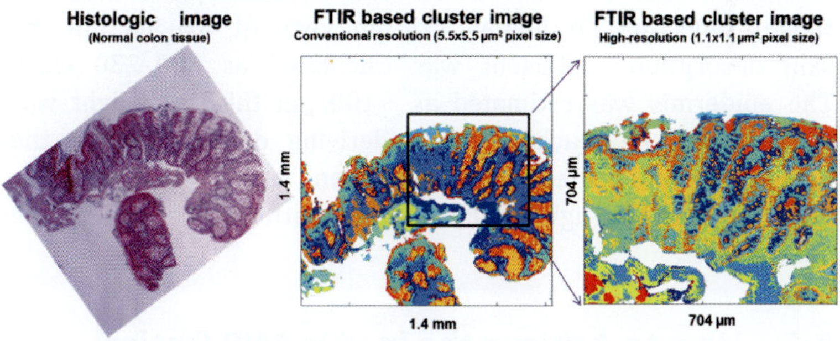

Figure 8.14 State-of-the-art mid-infrared spectral images of excised normal colon tissue compared to a conventional histopathological micrograph of the H&E stained colon tissue (*Images from N. Stone, University of Exeter, UK, Co-Investigator MINERVA*) [58].

8.5.3 MIR Light Coherent Imaging

Guo et al. [59] have used modest power MIR lasers for diffuse reflectance imaging of biologic tissue. Coherent imaging of a plant-leaf using a homebuilt 4.6 μm laser (average power 2.5 mW, 1000 ns pulse width, 10% duty cycle) and an imaging (focal plane array (FPA), see Section 8.6.1.2 and Fig. 8.18) MCT detector enabled not only surface imaging of the leaf but also confocal focusing enabling sub-surface imaging not achievable using visible light. This is because visible light is scattered far more than MIR light; on the other hand, MIR light is absorbed more than visible light. In addition, imaging of a gold-patterned substrate under a 5.4 μm laser (average power 2 mW, 600 ns pulse width and 1.5% duty cycle) was demonstrated through 150 μm depth of blood or water. Guo et al. [59] also suggest that MIR light is able to carry a tissue molecular absorption signal long before it suffers phase scrambling (in direct contrast to visible light) which has implications for prospects, for instance, for holography to re-construct 3-D MIR images. MIR chalcogenide glass fiber lasers could potentially be applied in diffuse reflection imaging of biologic tissue.

More recent work [45] employed a quantum cascade laser (tunable, pulsed (through duty cycle) at 100 kHz, 500 ns pulses, 100 mW peak power, 5 mW average power and estimated power density 70 mW cm^{-2} source covering 8.2–10.1 μm wavelength range) reflecting off human skin in vivo. At 8.6–9.6 μm the skin absorption coefficient was calculated as 200–230 cm^{-1}. The epidermis was estimated as ~100 μm thick and light was suggested to be reaching the underlying dermis because the detected light varied with detection angle, attributed to interaction of the light with collagen scattering centers in the dermal layer.

8.6 How to Achieve the in vivo MIR Optical Biopsy

As shown in the previous section (Section 8.5), MIR light molecular spectral tissue imaging is at present being carried out solely on *excised* external (skin) tissue and *excised* internal (deep) tissue which has been micro-tombed. As highlighted, excellent progress has been made on in vitro and ex vivo MIR spectral imaging over the last decade, but in order to develop MIR light spectral imaging in vivo, bright, portable MIR fiber-compatible sources are needed and MIR optical fiber to rout the light. In Section 8.6, progress made towards achieving these under-pinning technologies and towards bringing MIR light molecular spectral imaging in vivo into the clinic is described here. In Section 8.6.1, MIR bulk and fiber optical component availability, and working in the MIR region, are introduced. Established, emerging and future bright MIR light sources are discussed in Section 8.6.2.

8.6.1 MIR Optical Components, Circuits and Detectors

8.6.1.1 MIR optical components and circuits

For clinical application of MIR molecular spectral imaging, there is a need to rout MIR light spanning approximately the 3 to ≥10 μm wavelength range. Chalcogenide glass fiber has been demonstrated to exhibit low optical loss across this MIR window (see Section 8.4, Figs. 8.7(b),(c)) and [33, 35].

However, for constructing MIR optical circuits, there is a current dearth of MIR "cots" (commercial-off-the-shelf), and customized, linear and non-linear optical components (no MIR narrowline fiber lasers, no room-temperature, sensitive MIR detectors, no MIR cheap FPAs (focal-plane array detectors) and lack of range of tailored: MIR fiber-pigtails, MIR fiber couplers, MIR lenses, MIR mirrors, MIR AR (anti-reflection) coatings, MIR filters, MIR connectors, MIR optical fiber, etc.). Setting up a MIR photonics laboratory is therefore not as straight forward as working in the NIR spectral region. Also, free-space MIR optical circuits encounter the two MIR atmospheric windows (see Section 8.1). Moreover, MIR optical circuits are vulnerable to blackbody radiation interference; standard room-temperature blackbody radiation interference peaks at ~10 μm wavelength.

8.6.1.2 MIR detectors

MCT detectors are currently the most sensitive and widely used MIR detector, but are weak in comparison to NIR and visible detection (see Fig. 8.15). MCT FPA detectors are costly but becoming more widely available and are still under yet further development at this time. MCT FPAs comprise a pixelated grid of perhaps 128 × 128 pixels, each pixel being of a few microns' *x–y* dimensions. Useful for fast MIR imaging (replacing the slower rastoring of MIR *x–y* mapping), each pixel can record a full MIR spectrum averaged over its *x–y* plane area (Fig. 8.18).

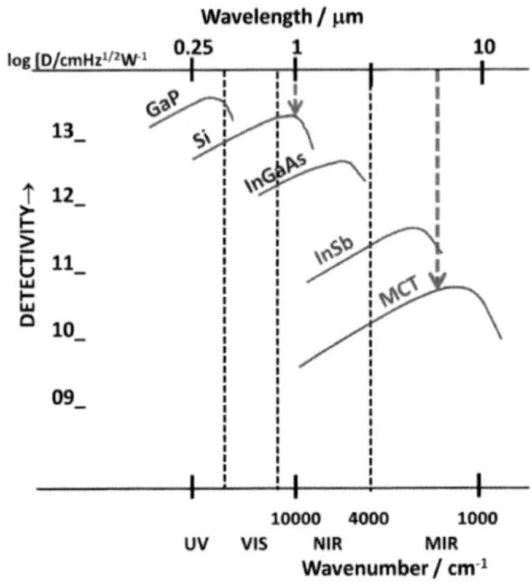

Figure 8.15 Comparison of the detectivity of common detectors used in the UV (ultraviolet), VIS (visible), NIR (near-infrared) and MIR (mid-infrared) regions of the electromagnetic spectrum, highlighting the relative weakness of the MIR MCT detector. Adapted from *Bruker Handbook*.

8.6.2 MIR Light Sources: Traditional, Emerging and New

The traditional MIR broadband sources, which are in common use, are still based on heated ceramic filaments, reminiscent of the use of candles for the main means of visible illumination in the 18th century. The Globar© (or Nernst filament) blackbody source, which is ubiquitous throughout infrared spectroscopy and readily available, is a weak light source. Max Diem, as a leader in the field of biomedical spectroscopy, remarked to the author: "*I've been fighting against the lack of mid-infrared brightness all my working life.*"

The brightness of a light source can be defined in terms of power per unit area and unit angle; spectral brightness is that portion of brightness lying within a defined spectral bandwidth (see Fig. 8.16).

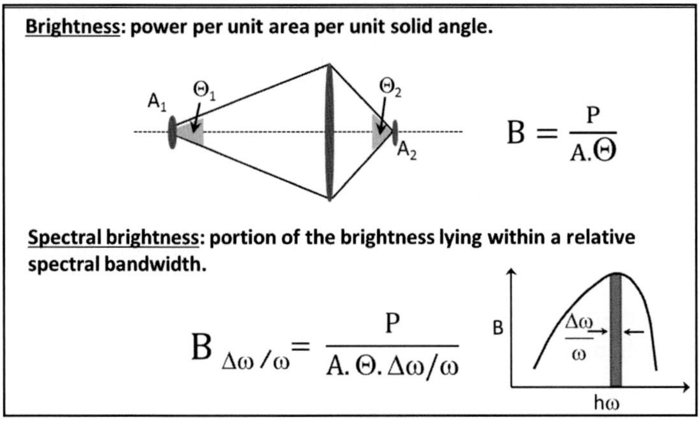

Figure 8.16 Brightness of a light source is defined here as power per unit area and unit solid angle; spectral brightness is that portion of brightness lying within a defined spectral bandwidth. Adapted from Diamond-Light Synchrotron, UK, Web site 2014.

To date, due to the lack of brightness of the common blackbody MIR wideband sources, MIR light molecular spectral imaging of tissue has been forced to be carried out on the bench-top, with source/analyte/detector in close proximity in order to have enough MIR light photons to complete the optical circuit (Fig. 8.17, upper scheme).

MIR microscopy is a popular adjunct to MIR FT bench-top spectrometers to achieve so-called FT-MIR micro-spectroscopy. This approach, at high magnification, can achieve MIR high-resolution, spectral-mapping of excised biologic tissue (see Fig. 8.14, [58]), exfoliated cells and, more recently, live cells in vitro [60]. FT-MIR micro-spectroscopy imaging is made possible by using a MCT FPA detector in the FT-MIR micro-spectroscopy setup (see Fig. 8.18).

FT-MIR microspectroscopy spectral imaging is recognized as non-destructive, label-free and highly sensitive and is already widely applied in biomedical research and in particular for cancer research and diagnosis on excised tissue.

The MIR diffraction-limited spatial resolution, *d*, according to Abbe [61]:

$$d = \lambda/2\text{NA}, \tag{8.4}$$

where λ is the wavelength of operation and NA is the numerical aperture in the microscope, assuming at most an NA ~1 then the resolution is at best ~1.5–6 μm for the MIR wavelength range ~3–12 μm. When using a Globar© blackbody-type source, the lack of brightness limits resolution further, but over-sampling and data-processing can help mitigate this problem. In summary, to date the lack of brightness of MIR wideband blackbody light sources has restricted the application of MIR light molecular spectral mapping and imaging to be carried out on the bench-top and only on excised biologic tissue (Fig. 8.17, upper scheme).

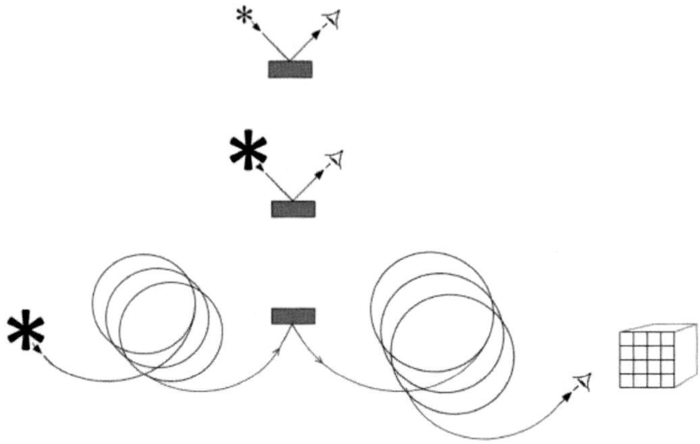

Figure 8.17 Demonstrating the need for bright, wide band, MIR light sources, and MIR fiber routing, to achieve the in vivo MIR optical biopsy. *Upper scheme*: current common setup for bench-top MIR spectroscopy based on a weak, wideband MIR blackbody source (Globar©, Nernst filament type blackbody) and weak MCT detection in close proximity to the analyte and to each other; used to analyze *excised* tissue. *Middle scheme*: a bright, wideband MIR synchrotron source: but this is non-portable and, perforce, *excised* tissue must be brought to it. *Lower scheme*: a future view of MIR spectroscopic molecular imaging, *e.g. for use in the medical clinic* using new bright portable wideband sources (under development, e.g. MIR optical fiber supercontinuum generation; arrayed quantum cascade lasers), 2-D pixel arrayed focal plane detection (becoming more commonly available), with 3-D depth imaging (future) using MIR optical fiber (already available as low loss) for routing the MIR light to where it is needed.

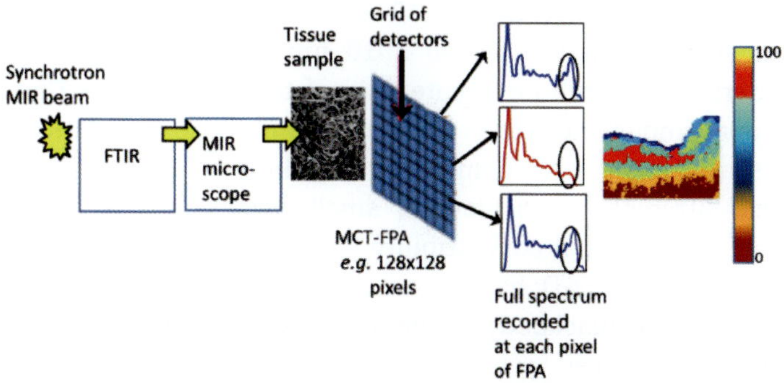

Figure 8.18 Bench-top setup, and systematic methodology, for FT-MIR spectral imaging of excised biologic tissue as currently carried out. At the moment, MIR blackbody source light, or a synchrotron-generated MIR beam (depicted here), is passed through a FT-MIR spectrometer. A MCT focal plane array (FPA) detector captures spectral images of excised biologic tissue: a full spectrum is recorded at each pixel. Spectral sets acquired are treated statistically to yield molecular discrimination across the MIR tissue image.

Synchrotron-generated MIR light, such as the B22 beamline at the Diamond Light Synchrotron (Oxfordshire, England) exhibits high brightness for molecular identification by MIR spectroscopy. Figure 8.19(a) demonstrates that the spectral brightness across the MIR spectrum of a synchrotron-generated beam can be two to three orders of magnitude greater than a blackbody type MIR source. The synchrotron-generated MIR beam is also used for FT-MIR microspectroscopy and MIR spectral imaging with a MCT FPA (Fig. 8.18); however, synchrotron beam-time is necessarily limited in user-access. Synchrotron MIR radiation high brightness sources are providing improved data quality, higher signal-to-noise and shorter acquisition times [62]. Despite those advantages, analytes must, perforce, be brought to the synchrotron. In summary, the non-portability of synchrotron-generated wideband MIR light has restricted the application of MIR light molecular spectral mapping and imaging to date to be carried out only on excised biological tissue brought to the synchrotron (Fig. 8.17, middle scheme).

It is important to note that a synchrotron-generated MIR beam is a parallel beam of a few mm diameter and therefore is capable of being focused down to a small aperture, unlike the 3-D radiant blackbody MIR sources. From Fig. 19(b), the synchrotron beam focused to a 10 μm aperture achieves over 3 orders of magnitude greater signal-to-noise than a Globar© [63] and, importantly, then becomes compatible with MIR fiber-optics and capable of being launched into a MIR optical fiber source. Attempts to achieve a small bright aperture from a Globar© for potential compatibility with fiber-optics (e.g. abutting a diamond rod [64]) have proved futile, due to the associated étendue of a blackbody radiant source emitting over three dimensions in space.

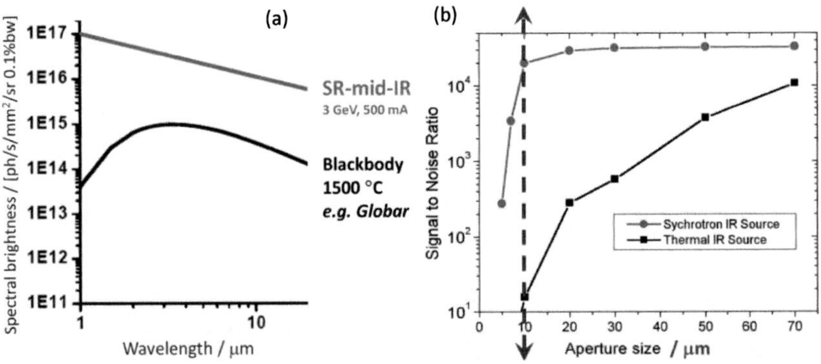

Figure 8.19 (a) Synchrotron radiation (SR): wide-band MIR emission is 100 to 1000 times brighter than the Globar© backbody type of blackbody source. Adapted from Diamond-Light Synchrotron, UK, Web site 2014. (b) SR MIR emission, being a pencil-beam may be focused to a small aperture and offers an advantage over the blackbody type source. Adapted from [63].

Quantum cascade lasers (QCLs) are emerging MIR lasers. They are relatively narrow band, spanning maybe ~250 nm. They are being developed for sensing of molecular gases and solids, including in stand-off format. MIR molecular spectral sensing inherently is direct, so neither labeling nor fluorescent tag are required; it is high contrast, can be quantitative and is widely applicable. Chang et al. [65] have demonstrated a lab-on-a-chip evanescent wave detection of cocaine using a germanium waveguide on silicon substrate hooked up to a QCL operating at 5.8 μm (see Fig. 8.20). However, this type of detection is non-unique (you have to know that cocaine is the only molecular species potentially present in order to confirm its presence) and to be unique requires arrayed QCLs to cover the entire MIR range and take advantage of the rich spectral absorption features which together provide a comprehensive, conclusive molecular fingerprint of cocaine. Microspectroscopy is under development using arrayed QCLs.

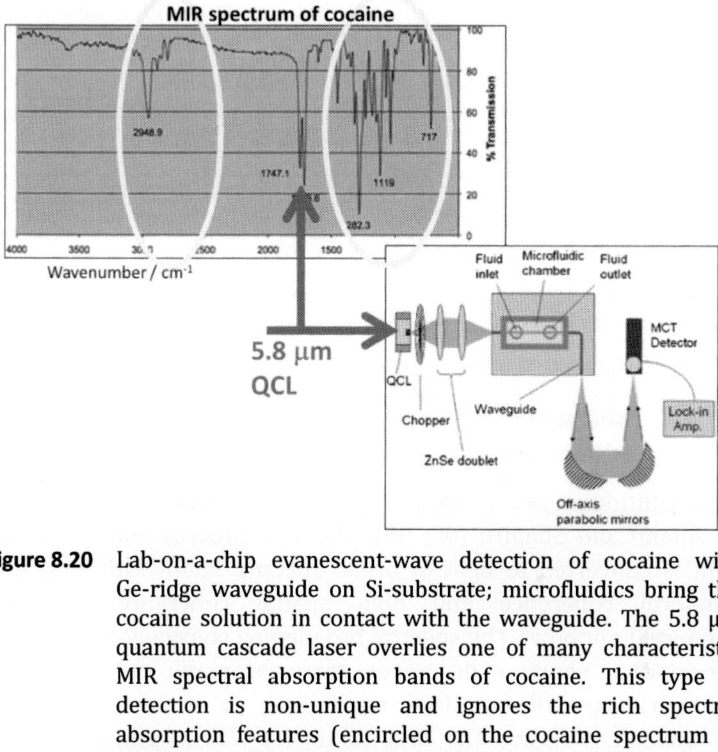

Figure 8.20 Lab-on-a-chip evanescent-wave detection of cocaine with Ge-ridge waveguide on Si-substrate; microfluidics bring the cocaine solution in contact with the waveguide. The 5.8 μm quantum cascade laser overlies one of many characteristic MIR spectral absorption bands of cocaine. This type of detection is non-unique and ignores the rich spectral absorption features (encircled on the cocaine spectrum in the figure) which, together, would indeed provide a unique molecular fingerprint of the cocaine. Adapted from [65].

In summary, because of the need to be arrayed to provide wideband MIR coverage, key to MIR medical spectral imaging of biologic tissue to distinguish disease, QCLs have the disadvantage of being narrowline. In contrast, MIR supercontinuum (SC) fiber lasers (see Section 8.6.3) are wideband and could cover the MIR spectral region for imaging to distinguish disease, can be at least as bright as synchrotron generated MIR light, have good beam quality, are readily integratable with fiber-optics and have already been demonstrated [66, 67]. QCLs exhibit poor beam quality are not readily integrated with MIR fiber-optics, can only be pulsed on a duty cycle and therefore are not suited for pumping fiber MIR SC generation (SCG). Narrowline direct-emission fiber lasers, on the other hand, have excellent beam quality, can be pulsed and are readily integratable with fiber-optics but have only been demonstrated at <4 μm wavelength in the MIR region. Narrowline direct-emission fiber lasers are now needed at longer wavelengths than this in order to pump MIR SCG fiber lasers to achieve a compact wideband all-fiber MIR SCG solution capable of coverage such as to service cancer diagnosis through MIR medical imaging of tissue (Fig. 8.17 lower scheme). Optical Parametric Oscillator MIR lasers, although satisfying most of the practical requirements for pumping fiber MIR SC, are bulky and costly.

8.6.3 Progress on MIR Fiber Lasers: MIR Supercontinuum Generation

8.6.3.1 MIR supercontinuum generation (SCG) wideband fiber lasers

Supercontinuum generation (SCG) in bulk glasses was first reported by Alfano and Shapiro [68, 69]. SCG is a process whereby laser light is converted to light of broad spectral bandwidth with low temporal coherence and high spatial coherence—a "laser generated rainbow". The spectral broadening is usually achieved by propagating short, high power, monochromatic optical pulses through an optically non-linear medium.

Russell [70] was the first to report SCG in silica photonic crystal fibers (PCFs). Silica glass PCF for VIS and NIR SCG sources are now commercially available and can exhibit of the order 10^6 × the brightness of an incandescent visible source (Fig. 8.21). The physical processes behind fiber SCG depend on pump pulse duration, peak power and repetition rate, pump wavelength, fiber chromatic dispersion and, usually, fiber length. For fs pulses, spectral broadening can be dominantly caused by self-phase modulation (SPM). In the anomalous dispersion regime, the combination of self-phase modulation and dispersion can lead to complicated soliton dynamics, including the splitting-up of higher-order solitons into multiple fundamental solitons (soliton fission). On pumping SC with ps, ns pulses, or even with CW (continuous wave) multiwatt radiation, then Raman scattering and four-wave mixing can be important. SC sources for the MWIR 3–5 μm wavelength range are, right now, an emerging technology. Figure 8.21 shows an early result in pumping a chalcogenide sulfide glass optical fiber with a fiber seed laser of wavelength 2.5 μm: the MIR SC is shown curtailed at ~5 μm wavelength (see Table 8.2) due, the Authors said, to the innate multiphonon edge absorption of the sulfide-based glass fiber used [71].

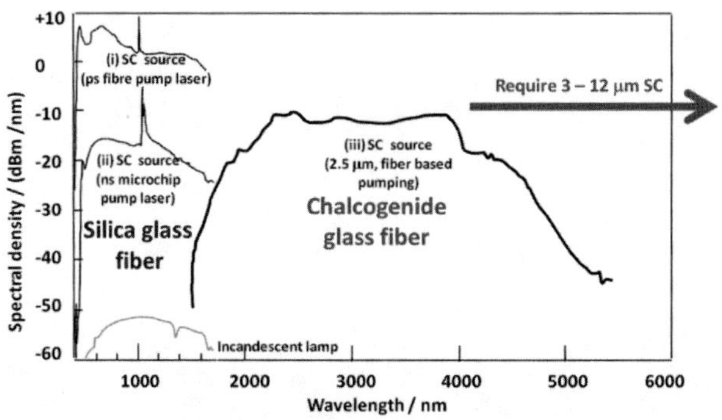

Figure 8.21 Supercontinuum (SC) generated broadband light in the near-infrared (NIR) and mid-infrared (MIR) reported prior to 2014 replotted for comparison of brightness and output spectral range of various broadband light sources. Plots (i) and (ii) are based on silica-glass fiber in the visible and NIR spectral regions and are adapted from [70]. Plot (iii) is based on chalcogenide glass fiber in the MIR spectral region and adapted from [71]. The incandescent lamp plot is adapted from [70]. MIR SCG spanning 3 to 12 μm is required in order to be of use in collecting MIR spectra of tissue for disease diagnosis—as indicated by the horizontal arrow. (Adapted from [17]; see Table 8.2.)

MIR transparent chalcogenide glasses exhibit very high Kerr coefficients and optical nonlinearities up to *circa* 1000× that of silica glass [39], suggesting the possibility of short MIR SCG fiber and waveguide devices. As depicted in Fig. 8.21, MIR SC covering the whole span of 3 to ≥10 μm wavelength are ideally required for collecting MIR spectra of tissue for disease diagnosis because of the subtle nature of the spectral changes in diseased tissue with respect to normal tissue, and the need to collect as wide-bandwidth MIR spectral signature as possible to maximize the sensitivity and specificity of the medical diagnosis.

Table 8.2 MIR SCG chalcogenide fiber sources reported, arranged in reverse order output wavelength-span

MIR SCG fiber design		• SIF, NA = 1.0 • core/clad. = • AsSe/GeAsSe • core: $\varphi \sim 16$ µm • 85 mm long		• SIF, NA = 1.3 • core/clad. = • GeAsSe/GeAsS • core: $\varphi = 4.5$µm • 110 mm long	• “holey” MOF • air/AsSe • core: $\varphi = 4.5$µm • 180 mm long	• SIF, NA = 0.3 • core/cladding = • “AsS” • core: $\varphi = 10$ µm • 2 m long		• “holey” MOF • air/AsSe • core: $\varphi = 12$ µm • 2 m long
MIR SC wave-length span/µm		**1.4 to 13.3**	**1.5 to 11.7**	**1.8 to 10**	**1.7 to 7.5**	**1.65 to 4.78 #**	**1.9 to 4.8 #**	**3.65 to 4.9 #**
MIR SC average power out/W		**“few” 100 µW**		**“few” mW**	**15.6 mW**	**140 mW**	**565 mW**	**5 mW**
Pump laser	**Type**	OPA		OPA	OPA	Raman-shifted Er^{3+} SiO_2 mode-locked fiber		Nd-YAG micro-chip pumped PPLN
	Center wave-length/µm	6.3	4.5	4.0	4.4	2.5	2.45	3.82
	Pulse width/fs	100		330 ± 15	320	—	—	—
	Pulse rate/kHz	1		21,000	21,000	560	10,000	20
	Peak power/kW	2290	1250	~3	5.2	>10	>3.5	>2
	Av. pump power	~760 µW	~350 µW	40 mW	100 mW	—	1.4 W	30 mW
Year [reference]		2014 [66]		2015 [67]	2015 [80]	2011 [71]	2012 [81]	2014 [82]

Key: Av. = average; clad. = cladding; MOF = microstructured optical fiber; OPA = optical parametric amplifier; Ref. = reference; φ = diameter; (#) = –20 dB point; — not stated.

The idea was first mooted in 2011 [17] to pump at a longer wavelength in a lower phonon energy chalcogenide glass fiber than [71] (Fig. 8.21) in order to try to get out to an ideal 3 to ≥10 μm MIR SCG for application in in vivo MIR spectral imaging of tissue. In 2013, numerical modeling, led by Ole Bang of DTU (Denmark), was the first to show the feasibility of generating a 3–12 μm fiber SC to span the MIR absorption bands of key biomolecules (Fig. 8.22) [72].

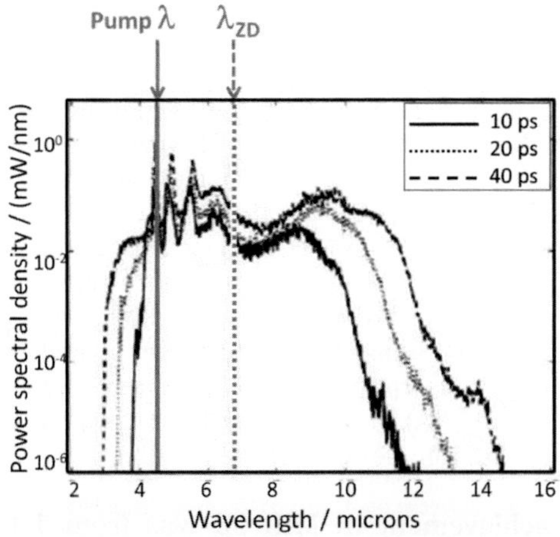

Figure 8.22 Numerical demonstration of 3–12 μm SCG in large-core step-index chalcogenide fibers pumped at 4.5 μm. Adapted from [72].

Refinement of the modeling data [73] was made possible by having to hand: characterized, specially designed, and engineered MIR fiber of exceptionally high NA (numerical aperture) around 1, which was dispersion optimized and whose loss had been measured (supposed <2.5 dB/m across 3.3–9.4 μm wavelength region) [74] (see Fig. 8.23). The modeling assumed pumping the fiber MIR SC with a mode-locked rare earth ion doped chalcogenide glass fiber laser with emission at 4.5 μm wavelength (not experimentally developed at this time), and with a repetition rate of 4 MHz and 50 ps long pulses, having a peak power of 4.7 kW. Fibers of 8 and 10 μm core diameters in the numerical modeling generated a MIR SC out to 12.5 and 10.7 μm, respectively, in less than 2 m of fiber when pumped with 0.75 and 1 kW, respectively, and power converted into the 8–10 μm band was 7.5 and 8.8 mW, respectively.

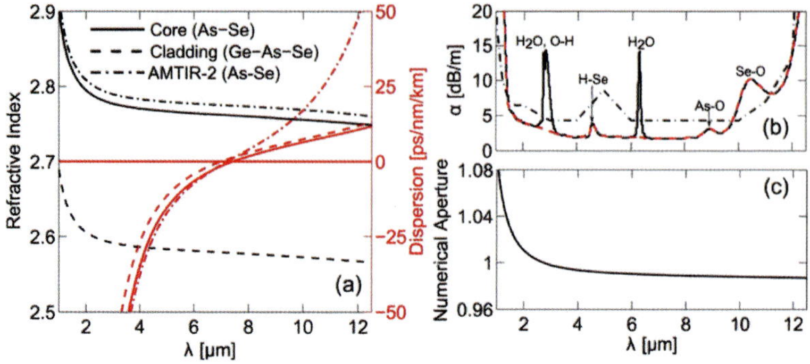

Figure 8.23 Optical properties of Se chalcogenide glasses, and fiber, used for numerical modeling and experimental measurement of fiber MIR SCG of record span (see Fig. 8.25) [66]. (a) Refractive index (black) and dispersion (red) of bulk As-Se (solid curves) and bulk Ge-As-Se (dashed curves) [74] and bulk AMTIR-2 glass (As-Se dash-dotted) [79]. (b) Large core fiber loss with absorption bands indicated (solid black), and idealized loss without O-H and H_2O bands at 2.9 and 6.3 μm (dashed red curve) as well as AMTIR-2 bulk material absorption (dotted black curve) [79]. (c) Numerical aperture (NA) of a SIF (step-index fiber) with As-Se as the core glass and Ge-As-Se as the cladding glass [74].

A major achievement in MIR SC was from 1.4 to 13.3 μm spectral range in fiber in [66]. This was the first experimental demonstration truly to reveal the potential of MIR fibers to emit across the MIR molecular "fingerprint region" and a key first step towards portable, broadband MIR sources for in vivo cancer diagnosis in a clinical setting, as highlighted in [17, 66, 75]. In detail: it was demonstrated experimentally that launching intense, ultra-short fs pulses, with a central wavelength of either 4.5 or 6.3 μm, into short pieces of ultra-high NA (numerical-aperture), (all-solid) chalcogenide glass, step-index-fiber (SIF) generated record MIR SCG, spanning 1.5 to 11.7 μm and 1.4 to 13.3 μm, respectively (Fig. 8.25(a),(b), respectively, and Table 8.2). Output MIR SCG power, for fs pumping of the fiber at a few pW average power, was a few 100s μW. Corkum had previously recorded in 1985 a 3–14 μm MIR SC span for ps pumping of semiconductors using a SHG (second harmonic generation) CO_2 laser 5 μm pump [76].

Seeding of the fiber record MIR SC was with a non-collinear and difference frequency generation unit pumped by an optical parametric amplifier (Fig. 8.24), rather than seeding with the MIR fiber laser pump as in the modeling. This is because, to date, such a fiber laser pump has not yet been realized [77, 78]. As pointed out recently [67], an important practical factor is that the power required to generate the MIR SC should be relatively low (a few kW) and compact, high-repetition-rate (20–100 MHz) pump sources with 10s or 100s mW of average power should be instead employed. Importantly also, this should give compatibility of the fiber MIR SCS (SC source) with FT-IR spectrometer operation, whose more slowly moving interferometer would not then "see" the time-constant of the pump and MIR SC emission. Towards this aim, recent work pumping a 110 mm high NA chalcogenide SIF with a relatively low peak pulse power of ~3000 W and 330 fs duration pulses, gave rise to a MIR SCG spanning ~1.8–10 μm (Table 8.2). The pump was a customized fs OPA. Importantly, this allowed a MIR SC of a few mW average power to be generated of an output power, as the authors stated: "similar output power to that of a typical MIR beamline of a synchrotron". This recent work represents significant steps now taken towards achieving the in vivo MIR optical biopsy because bright, wideband fiber-compatible MIR SCSs [66, 67] and low loss MIR routing fiber [35 and references therein] have been demonstrated. Table 8.2 summarizes the spectral range of MIR SCG chalcogenide fiber sources to date, with details of average output power and pump arrangement.

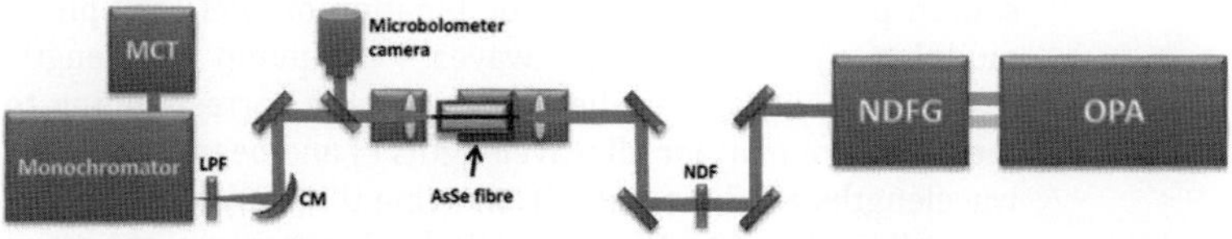

Figure 8.24 Setup to achieve record MIR SCG [66]: A non-collinear difference frequency generation (NDFG) unit pumped by an optical parametric amplifier (OPA) was used to produce the MIR pump. The output was free-space-coupled into the chalcogenide selenide fiber and subsequently collimated by aspheric lenses. A concave mirror was placed before the monochromator to prevent beam-clipping and compensate for chromatic aberration. Proper coupling to the core was verified by near-field imaging using a micro-bolometer camera. BD-2, black-diamond-2 aspheric lenses; NDF, neutral density filter; CM, concave mirror; LPF, long-pass filter.

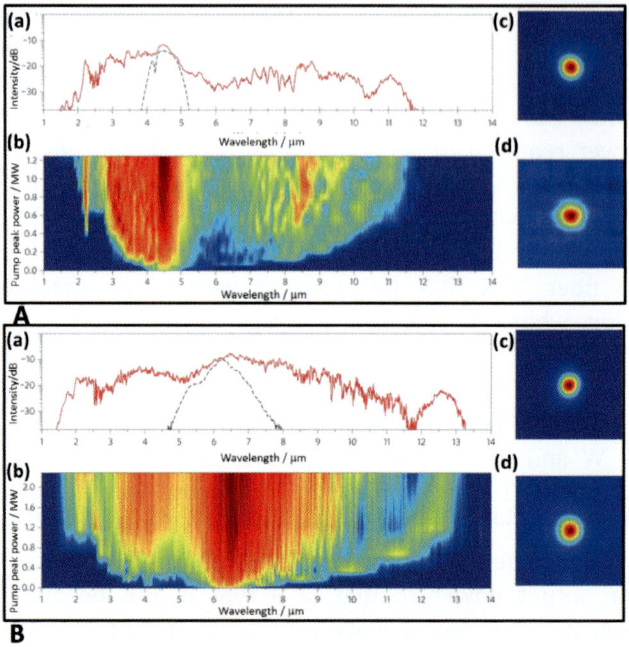

Figure 8.25 (A) Details of record MIR SCG [66]: (a) Input pump spectrum (dashed line) and spectral profile at maximum pump power (solid line), showing relatively flat SC (2.08–10.29 μm @ –20 dB from signal peak) with distinct soliton peaks > ZDW (zero dispersion wavelength) of ~5.83 μm, especially at 11 μm. (b) Spectral evolution with increasing pump peak power, showing gradual redshift of distinct soliton peaks > ZDW, and a combination of SPM (self-phase modulation) and dispersive waves < the pump wavelength. (c,d) Fiber output near-field beam profile corresponding to the spectrum in a) for all wavelengths c) and beam profile for wavelengths > 7.3 μm only d), showing that long wavelengths are still confined to the core. (B) (a) Input pump spectrum (dashed line) and spectral profile at maximum pump power (solid line), showing a broad, flat SC (1.64–11.38 μm @ –20 dB) followed by a strong spectral peak extending the spectrum all the way to 13.3 μm. (b) Spectral evolution with increasing pump peak power, showing gradual redshift of a distinct spectral peak at the long-wavelength edge and corresponding formation and blueshift of dispersive waves. (c,d) Fiber output near-field beam profile corresponding to spectrum in (a) for all wavelengths (c) and beam profile for wavelengths >7.3 μm only (d), showing that long wavelengths are still confined to the core.

8.6.3.2 MIR Narrowline Direct-Emission Fiber Lasers

MIR narrowline direct-emission fiber lasers based on rare earth (RE)-ion doped MIR-transmitting glass-hosts are of interest for cutting/welding soft materials, such as polymers, and also biologic tissue—potentially offering new wavelengths for medical laser surgery. Such discrete fiber lasers are suitable for coherent imaging (see Section 8.5.3). Also, as the MIR spectral region encompasses the fundamental vibrational absorptions of molecular materials, so a MIR fiber laser tuned to a specific molecular vibrational absorption has the potential to act as a direct molecular sensor (with the caveats of specificity as expressed earlier for QCLs, see Section 8.6.2). The large extinction coefficients of fundamental vibrational absorption bands in the MIR, mean such molecular sensors will lead to good contrast, and so good sensitivity, and could be quantitative. Importantly, MIR fiber direct emission narrowline lasers are required for pumping MIR fiber supercontinuum generation for an all-fiber compact, wideband, MIR SC molecular sensing with broad applicability across many sectors including for in vivo early cancer diagnosis [17].

However, no MIR narrowline fiber laser has yet been demonstrated at ≥4 μm wavelength [77, 78]; chalcogenide glasses are attractive hosts for RE-ions, which provide the MIR emission. The chalcogenide glasses have low phonon energies, giving transparency at both pump and emission MIR wavelengths, and show sufficient solubility for RE-ions. In 1996 a Nd^{3+}-ion doped *oxy*-chalcogenide glass Ga:La:S:O fiber was demonstrated to lase in the NIR (1.08 μm) [83]. But the high phonon energy of the *oxy*-chalcogenide host limited lasing to the NIR [77]. In contrast, RE-ion doped chalcogenide glasses are considered one of the most promising materials for MIR direct-emission narrowline fiber laser sources due to their very low phonon energy. Building on the comprehensive studies of rare earth ion emission in chalcogenide bulk glasses and/or fiber of the groups of Sanghera [85], Heo [86, 87], Tanabe [88], Adam [89], Frumar [90] and Aitken and Quimby et al. [91, 92] and also the modeling work of Prudenza et al. [93], Hu et al. [94], Quimby and Shaw [95] and Sujecki et al. [96], recently, for the first time, Pr^{3+} PL in selenide chalcogenide glass SIF of small core diameter has been reported [78, see Fig. 8.26]. Encouragingly we have found that the long lifetimes of several ms measured in RE-ion doped chalcogenide bulk glasses are maintained in small-core fiber [78], demonstrating that the RE ion dopant local environment survived the thermal cycling required to make fiber and glass devitrification did not occur. However, it is thought that a large population of the excited ions in chalcogenide bulk glass and fiber undergo a fast decay due to an additional mechanism of non-radiative depopulation of the excited state in the MIR that has not been properly recognized before: that of impurity multiphoton relaxation and which is suggested to be a key player in the non-radiative decay of MIR excited states, perhaps responsible for precluding MIR fiber lasing to date [77].

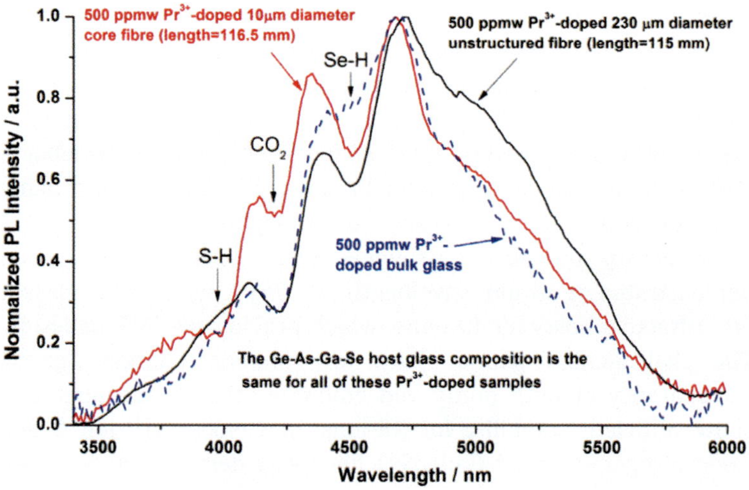

Figure 8.26 MIR PL spectra of a 500 ppmw Pr^{3+}-doped 10 μm diameter core fiber (red continuous solid curve), a 500 ppmw Pr^{3+}-doped 230 μm diameter unclad/intermediate fiber (black continuous solid curve) and a 500 ppmw Pr^{3+}-doped bulk glass (dashed curve; the bulk datum was taken from [97]). 50.2 mW pump power at 1550 nm wavelength was applied in the collection of these spectra [78].

8.7 Summary and Future Prospects

This chapter has attempted to convey the impressive strides seen in MIR spectral molecular imaging of excised tissue of many types ranging from external (skin) tissue to internal (deep) tissue; only a small fraction of the published work in this area has been referred to. It has been analyzed why it is that Raman spectral molecular imaging (sister technique of MIR spectral imaging) has made such good recent progress in in vivo application and that the only issues precluding similar application of MIR spectral imaging in vivo, and the MIR optical biopsy, have been the lack to date of bright MIR sources and that readily available silica optical fiber cannot rout MIR light. The solution proposed is chalcogenide glasses for passive routing and achieving portable, bright MIR fiber sources: both SC and narrowline MIR fiber pump lasers.

The recent demonstration of very wideband MIR SC [66] of high spatial coherence at low peak pulse pump power in optically non-linear chalcogenide glass optical fiber is braced to impact biomedical MIR spectral imaging. This represents a first important step towards portable, bright and wideband MIR-SC sources suitable for MIR spectral imaging in vivo, in real-time, in the clinic.

Add to this the future development of rare earth ion doped chalcogenide glass, narrowline, direct-emission fiber lasers to pump MIR-SC and we have a potentially elegant, compact, all-fiber solution to satisfy the need for bright mid-infrared broadband sources to accomplish in vivo sensing of tissue through remote mid-infrared spectroscopy of vibrational molecular signatures of tissue.

The MIR chalcogenide fiber carrying the light to/from tissue in vivo should ideally sense the tissue in reflection, to avoid direct contact of the chalcogenide glass material with the human body, while obviating limitations associated with evanescent field spectroscopy (spectral depth acquisition varies with wavelength) and the inability to carry out transmission spectroscopy; transflection could be possible in vivo where the underlying tissue is stratified. Fully encased, compact, antireflection-coated, or -patterned, end-fiber-faced, chalcogenide fiber systems equipped with mid-infrared-transmitting non-toxic diamond or fluoride windows are envisaged, which could be contacting or non-contacting with tissue, as required. A probe head based on reflectance could take the form of a coherent chalcogenide glass fiber bundle working at the diffraction limit for cellular resolution, if light sources bright enough can be fabricated. The preferred detecting system is a MCT FPA for imaging. Add to this, the utility of new rare-earth-doped chalcogenide glass fiber lasers for fiber surgery at new wavelengths resonant with particular tissue molecular absorption and also for coherent imaging using a mid-infrared fiber laser at more modest power. This suggests a complete MIR fiber optic endoscopic system for real-time diagnosis, medical decision-taking, and treatment planning, and enactment, of deep tissue cancers. To repeat from earlier, a remote mid-infrared probe guided by the surgeon's hand carrying out imaging and spectral mapping in situ during surgery to locate malignant tissue leading to its precise excision at its extreme borders to cellular tolerances, to preserve anatomic features as far as possible, in one go during the surgery would cut medical consultant time and unnecessary patient suffering.

This chapter has described future in vivo mid-infrared medical surface probes and endoscopy and is also a vehicle to showcase the potential of chalcogenide glass fiber optic systems for mid-infrared remote spectroscopic mapping, coherent imaging and laser processing with potential for application in other sectors, like manufacturing, chemical processing, energy production/use, security, "green" environmental monitoring, food production, pharmaceuticals, and so on to give real-time feedback on molecular signatures and morphology from MIR imaging to inform decision-taking.

Thus, MIR fiber photonics is a disruptive technology for real-time molecular sensing in monitoring environmental quality (sensing pollutants, e.g. exhaust gases from industrial plant, cars, jet-engines), energy-efficiency of fossil fuels (CO/CO_2 balance), security (e.g. in stand-off detection of narcotics, explosives) and food (authenticity and adulteration) and drink security (e.g. beer quality!), in agriculture (e.g. sensing ethylene gas in tomato ripening; monitoring nutrients and leaf-hydration in vertical-farming), aquaculture (fish-farming: sensing nutrients and bacterial contamination) and real-time monitoring of manufacturing and chemical processing (e.g. distributed fiber-optic MIR sensing enabling process control), including of oil fractionates/products; machining and welding of soft materials and MIR "radar" such as for collision avoidance. So, in many walks of life the application of MIR spectral sensing could enable greater control to be had in real-time over a diversity of processes. MIR laser machining of soft materials is enabled, including of polymeric materials and biologic tissue, analogous (power-scaled to requirements) to the more established technologies of NIR kW fiber laser machining and welding of hard materials.

Some of the specific developments in chalcogenide glasses now needed to realize mid-infrared photonic systems with such capability are as follows. First, to continue to develop MIR supercontinuum broadband efficient sources of high output power, to cover the molecular "fingerprint" spectral region. Also, the continued development of intensive purification of suitable chalcogenide glasses is required to give clean, low loss transparent windows across the mid-infrared. For new MIR fiber laser narrowline sources, a better understanding is required of the active dopant local environment to minimize unwanted non-radiative decay. Lower optical loss fiber of the complex chalcogenide glass compositions to solubilize the active dopants is required; understanding of glass stability to devitrification has been reported [98, 99]. Better understanding of any impact of host chalcogenide glass native defects on the dopant fluorescence is needed. Finally, the development of chalcogenide glass photonic systems in the future must go hand-in-hand with the end-user to tailor mid-infrared photonic systems fit for purpose. Continued development of chemometrics to handle large MIR hyperspectral data sets is required. There is a need to continue to refine stratification of MIR biomedical spectra of tissue with sophisticated application of confounding factors, e.g. DNA genome data.

Finally, to realize MIR sensing and imaging, including for the in vivo MIR optical biopsy, then the development and deployment of chalcogenide glass MIR fiber-optic devices and systems (although not for long haul) over the next while could mirror the complexity and versatility of silica glass fiber optics developed towards the end of the 20th century for NIR telecommunications and NIR fiber lasing [100].

Acknowledgments

The author gratefully acknowledges UK Fellowships, in the application of MIR optical fibers in cancer detection, awarded from The Royal Academy of Engineering/Leverhulme Trust, UK, 2007–2008, and Medical Research Council (G0701869), UK, 2008–2009, including a short sabbatical stay at the Department of Materials Science and Engineering, University of Arizona, USA, kindly hosted by Prof. Joe Simmons and Dr. Cate Simmons and visits to Prof. Max Diem at Northeastern University, Boston, USA and Prof. H. H. Mantsch at Ottawa, Canada, and accompanying Dr. Sandeep (Sunny) Varma at Queen's Medical Centre, University of Nottingham, UK, during cancer clinics and Moh's micrographic surgery. Also acknowledged is UK funding from: NEAT (New and Emerging Applications of Technology; FSG028), 2008–09; the STFC: 2010, 2011, 2012, the Royal Society (JP100296), 2010–2012, and support from the European Commission, 2012–2016, through Framework Seven (FP7) project: MINERVA: MId- to NeaR-infrared spectroscopy for improVed medical diAgnostics (317803; www.minerva project.eu). I sincerely thank my University of Nottingham Colleagues for many enjoyable and stimulating conversations and dedicate this chapter to my Husband Martin for his patience, understanding and support.

References

1. Herschel, W. (1800) Experiments on the refrangibility of the invisible rays of the sun, *Philos. Trans. R. Soc. Lond.*, **90**, pp. 284–292.
2. Savage, J. H. (1985) *Infrared Optical Materials and their Antireflection Coatings* (Adam Hilger Ltd. Bristol and Boston) Figure 1.1. p. 2.
3. Eiseberg, D., and Kauzmann, W. (1969) *Structure and Properties of Water* (Oxford University Press, London) p. 7.
4. France, P. W., Drexhage, M. G., Carter, S. F., and Wright, J. V. (1990) *Fluoride Glass Optical Fibers* (Blackie, London) p. 58.
5. (a) See for instance: (b) Herzberg, G. (1945) *Infrared and Raman Spectra of Polyatomic Molecules* (D Van Nostrand, New York).
6. Diem, M., Boydston-White, S., and Chiriboga, L. (1999) Infrared spectroscopy of cells and tissues: Shining light onto a novel subject, *Appl. Spectrosc.*, **53**(4), pp. 148A–161A.
7. Kretlow, A., Kneipp, J., Lasch, P., Beekes, M., Miller, L., and Naumann, D. (2011) *Biomedical Applications of Synchrotron Infrared Micro-Spectroscopy: A Practical Approach*, Moss, D., ed., RSC Analytical Spectroscopy Monographs, Compiled from Chapter 11.
8. Lucassen, G. W., van Veen, G. N., A., and Jansen, J. A. (1998) Band analysis of hydrated human skin stratum corneum attenuated total

reflectance Fourier Transform infrared spectra in vivo, *J. Biomed. Opt.*, **3**(3), pp. 267–280.

9. Yu, S., and Prendergast, F. G. (1997) Water (H_2O and D_2O) molar absorptivity in the 1000–4000 cm^{-1} range and quantitative infrared spectroscopy in aqueous solutions, *Anal. Biochem.*, **248**, pp. 234–245.
10. Alison, M., and Sarraf, C. (1997) *Understanding Cancer: From Basic Science to Clinical Practice, Postgraduate Medical Science* (Cambridge University Press, Cambridge).
11. Maddams, J., Utley, M., and Moller, H. (2012) Projections of cancer prevalence in the UK, 2010–2040, *Br. J. Cancer*, **107**, pp. 1195–1202.
12. Cancer Research UK (CRUK): A leading UK registered cancer which funds cancer research. www.cancerresearchuk.org. [12a] 'All cancers combined' Feb. 2015, CRUK.
13. Seddon, A. B. (2013) Mid-infrared (IR) – a hot topic, the potential for using mid-IR light for non-invasive, early detection of skin cancers in vivo, *Phys. Status Solidi (b)*, **250**(5), pp. 1020–1027.
14. Altman, D. G., and Bland, J. M. (1994), Diagnostic tests 1: Sensitivity and specificity, *Br. Med. J.*, **308**, p. 1552.
15. NHS England/Eastern Cancer registration and Information Centre/ Commons' Public Accounts Committee (2009)–reported in the Sunday Times, 4 May 2014.
16. Yao, K. (2013) The endoscopic diagnosis of early gastric cancer, *Ann. Gastroenterol.*, **26**(1), pp. 11–22.
17. Seddon, A. B. (2011) A prospective for new mid-infrared medical endoscopy using chalcogenide glasses, *Int. J. Appl. Glass Sci.*, **2**(3), pp. 177–191.
18. Workshadowing of Consultant Dermatologist and Surgeon Dr Sandeep Varma at the Queen's Medical Centre NHS Hospital, by Prof. AB Seddon, Faculty of Engineering, University of Nottingham, UK, under the auspices of National Health Service permission.
19. Kanter, E. M., Keller, M. D., and Mahadevan-Jansen, A. (2006) Raman spectroscopy for cancer diagnosis, *Spectroscopy*, **21**(9), pp. 33–38.
20. Raman, V. (1930) The molecular scattering of light, *Nobel Lecture*, 11 Dec.
21. Alfano, R. R., Liu, C. H., Sha, W. L., Zhu, H. R., Akins, D. L., Cleary, J., Prudente, R., and Cellmer, E. (1991) Human breast tissues studied by IR Fourier transform Raman spectroscopy, *Lasers Life Sci.*, **4**, p. 23.

22. Alfano, R. R., Liu, C. H., and Glassman, W. S. (1993) Method for determining if a tissue is a malignant tissue, a benign tissue or a normal or benign tissue using Raman spectroscopy, U. S., Patent 5261410.

23. Haka, A. S., Shafer-Peltier, K. E., Fitzmaurice, M., Crowe, J., Dasari, R. R., and Feld, M. S. (2005) Diagnosing breast cancer by using Raman spectroscopy, *Proc. Natl. Acad. Sci.*, **102**(35), pp. 12371–12376.

24. Manoharan, R., Shafer, K., Perelman, L., Wu, J., Chen, K., Deinum, G., Fitzmaurice, M., Myles, J., Crowe, J., Dasari, R., and Feld, M. (1998) Raman spectroscopy and fluorescence photon migration for breast cancer diagnosis and imagings, *Photochem. Photobiol.*, **67**(1), pp. 15–22.

25. Krafft, C., and Sergo, V., Biomedical applications of Raman and infrared spectroscopy to diagnose tissues (2006) *Spectroscopy*, **20**, pp. 195–218.

26. Kong, K., Kendall, C., Stone, N., and Notingher, I. (2015) Raman spectroscopy for medical diagnostics—from in vitro biofluid assays to in vivo cancer detection, *Adv. Drug Deliv. Rev.*, **89**, pp. 121–134.

27. Salzer, R., and Siesler, H. W. (2014) *Infrared and Raman Spectroscopic Imaging*, 2nd ed. (Wiley-VCH, Wenheim, Germany).

28. (a) Diem, M., Griffiths, P. R., and Chalmers, J. M. (2008) *Vibrational Spectroscopy for Medical Diagnosis* (Wiley, Chichester, England). (b) Diem, M. (2015) *Modern Vibrational Spectroscopy and Micro-Spectroscopy* (John Wiley Sons Ltd.) published on-line June 2015, DOI: 10.1002/9781118824924.

29. Shafer-Peltier, K. E., Haka, A. S., Fitzmaurice, M., Crowe, J., Dasari, R. R., and Feld, M. S. (2002) Raman microspectroscopic model of human breast tissue: Implications for breast cancer diagnosis in vivo, *J. Raman. Spectrosc.*, **33**, pp. 552–563.

30. Zhou, Y., Liu, C. H., Zhou, L., Zhu, Z., Liu., Y., Zhang, L., Boydston-White, S., Cheng, G., Bidyut, D., Pu, Y., and Alfano, R. R. (2015) Resonant Raman spectra of grades of human glioma tumors reveal the content of tryptophan by the 1588 cm^{-1} mode, *Proc. SPIE*, **9318**, pp. 931810-1-7.

31. Kanter, E. M., Vargis, E., Majumder, S., Keller, M. D., Woeste, E., Rao, G. G., and Mahadevan-Jansen, A. (2009) Application of Raman spectroscopy for cervical dysplasia diagnosis, *J. Biophoton.*, **2**(1–2), pp. 81–90.

32. Shephard, J., MacPherson, W., Maier, R., Jones, J., Hand, D., Mohebbi, M., George, A., Roberts, P., and Knight, J. (2005) Single-mode mid-IR

guidance in a hollow-core photonic crystal fiber, *Opt. Express*, **13**(18), pp. 7139–7144.

33. Sanghera, J. S., Nguyen, V. Q., Pureza, P. C., Milos, R. E., Kung, F. H., and Aggarwal, I. D. (1996) Fabrication of long lengths of low loss transmitting $As_{40}S_{60\text{-}x}Se_x$ glass fibers, *J. Lightwave Technol.*, **14**(5), pp. 743–748.

34. Snopatin, G. E., Shiryaev, V. S., Plotnichenko, V. G., Dianov, E. M., and Churbanov, M. F. (2009) High purity chalcogenide glasses for fiber optics, *Inorg. Mater.*, **45**(13), pp. 1439–1460.

35. Tang, Z., Shiryaev, V. S., Furniss, D., Sojka, L., Sujecki, S., Benson, T. M., Seddon, A. B., and Churbanov, M. F. (2015) Low loss Ge-As-Se chalcogenide glass fiber, fabricated using extruded preform, for mid-infrared photonics, *Opt. Mater. Express*, **5**(8), pp. 1722–1737.

36. Seddon, A. B. (1995) Chalcogenide glasses, their preparation, properties and applications, *J. Non-Cryst. Solids*, **184**, pp. 44–50.

37. Zachery, A., and Elliott, S. R. (2003) Optical properties and applications of chalcogenide glasses, *J. Non-Cryst. Solids*, **330**, pp. 1–12.

38. Sanghera, J. S., Shaw, L. B., and Aggarwal, I. D. (2009) Chalcogenide glass-fiber-based mid-IR sources and applications, *IEEE J. Sel. Top. Quantum Electron.*, **15**(1), pp. 114–119.

39. Zackery, A., and Elliott, S. R. (2007) *Optical Nonlinearities in Chalcogenide Glasses and their Applications* (Springer Series: Optical Sciences 135, New York, USA).

40. *Chalcogenide Glasses Preparation, Properties and Applications* (2013) Adam J.-L., and Zhang, X., eds. (Woodhead Publishing Series in Electronic and Optical Materials, Number 44, Elsevier).

41. Miya, T., Terunuma, Y., Hosaka, T., and Mujashita, T. (1979) Ultimate low loss single-mode fiber at 1.55 μm, *Electron. Lett.*, **15**(4), pp. 106–108.

42. Baker, M. J., Trevisan, J., Bassan, P., Bhargava, R., Butler, H. J., Dorling, K. M., Fielden, P. R., Fogarty, S. W., Fullwood, N. J., Heys, K. A., Hughs, C., Lasch, P., Martin-Hirsch, P. L., Obinaju, B., Socalingum, G. D., Sule-Suso, J., Strong, R. J., Walsh, M. J., Wood, B. R., Gardner, P., and Martin, F. L. (2014) Using Fourier transform IR spectroscopy to analyze biological materials, *Nat. Protocols*, **9**(8), pp. 1771–1791.

43. Sordillo, L., Pu, Y., Pratavieria, S., Budansky, Y., and Alfano, R. R. (1991) Deep optical imaging of tissue using the second and third near-infrared spectral windows, *J. Biomed. Opt.*, **19**(5), pp. 056004-1–056004-6.

44. Shi, L., Sordillo, L. A., Rodríguez-Contreras, A., and Alfano, R. R. (2016) Transmission in near-infrared optical windows for deep brain imaging, *J. Biophoton.*, **9**(1–2), pp. 38–43.

45. Michel, A. P. M., Liakat, S., Bors, K., and Gmachl, C. F. (2013) In vivo measurement of mid-infrared light scattering from human skin, *Biomed. Opt. Express,* **4**(4), pp. 520–530.

46. McIntosh, L. M., Jackson, M., Mantsch, H. H., Stranc, M. F., Pilavdzic, D., and Crowson, A. N. (1999) Infrared spectra of basal cell carcinomas are distinct from non-tumor-bearing skin components, *J. Invest. Dermatol.,* **112**, pp. 951–956.

47. Mostaço-Guidolin, L. B., Murakami, L. S., Nomizo, A., and Bachmann, L. (2009) Fourier transform spectroscopy of skin cancer cells and tissues, *Appl. Spect. Rev.,* **44**, pp. 438–455.

48. Hammody, Z., Argov, S., Sahu, R. K., Cargnano, E., Moreh, R., and Mordechai, S. (2008) Distinction of malignant melanoma and epidermis using IR micro-spectroscopy and statistical methods, *Analyst,* **133**(3), pp. 372–378.

49. Ly, E., Piot, O., Durlach, A., Bernard, P., and Manfait, M. (2009) Differential diagnosis of cutaneous carcinomas by infrared spectral micro-imaging combined with pattern recognition, *Analyst,* **134**, pp. 1208–1214.

50. Lasch, P., Haensch, W., Naumann, D., and Diem, M. (2004) Imaging of colorectal adenocarcinoma using FT-IR microspectroscopy and cluster analysis, *Biochim. Biophy. Acta,* **1688**, pp. 176–186.

51. Kendall, C., Isabelle, M., Bazant-Hegemark, F., Hutchings, J., Orr, L., Babrah, J., Baker, R., and Stone, N. (2009) Vibrational spectroscopy: A clinical tool for cancer diagnostics (Critical review), *Analyst,* **134**(6), pp. 1029–1045.

52. Krafft, C., Steiner, G., Beleites, C., and Salzer, R. (2009) Disease recognition by infrared and Raman spectroscopy (Review article), *J. Biophoton.,* **2**(1–2), pp. 13–28.

53. Bellisola, G., and Sorio, C. (2012) Infrared spectroscopy and microscopy in cancer research and diagnosis (Review article), *Am. J. Cancer Res.,* **2**(1), pp. 1–21.

54. World Cancer Research Fund International, http://www.wcrf.org/int/cancer-facts-figures/data-specific-cancers/cervical-cancer-statistics, consulted October 2015.

55. Schubert, J. M., Bird, B., Papamarkakis, K., Miljkovic, M., Bedrossian, K., Laver, N., and Diem, M. (2010) Spectral cytopathology of cervical samples: detecting cellular abnormalities in cytologically normal cells, *Lab. Invest.,* **90**(7), pp. 1068–1077.

56. Wood, B. R., Chiriboga, L., Yee, H., Quinn, M. A., McNaughton, D., and Diem, M. (2004) Fourier transform infrared (FTIR) spectral mapping

of the cervical transformation zone and dysplastic squamous epithelium, *Gynecol. Oncol.*, **93**(1), pp. 59–68.

57. Nallala, J., Piot, O., Diebold, M.-D., Gobinet, C., Bouché, O., Manfait, M., and Sockalingum, G. D. (2013) Infrared imaging as a cancer diagnostic tool: Introducing a new concept of spectral barcodes for identifying molecular changes in colon tumors, *Cytometry Part A*, **83A,** pp. 294–300.

58. Nallala, J., Lloyd, G. R., Shephard, N., and Stone, N. (2016) High-resolution FTIR imaging of colon tissues for elucidation of individual cellular and histopathological features, *Analyst*, **141,** pp. 630–639.

59. Guo, B., Wang, Y., Peng, B., Zhang, H. L., Luo, G. P., Le, H. Q., Gmachl, C., Sivco, D. L., Peabody, M. L., and Cho, A. Y. (2004) Laser-based mid-infrared reflectance imaging of biological tissues, *Opt. Exp.*, **12**(12), pp. 208–219.

60. Fale, P. L., Altharawi, A., and Chan, K. L. A. (2015) In situ Fourier transform infrared analysis of live cells' response to doxorubicin, *Biochim. Biophys. Acta,* **1853,** pp. 2640–2648.

61. Abbe, E. (1882) The relation of aperture and power in the microscope (continued), *J. R. Microscopical Soc.*, **2**(4), pp. 460–473.

62. Miller, L. M., and Dumas, P. (2010) From structure to cellular mechanism with infrared micro-spectroscopy, *Curr. Opin. Struct. Biol.*, **20,** pp. 649–656.

63. Holman, H.-Y. N., Martin, M. C., and McKinney, W. R. (2003) Tracking chemical changes in a live cell: Biomedical applications of SR-FTIR spectromicroscopy, *Spectroscopy,* **17**(2–3), pp. 139–159.

64. Jonas, R. E., and Braiman, M. S. (1993) Efficient source-to-fiber coupling method using a diamond rod: theory and application to multi-mode evanescent-wave IR absorption spectroscopy, *Appl. Spectrosc.*, **47**(11), pp. 1751–1759.

65. Chang, Y.-C., Wagli, P., Paeder, V., Homsy, A., Hvozdara, L., Van der Wal, P., Di Francesco, J., de Rooij, N. F., and Herzig, H. P. (2012) Cocaine detection by a mid-infrared waveguide integrated with a microfluidic chip, *Lab Chip,* **12,** pp. 3020–3023.

66. Petersen, C. R., Møller, U., Kubat, I., Zhou, B., Dupont, S., Ramsay, J., Benson, T., Sujecki, S., Abdel-Moneim, N., Tang, Z., Furniss, D., Seddon, A., and Bang, O. (2014) Mid-infrared supercontinuum covering the 1.4–13.3 μm molecular fingerprint region using ultra-high NA chalcogenide step-index fibre, *Nat. Photonics,* **8,** pp. 830–834. Highlighted in Nat. Photonics News & Views, see [75].

67. Yu, Y., Zhang, B., Gai, X., Zhai, C., Qi, S., Guo, W., Yang, Z., Wang, R., Choi, D.-Y., Madden, S., and Luther-Davies, B. (2015) 1.8–10 μm mid-infrared supercontinuum generated in a step-index chalcogenide fiber using low peak pump power, *Opt. Lett.*, **40**(6), pp. 1081–1084.

68. Alfano, R. R., and Shapiro, S. L. (1970) Emission in the region 4000 to 7000 Å via four-photon coupling in glass, *Phys. Rev. Lett.*, **24**, p. 584.

69. Alfano, R. R., and Shapiro, S. L. (1970) Observation of self-phase modulation and small-scale filaments in crystals and glasses, *Phys. Rev. Lett.*, **24**, pp. 592–594.

70. Russell, P. St. J. (2006) Photonic-crystal fibers, *J. Lightwave Technol.*, **24**(12), pp. 4729–4749.

71. Shaw, L. B., Gattass, R. R., Sanghera, J., and Aggarwal, I. (2011) All fiber mid-IR supercontinuum source from 1.5 to 5 μm, *Fibre Lasers VIII Technology Systems and Applications, Photonic West, San Francisco, CA, USA,* SPIE, pp. 7924–7924.

72. Agger, C., Kubat, I., Møller, U., Moselund, P. M., Petersen, C., Napier, B., Seddon, A., Sujecki, S., Benson, T., Farries, M., Ward, J., Lamrini, S., Scholle, K., Fuhrberg, P., and Bang, O. (2013) Numerical demonstration of 3–12 μm supercontinuum generation in large-core step-index chalcogenide fibers pumped at 4.5 μm, *Nonlinear Optics 2013, Hawaii, USA, 21–26 July 2013, ISBN: 978-1-55752-977-0, paper NW4A.09, doi:10.1364/NLO.2013.NW4A.09, Nonlinear Optics, OSA Technical Digest (online),* Optical Society of America.

73. Kubat, I., Agger, C. S., Møller, U., Seddon, A. B., Tang, Z., Sujecki, S., Benson, T. M., Furniss, D., Lamrini, S., Scholle, K., Fuhrberg, P., Napier, B., Farries, M., Ward, J., Moselund, P. M., and Bang, O. (2014) Mid-infrared supercontinuum generation to 12.5 μm in large NA chalcogenide step-index fibres pumped at 4.5 mm, *Opt. Express*, **22**(16), pp. 19169–19182. This paper was selected for *Nature Photonics Highlights*: Maragkou, M. (2014) Reaching the Mid-infrared light, *Nat. Photonics*, **8**, p. 746, doi:10.1038/nphoton.2014.231, Published online 01 October 2014.

74. H. G. Dantanarayana, H. G., Moneim. N. M., Tang, Z., Sojka, L., Sujecki, S., Furniss, D., Seddon, A. B., Kubat, I., Bang, O., and Benson, T. M. (2014) Refractive index dispersion of chalcogenide glasses for ultra-high numerical-aperture fiber for mid-infrared supercontinuum generation, *Opt. Mater. Express*, **4**, pp. 1444–1455.

75. Steinmeyer, G., and Skibina, J. S. (2014) Entering the mid-infrared: The demonstration of chalcogenide fibre-based supercontinuum sources that reach beyond a wavelength of ten micrometres is set

to have a major impact on spectroscopy and molecular sensing. Entering the mid-infrared, *Nat. Photonics*, **8**, 814–815.

76. Corkum, P. B., Ho, P. P., Alfano, R. R., and Manassah, J. T. (1985) Generation of infrared supercontinuum covering 3–14 μm in dielectrics and semiconductors, *Opt. Lett.*, **10**(12), pp. 624–626.
77. Seddon, A. B., Tang, Z., Furniss, D., Sujecki, S., and Benson, T. M. (2010) Progress in rare-earth-doped mid-infrared fiber lasers, *Opt. Express*, **18**(25), pp. 26704–26719.
78. Tang, Z., Furniss, D., Fay, M., Sakr, H., Sójka, L., Neate, N., Weston, N., Sujecki, S., Benson, T. M., and Seddon, A. B. (2015) Mid-infrared photoluminescence in small-core fiber of praseodymium-ion doped selenide-based chalcogenide glass, *Opt. Mater. Express*, **5**(4), pp. 870–886.
79. Amorphous Materials Inc., Chalcogenide glasses, available at http://www.amorphousmaterials.com/products/ (accessed 2013).
80. Møller, U., Yu, Y., Kubat, I., Petersen, C. R., Gai, X., Brilland, L., Méchin, D., Caillaud, C., Troles, J., Luther-Davies, B., and Bang, O. (2015) Multi-milliwatt mid-infrared supercontinuum generation in a suspended core chalcogenide fiber, *Opt. Express*, **23**(3), pp. 3282–3291.
81. Gattass, R. R., Shaw, B. L., Nguyen, V. Q., Pureza, P. C., Aggarwal, I. D., and Sanghera, J. S. (2012) All-fiber chalcogenide-based mid-infrared supercontinuum source, *Opt. Fiber Technol.*, **18**, pp. 345–348.
82. Gattass, R. R., Shaw, B. L., and Sanghera, J. S. (2014) Microchip laser mid-infrared supercontinuum laser source based on an As_2Se_3 fiber, *Opt. Lett.*, **39**(12), pp. 3418–3420.
83. Schweizer, T., Hewak, D. W., Jensen, T., and Huber, G. (1996) Rare earth doped chalcogenide glass laser, *Electron. Lett.*, **32**(7), pp. 666–667.
84. Jackson, S. D. (2012) Towards high-power mid-infrared emission from a fibre laser, *Nat. Photonics*, **6**(7), pp. 423–431.
85. Shaw, L. B., Cole, B., Thielen, P. A., Sanghera, J. S., and Aggarwal, I. D. (2001) Mid-wave IR and long-wave IR laser potential of rare-earth doped chalcogenide glass fiber, *IEEE J. Quant. Electron.*, **48**(9), pp. 1127–1137.
86. Shin, Y. B., and Heo, J. (1999) Mid-infrared emissions and multiphonon relaxation in Dy^{3+}-doped chalcohalide glasses, *J. Non-Cryst. Solids*, **253**, pp. 23–29.
87. Heo, J. (2007) Emission and local structure of rare-earth ions in chalcogenide glasses, *J. Non-Cryst. Solids*, **353**, pp. 1358–1363.

88. Tanabe, S. (1999) Optical transitions of rare earth ions for amplifiers: How the local structure works in glass, *J. Non-Cryst. Solids*, **259**, pp. 1–9.

89. Moizen, V., Nazabal, V., Troles, J., Houizot, P., Adam, J.-L., Doualan, J.-L., Moncorgé, R., Smektala, F., Cadret, G., Pitois, S., and Canat, G. (2008) Er^{3+}-doped GeGaSbS glasses for mid-IR fiber laser applications: Synthesis and rare earth spectroscopy, *Opt. Mater.*, **31** pp. 39–46.

90. Němec, P., and Frumar, M. (2008) Compositional dependence of spectroscopic parameters of Dy^{3+} ions in Ge-Ga-Se glasses, *Mater. Lett.*, **62** pp. 2799–2801.

91. Quimby, R. S., and Aitken, B. G. (2003) Multiphonon gap law in rare-earth doped chalcogenide glass, *J. Non-Cryst. Solids*, **320**, pp. 100–112.

92. Aitken, B. G., Ponader, C. W., and Quimby, R. S. (2002) Clustering of rare earths in GeAs sulfide glass, *C. R. Chim.*, **5**, pp. 865–872.

93. Prudenzano, F., Mescia, L., Alligretti, L. A., de Sario, M., Palmisano, T., Smektala, F., Moizan, V., Nazabal, V., and Troles, J. (2009) Design of Er^{3+}-doped chalcogenide glass laser for MID-IR application, *J. Non-Cryst. Solids* **355**, pp. 1145–1148.

94. Hu, J., Menyuk, C. R., Wei, C., Shaw, L. B., Sanghera, J. S., and Aggarwal, I. D. (2015) Highly efficient cascaded amplification using Pr^{3+}-doped mid-infrared chalcogenide fiber amplifiers, *Opt. Lett.*, **40**(16), pp. 3687–3690.

95. Quimby, R. S., Shaw, L. B., Sanghera, J. S., and Aggarwal, I. (2008) Modeling of cascade lasing in Dy: Chalcogenide glass fiber laser with efficient output at 4.5 μm, *IEEE Photonics Lett.*, **20**(2), pp. 123–125.

96. Sójka, L., Tang, Z., Zhu, H., Beres-Pawlik, E., Furniss, D., Seddon, A. B., Benson, T. M., and Sujecki, S. (2012) Study of mid-infrared laser action in chalcogenide rare earth doped glass with Dy^{3+}, Pr^{3+} and Tb^{3+}, *Opt. Mater. Express*, **2**(11), pp. 1632–1640.

97. Sakr, H., Furniss, D., Tang, Z., Sójka, L., Moneim, N., Barney, E., Sujecki, S., Benson, T. M., and Seddon, A. B. (2014) Superior photoluminescence (PL) of Pr^{3+}-In compared to Pr^{3+}-Ga selenide bulk glasses and PL of optically-clad fiber, *Opt. Express*, **22**(18), pp. 21236–21252.

98. Tang, Z., Furniss, D., Neate, N. C., Barney, E. Benson, T. M., and Seddon, A. B. (2016) Dy^{3+}-doped selenide chalcogenide glasses: Influence of Dy^{3+} dopant-additive and containment, *J. Am. Ceram. Soc.*, **99**(7), pp. 2283–2291.

99. Tang, Z., Furniss, D., Fay, M., Neate, N. C., Cheng, Y., Barney, E., Sójka, L., Sujecki, S., Benson, T. M., and Seddon, A. B. (2014) First identification of rare earth oxide nucleation in chalcogenide glasses, implications for fabrication of mid-infrared active fibers, *J. Am. Ceram. Soc.,* **97**(2), pp. 432–441.

100. Benson, T. M. (2011) University of Nottingham, UK, personal communication.

Chapter 9

Light Propagation in Turbid Tissue-Like Scattering Media

Alexander Bykov,[a] Alexander Doronin,[b] and Igor Meglinski[a]

[a]*Optoelectronics and Measurement Techniques Laboratory, University of Oulu, P.O. Box 4500, 90014 Oulu, Finland*
[b]*Computer Graphics Group, Department of Computer Science, Yale University, P.O. Box 208285, New Haven, CT, USA*

alexander.bykov@oulu.fi, alexander.doronin@yale.edu, igor.meglinski@oulu.fi

This chapter is an overview of the theories and derived models routinely employed to describe light transport in turbid tissue-like scattering media. The relative merits and drawbacks of the models are discussed, and commonly used terms are highlighted. For the main purpose of the further development of a comprehensive computational model of light-tissue interaction, we present an approach for modeling coherent polarized light propagation in turbid scattering media. The temporal coherence of light, linear and circular polarization, interference, and the helicity flip of circularly polarized light due to diffuse reflection are taken into account.

Deep Imaging in Tissue and Biomedical Materials: Using Linear and Nonlinear Optical Methods
Edited by Lingyan Shi and Robert R. Alfano

ISBN 978-981-4745-88-8 (Hardcover), 978-1-315-20655-4 (eBook)
www.panstanford.com

9.1 Introduction

Since the development of laser in 1960, optical diagnostic modalities strongly influence various scientific fields, including automotive, materials science, and biomedicine. These techniques offer a strong advantage for medical diagnostics compare to many other diagnostic approaches as the use of non-ionizing low-intensity visible and infrared optical radiation (typically, radiant power < 5 mW for exposure duration of 0.25 s, as defined by UK and US standards) is not harmful.

Conceptual design of particular optical diagnostic systems for biomedical application requires careful selection of various technical parameters, including wavelength, coherence, polarization and intensity profile of incident light, sensitivity of the detector, size, and the geometry and mutual position of source and detector, etc. In this point of view for the selection and optimization these parameters, a theoretical model that can comprehensively describe propagation of light in the biological tissues is required.

The description of optical radiation propagation within the random scattering media, like a biological tissue, is based on the radiative transfer equation that forms a basis of several popular solutions considered in this chapter. To describe a propagation of polarized electromagnetic wave within a dispersed random medium we apply a Monte Carlo based computational approach with an adopted vector formalism. We believe this approach has a privilege in simulation of propagation of coherent linearly and/or circularly polarized light in turbid tissue-like scattering medium.

9.2 Light Scattering Characteristics of Biotissues

Light scattering in biotissues is caused by spatial inhomogeneity of their structure at cellular and subcellular levels. Such an inhomogeneity results from the existence of variations in the refractive index δn. The main scatterers are the cellular membranes, nuclei, mitochondria and other organelles [1]. The relations between the sizes of the scatterers and the values of δn lead to an anisotropic scattering of probing radiation.

Light absorption in biological tissues arises due to the presence in their structure of natural chromophores (mainly hemoglobin and melanin). However, within the so-called "diagnostic window" the absorption is rather low. This fact leads to an increase of the penetration depth of the probing radiation into the tissue up to 1–10 mm depending on optical window in NIR wavelength from 700 to 2500 nm (see Chapter 5). Until recently, utilization of the longer NIR wavelengths (>1300 nm) was complicated due to the lack of corresponding laser sources and detectors. However, recent study [2] has introduced the "golden window" covering from 1600 to 1870 nm as an optimal window for light penetration in brain tissue for deep brain imaging due to a reduction in scattering and minimal absorption within this spectral band.

The main parameters describing the process of light propagation inside scattering medium are the absorption coefficient, the scattering coefficient and the phase function of the scatterer. The absorption coefficient $\mu_a(\lambda)$ is the reciprocal of the distance at which a monochromatic collimated beam is attenuated due to the absorption in *e* times. Similarly, the scattering coefficient $\mu_s(\lambda)$ is the reciprocal of the distance at which a monochromatic collimated beam is attenuated by scattering in *e* times. On the other hand, the scattering coefficient is the value which characterizes the mean number of scattering events during the propagation of a photon along the initial direction per unit length. The total value of μ_a and μ_s is called the attenuation (extinction) coefficient μ_t.

The phase function $p(\vec{\Omega}, \vec{\Omega}')$ describes the scattering properties of the medium and represents the probability density function for the scattering of a photon from the initial direction $\vec{\Omega}$ to the new direction $\vec{\Omega}'$ (see Fig. 9.1). In other words, this function characterizes the elementary act of scattering. To characterize average direction of the scatted light, the anisotropy factor g is introduced as the mean value of the cosine of the scattering angle θ:

$$g = \overline{\cos(\theta)} = \frac{\oiint_S \cos(\theta)(\vec{P}_s \cdot \vec{n})dS}{\oiint_S (\vec{P}_s \cdot \vec{n})dS}, \qquad (9.1)$$

where $\vec{P}_s$ [W/m²] is the scattered irradiance defined as the energy flux of the scattered wave; $\vec{n}$ is the unit vector normal to the sphere S surrounding the particle.

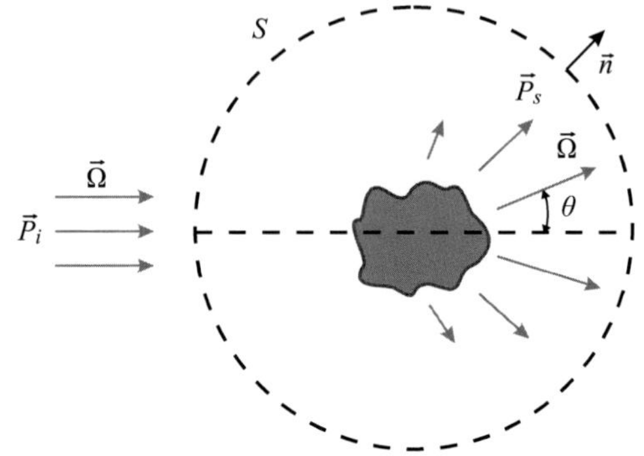

Figure 9.1 Scattering of the light by an arbitrary shaped particle surrounded by the sphere S. $\vec{P}_i$ and $\vec{P}_s$ are the incident and scattered irradiance defined as the energy flux of the incident or scattered wave, respectively.

If the scattering is symmetrical relatively to the direction of the incident wave then the phase function depends only on the angle between the directions $\vec{\Omega}$ and $\vec{\Omega}'$, i.e., $p(\vec{\Omega},\vec{\Omega}')=p(\theta)$. For such a function, the normalization requirement is as follows:

$$\int_0^{\pi} p(\theta)\cdot 2\pi\cdot\sin(\theta)d\theta = 1 \tag{9.2}$$

In many practical cases the phase function is well approximated by an analytic Henyey–Greenstein (9.3) or Gegenbauer Kernel (9.4) phase function [3]:

$$p(\theta)=\frac{1}{4\pi}\cdot\frac{1-g^2}{(1+g^2-2g\cos\theta)^{3/2}}, \tag{9.3}$$

$$p(\theta)=\frac{\alpha g}{\pi}\cdot\frac{(1-g^2)^{2\alpha}}{[(1+g)^{2\alpha}-(1-g)^{2\alpha}](1+g^2-2g\cos\theta)^{1+\alpha}}. \tag{9.4}$$

The latter is often used to represent the scattering properties of a single erythrocyte [4]. The value of g varies from −1 to 1, $\alpha > -0.5$. The case when g = 0 corresponds to isotropic (Rayleigh) scattering, $g = 1$ – to full forward scattering (Mie scattering on large particles), $g = -1$ – to full backward scattering. It is also seen that Henyey–Greenstein phase function is a special case of the Gegenbauer kernel phase function with $\alpha = 0.5$.

Photons propagating through a scattering medium usually are divided into three main groups: (a) ballistic (coherent, forward-scattered) photons that arrive first by traveling over the shortest path, (b) snake (or quasi-coherent) photons that arrive next after ballistic photons by traveling over relatively short paths, and (c) diffusive (incoherent) photons. The separation of the ballistic and snake components from the diffusive component of light propagating in turbid media can be performed by time resolved technique [5]. Generally, the diffusive scattered photons represent the major part of the optical signal in highly scattering media [6]. They travel over a much longer distance than the ballistic or snake photons. However, due to the long random path within the scattering media, diffuse photons carry distorted information about the inner structure of the medium unlike the ballistic and snake photons. Crucial role for the formation of the scattering regime plays the scattering anisotropy of the turbid media described by the factor g. Thus, the snake component in thick turbid media with low g value ($g \sim 0$) is considerably small whereas high g value (e.g., $g > 0.995$) provides a significant contribution of those photons.

9.3 Radiative Transfer Equation

Exact mathematical description of light propagation in optically inhomogeneous medium can be performed with the help of the radiative transfer theory (RTT). This theory is valid for an ensemble of scatterers which are sufficiently distant from each other. RTT is successfully applied in solving some practical problems of the optics of biotissues [7–10]. Classical RTT considers the wave field as a superposition of incoherent radiation beams. The assumption of total mutual incoherence of the fields allows for summation of the average intensities of the beams without taking into account phase relations. Thus, RTT does not include diffraction effects. The basic equation of RTT for a monochromatic light has the form [11]:

$$\frac{1}{c}\frac{\partial I(\vec{r},\vec{\Omega},t)}{\partial t}+\vec{\Omega}\nabla\cdot I(\vec{r},\vec{\Omega},t)+(\mu_s+\mu_a)I(\vec{r},\vec{\Omega},t)$$
$$=\mu_s\int_{4\pi} I(\vec{r},\vec{\Omega},t)p(\vec{\Omega},\vec{\Omega}')d\Omega'+Q(\vec{r},\vec{\Omega},t), \qquad (9.5)$$

where $I(\vec{r},\vec{\Omega},t)$ $[\mathrm{W}\cdot\mathrm{m}^{-2}\cdot\mathrm{sr}^{-1}]$ is the radiation intensity at the point $\vec{r}$ in the direction $\vec{\Omega}$; $p(\vec{\Omega},\vec{\Omega}')$ represents the scattering phase function; c is the speed of light in the medium; $Q(\vec{r},\vec{\Omega},t)$ – the source term.

If the process of radiation transfer is considered in a region G and ∂G is its border then the boundary conditions can be written in general form [12]:

$$I(\vec{r},\vec{\Omega})\Big|_{(\vec{\Omega}\cdot\vec{n})<0}=q(\vec{r},\vec{\Omega})+\int_{\partial G} d\vec{r}'\int_{(\vec{\Omega}\cdot\vec{n})>0} R(\vec{r}',\vec{\Omega}',\vec{r},\vec{\Omega})I(\vec{r}',\vec{\Omega}')d\vec{\Omega}', \qquad (9.6)$$

where $\vec{r},\vec{r}'\in\partial G$; $\vec{n}$ is the external normal to ∂G; $R(\vec{r}',\vec{\Omega}',\vec{r},\vec{\Omega})$ is the reflection coefficient; $q(\vec{r},\Omega)$ represents the radiation of the external sources. If the region G contains the surfaces reflecting or refracting the light, the analogous conditions should be applied to each of them.

The exact analytical solutions of the radiative transfer Eq. (9.5) exist only for a small number of special cases. However, there are approximate analytical solutions of time-dependent radiative transfer equation for an infinite uniform medium with an arbitrary phase function using cumulant expansion [13] including the polarization phenomena [14, 15].

In practice, the approximations of the radiative transfer equation and its numerical solutions are commonly used. The most popular ways to solve the radiative transfer equation numerically are the Monte Carlo technique [16, 17] and the iterative approach based on the discrete ordinates method using the expansions in spherical harmonics [18, 12].

9.4 Approximations of the Radiative Transfer

Due to complex boundary conditions for the majority of biological tissues the Eq. (9.5) is extremely complicated to be solved. Therefore, for certain cases (e.g., media with high anisotropy, media with strong/weak scattering, etc.) the approximate solutions are typically used. The extensively used approximations are briefly presented below.

9.4.1 Small-Angular Approximation

In case of scattering by a particle the size of which is large compared to the wavelength of the radiation and its relative refractive index is close to 1, the scattering phase function differs from zero only at small angles in the direction of radiation propagation (medium with high anisotropy factor $g > 0.9$). This fact allows one to significantly simplify the radiative transfer equation [19]. In this approximation, the transport equation has an exact, albeit complicated solution. Small-angle approximation is valid for relatively low optical thickness, as long as a beam is still rather collimated [20].

9.4.2 Diffuse Approximation

With an increase of the penetration depth in a scattering medium, the radiation becomes more isotropic. At a certain depth it is possible to assume that the radiation "forgets" its initial direction. Such a regime of light propagation is defined as diffusive and described by the diffusion equation, which is much simpler than the original equation of radiative transfer. The diffusion equation can be derived from the transport equation and for a homogeneous isotropic medium it has the following form [11]:

$$\frac{1}{c}\frac{\partial\Phi(\vec{r},t)}{\partial t} - D\nabla^2\Phi(\vec{r},t) + \mu_a\Phi(\vec{r},t) = S(\vec{r},t),$$
$$\Phi(\vec{r},t) = \int I(\vec{r},\vec{\Omega},t)d\vec{\Omega}, \tag{9.7}$$

where $\Phi(\vec{r},t)$ has the dimension of $[\mathrm{W\cdot m^{-2}}]$; $D = (3(\mu_a + \mu_s'))^{-1}$ is the photon diffusion coefficient; $\mu_s' = \mu_s(1 - g)$ is the reduced scattering coefficient; $S(\vec{r},t)$ represents the source function. The value of $l^* = (\mu_a + \mu_s')^{-1}$ is called the transport length and characterizes the distance at which the photon "forgets" its initial direction. Three commonly used boundary conditions have been investigated in literature, i.e., the partial-current boundary conditions [21], the zero-boundary conditions [22], and the extrapolated boundary conditions [23]. The boundary conditions and the matching conditions for the case when the medium under study consists of several layers could be found in [24, 25]. In practical applications such as measurements on biological tissues, the photon fluence rate Φ is not measured directly. The measurable quantity is the number of photons $R(\rho,t)$ that reach the surface per unit time and unit area at a given source- detector separation ρ and at a given time t. $R(\rho, t)$ is called reflectance and can be calculated as current across the medium boundary [26, 27]:

$$R(\rho,t) = -D\nabla\Phi(\rho,z,t)\cdot(-\vec{z})_{z=0}. \tag{9.8}$$

For the extrapolated boundary condition, the solution of the Eq. (9.7) for the semi-infinite homogeneous medium for the fluence rate within the medium is as follows [28]:

$$\Phi(\rho,z,t) = \frac{c}{(4\pi Dct)^{3/2}}\exp(-\mu_a ct) \times\left\{\exp\left(-\frac{(z-z_0)^2+\rho^2}{4Dct}\right)-\exp\left(-\frac{(z+z_0+2z_b)^2+\rho^2}{4Dct}\right)\right\} \tag{9.9}$$

The first term in (9.9) arises due to a point source at $z_0 = (\mu_a + \mu_s)^{-1}$ that results from the perpendicularly incident collimated light, and the second one is due to a negative image source at $-z_0 - 2z_b$. The fluence rate is set to zero at the extrapolated boundary located at z_b:

$$z_b = 2D\cdot(1+R_{eff})/(1-R_{eff}), \tag{9.10}$$

where R_{eff} represents the fraction of photons that have been internally diffusely reflected at the boundary. There are also analytical solutions of the diffusion equation for a two-layer and N-layer medium [29, 30], however, they are rather complicated.

The diffusion theory is a valid approximation in the case when the following conditions are fulfilled: (a) the source of radiation is isotropic or the distance between the source and the detector is much larger than l^*; (b) $\mu_a << \mu_s'$; c) if the intensity of the radiation source is modulated with a frequency f then the modulation frequency should satisfy the relation $c\mu_s'/2\pi f >> 1$ [31]. Thus, the diffusion approximation is valid only for media with a low anisotropy factor, at the same time the scattering should predominate over the absorption. For most of biotissues the anisotropy factor is in the range of 0.7–0.9 and for blood it may reach the value of 0.990 [1] which significantly limits the applicability of the diffusion approximation for such objects. The diffusion approximation is also inapplicable in the regions close to the surface where the low scattered photons dominate and the diffusion regime is not yet established.

9.4.3 Other Methods and Approximations

In addition to the small-angle and diffusion approximation there are some other methods available for a solution of the radiative transfer equation, e.g., first-order approximation, two- and multi-flux approximations. First-order approximation of the radiative transfer equation is valid for optically thin and weakly scattering media [11]. If the medium is illuminated by a diffuse radiation source and its scattering coefficient is high enough to establish the diffusive scattering regime then in this case the two-flux Kubelka–Munk theory becomes applicable [11, 32]. Four-flux theory should be used for the collimated light beam [11]. There is also a seven-flux theory which considers the scattering of a laser beam in a semi-infinite 3D configuration [11].

9.5 Monte Carlo Simulations

Due to the complex structure of biological tissues and high scattering of light therein, the analytical approaches describing radiative transfer are impractical. Thus, stochastic alternative, such as Monte Carlo (MC) method [16] is ideally suited for modeling of optical radiation propagation in complex scattering media. The method is based on multiple computations of random trajectories of photons in the studied medium and subsequent analysis of the accumulated statistical data. The trajectory of each photon represents a sequence of free random steps between the scatterers of the medium and interactions with them. The simulation process consists of the following key steps: injection of the photon packet into medium, generation of the random pathlength, generation of scattering events, accounting for the absorption, definition of reflection/refraction at the medium boundary and definition of the detection conditions.

Both, deterministic method described above and MC method, have errors. However, the source of errors is different for each method. In the treatment of deterministic computational methods the computing errors are systematic. They arise from the discretization of the time, space, angle, energy, phase, and the approximate geometry. The present state of the art does not allow full representation of complicated three-dimensional geometries for deterministic transport methods. MC methods, on the other hand, can treat continuous energy, space, and angle, and hence avoid discretization errors. The errors in MC methods take the form of stochastic uncertainties. Estimation of the statistical uncertainty of MC results requires understanding of properties of random variables such as expectation values and their variances.

The conventional MC approach is comprehensively described elsewhere (see, e.g., [33, 34]), and considers the modeling of energy or intensity transfer through the medium. However, it does not take into account the wave nature of light, i.e., polarization, coherence, phase retardation, interference, and associated wave phenomena.

Recently a new generation of stochastic MC techniques has been developed and widely employed in the studies of coherent effects of multiple scattering, such as enhancement of coherent back-scattering (CBS) and decay of the temporal correlation function of scattered light [35, 36]. MC has been used to study depolarization of the backscattered electromagnetic field [37] and for calculation of low-coherence enhanced backscattering [38]. The MC approach has been extended and extensively used for imitation of 2D images of human skin obtained by optical coherence tomography (OCT) [39–41].

In this chapter, we consider the unified MC approach [42] for modeling of coherent polarized light propagation in a complex multiple scattering medium. The developed model utilizes Jones formalism specially adapted for the online object oriented MC (O3MC) approach developed earlier [42, 43]. The model takes into account the wave properties of light, including temporal coherence, polarization, interference and reflection/refraction at the medium boundary, as well as the helicity flip of the backscattered circularly and/or elliptically polarized light.

The conventional MC model of photon migration in a random scattering medium [33, 34] is based on simulation of a large number of photon packet trajectories ($\sim 10^6$–10^9) consisting of N_s scattering events (up to 10^2–10^3). Typically, MC is used to simulate diffuse reflectance, fluence rate distribution, sampling volume, reflectance spectra, etc. [16]. Depending on the computational performance of available facilities this procedure can be extremely time-consuming (in a range from hours to several days). MC approach can be effectively sped up allowing the results of simulations to be achieved in near real-time [42, 43]. This approach utilizes the Compute Unified Device Architecture (CUDA) parallel computing platform (introduced by NVIDIA Corporation) that allows simultaneous execution of thousands of lightweight parallel threads [44].

9.5.1 Theoretical Basis for Modeling of Coherent Polarized Light Propagation in Scattering Media

Polarization and coherence are the fundamental properties of light with respect to its wave nature. Various MC models have been developed in the past to simulate the propagation of polarized light in scattering media [37, 45, 46, 47–50]. To describe and track the polarization changes, recently developed MC approaches typically utilize Jones [51, 52] and/or Stokes–Mueller [53, 54] formalisms, but struggle with the limitations when modeling of coherent polarized light propagation in highly scattering media is required.

Light propagation in randomly inhomogeneous scattering media including coherent and interferential effects can be described by the Bethe–Salpeter equation [35, 55]:

$$\begin{aligned} G(\vec{r}_1,\vec{r}_n,t\,|\,\vec{k}_\text{i},\vec{k}_\text{s}) &= \mu_\text{s} p(\vec{k}_\text{s}-\vec{k}_\text{i})\delta(\vec{r}_n-\vec{r}_1) \\ &\quad + \mu_\text{s}\int p(\vec{k}_\text{s}-\vec{k}_{nj})\Lambda(R_{nj})G(\vec{r}_j,\vec{r}_1,t\,|\,\vec{k}_{nj},\vec{k}_i)d\vec{r}_j, \quad (9.11) \end{aligned}$$

where $G(\vec{r}_1,\vec{r}_n,t\,|\,\vec{k}_\text{i},\vec{k}_\text{s})$ is Green's function or propagator of the Bethe–Salpeter equation describing propagation of two complex-conjugated electromagnetic fields, delayed in time t, from the point of the first $\vec{r}_1$ to the final $\vec{r}_n$ scattering; $\vec{k}_i$ and $\vec{k}_s$ are the wave vectors of incident and scattered plain waves, respectively; $\Lambda(R) = R^{-2}\exp(-R/l)$ is the propagator describing the transfer of optical radiation between two successive scattering events, and stems from the product of two complex-conjugated Green functions of scalar field; distance $R = |\vec{r}_j - \vec{r}_{j-1}|$; l is mean free path; $p(\vec{k}_s - \vec{k}_i)$ is the phase function; vector $\vec{k}_{nj} = k(\vec{r}_n - \vec{r}_j)/|\vec{r}_n - \vec{r}_j|$ is the wave vector of radiation propagation from a point of the medium $\vec{r}_j$ to $\vec{r}_n$, $k = nk_0$, where $k_0 = 2\pi/\lambda$ is the wave number, λ is the wavelength and n is the refractive index of the medium. Iterating the Bethe–Salpeter Eq. (9.11) the series of scattering orders can be obtained as

$$\begin{aligned} G(\vec{r}_1,\vec{r}_n,t|\vec{k}_i,\vec{k}_s) = {} & \mu_s p(\vec{k}_s - \vec{k}_i)\delta(\vec{r}_n - \vec{r}_1) \\ & + \mu_s^2 p(\vec{k}_s - \vec{k}_{n1})\Lambda(R_{n1})p(\vec{k}_{n1} - \vec{k}_i) \\ & + \mu_s^3 p(\vec{k}_s - \vec{k}_{n3})\Lambda(R_{n2})p(\vec{k}_{n2} - \vec{k}_{21})\Lambda(R_{21})p(\vec{k}_{21} - \vec{k}_i) + \ldots \\ & + \mu_s^n p(\vec{k}_s - \vec{k}_{nj})\Lambda(R_{nj})p(\vec{k}_{nj} - \vec{k}_{jj-1})\Lambda(R_{jj-1})p(\vec{k}_{jj-1} - \vec{k}_{j-1j-2})\ldots \\ & \ldots\Lambda(R_{21})p(\vec{k}_{21} - \vec{k}_i), \end{aligned} \tag{9.12}$$

where the first term describes the single scattering, the second term describes two scattering events, etc. This series is usually illustrated by ladder diagrams [55] as shown in Fig. 9.2.

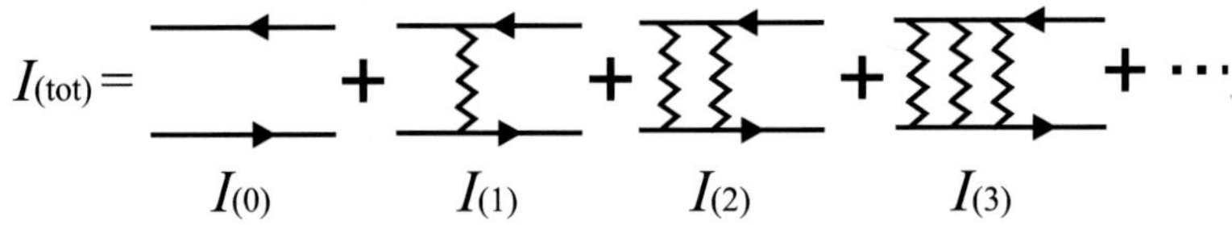

Figure 9.2 Schematic presentation of ballistic $I(0)$, single $I(1)$, double $I(2)$ and triply $I(3)$ scattered photons typically described as a series of ladder diagrams.

In a same manner MC describes the radiation propagation as a stochastic process consisting 1, 2 ... N_s scattering acts, and the addition of extra ladder section $\Lambda(R_{jj-1})p(\vec{k}_{jj-1} - \vec{k}_{j-1j-2})$ is obtained by simulating a photon packet which travels a certain distance to the next scattering event j.

9.5.1.1 Linearly polarized light

In the Jones approach, linearly polarized light is presented as follows:

$$\vec{E} = \begin{bmatrix} E_x \\ E_y \end{bmatrix} = \begin{bmatrix} E_{0x} e^{i\varphi_x} \\ E_{0y} e^{i\varphi_y} \end{bmatrix} = \begin{bmatrix} E_p \\ E_s \end{bmatrix}, \tag{9.13}$$

where E_{0x} and E_{0y} are the instantaneous x and y amplitudes of the wave propagating in the z direction, and coincident with p and s polarization (E_p and E_s, respectively), i is the imaginary unit, and φ is the phase. The propagation of a wave through the medium is then described by a 2 × 2 matrix [54]:

$$\vec{E}^{\text{out}} = J \cdot \vec{E}^{\text{in}}, \tag{9.14}$$

where J is the Jones matrix; $\vec{E}^{\text{in}}$ and $\vec{E}^{\text{out}}$ define, respectively, the polarization of incident beam and the beam emerging from a polarizing element that changes the state of polarization. If the polarized light propagates through multiple polarizing elements, the resulting Jones matrix is given by the product of each element (i.e., by multiple 2 × 2 matrix multiplications [56]). An application of the alternative Stokes-Muller formalism for the same problem requires multiplication of 4 × 4 matrices. The implementation of these approaches into MC modeling requires considerable allocation of memory and time. Here, the propagation of linear polarized light in a scattering medium is modeled by an analogy to the iterative procedure of the solution of the Bethe–Salpeter equation [35]. In this approach, the polarization state is described by a vector $\vec{P}$ undergoing a sequence of transformations after each scattering event. The polarization of incident light polarized along the x axis is defined by unit the vector $\vec{P}^{\text{in}} = [P_x, P_y, P_z] = [1, 0, 0]$. The output vector $\vec{P}^{\text{out}}$ is determined at the end of each photon trajectory by the sequential transformations of the initial polarization [35, 36]:

$$\vec{P}^{\text{out}} = \prod_{j=1}^{N_s} \left[\hat{I} - \frac{(\vec{r}_{j+1} - \vec{r}_j) \otimes (\vec{r}_{j+1} - \vec{r}_j)}{\left| \vec{r}_{j+1} - \vec{r}_j \right|^2} \right] \vec{P}^{\text{in}}, \tag{9.15}$$

where $\hat{I}$ is the unit fourh-rank and ⊗ indicates a direct product of two vectors (see Fig. 9.3). Finally, the corresponding Jones vector components E_x and E_y are defined as

$$
\begin{aligned}
E_x^2 &= W_j P_x^2 \Gamma_R^{N_s} \\
E_y^2 &= W_j P_y^2 \Gamma_R^{N_s}
\end{aligned} \tag{9.16}
$$

where W_j is the statistical weight of j-th photon packet (the initial weight of the photon packet is $W_0 = 1$) and $\Gamma_R = 2(1+\cos^2\theta)^{-1}$ is the Rayleigh factor. This relation is based on the optical theorem [57, 58], which relates the scattering cross-section and the scattering pathlength lj.

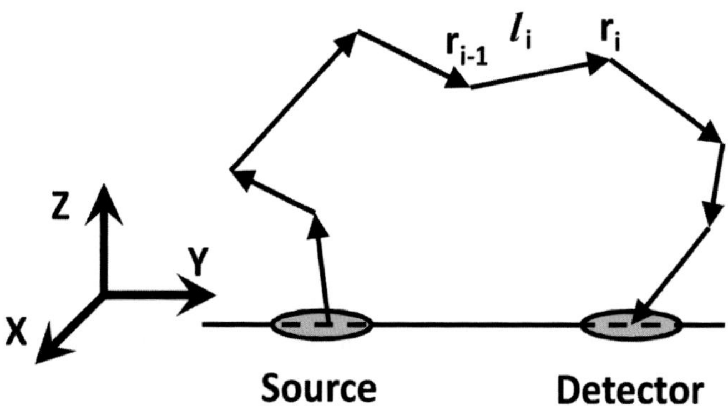

Figure 9.3 Typical random trajectory of a photon packet traveling from the source to the detector; $\vec{r}_j$ and $\vec{r}_{j+1}$ are the points of j and j + 1 scattering events, and l_j is the pathlength of the photon packets at j-th scattering event.

Total internal reflection and/or refraction at the medium boundary are taken into account by splitting the photon packet into the transmitted and reflected parts [59]. Thus, the statistical weight of the detected j-th photon packet that experienced M number of reflections/refractions at the medium boundary is defined as

$$
W_j = W_0\left[1 - R^{\text{in}}(\alpha)\right]\left[\prod_{f=1}^{M_j} R_f(\alpha)\right]\left[1 - R^{\text{out}}(\alpha)\right] \tag{9.17}
$$

where R^{in} and R^{out} are the Fresnel reflection coefficients at the medium/air boundary for the photon packets entering and leaving the medium, respectively; $R_f(\alpha)$ is the Fresnel reflection coefficient that accounts for all other reflections at the surface of the medium; α is an angle of incidence of the photon packet at the surface of the medium.

Absorption is taken into account according to the microscopic Beer-Lambert law [59] by attenuating the statistical weight W_j of the photon packet by the factor $\exp(-\mu_a \cdot L)$, where μ_a is the absorption coefficient and L is the photon packet pathlength.

Finally, for the detected photon packets (typically $N_{ph} = 10^7$–10^9) with taking into account interference and the coherent properties of light, the resulting intensities of co-polarized $I_{||}$ and cross-polarized $I_{\perp}$ light are defined as:

$$I_{||} = \frac{1}{N_{ph}} \sum_{i=1}^{N_{ph}} E_{x_i}^2 + \frac{2}{N_{ph}} \sum_{i=1}^{N_{ph}} \sum_{j=i+1}^{N_{ph}} E_{x_i} E_{x_j} \times \cos\left[\frac{2\pi}{\lambda}(L_i - L_j) + \pi(m_i - m_j)\right] \exp\left[-\left(\frac{L_i - L_j}{l_c}\right)^2\right],$$

$$I_{\perp} = \frac{1}{N_{ph}} \sum_{i=1}^{N_{ph}} E_{y_i}^2 + \frac{2}{N_{ph}} \sum_{i=1}^{N_{ph}} \sum_{j=i+1}^{N_{ph}} E_{y_i} E_{y_j} \times \cos\left[\frac{2\pi}{\lambda}(L_i - L_j)\right] \exp\left[-\left(\frac{L_i - L_j}{l_c}\right)^2\right], \tag{9.18}$$

where L_i and L_j are the individual total pathlengths of i-th and j-th photon packets, respectively; m is the number of displacements of E_x giving a correct representation of the nodality [60] for the i-th and j-th photon packets; l_c is the temporal coherence length of the incident radiation and λ is the wavelength of the incident light.

9.5.1.2 Circularly polarized light

A similar approach can be applied for modeling of circularly and elliptically polarized light propagation in turbid scattering media. Circularly polarized light is characterized by two orthogonal vectors, i.e., for the incident light with right handed circular polarization (RCP) each photon packet injected into the medium is assigned with two orthogonal vectors $\vec{P}_x^{\text{in}} = [1, 0, 0]$ and $\vec{P}_y^{\text{in}} = [0, 1, 0]$ with a phase shift $(\varphi_x - \varphi_y = \pi/2)$. Thus, the resulting components of RCP and left-handed circular polarization (LCP) at the end of each photon trajectory are, respectively, defined as follows:

$$\begin{bmatrix} E_{xx}^2 \\ E_{yy}^2 \end{bmatrix} = \begin{bmatrix} W_j P_{xx}^2 \Gamma_R^{N_s} \\ W_j P_{xx}^2 \Gamma_R^{N_s} \end{bmatrix}, \begin{bmatrix} E_{xy}^2 \\ E_{yx}^2 \end{bmatrix} = \begin{bmatrix} W_j P_{xy}^2 \Gamma_R^{N_s} \\ W_j P_{yx}^2 \Gamma_R^{N_s} \end{bmatrix}. \quad (9.19)$$

It should be also pointed out here, that when circularly or elliptically polarized light is reflected at the medium surface the state of polarization undergoes a helicity flip [61, 62]. The same goes for backscattered light, i.e., if the photon packet is scattered backwards (where $\theta > \pi/2$, θ is the scattering angle, see Fig. 9.1) this results in a RCP photon packet changing to LCP; if the photon packet scatters forward $(\theta < \pi/2)$ the helicity remains the same. This helicity flip phenomenon is of fundamental importance [63, 62]. Linear polarization possesses no such sense of the direction in which it travels. It was also demonstrated that the use of circularly polarized light for reflective target detection in highly scattering medium significantly improves the image contrast [62, 64].

Therefore, considering the propagation of circularly polarized light through a turbid medium where multiple scattering events occur, the photon packets which have backscattered an odd number of times will correspond to a reversal in helicity, and thus, contribute the cross polarized portion of the detected signal (I_{cross}). Correspondingly, the photon packets that have experienced an even number of backscattering events contribute to the co-polarized signal (I_{co}) and the handedness of light is unchanged with respect to the incident polarization. Thus, for total number of helicity flips m, either due to reflections on the medium boundary or back-scattering events $(\theta > \pi/2)$, and taking into account interference, the intensity of co-polarized light (that has preserved its helicity) is given by

$$I_{xx} = \frac{1}{N_{ph}} \sum_{i=1}^{N_{ph}} \left(E_{xx_i}^2 \left| \cos\frac{\pi m_i}{2} \right| + E_{yx_i}^2 \left| \sin\frac{\pi m_i}{2} \right| \right)$$
$$+ \frac{1}{N_{ph}} \sum_{i=1}^{N_{ph}} \sum_{j=i+1}^{N_{ph}} \left(E_{xx_i} \left| \cos\frac{\pi m_i}{2} \right| + E_{yx_i} \left| \sin\frac{\pi m_i}{2} \right| \right)$$
$$\times \left(E_{xx_j} \left| \cos\frac{\pi m_j}{2} \right| + E_{yx_i} \left| \sin\frac{\pi m_i}{2} \right| \right)$$
$$\times \cos\left[\frac{2\pi}{\lambda}(L_i - L_j) + \pi(m_i - m_j) \right] \exp\left[-\left(\frac{L_i - L_j}{l_c} \right)^2 \right],$$

$$I_{yy} = \frac{1}{N_{ph}} \sum_{i=1}^{N_{ph}} \left(E_{yy_i}^2 \left| \cos\frac{\pi m_i}{2} \right| + E_{xy_i}^2 \left| \sin\frac{\pi m_i}{2} \right| \right)$$
$$+ \frac{2}{N_{ph}} \sum_{i=1}^{N_{ph}} \sum_{j=i+1}^{N_{ph}} \left(E_{yy_i} \left| \cos\frac{\pi m_i}{2} \right| + E_{xy_i} \left| \sin\frac{\pi m_i}{2} \right| \right)$$
$$\times \left(E_{yy_j} \left| \cos\frac{\pi m_j}{2} \right| + E_{xy_i} \sin\frac{\pi m_i}{2} \right)$$
$$\times \cos\left[\frac{2\pi}{\lambda}(L_i - L_j) + \pi(m_i - m_j) \right] \exp\left[-\left(\frac{L_i - L_j}{l_c} \right)^2 \right]. \qquad (9.20)$$

Similar equations can be written for depolarized light, defined by I_{yx} and I_{xy}, respectively.

9.5.2 Results of Modeling of Polarized Light Propagation

To verify the presented approach MC simulations have been performed in backscattering geometry utilizing a semi-infinite scattering medium with μ_s = 30 mm^{-1}, μ_a = 0.001 mm^{-1} and different values of anisotropy factor g for 10^6 of photon packets. Table 9.1 shows results of MC simulations for linearly polarized light in isotropic scattering medium ($g = 0$) compared against the exact theoretical predictions and other methods.

Table 9.1 Comparison of the results of MC simulation for linearly polarized light in isotropic scattering medium with the exact solution by Milne and alternative modeling

Reference	$I_{\|\|}$	$I_{\perp}$	$I_{\|\|}/I_{\perp}$
Milne solution [35]	3.025	1.563	1.935
Kuzmin, et al. [35]	3.029	1.570	1.929
Developed MC [65]	3.023	1.565	1.931

MC stimulation yields the ratio between polarized and depolarized backscattered intensities $I_{||}/I_{\perp} = 1.931$ which is in an excellent agreement with the generalized Milne solution providing $I_{||}/I_{\perp} = 1.935$.

Results of simulation of $I_{||}$ (squares) and $I_{\perp}$ (circles) polarized intensities as a function of a number of scattering events are presented in Fig. 9.4a–c. Figure 9.4d shows the corresponding depolarization ratio DR = $(I_{||} - I_{\perp})/(I_{||} + I_{\perp})$ also known as polarization contrast. Strong depolarization is clearly observed after a several number of scattering orders (see Fig. 9.4d). Thus, it is clearly demonstrated that the scattering properties of the medium significantly affect depolarization of the detected signal and need to be accessed when developing a particular optical diagnostic system. Figure 9.5 presents the results of simulation of spatial distribution of $I_{||}$, $I_{\perp}$ and the depolarization ratio DR at the surface of the medium with various values of scattering anisotropy. Correspondingly, Fig. 9.6 presents the effective probing depths distribution (sampling volume) for $I_{||}$ and $I_{\perp}$ and their relative difference counted by analogy to DR. It can be clearly seen from Fig. 9.5 that $I_{||}$ significantly prevails over the intensity of the cross-polarized light $I_{\perp}$. Figure 9.6 shows that for the medium with high anisotropy of scattering ($g = 0.98$) the effective probing depth of $I_{||}$ exceeds the propagation depth for $I_{\perp}$. Therefore, with a significant increase of the photon packets' pathlengths within the medium, the residual polarization in the backscattered light becomes preserved due to a large relative contribution of the photon packets' trajectories with low orders of scattering. Thus, for the medium with isotropic scattering, the light is almost immediately depolarized. On the contrary, strong forward scattering preserves polarization within the tissue. When scattering is highly anisotropic which is common for many biological tissues, polarized light can be detected and analyzed at larger distances from the light source. These results are well agreed with the results of theoretical predictions [37]. As a result of multiple scattering, polarized light becomes almost completely depolarized that admits scalar field approach for light propagation in media. However, in the case of backscattering contributions of lower scattering orders, which are still partly polarized, results in that the detected backscattered light partly retains its initial polarization.

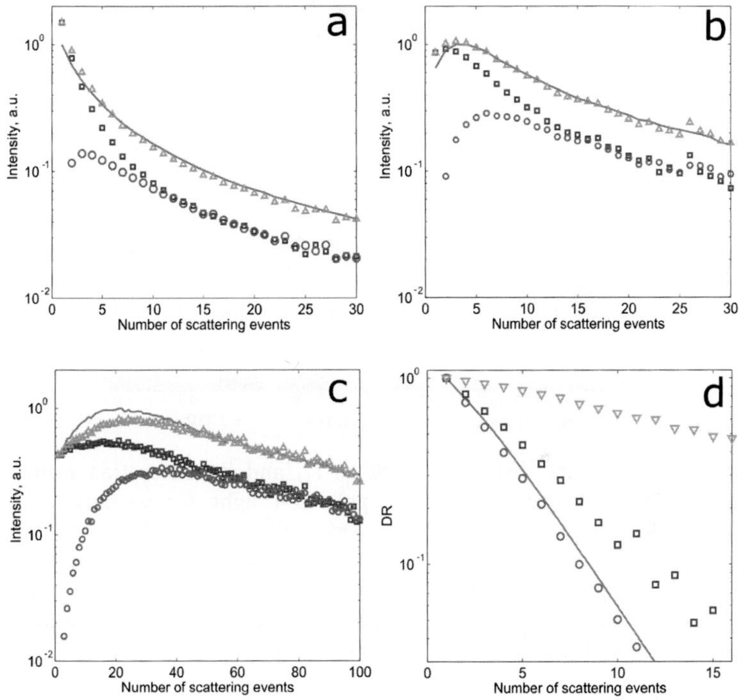

Figure 9.4 Results of MC modeling of $I_{||}$ (squares) and $I_{\perp}$ (circles) presented as a function of the number of scattering events for various anisotropy factors: (a) shows the results for $g = 0$; (b) depicts $g = 0.5$; (c) corresponds to $g = 0.9$; triangles represent $I_{||} + I_{\perp}$; solid line represents the total intensity of the scalar filed. Figure (d) demonstrates the comparison between the depolarization ratios for $g = 0$ (circles), $g = 0.5$ (squares), $g = 0.9$ (triangles) with the known theoretical solution (solid line).

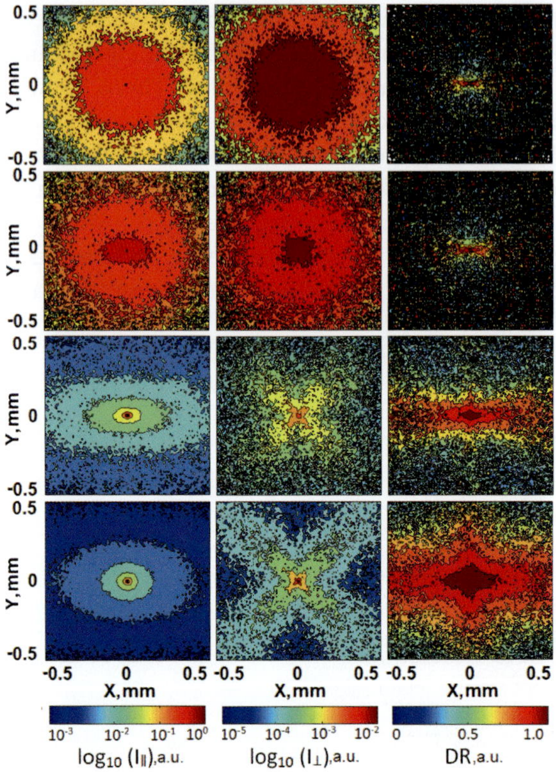

Figure 9.5 Surface distributions of $I_{||}$, $I_{\perp}$, and depolarization ratio of backscattered linearly polarized light for various values of scattering anisotropy (from the top g = 0, 0.5, 0.9, and 0.98). The simulations were performed for non-coherent light normally incident on the semi-infinite scattering medium with μ_s = 30 mm^{-1}, μ_a = 0.001 mm^{-1}, and detected over entire medium surface [65].

To validate the developed MC for circularly polarized light, MC simulations for co- (I_{co}) and cross- (I_{cross}) components have been performed for the exact backscattering detection geometry. A semi-infinite scattering medium with μ_s = 30 mm^{-1}, μ_a = 0.001 mm^{-1} has been employed, 10^{10} of photon packets were used for each MC simulation.

Results produced by the developed MC have been compared against the known analytical solutions and other models and are presented in Table 9.2.

Table 9.2 Comparison of the results of MC simulation for circularly polarized light in isotropic scattering medium with the exact solution by Milne and alternative modeling

Reference	I_{co}	I_{cross}	I_{co}/I_{cross}
Milne solution [35]	1.751	2.837	0.617
Kuzmin et al. [35]	1.758	2.841	0.618
Developed MC [65]	1.754	2.839	0.617

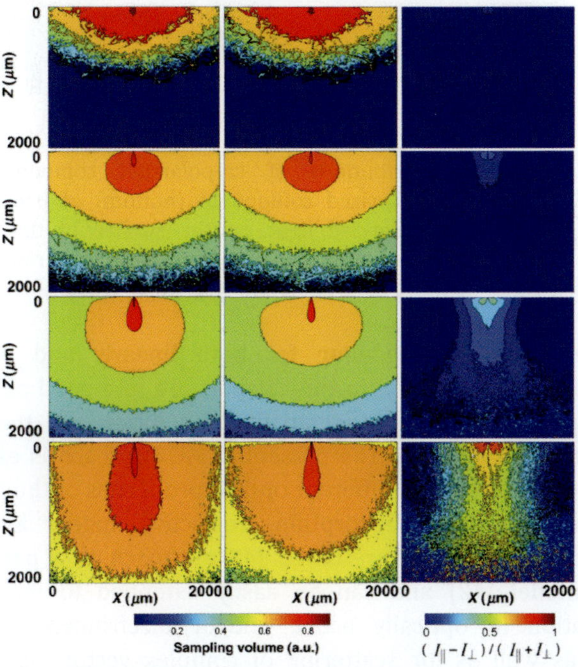

Figure 9.6 Corresponding sampling volumes for $I_{||}$, $I_{\perp}$, and spatial distribution of the depolarization ratio DR simulated for the same parameters as in Fig. 9.5 [65].

A good agreement has been observed between simulations results and analytical predictions. Milne solution yields $I_{||} = 1.751$, $I_{\perp} = 2.837$, $I_{||}/I_{\perp} = 0.617$, whereas the developed model yields $I_{||} = 1.754$, $I_{\perp} = 2.839$, $I_{||}/I_{\perp} = 0.617$.

Figure 9.7 shows the surface distribution of the portions of co-polarized I_{co} (RCP) and cross-polarized I_{cross} (LCP) intensities of backscattered light, and the helicity flip is clearly observed.

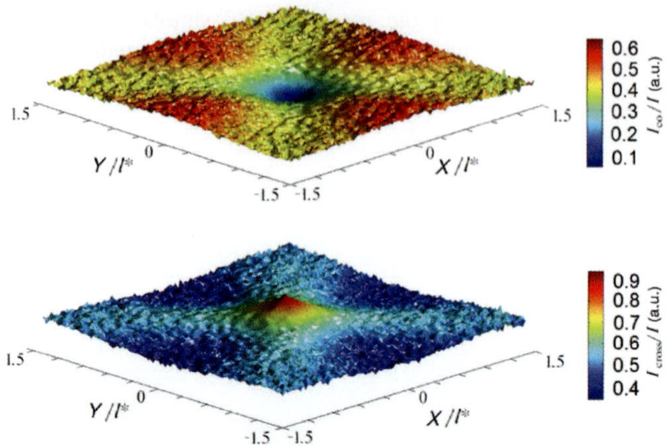

Figure 9.7 Spatial distributions of co-polarized component (top) and cross-polarized component (bottom) of backscattered circularly polarized light on the surface of the medium, $l^* = (\mu_s')^{-1}$; counted for a semi-infinite scattering medium ($\mu_s = 30$ mm^{-1}, $\mu_a = 0.001$ mm^{-1}, $g = 0.98$) [65].

The results of modeling for both linearly and circularly polarized light presented in this chapter are in a good agreement with the results of the exact analytical solution and the results of alternative modeling. The presented MC approach is extremely flexible to operate with different optical properties of the medium, including scattering, absorption, refractive index, anisotropy of scattering, etc. The presented MC approach is a part of the O3MC model [42] and can be easily extended to model light propagation in optically active and/or birefringent scattering media, as well as for scattering of complex vector laser beams in turbid tissue-like media.

9.6 Summary

In this chapter, the major models used for a description of optical radiation propagation in turbid tissue-like scattering media have been considered. Rigorous theoretical background behind the MC approach, chosen as a primary modeling tool, has been given. Versatile usage of the developed MC model has been exemplified by its application to the computational imitation of known effects associated with the polarization and coherent properties of multiple scattering of light. The developed MC model also takes into account the influence of total reflection and refraction at the medium boundary, helicity flip of circularly polarized light and interference. The results of modeling for both linearly and circularly polarized light are in good agreement with the results of known analytical solution and the results of alternative modeling. The presented computational approach can be easily extended for consideration of light propagation in optically active and/or birefringent scattering media, as well as for modeling of propagation complex vector laser beams in turbid tissue-like scattering media.

Acknowledgments

The authors acknowledge partial support provided by the Faculty of Information Technology and Electrical Engineering of the University of Oulu (Finland) and Academy of Finland (290596).

References

1. Tuchin, V. V. (2000). *Tissue Optics: Light Scattering Methods and Instruments for Medical Diagnostics* (SPIE Press, Bellingham, USA).
2. Shi, L., Sordillo, L. A., Rodríguez-Contreras, A., and Alfano, R. (2016). Transmission in near-infrared optical windows for deep brain imaging, *J. Biophotonics*, **9** (1–2), pp. 38–43.
3. Yaroslavsky, A. N., Yaroslavsky, I. V., Goldbach, T., and Schwarzmaier, H.-J. (1999). Influence of the scattering phase function approximation on the optical properties of blood determined from the integrating sphere measurements, *J. Biomed. Opt.*, **4**(1), pp. 47–53.
4. Hammer, M., Schweitzer, D., Michel, B., Thamm, E., and Kolb, A. (1998). Single scattering by red blood cells, *Appl. Opt.*, **37**(31), pp. 7410–7418.
5. Wang, L., Ho, P. P., Liu, G., Zhang, G., and Alfano, R. R. (1991). Ballistic 2-D imaging through scattering walls using an ultrafast optical Kerr gate, *Science*, **253**, pp. 769–771.
6. Cai, W., Das, B. B., Liu, F., Zevallos, M., Lax, M., and Alfano, R. (1996). Time-resolved optical diffusion tomographic image reconstruction in highly scattering turbid media, *Proc. Natl. Acad. Sci. U. S. A.*, **93**, pp. 13561–13564.

7. Abdoulaev, G. S., and Hielscher, A. H. (2003). Three-dimensional optical tomography with the equation of radiative transfer, *J. Electron. Imaging*, **12**, pp. 594–601.
8. Klose, A. D., Netz, U., Beuthan, J., and Hielscher, A. H. (2002). Optical tomography using the time-independent equation of radiative transfer—Part 1: Forward model, *J. Quant. Spectrosc. Radiat. Transf.*, **72**, pp. 691–713.
9. Klose, A. D., and Hielscher, A. H. (2002). Optical tomography using the time-independent equation of radiative transfer—Part 2: Inverse model, *J. Quant. Spectrosc. Radiat. Transf.*, **72**, pp. 715–732.
10. Klose, A. D., Ntziachristos, V., and Hielscher, A. H. (2005). The inverse source problem based on the radiative transfer equation in optical molecular imaging, *J. Comput. Phys.*, **202**, pp. 323–345.
11. Ishimaru, A. (1997). *Wave Propagation and Scattering in Random Media* (IEEE Press, New York, USA).
12. Lewis, E. E., and Miller, W. F. (1990). *Computational Methods of Neutron Transport* (American Nuclear Society, New York, USA).
13. Cai, W., Lax, M., and Alfano, R. R. (2000). Cumulant solution of the elastic Boltzmann transport equation in an infinite uniform medium, *Phys. Rev. E*, **61**, pp. 3871–3876.
14. Cai, W., Lax, M., and Alfano, R. R. (2000). Analytical solution of the polarized photon transport equation in an infinite uniform medium using cumulant expansion, *Phys. Rev. E*, **63**, p. 016606.
15. Cai, W., Ni, X., Gayen, S. K., and Alfano, R. R. (2006). Analytical cumulant solution of the vector radiative transfer equation investigates backscattering of circularly polarized light from turbid media, *Phys. Rev. E*, **74**, p. 056605.
16. Zhu, C., and Liu, Q. (2013). Review of Monte Carlo modeling of light transport in tissues, *J. Biomed. Opt.* **18**(5), p. 050902.
17. Kandidov, V. P., Militsin, V. O., Bykov, A. V., and Priezzhev, A. V. (2006). Application of corpuscular and wave Monte-Carlo methods in optics of dispersive media, *Quant. Electron.*, **36**(11), pp. 1003–1008.
18. Kuznetsov, V. S., Nikolaeva, O. V., Bass, L. P., Bykov, A. V., and Priezzhev, A. V. (2010). Mathematical simulation of ultrashort light pulse propagation in a highly scattering medium, *Math. Models. Comput. Simulation*, **2**(1), pp. 22–32.
19. Kokhanovsky, A. A. (1997). Small-angle approximations of the radiative transfer theory, *J. Phys. D Appl. Phys.*, **30**, pp. 2837–2840.

20. Swanson, N. L., Gehman, V. M., Billard, B. D., and Gennaro, T. L. (2001). Limits of the small-angle approximation to the radiative transport equation, *J. Opt. Soc. Am. A*, **18**, pp. 385–391.
21. Haskell, R., Svaasand, L., Tsay, T. T., Feng, T. C., and McAdams, M. (1994). Boundary conditions for the diffusion equation in the radiative transfer, *J. Opt. Soc. Am. A*, **11**(10), pp. 2727–2741.
22. Kienle, A., Lilge, L., Patterson, M., Steiner, R. H., and Wilson, B. (1996). Spatially resolved absolute diffuse reflectance measurements for noninvasive determination of the optical scattering and absorption coefficients of biological tissue, *Appl. Opt.*, **35**(13), pp. 2304–2314.
23. Kienle, A., and Patterson, M. (1997). Improved solutions of the steady-state and the time-resolved diffusion equations for reflectance from semi-infinite turbid medium, *J. Opt. Soc. Am. A*, **14**(1), pp. 246–254.
24. Keijzer, M., Star, W. M., and Storchi, P. R. (1988). Optical diffusion in layered media, *Appl. Opt.*, **27**, pp. 1820–1824.
25. Dehghani, H., Brooksby, B., Vishwanath, K., Pogue, B. W., and Paulsen, K. D. (2003). The effects of internal refractive index variation in near-infrared optical tomography: A finite element modelling approach, *Phys. Med. Biol.*, **48**, pp. 2713–2727.
26. Patterson, M., Wilson, B., and Wyman, D. (1991). The propagation of optical radiation in tissue I. Models of radiation transport and their application, *Las. Med. Sci.*, **6**, pp. 155–167.
27. Durduran, T., Yodh, A., Chance, B., and Boas, D. (1997). Does the photon-diffusion coefficient depend on absorption, *J. Opt. Soc. Am. A*, **14**(12), pp. 3358–3365.
28. Patterson, M., Chance, B., and Wilson, B. (1989). Time resolved reflectance and transmittance for the non-invasive measurement of tissue optical properties, *Appl. Opt.*, **28**(12), pp. 2331–2336.
29. Kienle, A., Patterson, M. S., Dognitz, N., Bays, R., Wagnieres, G., and van den Bergh, H. (1998). Noninvasive determination of the optical properties of two-layered turbid media, *Appl. Opt.*, **37**, pp. 779–791.
30. Liemert, A., and Kienle, A. (2009). Light diffusion in N-layered turbid media, *Proc SPIE*, **7369**, p. 73690F.
31. Guenther, B. D. (ed.) (2004). *Encyclopedia of Modern Optics* (Academic Press Inc., Oxford, UK).

32. Vargas, W. E., and Niklasson, G. A. (1997), Applicability conditions of the Kubelka-Munk theory, *Appl. Opt.*, **36**, pp. 5580–5586.

33. Prahl, S. A., Keijzer, M., Jacques, S. L., and Welch, A. J. (1989). A Monte Carlo model of light propagation in tissue, *SPIE Series*, IS **5**, pp. 102–111.

34. Jacques, S. L., and Wang, L. V. (1995). Monte Carlo modeling of light transport in tissues, in: *Optical Thermal Response of Laser Irradiated Tissue*, Welch, A. J., and van Gemert, M. J. C., eds. (Plenum, New York), pp. 73–100.

35. Kuzmin, V., and Meglinski, I. (2007). Coherent effects of multiple scattering for scalar and electromagnetic fields: Monte-Carlo simulation and Milne-like solutions, *Opt. Commun.*, **273**(2), pp. 307–310.

36. Kuzmin, V., and Meglinski, I. (2006). Numerical simulation of coherent back-scattering and temporal intensity correlations in random media (overview), *Quantum Electron.*, **36**(11), pp. 990–1002.

37. Rojas-Ochoa, L., Lacoste, D., Lenke, R., Schurtenberger, P., and Scheffold, F. (2004). Depolarization of backscattered linearly polarized light, *J. Opt. Soc. Am. A*, **21**(9), pp. 1799–1804.

38. Subramanian, H., Pradhan, P., Kim, Y. L., Liu Y., Li, X., and Backman, V. (2006). Modeling low-coherence enhanced backscattering using Monte Carlo simulation, *Appl. Opt.*, **45**(24), pp. 6292–6300.

39. Meglinski, I., Kirillin, M., Kuzmin, V., and Myllylä, R. (2008). Simulation of polarization-sensitive optical coherence tomography images by a Monte Carlo method, *Opt. Lett.*, **33**(14), pp. 1581–1583.

40. Kirillin, M., Meglinski, I., and Priezzhev, A. (2006). Effect of photons of different scattering orders on the formation of a signal in optical low-coherence tomography of highly scattering media, *Quantum Electron.*, **36**(11), pp. 990–1002.

41. Kirillin, M., Meglinski, I., Kuzmin, V., Sergeeva, E., and Myllylä, R. (2010). Simulation of optical coherence tomography images by Monte Carlo modeling based on polarization vector approach, *Opt. Express*, **18**(21), pp. 21714–21724.

42. Doronin, A., and Meglinski, I. (2011). Online object oriented Monte Carlo computational tool for the needs of biomedical optics, *Biomed. Opt. Express*, **2**(9), pp. 2461–2469.

43. Doronin, A., and Meglinski, I. (2012). Peer-to-peer Monte Carlo simulation of photon migration in topical applications of biomedical optics, *J. Biomed. Opt.*, **17**(9), p. 090504.

44. Kirk, D., and Hwu, W. (2010). *Programming Massively Parallel Processors: A Hands-on Approach* (Morgan Kaufmann, Burlington, USA).

45. Sawicki, J., Kastor, N., and Xu, M. (2008). Electric field Monte Carlo simulation of coherent backscattering of polarized light by a turbid medium containing Mie scatterers, *Opt. Express*, **16**(8), pp. 5728–5738.

46. Radosevich, A., Rogers, J. D., Capoglu, I. R., Mutyal, N. N., Pradhan, P., and Backman, V. (2012). Open source software for electric field Monte Carlo simulation of coherent backscattering in biological media containing birefringence, *J. Biomed. Opt.*, **17**(11), pp. 115001.

47. Wang, X., and Wang, L. (2001). Propagation of polarized light in birefringent turbid media: Time-resolved simulations, *Opt. Express*, **9**(5), pp. 254–259.

48. Wang, X., and Wang, L. (2002). Propagation of polarized light in birefringent turbid media: A Monte Carlo study, *J. Biomed. Opt.*, **7**(3), pp. 279–290.

49. Gangnus, S., Matcher, S., and Meglinski, I. (2004). Monte Carlo modeling of polarized light propagation in biological tissues, *Laser Phys.*, **14**(6), pp. 886–891.

50. Ramella-Roman, J., Prahl, S., and Jacques, S. (2005). Three Monte Carlo programs of polarized light transport into scattering media: Part I, *Opt. Express*, **13**(12), pp. 4420–4438.

51. Jones, R. C. (1941). A new calculus for the treatment of optical systems. I. Description and discussion of the new calculus, *J. Opt. Soc. Am.*, **31**(7), pp. 488–493.

52. Hurwitz, H., and Jones, R. C. (1941). A new calculus for the treatment of optical systems. II. Proof of three general equivalence theorems, *J. Opt. Soc. Am.*, **31**(7), pp. 493–499.

53. Gerald, A., and Burch, J. M. (1975). *Introduction to Matrix Methods in Optics* (John Wiley & Sons, London, UK).

54. Collett, E. (2005). *Field Guide to Polarization* (SPIE, Bellingham, USA).

55. Meglinski, I. V., Kuzmin, V. L., Churmakov, D. Y., Greenhalgh, D. A. (2005). Monte Carlo simulation of coherent effects in multiple scattering, *Proc. R. Soc. A*, **461**, pp. 43–53.

56. Yeh, P. (1982). Extended Jones matrix method, *J. Opt. Soc. Am.*, **72**(4), pp. 507–513.

57. Newton, R. G. (1976). Optical theorem and beyond, *Am. J. Phys.*, **44**(7), pp. 639–642.

58. Carney, P. S., Wolf, E., and Agarwal, G. S. (1997). Statistical generalizations of the optical cross-section theorem with application to inverse scatter, *J. Opt. Soc. Am. A*, **14**(12), pp. 3366–3371.

59. Churmakov, D., Meglinski, I., and Greenhalgh, D. (2002). Influence of refractive index matching on the photon diffuse reflectance, *Phys. Med. Biol.*, **47**(23), pp. 4271–4285.

60. Clarke, D., and Grainger, J. F. (1971). *Polarized Light and Optical Measurements* (Pergamon, New York, USA).

61. Born, M., and Wolf, E. (1986). *Principles of Optics: Electromagnetic Theory of Propagation, Interference and Diffraction of Light*, 6th ed. (Pergamon, London, UK).

62. Ni, X., and Alfano, R. (2004). Time-resolved backscattering of circularly and linearly polarized light in a turbid medium, *Opt. Lett.*, **29**(23), pp. 2773–2775.

63. Kim, Y., Pradhan, P., Kim, M. H., and Backman, V. (2006). Circular polarization memory effect in low-coherence enhanced back-scattering of light, *Opt. Lett.*, **31**(18), pp. 2744–2746.

64. Kartazayeva, S. A., Ni, X., and Alfano, R. R. (2005). Backscattering target detection in a turbid medium by use of circularly and linearly polarized light, *Opt. Lett.*, **30**(10), pp. 1168–1170.

65. Doronin, A., Macdonald, C., Meglinski, I. (2014). Propagation of coherent polarized light in turbid highly scattering medium, *J. Biomed. Opt.*, **19**(2), p. 025005.

Chapter 10

Overview of the Cumulant Solution to Light Propagation Inside a Turbid Medium and Its Applications in Deep Imaging Beyond the Diffusion Approximation

Min Xu,[a] Cai Wei,[b] and Robert R. Alfano[b]

[a]*Physics Department, Fairfield University,*
1073 N Benson Rd, Fairfield, Connecticut 06824, USA
[b]*Physics Department, The City College of New York 136th St & Convent Av.,*
New York, New York 10037, USA

mxu@fairfield.edu

Radiative transfer is a phenomenological theory describing the propagation of multiple scattering light in turbid media. It serves as the foundation of optical imaging where the optical properties of a turbid medium such as biological tissue are characterized by solving an inverse problem of how light propagates inside the medium. Radiative transfer in optical imaging is commonly approximated as diffusion due to the computational demand of the exact solution of the radiative transfer equation (RTE).

Deep Imaging in Tissue and Biomedical Materials: Using Linear and Nonlinear Optical Methods
Edited by Lingyan Shi and Robert R. Alfano

ISBN 978-981-4745-88-8 (Hardcover), 978-1-315-20655-4 (eBook)
www.panstanford.com

In this chapter, we will overview the cumulant solution to light propagation in a turbid medium and its applications in deep imaging beyond the diffusion equation. We will first review the analytical solution of the time-dependent RTE in an infinite uniform medium with an arbitrary light scattering phase function using cumulant expansion. Analytical expressions for the exact distribution in angle and the spatial cumulants at any angle, exact up to an arbitrary high order, *n*, of photons are derived. By a cutoff at the second cumulant order, a Gaussian analytical approximate expressions of the photon spatial distribution are obtained as a function of the direction of light propagation and time, whose center position and half-width are always exact at arbitrary time. The center of this distribution advances and the half-width grows in time, depicting the evolution of the particle migration from near ballistic, through snake-like, and into the final diffusive regime. We will then present strategies to incorporate the boundary conditions in the cumulant solution. The performance of the cumulant solution in an infinite and a semi-infinite medium is verified by Monte Carlo simulations. We will finally discuss particular applications of the cumulant solution to RTE for optical imaging beyond the diffusion approximation with examples in a transport forward model for optical imaging, cloud remote sensing, and early photon tomography for imaging targets deep inside breast and lung.

10.1 Introduction

Light has some amazing properties. It travels at ultimately the fastest speed and can go around corners by diffraction, Its wavelength covers a wide spectral range from gamma and X rays, UV, visible, NIR, MIR, IR, THz, microwaves, to radio waves. Light excites transitions by absorption of single or multiphotons from which the vibrations and other elementary or quasi particles can be probed. Radiative transfer describes light propagation in a turbid medium, where photons suffer multiple scattering by randomly distributed scatterers in the medium. The kinetic equation governing photon propagation is the classic Boltzmann transport equation. The search for an analytical solution of the time-dependent elastic Boltzmann transport equation has lasted for many years. Besides being considered as a classic problem in fundamental research in statistical dynamics, the study of analytical solution of this equation has applications in a broad variety of fields from astrophysics, to geophysics, remote sensing, and medicine. One classic approach to solve this equation is to transform the Boltzmann transport equation into a series of moment equations based on the angular moment expansion and to retain the leading orders [1–3]. To the second order, a diffusion equation is derived. One of the main advantages of the moment equations (and related methods such as the diffusion approximation) is that they are manifestly energy conservative. The diffusion approximation admits an analytical solution in a uniform medium of infinite, semi-infinite, slab and other geometries and has been widely used [4–6]. The diffusion approximation, however, fails at early times when the particle distribution is still highly anisotropic. Numerical approaches, including the Monte Carlo simulation [7–9], remain the main tools in solving the elastic Boltzmann equation, which are cumbersome tasks. To address these challenges, we have developed the cumulant solution to RTE which is an analytical solution of the classic elastic Boltzmann transport equation in an infinite uniform medium, with the photon's velocity $c\mathbf{s}$, where $\mathbf{s}$ is a unit vector of direction, and c is the (constant) speed in the medium [6, 10–20]. The cumulant method and its applications have been overlooked since Cai and Xu introduced this approach together with Melvin Lax and Robert Alfano despite its very appealing properties, possibly due to its mathematical complexity in derivation. We hope this Chapter may serve as the starting point where the readers will be able to grasp the key features of the cumulant solution to RTE and appreciate the power of this Cumulant method from some salient diverse applications in biophotonics and remote sensing.

In this approach we assume that the phase function depends only on the scattering angle. Under this assumption, we can handle an arbitrary phase function for obtaining the photon distribution (specific intensity), $I(\mathbf{r}, \mathbf{s}, t)$, as a function of time t, position $\mathbf{r}$ and direction $\mathbf{s}$, and the particle density distribution $N(\mathbf{r}, t)$. Specifically, the exact expression of the total angular distribution, $F(\mathbf{s}, \mathbf{s}_0, t)$, as a function of time in an infinite uniform medium is derived first where $\mathbf{s}_0$ is the incident direction of the beam at time $t = 0$. Based on this angular distribution, we then derive exact spatial cumulants of $I(\mathbf{r}, \mathbf{s}, t)$ up to an arbitrary order at any angle and time [11, 14, 18]. A cut-off at the second order of cumulants, $I(\mathbf{r}, \mathbf{s}, t)$ and $N(\mathbf{r}, t)$ can be expressed by Gaussian distributions [2, 5, 8], which have exact first cumulant (the position of the center of the distribution) and exact second cumulant (the half-width of the spread of the distribution). The solution has been extended to the case of polarized photon distribution although we will limit the discussion to scalar photons alone in this chapter [10, 16, 17]. By use of a perturbative method, the distribution in a weak heterogeneous scattering medium can be computed [6, 13, 14]. This establishes the foundation to use the cumulant solution to RTE as the forward model. The cumulant solution to RTE has been successfully applied in optical imaging beyond the diffusion approximation, for example, constructing a transport forward model for optical imaging, cloud remote sensing, and early photon tomography for imaging.

This Chapter is organized as follows: In Section 10.2, expressions of spatial cumulants of the distribution up to an arbitrary order are derived. In Section 10.3, the explicit Gaussian forms of the distribution (up to the second order cumulant) are presented; and numerical results of the solutions are shown. In Section 10.4, the photon transport forward model (PTFM) is first presented together with the strategies of extending the cumulant solution to RTE inside an infinite uniform medium to planar geometries. Applications of the cumulant solution of radiative transfer in areas of early time gated technology [21] where early photon tomography allows fluorescence detection of lung carcinomas and disease progression in mice in vivo [21], and remote sensing for determining size and density of water drop in cloud [20] are presented afterwards. This chapter is finally summarized in Section 10.5.

10.2 Derivation of Cumulants to an Arbitrary Order

The elastic Boltzmann kinetic equation of light for the distribution function $I(\mathbf{r}, \mathbf{s}, t)$ as a function of time t, position $\mathbf{r}$ and direction $\mathbf{s}$, in an infinite uniform medium, from a point pulse light source $\delta(\mathbf{r} - \mathbf{r}_0)\delta(\mathbf{s} - \mathbf{s}_0)\delta(t - 0)$ is given by:

$$\partial I(\mathbf{r},\mathbf{s},t)/\partial t + c\mathbf{s}\cdot\nabla_r I(\mathbf{r},\mathbf{s},t) + \mu_a I(\mathbf{r},\mathbf{s},t)$$
$$= \mu_s \int P(\mathbf{s},\mathbf{s}')[I(\mathbf{r},\mathbf{s}',t) - I(\mathbf{r},\mathbf{s},t)]d\mathbf{s}' + \delta(\mathbf{r}-\mathbf{r}_0)\delta(\mathbf{s}-\mathbf{s}_0)\delta(t-0), \tag{10.1}$$

where c is the speed of light, μ_s is the scattering rate, μ_a is the absorption rate, and $P(\mathbf{s}, \mathbf{s}')$ is the phase function, normalized to $\int P(\mathbf{s},\mathbf{s}')d\mathbf{s}' = 1$.

When the phase function depends only on the scattering angle in an isotropic medium, we can expand the phase function in Legendre polynomials,

$$P(\mathbf{s},\mathbf{s}') = \frac{1}{4\pi}\sum_l a_l P_l[\cos(\mathbf{s}\cdot\mathbf{s}')], \tag{10.2}$$

where a_l are coefficients determined by the phase function. A difficulty in solving Eq. (10.1) originates from the term

$$c\mathbf{s}\cdot\nabla_r I(\mathbf{r},\mathbf{s},t),$$

which couples components of $I(\mathbf{r}, \mathbf{s}, t)$ – spherical harmonics of different orders with each other. We first study the dynamics of the distribution in the direction space, $F(\mathbf{s}, \mathbf{s}_0, t) = \int I(\mathbf{r}, \mathbf{s}, t)d\mathbf{r}$, on a spherical surface of unit radius. The kinetic equation for $F(\mathbf{s}, \mathbf{s}_0, t)$, can be obtained by integrating Eq. (10.1) over the whole spatial space, $\mathbf{r}$. The spatial independence of μ_s, μ_a and $P(\mathbf{s}, \mathbf{s}')$ ensures translation invariance. Thus the integral of Eq. (10.1) obeys

$$\partial F(\mathbf{s},\mathbf{s}_0,t)/\partial t + \mu_a F(\mathbf{s},\mathbf{s}_0,t) + \mu_s[F(\mathbf{s},\mathbf{s}_0,t)$$
$$-\int P(\mathbf{s},\mathbf{s}')F(\mathbf{s}',\mathbf{s}_0,t)d\mathbf{s}'] = \delta(\mathbf{s}-\mathbf{s}_0)\delta(t-0) \tag{10.3}$$

since the integral of the gradient term over all space vanishes. In contrast to Eq. (10.1), upon expanding $F(\mathbf{s}, \mathbf{s}', t)$ in spherical harmonics, its components do not couple with each other. Therefore, it is easy to obtain the exact solution of Eq. (10.3):

$$F(\mathbf{s},\mathbf{s}_0,t) = \exp(-\mu_a t)\sum_l \frac{2l+1}{4\pi}\exp(-g_l t)p_l[\cos(\mathbf{s}\cdot\mathbf{s})], \tag{10.4}$$

where $g_l = \mu_s[1 - a_l/(2l + 1)]$. Two special values of g_l are: $g_0 = 0$, which follows from the normalization of $P(\mathbf{s}, \mathbf{s}')$ and $g_1 = c/l_{tr}$, where l_{tr} is the transport mean free path, defined by $l_{tr} = c/[\mu_s(1 - \langle\cos\theta\rangle)]$, where $\langle\cos\theta\rangle$ is the average of the cosine of the scattering angle θ. Equation (10.4) serves as the exact Green's function of light propagation in the velocity space. Since in an infinite uniform medium this function is independent of the source position, $\mathbf{r}_0$, the requirements for a Green's function are satisfied. Indeed, in an infinite uniform medium, this propagator determines the particle migration behavior, including its spatial distribution, because displacement is an integration of velocity over time. The distribution function $I(\mathbf{r}, \mathbf{s}, t)$ (the source is located at $\mathbf{r}_0 = 0$) is given by

$$I(\mathbf{r},\mathbf{s},t) = \langle\delta(\mathbf{r} - c\int_0^t \mathbf{s}(t')dt')\delta(\mathbf{s}(t)-\mathbf{s})\rangle, \tag{10.5}$$

where $\langle\ldots\rangle$ means the ensemble average in the velocity space. The first delta function imposes that the displacement is given by the path integral. The second delta function ensures the correct final value of direction. Equation (10.5) is an exact formal solution of Eq. (10.1), but cannot be evaluated directly. We make a Fourier transform for the first delta function in Eq. (10.5), then make a cumulant expansion, defined by [22]

$$\langle\exp(ikx)\rangle = \left(\exp\sum_{n=1}^{\infty}\langle x^n\rangle_c (ik)^n/n!\right) \tag{10.6}$$

where the first cumulant $\langle x^n\rangle_c = \langle x\rangle$ is the central position of x, $\langle x^2\rangle_c = \langle x^2\rangle - \langle x\rangle\langle x\rangle$ is the half-width of the distribution. $\langle x^3\rangle_c$ describes the skewness or asymmetry of the distribution, $\langle x^4\rangle_c$ describes the kurtosis or the difference from the Gaussian shaped bell form, etc. We obtain

$$I(\mathbf{r},\mathbf{s},t) = F(\mathbf{s},\mathbf{s}_0,t)\frac{1}{(2\pi)^3}\int d\mathbf{k}$$
$$\exp\left\{i\mathbf{k}\cdot\mathbf{r} + \sum_{n=1}^{\infty}\frac{(-ic)^n}{n!}\sum_{j_n}\ldots\sum_{j_1}k_{j_n}\ldots k_{j_1}\langle\int_0^t dt_n\ldots\int_0^t dt_1 T[\mathbf{s}_{j_n}(t_n)\ldots\mathbf{s}_{j_1}(t_1)]\rangle_c\right\}, \tag{10.7}$$

where T denotes time-ordered multiplication. In Eq. (10.7), the subscript "c" denotes the cumulant. As shown above, cumulants are just centered moments. Hence, if moments $\langle A^m \rangle$ (m = 1, 2, ..., n) have been calculated, $\langle A^m \rangle_c$ can be recursively obtained, and vice versa.

In the following, we derive the analytical expression for the moment ensemble average. Using a standard time-dependent Green's function approach, it is given by

$$\langle \int_0^t dt_n ... \int_0^t dt_1 T[\mathbf{s}_{j_n}(t_n) ... \mathbf{s}_{j_1}(t_1)] \rangle = \frac{1}{F(\mathbf{s}, \mathbf{s}^{(0)}, t)} \left\{ \int_0^t dt_n \int_0^{t_n} dt_{n-1} ... \int_0^{t_2} dt_1 \int d\mathbf{s}^{(n)} ... \right.$$

$$\left. \int d\mathbf{s}^{(1)} F(\mathbf{s}, \mathbf{s}^{(n)}, t - t_n) \mathbf{s}_{j_n}^{(n)} F(\mathbf{s}^{(n)}, \mathbf{s}^{(n-1)}, t_n - t_{n-1}) ... \mathbf{s}_{j_1}^{(1)} F(\mathbf{s}^{(1)}, \mathbf{s}^{(0)}, t_1 - 0) + \text{permut.} \right\} \tag{10.8}$$

where the word "permut." means all $n! - 1$ terms obtained by permutation of $\{j_i\}$, i = 1, ..., n, from the first term. An intuitive way to understand Eq. (10.8) is to use a basic concept in quantum mechanics that the left side of the equation is written in the Heisenberg representation while the right side of the equation is written in the Schrodinger representation. Equation (10.8) can also be understood as a Feymann path integral.

In Eq. (10.8), $F(s^{(i)}, s^{(i-1)}, t_i, t_{i-1})$ is given by Eq. (10.4). Since Eq. (10.4) is exact, Eq. (10.8) provides the exact n-th order moments of the distribution.

In Cartesian coordinates three components of s are s_x, s_y, s_z. For convenience in calculation, however, we will use the components of s on the base of spherical harmonics:

$$\mathbf{s} = [s_1, s_0, s_{-1}] = [Y_{11}(s), Y_{10}(s), Y_{1-1}(s)] = \left[-\frac{1}{2^{1/2}} \sin\theta e^{+i\varphi}, \cos\theta, \frac{1}{2^{1/2}} \sin\theta e^{-i\varphi} \right] \tag{10.9}$$

The recurrence relation of the spherical harmonics is given by

$$Y_{lm}(\mathbf{s}) Y_{1j}(\mathbf{s}) = \sum_i Y_{l+i,\, m+j}(\mathbf{s}) \langle l, 1, m, j \,|\, l+i, m+j \rangle \langle l, 1, 0, 0 \,|\, l, 0 \rangle, \tag{10.10}$$

where $i = \pm 1$, $\langle l, 1, m, j \,|\, l + i, m + j \rangle$ is the Clebsch–Gordan coefficients of angular momentum theory [23]. The orthogonality relation of spherical harmonics is given by

$$\int d\mathbf{s} Y_{l'm'}^*(\mathbf{s}) Y_{lm}(\mathbf{s}) = \frac{4\pi}{2l+1} \delta_{l,l'} \delta_{m,m'} \tag{10.11}$$

Using Eq. (10.10) and Eq. (10.11), the integrals over $d\mathbf{s}^{(n)}$ ... $d\mathbf{s}^{(1)}$ in Eq. (10.8) can be analytically performed. We obtain, when $\mathbf{s}_0$ is set to along $\mathbf{z}$, that

$$\langle \int_0^t dt_n \ldots \int_0^t dt_1 T[\mathbf{s}_{j_n}(t_n) \ldots \mathbf{s}_{j_1}(t_1)] \rangle$$

$$= \frac{1}{F(\mathbf{s}, \mathbf{s}^{(0)}, t)} \left\{ \sum_l Y_{\left[l, \sum_{m=1}^n j_m\right]}(\mathbf{s}) \sum_{i_n} \ldots \sum_{i_1} \frac{2\left(l - \sum_{m=1}^n i_m\right) + 1}{4\pi} \right.$$

$$\prod_{k=1}^n \left\langle l - \sum_{m=1}^{n-k+1} i_{n-m+1}, 1, \sum_{m=1}^{k-1} j_m, j_k \middle| l - \sum_{m=1}^{n-k} i_{n-m+1}, \sum_{m=1}^{k} j_m \right\rangle \left\langle l - \sum_{m=1}^{n-k+1} i_{n-m+1}, 1, 0, 0 \middle| \right.$$

$$\left. l - \sum_{m=1}^{n-k} i_{n-m+1}, 0 \right\rangle D^l_{i_n, \ldots, i_1}(t) + \text{permut.} \Bigg\} \tag{10.12}$$

with $i_f = \pm 1, f = 1, 2, \ldots n$, and

$$D^l_{i_n \ldots i_1}(t) = \exp(-\mu_a t) \left\{ \int_0^t dt_n \int_0^{t_n} dt_{n-1} \ldots \int_0^{t_2} dt_1 \right.$$
$$\exp[-g_l(t - t_n)] \exp[-g_{l-i_n}(t_n - t_{n-1})] \ldots$$
$$\left. \exp[-g_{l-\sum_{k=1}^n i_{n-k+1}}(t_1 - 0)] \right\} \tag{10.13}$$

Note that all ensemble averages have been performed. Equation (10.13) involves integrals of exponential functions, which can be analytically performed. Equation (10.13) includes all related scattering and absorption parameters, g_l, $l = 0, 1, \ldots$, and μ_a, that determines the time evolution dynamics. The final particle direction, $\mathbf{s}$, appears as the argument of the spherical harmonics $Y_{lm}(\mathbf{s})$ in Eq. (10.12). Substituting Eq. (10.13) into Eq. (10.12), and using a standard cumulant procedure, the cumulants as functions of angle $\mathbf{s}$ and time t up to an arbitrary n-th order can be analytically computed. The final position, $\mathbf{r}$, appears in Eq. (10.7), and its component can be expressed as $|\mathbf{r}|$ and Y_{1j}, $j = 1, 0, -1$. Then, performing a numerical three-dimensional inverse Fourier transform over $\mathbf{k}$, an approximate distribution function, $I(\mathbf{r}, \mathbf{s}, t)$, accurate up to the n-th order cumulant, is obtained.

10.3 Gaussian Approximation of the Distribution Function

By a cut-off at the second order cumulant, the integral over $\mathbf{k}$ in Eq. (10.7) can be analytically performed, which directly leads to a Gaussian spatial distribution, which will be displayed in Eq. (10.16). The exact first cumulant provides the correct center position of the distribution. The exact second cumulant provides the correct half-width of spread of the distribution. The expressions below are given in Cartesian coordinates with subscripts α, β = [x, y, z]. These expressions is obtained by use of the following unitary transform $\mathbf{s}_\alpha = \sum U_{\alpha j}\mathbf{s}_j$, j = 1, 0, −1, from Eq. (10.12) (up to the second order cumulant) which is based on $s_j = Y_{1j}(\mathbf{s})$, with

$$U=\begin{bmatrix} -2^{-1/2} & 0 & 2^{-1/2} \\ 2^{-1/2}i & 0 & 2^{1/2}i \\ 0 & 1 & 0 \end{bmatrix} \tag{10.14}$$

We set $\mathbf{s}_0$ along the z direction and denote $\mathbf{s}$ as (θ, ϕ). Our cumulant approximation to the distribution function reduces to

$$I(\mathbf{r},\mathbf{s},t)=\frac{F(\mathbf{s},\mathbf{s}_0,t)}{(4\pi)^{3/2}}\frac{1}{(\det B)^{1/2}}\exp\left[-\frac{1}{4}(B^{-1})_{\alpha\beta}(\mathbf{r}-\mathbf{r}^c)_\alpha(\mathbf{r}-\mathbf{r}^c)_\beta\right] \tag{10.15}$$

with the center of the packet (the first cumulant), denoted by $\mathbf{r}^c$, located at

$$r_z^c = G\sum_l A_l P_l(\cos\theta)[(l+1)f(g_l - g_{l+1}) + lf(g_l - g_{l-1})] \tag{10.16}$$

$$r_x^c = G\sum_l A_l P_l^{(1)}(\cos\theta)\cos\varphi[f(g_l - g_{l-1}) - f(g_l - g_{l+1})], \tag{10.17}$$

where $G = c\exp(-\mu_a t)/F(\mathbf{s}, \mathrm{s}_0, t)$, $A_l = (1/4\pi)\exp(-g_{lt})$, and

$$f(x)=[\exp(xt)-1]/x \tag{10.18}$$

r_y^c is obtained by replacing $\cos\theta$ in Eq. (10.18) by $\sin\theta$. In Eqs. (10.17), (10.18), $P_l^{(m)}(\cos\theta)$ is the associated Legendre function.

The square of the average spread width (the second order cumulant) is given by

$$B_{\alpha\beta} = cG\Delta_{\alpha\beta} - r_\alpha^c r_\beta^c/2, \tag{10.19}$$

where all the coefficients are functions of angle and time:

$$\Delta_{zz} = \sum_l A_l P_l(\cos\theta)\begin{bmatrix} \dfrac{l(l-1)}{2l-1}E_l^{(1)} + \dfrac{(l+1)(l+2)}{2l+3}E_l^{(2)} \\ +\dfrac{l^2}{2l-1}E_l^{(3)} + \dfrac{(l+1)^2}{2l+3}E_l^{(4)} \end{bmatrix} \tag{10.20}$$

$$\Delta_{xx,yy} = \sum_l \frac{1}{2}A_l P_l(\cos\theta)\begin{bmatrix} -\dfrac{l(l-1)}{2l-1}E_l^{(1)} - \dfrac{(l+1)(l+2)}{2l+3}E_l^{(2)} + \\ \dfrac{l(l-1)}{2l-1}E_l^{(3)} + \dfrac{(l+1)(l+2)}{2l+3}E_l^{(4)} \end{bmatrix}$$
$$\pm\sum_l \frac{1}{2}A_l P_l^{(2)}(\cos\theta)\cos(2\varphi)\begin{bmatrix} \dfrac{1}{2l-1}E_l^{(1)} + \dfrac{1}{2l+3}E_l^{(2)} - \\ \dfrac{1}{2l-1}E_l^{(3)} - \dfrac{1}{2l+3}E_l^{(4)} \end{bmatrix} \tag{10.21}$$

where (+) corresponds to Δ_{xx} and (–) corresponds to Δ_{yy},

$$\Delta_{xy} = \sum_l \frac{1}{2}A_l P_l^{(2)}(\cos\theta)\sin(2\varphi)\begin{bmatrix} \dfrac{1}{2l-1}E_l^{(1)} + \dfrac{1}{2l+3}E_l^{(2)} \\ -\dfrac{1}{2l-1}E_l^{(3)} - \dfrac{1}{2l+3}E_l^{(4)} \end{bmatrix} \tag{10.22}$$

$$\Delta_{xz} = \sum_l \frac{1}{2}A_l P_l^{(1)}(\cos\theta)\cos(\varphi)\begin{bmatrix} \dfrac{2(l-1)}{2l-1}E_l^{(1)} - \dfrac{2(l+2)}{2l+3}E_l^{(2)} \\ +\dfrac{1}{2l-1}E_l^{(3)} + \dfrac{1}{2l+3}E_l^{(4)} \end{bmatrix} \tag{10.23}$$

Δ_{yz} is obtained by replacing $\cos(\theta)$ in Eq. (3.10) by $\sin(\theta)$. In Eqs. (10.21– 10.24)

$$E_l^{(1)} = [f(g_l - g_{l-2}) - f(g_l - g_{l-1})]/(g_{l-1} - g_{l-2}) \tag{10.24}$$

$$E_l^{(2)} = [f(g_l - g_{l+2}) - f(g_l - g_{l+1})]/(g_{l+1} - g_{l+2}) \tag{10.25}$$

$$E_l^{(3)} = [f(g_l - g_{l-1}) - t]/(g_l - g_{l-1}) \tag{10.26}$$

$$E_l^{(4)} = [f(g_l - g_{l+1}) - t]/(g_l - g_{l+1}) \tag{10.27}$$

The photon density $N(\mathbf{r}, t)$ of the second cumulant solution is given by

$$N(\mathbf{r},t) = \frac{1}{(4\pi D_{zz}ct)^{1/2}} \frac{1}{4\pi D_{xx}ct} \exp\left[-\frac{(z-R_z)^2}{4D_{zz}ct}\right] \exp\left[-\frac{(x^2+y^2)}{4D_{xx}ct}\right] \exp(-\mu_a t) \tag{10.28}$$

with the mean position

$$R_z = c[1-\exp(-g_1 t)]/g_1 \tag{10.29}$$

The corresponding time-dependent diffusion coefficients are

$$D_{zz} = \frac{c}{3t}\left\{\begin{array}{l} \dfrac{t}{g_1} + \dfrac{3g_1 - g_2}{g_1^2(g_1 - g_2)}[1-\exp(-g_1 t)] + \dfrac{2}{g_2(g_1 - g_2)} \\ [1-\exp(-g_2 t)] - \dfrac{3}{2g_1^2}[1-\exp(-g_1 t)]^2 \end{array}\right\} \tag{10.30}$$

$$D_{xx} = D_{yy} = \frac{c}{3t}\left\{\begin{array}{l} \dfrac{t}{g_1} + \dfrac{g_2}{g_1^2(g_1 - g_2)}[1-\exp(-g_1 t)] \\ -\dfrac{1}{g_2(g_1 - g_2)}[1-\exp(-g_2 t)] \end{array}\right\} \tag{10.31}$$

We see that, at very early time, the diffusion coefficients is near zero, and the center of motion moves with speed c. This stage is ballistic-like mode. With increase of time, the motion of center slow down and the diffusion coefficients increase from zero. This stage of photon migration is often called a "snack-like mode." At large times, the center of motion tends toward $1l_{\text{tr}}$ and diffusion coefficients to $l_{\text{tr}}/3$, which is the result of the diffusion mode. Figure 10.1 shows the moving center of photons,

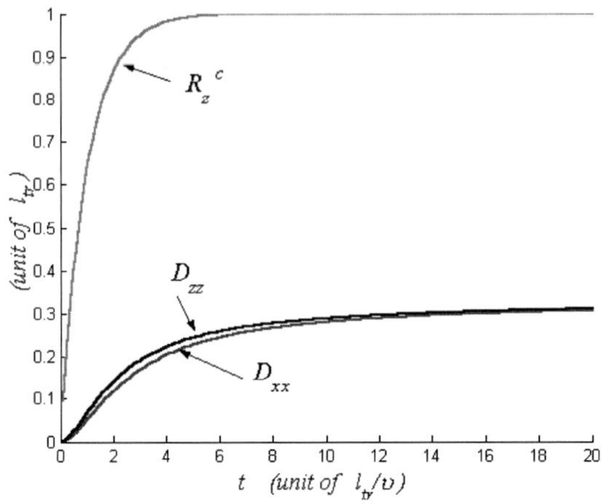

Figure 10.1 The moving center of photons, R_z^c, and the diffusion coefficients, D_{zz} and D_{xx}, as function of time, where g_l are calculated by Mie theory, assuming water drops with $a/\lambda = 1$, with a the radius of droplet and λ the wavelength of light, and the index of refraction $m = 1.33$.

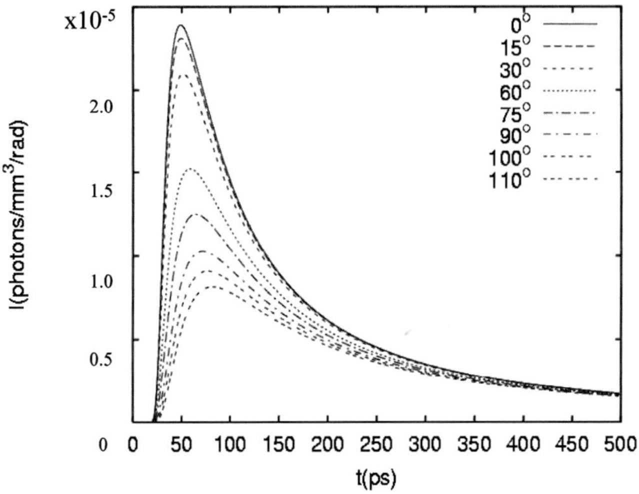

Figure 10.2 The time-resolved profile of light at different angles measured on a detector 10 mm from the source in the incident direction. The parameters are l_{tr} = 2 mm, l_a = 300 mm, the phase function is computed using Mie theory for polystyrene spheres in water, with diameter d = 1.11 μm, and the wavelength of laser source λ = 625 nm, which gives the g-factor g = 0.926.

R_{z}^{c}, and the diffusion coefficients, D_{zz} and D_{xx}, as function of time. Figures 10.2 and 10.3 show the time-resolved intensity profile of light scattered into different forward and backward directions computed with Eq. (10.15). Figure 10.3 shows the second order cumulant solution for backscattered light agrees well with Monte Carlo simulations. Figure 10.4 compares various orders of cumulant solution for light transmission into the forward direction to Monte Carlo simulations.

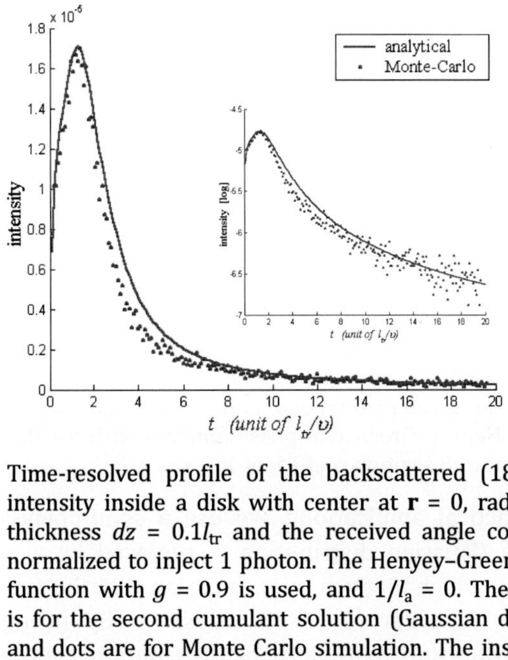

Figure 10.3 Time-resolved profile of the backscattered (180°) photon intensity inside a disk with center at $\mathbf{r} = 0$, radius $R = 1l_{tr}$, thickness $dz = 0.1l_{tr}$ and the received angle $\cos\theta = 0.001$, normalized to inject 1 photon. The Henyey–Greenstein phase function with $g = 0.9$ is used, and $1/l_a = 0$. The solid curve is for the second cumulant solution (Gaussian distribution), and dots are for Monte Carlo simulation. The insert diagram is the same result drawn using a log scale for intensity.

The analytical solution obtained, although it has exact center and half-width, is not satisfied in two aspects. First at very early times, $\exp(-g_l t) \to 1$, for all l, hence one cannot ensure summation over l to be convergent. Second, particles at front edge of the Gaussian distribution travel faster than speed c. thus violating causality.

In order to make the summation over l convergent, we separate the ballistic component from the total distribution and the compute the cumulants for the scattered component. By use of the Legendre expansion of delta function $\delta(\mathbf{s}-\mathbf{s}_0)$, the first and second cumulant expressions of the scattered component can be easily obtained.

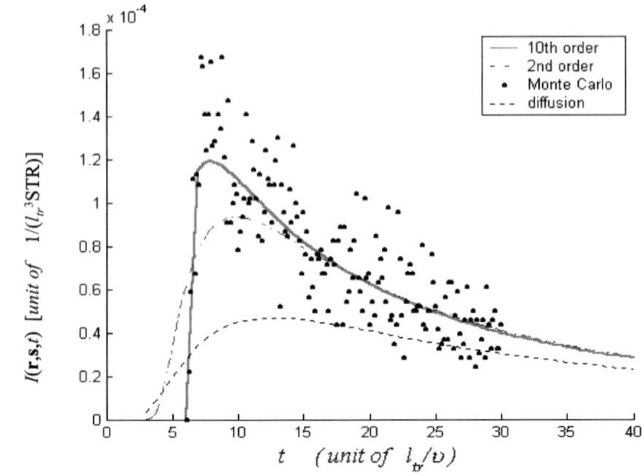

Figure 10.4 Time-resolved profile of transmission light in an infinite uniform medium, computed using the 10th order of cumulants solution (solid curve), the second cumulant solution (dotted curve) and the diffusion approximation (thick dotted curve), comparing with that of Monte Carlo simulation (discrete dots). The detector is located at $z = 6l_{tr}$ from source along the incident direction, and received direction is $\theta = 0$. The Henyey–Greenstein phase function with $g = 0.9$ is used, and the absorption coefficient $1/l_a = 0$.

For practical applications, we use a semi-phenomenological model. The Gaussian distribution is replaced by a new shaped distribution, which is given by a Gaussian distribution, multiplied by a factor in order to ensure causality. By fitting the exact center position and half-width of distribution computed by the cumulant solution, the related parameters of the new shaped distribution can be determined. The formulation for the above improvement of cumulant solution is given in [11, 24].

10.4 Applications of the Cumulant Solution of Radiative Transfer

10.4.1 Transport Forward Model for Optical Imaging

A perturbative method can be used to obtain the forward model when weak inhomogeneities are introduced in the otherwise uniform medium. Making a perturbation expansion of the radiative transfer Eq. (10.1) to the first order Born approximation, the change in photon distribution from, $I^{(0)}(\mathbf{r},\mathbf{s},t)$, that of the uniform background, is derived as

$$\delta I(\mathbf{r},\mathbf{s},t\,|\,\mathbf{r}_0,\mathbf{s}_0,t_0) = \int dt' \int d\mathbf{r}' \int d\mathbf{s}' I^{(0)}(\mathbf{r}',-\mathbf{s}',t-t'\,|\,\mathbf{r},-\mathbf{s})$$
$$\times\{\int d\mathbf{s}''\,\delta\big[\mu_s P\big](\mathbf{s}',\mathbf{s}'',\mathbf{r}')I^{(0)}(\mathbf{r}',\mathbf{s}'',t-t_0\,|\mathbf{r}_0,\mathbf{s}_0)$$
$$-\big[\delta\mu_s(\mathbf{r}')+\delta\mu_a(\mathbf{r}')\big]I^{(0)}(\mathbf{r}',\mathbf{s}',t-t_0\,|\,\mathbf{r}_0,\mathbf{s}_0)\} \tag{10.32}$$

where $\delta\mu_a$, $\delta\mu_s$, and $\delta[\mu_s P]$ are changes of the absorption coefficient, the scattering coefficient and the angular dependent differential scattering coefficient, from the background to the inhomogeneity, respectively. The optical reciprocity theorem is used in obtaining Eq. (10.33). Equation (10.15) can be used to evaluate $\delta I(\mathbf{r},\mathbf{s},t)$ straightforwardly.

One particularly attractive approach is first to approximate $I^{(0)}(\mathbf{r},\mathbf{s},t)$ in terms of the photon density $N^{(0)}(\mathbf{r},t)$. Specifically, we will assume the photon distribution in an infinite uniform medium, $I^{(0)}(\mathbf{r},\mathbf{s},t)$, to take the form:

$$I^{(0)}(\mathbf{r},\mathbf{s},t\,|\,\mathbf{r}_0,\mathbf{s}_0,t_0) = N^{(0)}(\mathbf{r},t\,|\mathbf{r}_0,\mathbf{s}_0,t)F^{(0)}(\mathbf{s},t\,|\,\mathbf{s}_0,t_0)$$
$$-\frac{3}{4\pi}D(t-t_0)\mathbf{s}\cdot\nabla_r N^{(0)}(\mathbf{r},t\,|\mathbf{r}_0,\mathbf{s}_0,t), \tag{10.33}$$

where $N^{(0)}(\mathbf{r},t/\mathbf{r}_0,\mathbf{s}_0,t_0)$ is photon density for a point pulse propagating along $\mathbf{s}_0$ at position $\mathbf{r}_0$ and time t_0 in an infinite uniform medium, $F^{(0)}(\mathbf{s},t/\mathbf{s}_0,t_0)$ is the known exact photon distribution in light direction space (Eq. 10.4), and $D(t-t_0)$ is the time-dependent diffusion coefficient. The diffusion coefficient $D(t)$ is taken to be an average of the time-dependent semi-major axis D_{zz} and semi-minor axes $D_{xx} = D_{yy}$ of the diffusion coefficient ellipsoid in Eqs. (10.31–10.32),

$$D(t) = \frac{D_{xx}+D_{yy}+D_{zz}}{3} = \frac{c}{3t}\left\{\frac{t}{g_1} - \frac{1}{g_1^2}[1-\exp(-g_1 t)] - \frac{1}{2g_1^2}[1-\exp(-g_1 t)]^2\right\} \tag{10.34}$$

The photon density is adapted from Eq. (10.29) to be:

$$N(\mathbf{r},t\,|\,\mathbf{r}_0,\mathbf{s}_0,t_0)=\frac{1}{[4\pi D(t-t_0)t]^{3/2}}\exp\left[-\frac{(\mathbf{r}-\mathbf{r}_0-\mathbf{s}_0\Delta(t-t_0))^2}{4D(t-t_0)(t-t_0)}\right] \quad (10.35)$$
$$\exp(-\mu_a(t-t_0)),$$

where $\Delta(t) = c[1 - \exp(-g_1 t)]/g_1$ is the average center of photons which moves with speed c initially and stops at $c/g_1 = l_{tr}$ in the long time limit.

Expanding $\delta[\mu_s P]$ in Legendre polynomials and substituting Eq. (10.32) into Eq. (10.31), the integrations over angular variables in Eq. (10.31) can be analytically performed. We obtain:

$$\delta I(\mathbf{r},\mathbf{s},t\,|\,\mathbf{r}_0,\mathbf{s}_0,t_0) = -\frac{1}{4\pi}\int dt' \int d\mathbf{r}'\delta\mu_a(\mathbf{r}')$$
$$N^{(0)}(\mathbf{r}',t-t'\,|\,\mathbf{r},-\mathbf{s})N^{(0)}(\mathbf{r}',t'-t_0\,|\,\mathbf{r}_0,\mathbf{s}_0)$$
$$+\frac{3}{4\pi}\int dt' \int d\mathbf{r}' D(t-t')D(t'-t_0)$$
$$[\delta\mu_a(\mathbf{r}')+\delta\mu_s'(\mathbf{r}')]\,\nabla_{r'}N^{(0)}(\mathbf{r}',t-t'\,|\,\mathbf{r},-\mathbf{s})$$
$$\bullet\,\nabla_{r'}N^{(0)}(\mathbf{r}',t'-t_0\,|\,\mathbf{r}_0,\mathbf{s}_0)+\frac{3}{4\pi}$$
$$\int dt' \int d\mathbf{r}' D(t-t')[\delta\mu_a(\mathbf{r}')+\delta\mu_s'(\mathbf{r}')]\exp(-g_1 t')$$
$$\times\{N^{(0)}(\mathbf{r}',t'-t_0\,|\,\mathbf{r},-\mathbf{s})\mathbf{s}\bullet\nabla_{r'}N^{(0)}(\mathbf{r}',t-t'\,|\,\mathbf{r}_0,\mathbf{s}_0)$$
$$-\mathbf{s}_0\bullet\nabla_{r'}N^{(0)}(\mathbf{r}',t-t'\,|\,\mathbf{r},-\mathbf{s})N^{(0)}(\mathbf{r}',t'-t_0\,|\,\mathbf{r}_0,\mathbf{s}_0) \quad (10.36)$$

after neglecting fast decaying terms involving $\exp(-2glt)$ for $l > 1$. One salient feature of Eq. (10.36) is that it no longer depends on the higher order moments of the phase function except for its first order through g_1 and $\mu_s' = 1/l_t = g_1/c$ called the reduced scattering coefficient. Equation (10.36), termed as the PTFM, obeys the reciprocal relation. In the long time limit, the term in Eq. (10.36) containing the exponential decay factor $\exp(-g_1 t)$ can be neglected, the change in photon density, $4\pi\delta I(\mathbf{r}, \mathbf{s}, t)$, in PTFM, reduces to the diffuse imaging model (Eq. 14 in [25]).

To see the difference between PTFM and the diffusion models, we consider a point photon pulse $\delta(\mathbf{r})\delta(\mathbf{s} - \hat{z})\delta(t)$ propagating inside an infinite scattering turbid medium with its absorption coefficient $\mu_a = 0$, reduced scattering coefficient $\mu_s' = 1\ \text{mm}^{-1}$, anisotropy $g_1 = 0.9$ and refractive index $n = 1.33$. These optical properties are similar to those of a typical breast tissue. A Henyey-Greenstein phase function [26] is adopted in the following calculations.

The time-resolved profiles of transmission $I^{(0)}(\mathbf{r}, \hat{z}, t)$ at position $\mathbf{r}$ = (0,0,5) mm and back-scattering $I^{(0)}(\mathbf{r}, \hat{z}, t)$ at position $\mathbf{r}$ = (0,2,0) mm are shown in Figs. 10.5(a,b) in solid curves, respectively. For comparison, the photon density divided by 4π from the original diffusion model (ODM) in dashed curve and the center-moved diffusion model (CDM) in dash-dot curve are also plotted.

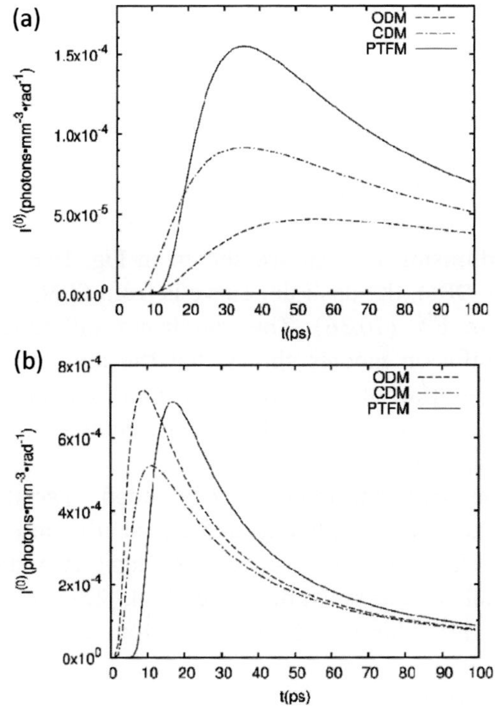

Figure 10.5 The photon distribution profiles (a) at position $\mathbf{r}$ = (0,0,5) mm and (b) at position $\mathbf{r}$ = (0,2,0) mm. The dashed curve is from the original diffusion model (ODM), the dash-dot line is from the center-moved diffusion model (CDM), and the solid curve is from the photon transport forward model (PTFM).

In a transmission case (Fig. 10.5a), the peak of photon intensity in ODM arrives too late compared to the experimental result, while in CDM, the artificial adjustment of the source position for a l_{tr} leads to an arrival of photons faster than light speed. The intensity at forward directions from PTFM is stronger than that of diffusion models, which indicates a certain anisotropic angular distribution remains even at a distance of $5l_{\text{tr}}$ from the source.

In the case of back scattering (Fig. 10.5b), photons diffuse from the origin (0,0,0) to (0,2,0) mm with a constant diffusion coefficient in ODM, while photons diffuse from the adjusted source position (0,0,1) mm to (0,2,0) mm with the same constant diffusion coefficient in CDM. The photons in PTFM are back-scattered to (0,2,0) mm later than those in ODM and CDM because the center of photons moves forward along the positive z-direction and diffuse from the moving center with a gradually increasing diffusion coefficient from 0 to $l_t/3$ in PTFM.

Consider a forward model with a scattering inhomogeneity of unit volume placed at position (0,0,2) mm. The time-resolved profiles of $-\delta I(\mathbf{r}, \hat{z}, t)$ at position $\mathbf{r}$ = (0,0,5) mm and $\delta I(\mathbf{r}, \hat{z}, t)$ at position $\mathbf{r}$ = (0,2,0)mm from the Eq. (10.36) and those from diffusion models are shown in Fig. 10.6. The dashed curve is from ODM, the dash-dot curve is from CDM, and the solid curve is from Eq. (10.36). The significant difference between PTFM and diffusion models shows that the non-diffusive nature of photon migration at early times is important and cannot be neglected when the separation of any pair of source, inhomogeneity and detector is small.

When the scattering medium is bounded, special conditions are needed to set the photon density at the interfaces. The reflection at the interface re-injects the light into the medium. Using a partial current technique, Zhu et. al. [27] showed that the boundary condition for a semi-infinite medium can be written as

$$\left[N^{(0)} - z_e \frac{\partial N^{(0)}}{\partial z} \right]_{z=0} = 0 \tag{10.37}$$

at the interface $z = 0$ where the extrapolation length z_e measures the distance outside the medium where the energy density from the diffusion approximation vanishes linearly. The extrapolated-boundary condition has been successfully employed for planar geometries such as a slab or a semi-infinite medium in diffuse imaging, in which the photon density is set equal to zero at an extrapolated boundary located a distance z_e outside the turbid medium [28]. The method of images is used to obtain the Green's function in such bounded media. The same technique can be applied here to the Green's function $N^{(0)}(\mathbf{r}, t/\mathbf{r}_0, \mathbf{s}_0, t_0)$.

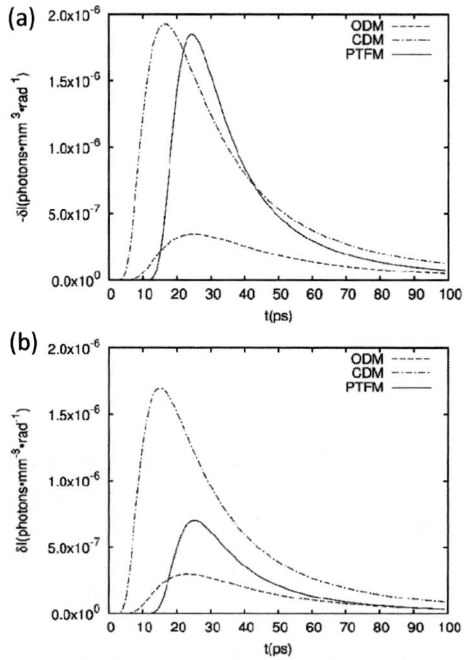

Figure 10.6 The change of photon distribution function (a) $-\delta I(\mathbf{r}, \hat{z}, t)$ at position $\mathbf{r}$ = (0,0,5)mm and (b) $\delta I(\mathbf{r}, \hat{z}, t)$ at position $\mathbf{r}$ = (0,2,0)mm. The dashed line is from ODM, the dash-dot line is from CDM, and the solid curve is from PTFM.

Keeping in mind that the source approaches gradually and stops finally at $\mathbf{r}_0 + \mathbf{s}_0 l_{tr}$ on average with the increase of time, the image of the incident point source at $(x_0, y_0, z_0 \geq 0)$ propagating along the positive z-axis inside a semi-infinite medium with its interface at z = 0 is a negative one at $(x_0, y_0, -z_0 - 2z_e - 2l_{tr})$ propagating along the same direction (Fig. 10.7). At early times, both the source and its image have not arrived at the extrapolated boundary and their contributions at the extrapolated boundary can be neglected. When the time increases, the contributions at the extrapolated boundary from both the source and image tend to cancel each other as both approach their final stops (shadow spots in Fig. 10.7). The shadow spots just represent the positions of the source and its image in the center-moved diffusion approximation. Green's function of a semi-infinite medium given by

$$N^{(0)}_{\text{semi}}(\mathbf{r},t\,|\,\mathbf{r}_0,\mathbf{s}_0) = N^{(0)}(\mathbf{r},t\,|\,x_0,y_0,z_0,\mathbf{s}_0) - N^{(0)}(\mathbf{r},t\,|\,x_0,y_0,-z_0-2z_e-2l_{\text{tr}},\mathbf{s}_0) \quad (10.38)$$

thus approximately satisfies the extrapolated-boundary condition.

The same procedure can be easily applied to a slab with its extrapolated boundaries at $z = 0$ and $z = L$. The images of an incident source at (x_0, y_0, z_0) with $0 \le z_0 \le L$ propagating along positive or negative z-axis ($s_z = \pm 1$) are a set of positives images at $(x_0, y_0, z_0 + 2m\text{L})$ and a set of negative ones at $(x_0, y_0, -z_0 - 2m\text{L} - 2s_z l_t)$, all propagating along the same direction as the source ($-\infty < m < \infty$ is integer).

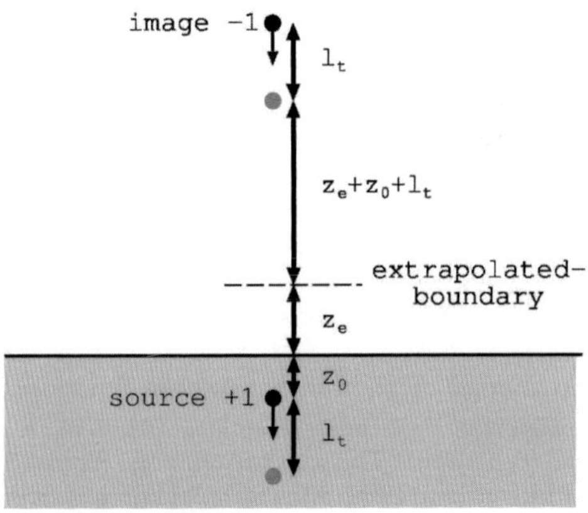

Figure 10.7 The incident source at position $(x_0, y_0, z_0 \ge 0)$ and its image source at $(x_0, y_0, -z_0 - 2z_e - 2l_{\text{tr}})$ propagating along the positive z-axis in a semi-infinite medium ($z \ge 0$) with its interface at $z = 0$. The source and its image move from their original positions (dark spots) to their final stops (shadow spots) at later times.

Further details for photon transport forward model, photon migration and radiative transfer tomography can be found in [6, 13, 14].

10.4.2 Early Photon Tomography (EPT)

M. K. Niedre et al. have developed early-photon technology for performing high-fidelity fluorescence tomography in living tissues (see Fig. 10.8) [21]. The theoretical model of the analytical cumulant solution [Eqs. (10.29)–(10.31)] is applied in this subject, that allowed accurate modeling of light propagation in tissue at early time for early photon tomography of fluorescence detection of lung carcinomas and disease progression in mice in vivo.

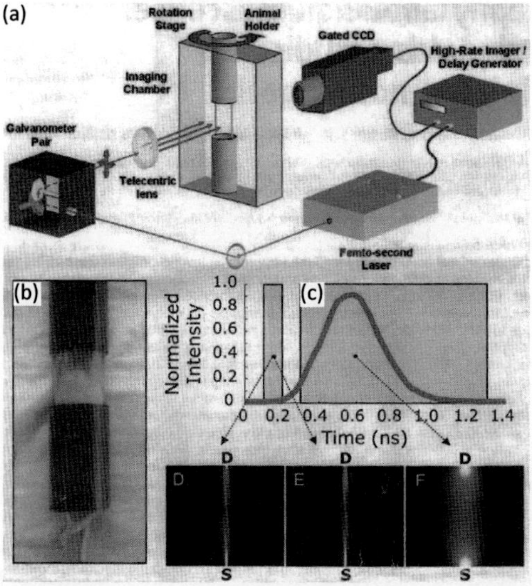

Figure 10.8 The early photon tomography (EPT) approach. (a) Experimental setup. Early arriving photons are detected by using the gated intensified CCD system. The animal is placed in the custom built cylindrical carbon fiber tube (b) for scanning. As photons propagate through the diffusive medium, they disperse temporally (c). The experimentally measured Green's function describing the path of photons propagating between a source (s) and detector (d) pair, measured at early time gate (d) and the early-photon forward model used in the reconstructions (e). This model was calculated by using a normalized cumulant approximation to the Boltzmann transport equation and agreed well with measured weight function. The weight function for diffuse (CW) photons for the same geometry is also shown for comparison (f). The radial spread of early arriving photons is drastically reduced, thereby allowing the excellent resolution possible with the EPT (adapted from Fig. 1 of Ref. [21]).

EPT uses early arriving photons, that is, photons emitted from an ultrafast laser source (i.e., pulse width < 1 ps) that propagate through tissue and are the first to arrive at time-gated detector placed at a distance away from source. The scattering of these early photons is strongly biased in the forward direction and correspondingly they experience a lower number of total scattering events. As a result, they preferentially propagate along significantly less diffusive paths connecting the source and the detector versus un-gated photons and can therefore be used to significantly improve image resolution (see Fig. 10.9).

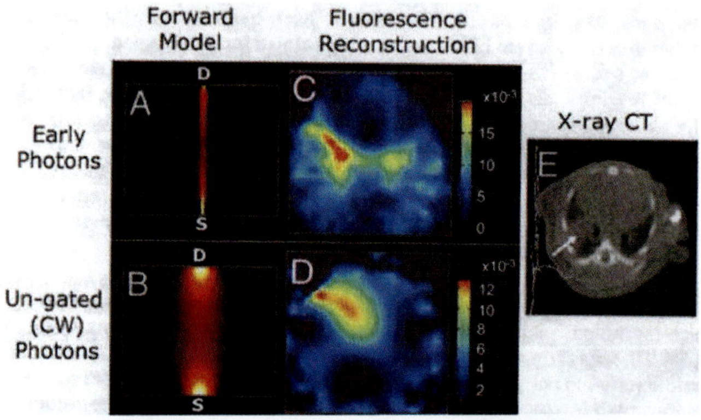

Figure 10.9 Fluorescence reconstructions obtained by using early arriving photons compared with ungated (continuous-wave) photons in mouse with an LLC tumor, 10 days after injection. Forward model for early arriving photons and ungated photons are shown (a and b), as well as the fluorescence reconstructions for the axial slice (c and d), and the corresponding x-ray CT axial slice (e). The use of early photons enables fluorescence reconstructions with superior resolution and localization of the activated fluorophore in the right lobe of the lung, as well as on the contralateral side compared with the ungated reconstruction (adapted from Fig. 5 in Ref. [21]).

10.4.3 Non Medical Use of Retrieving Parameters of Water Cloud from CALIPSO Data

The balance of the earth's radiation budget depends on various factors, including microphysical properties of cloud, such as density and size distribution of droplets in cloud, and the thickness of the cloud layer. The active measurement, pursued by the NASA CALIPSO mission, uses laser light pulses from lidars aboard a satellite to probe the atmosphere, and collects the time-resolved backscattered signal from each single measurement. Since being launched in 2006, CALIPSO has generated huge amounts of data, which are temporal profiles of 532 nm and 1064 nm pulses of light backscattered from earth atmosphere. The size and density of scattering entities in earth atmosphere determine the shape and intensity of the temporal profile of *backscattered* pulses, especially, for early arriving photons.

Difficulty in developing approaches to retrieve dense water cloud parameters from optical data stems from the paucity of adequate theoretical formalism to account for multiple scattering of photons that is prevalent given the high concentration of randomly distributed water droplets inside dense cloud. The approaches [29–32] to address the problem of multiple scattering may include numerical solutions of the radiative transfer equation or Monte Carlo simulation of photon migration in cloud. However, these are cumbersome, computation-time intensive, and are not suitable as forward models for retrieving the cloud parameters from measured data.

The analytical solution of radiative transfer equation based on the cumulant expansion provides a tool, which can compute the photon density distribution function rapidly and accurately taking full consideration of multiple scattering events. For the backscattering case, our results are shown to be in good agreement with Monte-Carlo simulation results.

The phase function of single scattering of light from a water drop, determined by Mie formula, is related to a/λ, where a is radius of the drop and λ is the wavelength of light. The cumulant solution can be used as a forward model for retrieving the density and the average size of water drops in cloud from CALIPSO data [20].

In a finite medium with boundaries, photons that pass through the boundary may not go back into medium again, while in an infinite medium photons can be scattered back into the region. Only those photons that leak out, scatter back into the finite medium from outside, and are scattered out again, account for the difference between the solution in an infinite medium and that in a finite medium. The boundary condition, hence, mainly affects the photon distribution at a later time rather than the early time.

An approximate method, similar to extending the solution of the diffusion equation from an infinite medium to semi-infinite and slab geometries is used [33]. The technique involves adding image-sources to satisfy the "extrapolated" boundary condition. For semi-infinite geometry, it requires the distribution of photons to be zero at an extrapolated plane, $z = -z_e$ with $z_e = \alpha l_{\text{tr}}$, and $\alpha \sim 0.7$ for the vacuum boundary. The physical interface is at $z = 0$. A virtual negative source is added to the original source at $z = -2\alpha l_{\text{tr}}$. At early times, the contribution from the virtual source is negligible. The boundary effect becomes essential at later times. Comparison with Monte Carlo simulation shows that this approach of extension to the semi-infinite geometry is quantitatively sound [6].

In the CALIPSO mission, light backscattering data is collected in the following manner. The satellite flies in an orbit at a height of ~700 km above the earth. The lidars on CALIPSO use linearly polarized laser beams to probe the atmosphere. Nanosecond-duration laser light pulses of wavelength 532 and 1064 nm are injected when satellite flies through different positions of earth. The field of view (FOV) of 130 μrad corresponds to an area of radius $R_{\text{b}} \approx 0.05$ km on the surface of cloud, from which backscattering signal is collected, and will be referred to as "footprint" area in subsequent discussion. For each laser pulse, a group of L1B data, which is the total calibrated attenuated backscattering rate β_i (in unit of $\text{km}^{-1}\ \text{SR}^{-1}$) versus altitude h_i (in unit of km) is collected. Data represent the backscattered intensity of light collected through the footprint area of cloud surface as a function of time (in unit of μsec). Time scale is obtained by mapping the change of altitudes to the round-trip time $\Delta t = 2\Delta h/c$, with c the speed of light.

Figure 10.11 shows image of CALIPSO data when satellite passed through North America on August 7, 2007. Each of 56,295 horizontal positions represents a location on the earth by its longitudinal and latitudinal coordinates, while each of the 583 vertical positions represents the altitude measured from the earth surface. Strength of the attenuated backscattering (in unit of $\text{km}^{-1}\text{SR}^{-1}$) is represented by pseudo color according to the scaled color bar shown in the figure. Each of these vertical positions correspond to a point on the temporal profile data based on the $\Delta t = 2\Delta h/c$ mapping. The white colored points represent the strongest signals backscattered from dense water cloud, while the white-gray colored points represent the weaker signals from rarer cloud. These signals are several orders of magnitude stronger than the ambient background and signals from other particles in the earth atmosphere. The dark-blue color under cloud indicates the strong decay of the tail part of time resolved signals. The numerical values of measured temporal profiles may be derived from the corresponding hdf-data file.

A piece of cloud shown inside the square frame in Fig. 10.10 is chosen for fitting data to obtain the parameters of cloud. Figure 10.11 shows some examples of fitting. In Figs. 10.11(a–c) the points represent CALIPSO data for the wavelength λ = 532 nm, and the solid curves are theoretical fits using the analytical model. Table 10.1 lists the parameters corresponding to Figs. 10.11(a–c) at different positions on earth, including satellite's flying UTC time, the longitudinal and latitudinal positions, and the fitted parameters of cloud: the transport mean free path l_{tr}, the effective radius of water drops r_d, the anisotropy factor g, the scattering cross section σ_s, the scattering length l_s, the density of drops ρ, and the altitude of top level of major cloud h_{top}, respectively.

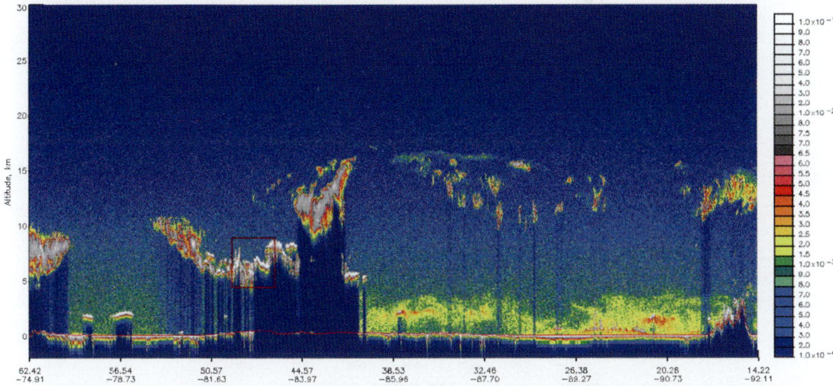

Figure 10.10 Image of CALIPSO L1B data is shown when satellite passed through North America on August 7, 2007. The horizontal points represent positions on the earth by longitudinal and latitudinal coordinates. The vertical levels represent altitude counted from the earth surface. Strength of the attenuated backscattering rate (in unit of km^{-1} SR^{-1}) is represented by pseudo color on the image-map, according to the scale shown in the color bar.

Table 10.1

Figures	10.11a	10.11b	10.11c
UTC time	70807.327223	70807.327323	70807.327017
Latitude (°)	48.139153	47.623989	49.192322
Longitude (°)	−82.636620	−82.838814	−82.212204
l_{tr} (km)	0.1516	0.118	0.334
r_d (μm)	3.0	1.7	0.5
g-factor	0.8397	0.7923	0.8508
σ_s (μm^2)	67.628	18.79	3.0716
l_s (km)	0.02428	0.0245	0.04983
ρ (10^9/m^3)	0.609	2.171	6.532
h_{top} (km)	6.721	6.826	5.260

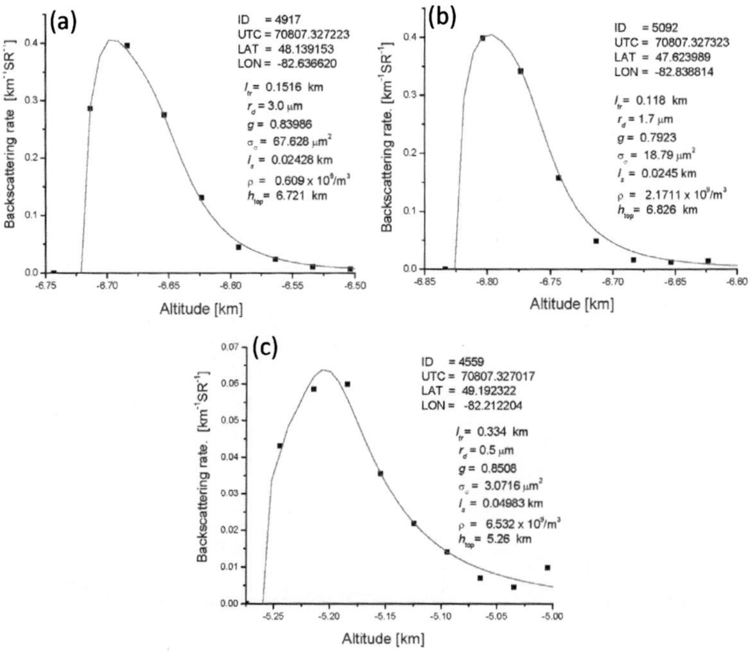

Figure 10.11 (a–c) Fitting of CALIPSO data recorded using 532 nm light (represented by black squares) to the theoretical formalism (shown by solid line) at three different positions on earth for extraction of cloud parameters. The parameters are listed in Table 10.1, and indicated on respective figures.

10.2 Summary

In this chapter, we have reviewed the analytical solution of the time-dependent RTE in an infinite uniform medium with an arbitrary light scattering phase function using cumulant expansion and some applications in imaging with multiply scattered light beyond the diffusion approximation. Analytical expressions have been derived for the exact distribution in angle and the spatial cumulants at any angle, exact up to an arbitrary high order. A simple Gaussian analytical approximate expressions of the photon spatial distribution have also been obtained, whose center position and half-width are always exact at arbitrary time. This provides a clear analytical picture that the center of photon distribution advances and the half-width grows in time, showing the evolution of the particle migration from near ballistic, through snake-like, and into the final diffusive regime. The cumulant solution has one unique advantage comparing with other approximation techniques of solving RTE that truncation of the cumulant expansion at order n is exact at that order and cumulants up to and including order n remain unchanged when contributions from higher orders are added. The cumulant solution to RTE is particularly suited to understand and model the temporal migration of photons inside a turbid medium. Three applications of the cumulant solution for achieving optical imaging beyond diffusion in PTFM, early photon tomography (EPT) and cloud sensing from lidar data have been discussed as examples. We expect the cumulant solution will find more applications in biophotonics for optical imaging and remote sensing with time-resolved data in general.

Acknowledgments

We thank ARO, DOD, NASA, and internal CUNY funds which partly support this research. We also thank Prof. Melvin Lax for his deep insights and exemplary teaching.

References

1. S. Chandrasekhar, *Radiative Transfer*. Dover, New York, 1960.
2. A. Peraiah, *An Introduction to Radiative Transfer: Methods and Applications in Astrophysics*. Cambridge University Press, 2002.

3. R. Wehrse and W. Kalkofen, Advances in radiative transfer, *Astron. Astrophys. Rev.*, vol. 13, no. 1–2, pp. 3–29, 2006.
4. R. Aronson, Boundary conditions for diffusion of light, *J. Opt. Soc. Am. A. Opt. Image Sci. Vis.*, vol. 12, no. 11, pp. 2532–2539, 1995.
5. D. J. Cuccia, F. Bevilacqua, A. J. Durkin, F. R. Ayers, and B. J. Tromberg, Quantitation and mapping of tissue optical properties using modulated imaging, *J. Biomed. Opt.*, vol. 14, no. 2, p. 024012, Jan. 2009.
6. M. Xu, W. Cai, M. Lax, and R. R. Alfano, Photon-transport forward model for imaging in turbid media, *Opt. Lett.*, vol. 26, no. 14, pp. 1066–1068, Jul. 2001.
7. S. V. Patwardhan, A. P. Dhawan, P. A. Relue, and M. Carlo, Monte Carlo simulation of light-tissue interaction: Three-dimensional simulation for trans-illumination-based imaging of skin lesions, *IEEE Trans. Biomed. Eng.*, vol. 52, no. 7, pp. 1227–1236, 2005.
8. M. Xu, Electric field Monte Carlo simulation of polarized light propagation in turbid media, *Opt. Express*, vol. 12, no. 26, pp. 6530–6539, Dec. 2004.
9. J. Sawicki, N. Kastor, and M. Xu, Electric field Monte Carlo simulation of coherent backscattering of polarized light by a turbid medium containing Mie scatterers, *Opt. Express*, vol. 16, no. 8, pp. 5728–5738, Apr. 2008.
10. W. Cai, X. Ni, S. K. Gayen, and R. R. Alfano, Analytical cumulant solution of the vector radiative transfer equation investigates backscattering of circularly polarized light from turbid media, *Phys. Rev. E*, vol. 74, no. 5, p. 056605, Nov. 2006.
11. W. Cai, M. Xu, and R. R. Alfano, Analytical form of the particle distribution based on the cumulant solution of the elastic Boltzmann transport equation, *Phys. Rev. E*, vol. 71, no. 4, p. 041202, Apr. 2005.
12. M. Xu, W. Cai, and R. R. Alfano, Multiple passages of light through an absorption inhomogeneity in optical imaging of turbid media, *Opt. Lett.*, vol. 29, no. 15, pp. 1757–1759, Aug. 2004.
13. W. Cai, M. Xu, and R. R. Alfano, Three-dimensional radiative transfer tomography for turbid media, *IEEE J. Sel. Top. Quantum Electron.*, vol. 9, no. 2, pp. 189–198, Mar. 2003.
14. M. Xu, W. Cai, M. Lax, and R. R. Alfano, Photon migration in turbid media using a cumulant approximation to radiative transfer, *Phys. Rev. E*, vol. 65, no. 6, p. 066609, Jun. 2002.

15. W. Cai, M. Xu, M. Lax, and R. R. Alfano, Diffusion coefficient depends on time, not on absorption, *Opt. Lett.*, vol. 27, no. 9, pp. 731–733, May 2002.

16. W. Cai, M. Lax, and R. R. Alfano, Analytical solution of the polarized photon transport equation in an infinite uniform medium using cumulant expansion, *Phys. Rev. E*, vol. 63, no. 1 Pt 2, p. 016606, Jan. 2001.

17. W. Cai, M. Lax, and R. R. Alfano, Analytical solution of the polarized photon transport equation in an infinite uniform medium using cumulant expansion, *Phys. Rev. E*, vol. 63, no. 1, p. 016606, Dec. 2000.

18. W. Cai, M. Lax, and R. Alfano, Cumulant solution of the elastic Boltzmann transport equation in an infinite uniform medium, *Phys. Rev. E*, vol. 61, no. 4 Pt A, pp. 3871–3876, Apr. 2000.

19. A. Liemert and A. Kienle, Infinite space Green's function of the time-dependent radiative transfer equation, *Biomed. Opt. Express*, vol. 3, no. 3, pp. 543–551, Mar. 2012.

20. W. Cai and S. K. Gayen, Retrieval of water cloud characteristic from active sensor data using the analytical solution of radiative transfer equation, *J. Quant. Spectrosc. Radiat. Transf.*, vol. 111, no. 1, pp. 226–233, Jan. 2010.

21. M. J. Niedre, R. H. de Kleine, E. Aikawa, D. G. Kirsch, R. Weissleder, and V. Ntziachristos, Early photon tomography allows fluorescence detection of lung carcinomas and disease progression in mice in vivo, *PNAS*, vol. 105, no. 49, pp. 19126–19131, Dec. 2008.

22. S. K. Ma, *Statistical Physics*. World Scientific, Philadelphia, 1985.

23. D. M. Brink and G. R. Satchler, *Angular Momentum*. Clarendon Press, Oxford, 1962.

24. M. Lax, W. Cai, and M. Xu, *Random Processes in Physics and Finance*. Oxford University Press, 2006.

25. S. R. Arridge, Photon-measurement density functions. Part I: Analytical forms, *Appl. Opt.*, vol. 34, no. 31, pp. 7395–7409, Nov. 1995.

26. L. C. Henyey and J. L. Greenstein, Diffuse radiation in the Galaxy, *Astrophys. J.*, vol. 93, p. 70, Jan. 1941.

27. J. X. Zhu, D. J. Pine, and D. A. Weitz, Internal reflection of diffusive light in random media, *Phys. Rev. A*, vol. 44, no. 6, pp. 3948–3959, Sep. 1991.

28. M. S. Patterson, B. Chance, and B. C. Wilson, Time resolved reflectance and transmittance for the non-invasive measurement of

tissue optical properties, *Appl. Opt.*, vol. 28, no. 12, pp. 2331–2336, Jun. 1989.

29. A. A. Kokhanovsky, Satellite-based retrieval of ice cloud properties using a semianalytical algorithm, *J. Geophys. Res.*, vol. 110, no. D19, p. D19206, 2005.
30. B. A. Baum, D. P. Kratz, P. Yang, S. C. Ou, Y. Hu, P. F. Soulen, and S.-C. Tsay, Remote sensing of cloud properties using MODIS airborne simulator imagery during SUCCESS: 1. Data and models, *J. Geophys. Res.*, vol. 105, no. D9, p. 11767, May 2000.
31. T. Nakajima and M. D. King, Determination of the optical thickness and effective particle radius of clouds from reflected solar radiation measurements. Part I: Theory, *J. Atmos. Sci.*, vol. 47, no. 15, pp. 1878–1893, Aug. 1990.
32. A. A. Kokhanovsky, P. J. McBride, K. S. Schmidt, and P. Pilewskie, The Determination of cloud optical thickness and effective particle size from measurements of transmitted solar diffuse light, *IEEE Geosci. Remote Sens. Lett.*, vol. 10, no. 6, pp. 1512–1516, Nov. 2013.
33. M. Lax, V. Nayaramamurti, and R. C. Fulton, Classical diffusion photon transport in a slab, in *Laser Optics of Condensed Matter* (J. L. Birman, H. Z. Cummins, and A. A. Kaplyanskii, eds.), Plenum, New York, 1987, pp. 229–237.

Chapter 11

Deep Imaging of Prostate Cancer Using Diffusion Reconstruction of Banana Paths with Near Infrared Prostatoscope Analyzer

Yang Pu,[a] Wubao Wang,[b] Min Xu,[c] James A. Eastham,[d] and Robert R. Alfano[b]

[a]*Departments of Electrical Engineering*
The City College of the City University of New York,
160 Convent Avenue, New York, New York 10031, USA
[b]*Department of Physics, The City College of the City University of New York,*
160 Convent Avenue, New York, New York 10031, USA
[c]*Department of Physics, Fairfield University, Fairfield, Connecticut 06824, USA,*
[d]*Department of Urology, Sidney Kimmel Center for Prostate and Urologic Cancers,*
Memorial Sloan-Kettering Cancer Center, New York, New York 10065, USA
ralfano@ccny.cuny.edu

11.1 Introduction: Screening Cancer Using Light

Optical spectroscopy and imaging using near-infrared (NIR) light provide powerful tools for non-invasive screening and clinical detection of diseased tissues located on surface, subsurface and even in the areas of deep organs [1]. These techniques are

Deep Imaging in Tissue and Biomedical Materials: Using Linear and Nonlinear Optical Methods
Edited by Lingyan Shi and Robert R. Alfano

ISBN 978-981-4745-88-8 (Hardcover), 978-1-315-20655-4 (eBook)
www.panstanford.com

capable of quantitative reconstructions of tissue absorption and scattering properties. Optical images can map in vivo difference of certain bio-markers between normal and cancerous tissues, for example, water H_2O content, oxygenated hemoglobin (HbO_2) and deoxygenated hemoglobin (HbO), to differentiate them. The key advantages of optical approaches are the ability to monitor multiple independent optical reporters simultaneously in vivo based upon wavelength or lifetime, the absence of radioactive intermediates, and the relative simplicity of the imaging hardware as compared to magnetic resonance imaging (MRI) and positron emission tomography (PET) equipment, which makes it unmatched in comparison with any other in vivo imaging techniques [1].

The major disadvantage of optical imaging approaches is that high scattering of biological tissue causes most photons diffused which blurs images. The very few percentage of ballistic and snake photons make direct imaging practically only in surface layer of tissue [2]. Scientists have to explore optical tomography methods and/or inverse image reconstruction approaches to locate the three dimension (3D) position of abnormal tissue or recovers the 3D spatial distribution information of optical parameters of the whole tissue [3]. When light transports in a highly scattering medium such as tissue, the opaque and blurs transmission or backscattering reflectance images are rendered to us. The behavior of diffusion essentially depends on optical properties of the tissue medium which are scattering coefficient (μ_s), anisotropy factor (g) and absorption coefficient (μ_a) [4]. Since the transmittance and backscattered images are acquired from those photons surviving passage from the tissue, which bear clues about their voyage path and the optical coefficients of the tissue [4], the measured light intensity distribution on the boundary of the turbid medium can be used to generate a map of the μ_s and μ_a of the whole tissue medium using an inversion algorithm. The location of the abnormality should be determined as diseased tissue using spectral imaging techniques.

Since the inversion problem is ill posed, the achievement of a unique solution depends on theoretical model and the applied numerical methods as well. Many algorithms were developed and tested in high scattering medium or tissue [3, 5–9]. In particular, Xu et al. introduce the method of independent component analysis

(ICA) from information theory [10, 11] to locate absorptive and scattering inhomogeneities embedded in a thick turbid medium and demonstrate the efficacy in simulated data [6], intralipid scattering medium [11] and human prostate [12] and breast tissues [13]. This information theory-inspired approach is referred as optical imaging using independent component analysis (OPTICA) [14]. OPTICA associates directly the independent components (ICs) to Green's functions responsible for light propagation in the turbid medium from the inhomogeneities to the source and the detector, and therefore the retrieved independent components can be used to locate and characterize the inhomogeneities [6].

This chapter will review and provide a noninvasive optical imaging technique for detecting and 3D-locating cancerous sites in prostate from viewing through the rectum. We have improved the OPTICA-based inverse image reconstruction algorithm specifically for the application of backscattering configuration. The difference between cancerous and normal prostate tissue can be enhanced using a smart receptor-targeted contrast agent (i.e., Cytate) [12]. The scanning polarization imaging acquires experimental data by sequentially scanning a polarized illuminating light beam at different locations of a prostate cancerous tissue sandwiched by two large pieces of prostate normal tissues, and recording the distribution of light intensity backscattered from the prostate samples using a CCD camera. The ICA-based inverse reconstruction algorithm specifically for the application of backscattering configuration is used to locate the 3D positions of inter abnormal prostate tissue from the recorded array of the 2D images. The retrieved independent component (IC) corresponds to the projection of the Green's function of light propagating from the object to the boundary of the tissue medium. The reflection (backscattering) geometry which is most amenable for probing targets hidden in the subsurface of a turbid medium poses the following major difficulties than the transmission geometry: First, the estimation of the clean image of the homogeneous host medium from the measured data set obtained at an array of scanning positions cannot be done simply by averaging on the measured data set. The subsurface, inhomogeneities affect the backscattering images appreciably. Second, the commonly used diffusion model in optical imaging breaks down in general for probing the subsurface, and computational expensive methods have to be used to model backscattering light at a close source-detector separation [11]. Third, the unknown surface property of the biological sample such as superficial structure and roughness impacts the spatial distribution of the backscattering light. All of these issues are critical in analyzing images measured in a backscattering geometry. We addressed this challenge by improving OPTICA in two procedures: (1) synthesizing a "clean" background image (CBI) of a nearly homogeneous host medium numerically from the experimental scanning images of the host medium with embedded target(s) recorded by a large frame CCD camera; (2) numerically marching the target(s) to the surface until matching the retrieved independent component, incorporating both the beam profile and the surface property of the sample [3]. The 3D locations of small piece of Cytate-stained cancerous prostate tissue hidden inside the large host normal prostate tissue were used to verify our new OPTICA algorithm by sets of 2D backscattering images recorded by our specifically developed prostatoscope analyzer scanning imaging unit.

11.2 Theoretical Formalism

The theory of OPTICA was first proposed by Xu et al. [6]. ICA, the core of OPTICA, is a solution to the *blind source separation* problems, which is a class of the problems that no precise knowledge is available on neither the mixing channels nor the sources [6]. Typically the observations are the output of a set of sensors, where each sensor receives a set of mixed source signals [15]. The algorithm of OPTICA is based on that ICA of the perturbations in the spatial intensity distributions provides the corresponding independent intensity distributions (IID) on the detector planes [15].

11.2.1 Clean Image Synthesis

Fast inversion algorithms require as input the scattering field due to the sought-after targets. This scattering field is the difference between the measured intensity distribution of backscattering light and a clean image of the homogeneous host medium without any embedded targets. The image for a target-free host medium is not available in most applications and must be generated from the measured data set of the medium with targets embedded inside. Denote the recorded images $I(\boldsymbol{\rho}_\mathrm{d}, \boldsymbol{\rho}_\mathrm{s})$, where $\boldsymbol{\rho}_\mathrm{d}$ covers the whole 2D array (for example, $\boldsymbol{\rho}_\mathrm{d}$ enumerates all pixels on a CCD image) for a series of beam scanning positions at $\boldsymbol{\rho}_\mathrm{s}$ on the sample surface ($z_\mathrm{d} = z_\mathrm{s} = 0$). One crude method to obtain the image of the host medium is simply the average of all the array images after shifting the scanning position to the origin. This results in, however, only a distorted version of the real background image. To illustrate this point, assume one thin slice absorptive object $\delta\mu_\mathrm{a}$ of volume ΔV is located at $\mathbf{r}' = (\boldsymbol{\rho}', z')$ with the extension $\Delta z << 1$ along the axial direction ($\delta\mu_\mathrm{a}$ is constant within ΔV and 0 outside), denote the exact Green's function for light propagation in the host medium G_0 and the intensity of the incident beam I_0, the measured array image is given by

$$I(\boldsymbol{\rho}_\mathrm{d}, \boldsymbol{\rho}_\mathrm{s}) = I_0 G_0(\mathbf{r}_\mathrm{d}, \mathbf{r}_\mathrm{s}) - I_0 \Delta V \int G_0(\mathbf{r}_\mathrm{d}; \boldsymbol{\rho}', z')\, \delta\mu_\mathrm{a}(\boldsymbol{\rho}', z') \\ G_0(\boldsymbol{\rho}', z'; \mathbf{r}_\mathrm{s})\, d^2\boldsymbol{\rho}'. \qquad (11.1)$$

We have set the speed of light to be unity for clarity. Scattering targets can be treated in a similar fashion.

The average of shifted array images (shifting the scanning position for each image $\boldsymbol{\rho}_\mathrm{s}$ to the origin 0) gives

$$\bar{I}(\boldsymbol{\rho}_\mathrm{d}, 0) = I_0 G_0(\mathbf{r}_\mathrm{d}, 0) - h_1(\boldsymbol{\rho}_\mathrm{d}, 0; z'), \qquad (11.2)$$

where the error term

$$h_1 \equiv \frac{1}{N_\mathrm{s}} I_0 \Delta Z \int G_0(\mathbf{r}_\mathrm{d}; \boldsymbol{\rho}, z') u(\boldsymbol{\rho}', z') G_0(\boldsymbol{\rho}', z'; 0) d^2\boldsymbol{\rho}' \\ = \frac{1}{A_\mathrm{s}} I_0 \delta\mu_\mathrm{a} \Delta V \int G_0(\mathbf{r}_\mathrm{d}; \boldsymbol{\rho}', z') G_0(\boldsymbol{\rho}', z'; 0) d^2\boldsymbol{\rho}', \qquad (11.3)$$

with $u(\boldsymbol{\rho}', z') = \sum_{\rho_\mathrm{s}} \delta\mu_\mathrm{a}\, (\boldsymbol{\rho}' - \boldsymbol{\rho}_\mathrm{s}; 0)$ is the superposition of the absorber at all possible shifted positions, N_s is the total number of scanning positions, and $A_\mathrm{s} \equiv N_\mathrm{s} a_\mathrm{s}^2$ is the total area of the scanning grid with step size a_s, provided that the scanning area A_s is sufficiently large and the pitch a_s is fine enough such that the scanning grid samples well Green's function on the $z = z'$ plane.

When a total N slices of strength $\delta\mu_j\Delta V_j$ are located on planes ($j = 1, 2, \ldots, N$), the averaging over the shifted images yields

$$\overline{I}(\boldsymbol{\rho}_\mathrm{d}, 0) = I_0 G_0(\mathbf{r}_\mathrm{d}, 0) - h(\boldsymbol{\rho}_\mathrm{d}, 0), \tag{11.4}$$

where $h = \Sigma_j h_1(\boldsymbol{\rho}_\mathrm{d}, 0; z_j)$ and the difference image between the dirty images and $\overline{I}$ can be written as

$$\Delta I(\boldsymbol{\rho}_\mathrm{d}, \boldsymbol{\rho}_\mathrm{s}) = \Sigma I_0 \int G_0(\mathbf{r}_\mathrm{d}; \boldsymbol{\rho}', z')\delta\mu_{aj}(\boldsymbol{\rho}', z')G_0(\boldsymbol{\rho}', z'; \mathbf{r}_\mathrm{s})d^2\boldsymbol{\rho}' d^2 z' + h(\boldsymbol{\rho}_\mathrm{d} - \boldsymbol{\rho}_\mathrm{s}, 0). \tag{11.5}$$

Here the error term h in Eqs. (11.4) and (11.5) only depends on $\boldsymbol{\rho}_\mathrm{d} - \boldsymbol{\rho}_\mathrm{s}$ and is shared for the difference images $\Delta I(\boldsymbol{\rho}_\mathrm{d}, \boldsymbol{\rho}_\mathrm{s})$ obtained at all scanning positions $\boldsymbol{\rho}_\mathrm{s}$. Hence a simple estimation of h is provided by averaging all shifted difference images which are minimally perturbed by the embedded objects, such as,

$$h(\boldsymbol{\rho}_\mathrm{d} - \boldsymbol{\rho}_\mathrm{s}, 0) = \frac{1}{N_B} \sum_{\rho_s \in B} \Delta I(\boldsymbol{\rho}_\mathrm{d} - \boldsymbol{\rho}_\mathrm{s}, 0) \tag{11.6}$$

where B denotes the perimeter of the scanning grid which contains a total N_s scanning positions. After h has been obtained, the clean image of the host medium and the difference between the dirty images and the clean one are given by $I^c = I(\boldsymbol{\rho}_\mathrm{d}, 0) + h(\boldsymbol{\rho}_\mathrm{d}, 0)$ and $\Delta I^c = \Delta I(\boldsymbol{\rho}_\mathrm{d}, \boldsymbol{\rho}_\mathrm{s}) - h(\boldsymbol{\rho}_\mathrm{d} - \boldsymbol{\rho}_\mathrm{s}, 0)$, respectively.

11.2.2 Numerical Target Marching

After the clean difference images $\Delta I^c(\boldsymbol{\rho}_\mathrm{d}, \boldsymbol{\rho}_\mathrm{s})$ have been obtained, independent component analysis can be used to unmix the signal arising from individual targets and the independent intensity distribution due to the j-th target ($j = 1, 2, \ldots, N$). The j-th independent component constitutes of the projections of Green's functions, $G_0(\mathbf{r}_\mathrm{s}, \mathbf{r}_j)$ and $G_0(\mathbf{r}_\mathrm{d}, \mathbf{r}_j)$, on the source and detector plane, respectively [15]. In the reflection geometry, the intensity of the backscattering light under illumination of an incident beam is strongly peaked at the point of incidence with a narrow width. There is no analytical solution for Green's function in this case and numerical simulations such as Monte Carlo methods have been commonly used instead to generate Green's function [16]. To overcome these difficulties, we approximate light propagation from $\mathbf{r}_j$ to $\mathbf{r}_\mathrm{d}$, by a two-step process: forward propagating light first bouncing back by the bottom semi-infinite medium ($z > z_j$), and then diffusing from the layer $z = z_j$ to the top surface $z = z_d$. This approximation is justified as the probing beam still retains its forward direction at the shallow depth ($|z_j - z_s| \sim l_t$) whereas the propagation direction of the bounced back light is randomized within the backward hemisphere, and light backscattered without interaction first with the bottom semi-infinite medium is negligible. The main *advantage* of this approximation is that the light propagator for the first process of light backscattering from the host medium, i.e., Green's function $G_0(\mathbf{r}_d, 0)$, is exactly the clean image, $\bar{I}^c$, which has been computed directly from the measured data set as outlined above without assumption of any particular light propagation model in the medium. Another advantage of using the clean image from experimentally measured data set as Green's function $G_0(\mathbf{r}_d, 0)$ is that this light propagator automatically incorporates the blurring effect due to the *unknown* superficial structure and surface roughness of a biological sample.

The projection of Green's functions for the j-th target, $G_0(\mathbf{r}_d, \mathbf{r}_j)$, on the detector plane can now be rewritten as

$$G_0(\mathbf{r}_d, \mathbf{r}_j) = \frac{1}{2\pi} \int G_0(\boldsymbol{\rho}' - \boldsymbol{\rho}_j, 0) \frac{\partial g(\boldsymbol{\rho}_d - \boldsymbol{\rho}', z_d, z_j)}{\partial z_j} d^2\boldsymbol{\rho}' \qquad (11.7)$$

using the first Rayleigh–Sommerfeld integral [17] where the integration is performed over the $z = z_j$ plane and g is Green's function inside an infinite homogeneous medium for a diffuse photon density wave [18]. Performing Fourier transform over the lateral coordinates, a simple relationship is obtained:

$$G_{0_j}(\mathbf{r}_d, \mathbf{r}_j) = \frac{1}{2\pi} G_0(\mathbf{q}, 0) \frac{\partial g(\mathbf{q}, z_d, z_j)}{\partial z_j} \qquad (11.8)$$

Here $G_{0_j}(\mathbf{q})$ is the Fourier transform of the j-th independent intensity on the detector plane, $G_0(\mathbf{q}, 0)$ is the Fourier transform of the clean image, $\bar{I}^c$, and $g(\mathbf{q}, z_d, z_j)$ is the Fourier transform of $g(\mathbf{q}, z_d, z_j)$ representing the propagation of a plane wave of spatial modulation frequency $\mathbf{q}$ from the $z = z_j$ plane to the surface $z = z_d$.

To incorporate the detection condition where only photons escaping the medium in the normal direction is being detected, we first use the optical reciprocal property [19] and consider instead for a normally incident beam at $\mathbf{r}_d$ migrating to the $z = z_j$ plane. In the center-moved diffusion model [20, 21], $g(\boldsymbol{\rho}, z_d, z_j) = \exp(-kr)/r$ where $k = \sqrt{(\mu_{a_0} - i\omega)/D_0}$ is the attenuation coefficient for incident beam of intensity modulation frequency ω, $D_0 = l_t/3$ is the diffusion coefficient, l_t is the transport mean free path, μ_{a0} is the absorption coefficient for the host medium, $r = \sqrt{\rho^2 + (z_j - z_d^*)^2}$, and $z_d^* = z_d + l_t$ is the location of the effective source by displacing the incident point of the collimated beam one l_t along the incident direction. Green's function g in the Fourier space is given by $g(\rho, z_d, z_j) = 2\pi \exp(-Q(|z_j - z_d^*|)/Q$ and Eq. (11.8) is simplified to

$$G_{0_j}(\mathbf{q}) = G_0(\mathbf{q}, 0) \exp(-Q(|z_j - z_d^*|)), \tag{11.9}$$

where $Q \equiv \sqrt{q^2 + k^2}$.

Equation (11.9) is the main expression for numerical target marching. The deeper the target is hidden inside the medium, the stronger damping of the high spatial frequency details in the independent intensity distribution owing to that target. The degree to which the independent intensity distribution flattens comparing to the clean image provides the basis to obtain the depth of the target.

The position of the j-th target is obtained by fitting $G_0(\mathbf{r}_d, \mathbf{r}_j)$ from inverse Fourier transform of Eq. (11.9) to the retrieved independent intensity distribution on the detector plane owing to that target. $G_0(\mathbf{r}_d, \mathbf{r}_j)$ needs not to be convoluted with the incident beam profile before fitting as beam profile affects $G_0(\mathbf{r}_s, \mathbf{r}_j)$ alone.

The graphical illustration of the OPTICA algorithm is shown in Fig. 11.1. Figure 11.1a shows the pathway of backscattered light in one direction and the formation of a backscattering image of the host scattering medium without guest object(s) acquired from photons surviving back from the medium, which carry the information of optical coefficients of the medium. Figure 11.1b exhibits the voyage of the backscattered light in one direction and their intensity distribution on the source-detector plane for the host medium with one embedded guest object. The light intensity distribution is affected by the photons having interaction with the guest object and the optical coefficients of the tissue. If the difference of the intensity distribution of images recorded at those two different conditions is detectable, ICA can be applied to retrieve the information of the object (IC). A more real-world situation (multiple guest objects) is shown by Fig. 11.1c, in which the 2D intensity distribution of a backscattering image is contributed by photons having interaction with the multiple guest objects. The OPTICA can be used to recover each IC (target) from the mixture perturbation. This can be achieved by using the approach of multiple detectors (or CCD camera) and multiple sources (or scanning the illuminated laser source at different positions on samples in our case) as shown in Fig. 11.1d. Each recovered IC is used to obtain the 3D localization of the corresponding target by marching the propagation of the scattered light from the target to the surface until matching the retrieved IC. The OPTICA algorithm has high sensitivity to detect a small-sized target with detectable contrasts to the host medium because of its capability to separate the signal of each IC from huge background noise of the host medium.

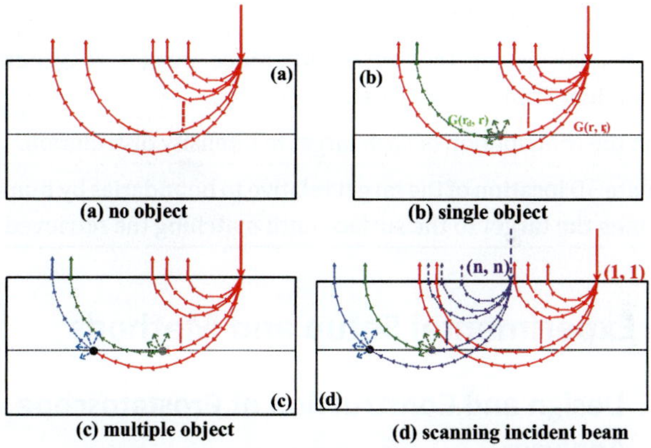

Figure 11.1 Graphical illustration of the OPTICA algorithm: (a) The formation of a backscattering image of host medium without embedded guest object(s); (b) the formation of the backscattering image of host medium with one guest object, which has detectable perturbation of intensity distribution from the image of the medium without object(s); (c) the formation of a backscattering image of the host medium with multiple guest objects. The image is formed by mixing diffused photons with information of multiple objects; (d) mechanism of multiple sources and multiple detectors by using CCD camera and scanning an illuminated laser source at different positions on samples, which acquires information needed for ICA.

ICA, the core of the OPTICA algorithm, can be used to extract intensity distribution contributed by each IC from the total perturbations of the spatial intensity distributions on the detector plane caused by guest objects [6, 13]. The procedures of the scanning imaging and OPTICA approach for obtaining 3D locations of the guest objects in the host media can be described as six steps shown in Table 11.1.

Table 11.1 The steps for obtaining 3D locations of the guest objects using scanning imaging and OPTICA approach

1.	Record a set of 2D images with embedded inhomogeneities at different illuminated scanning positions.
2.	Generate a "clean" background image from the "dirty" background image produced by selecting and averaging the minimally perturbated images.
3.	Calculate the difference images between the recorded images and the "clean" background image.
4.	Do ICA analysis for the set of difference images to generate the Independent Components (ICs).
5.	Obtain the contribution of each target to intensity distribution.
6.	Estimate 3D location of the target relative to boundaries by numerically marching the target to the surface until matching the retrieved IC.

11.3 Experimental Setup and Methods

11.3.1 Design and Construction of Prostatoscope Analyzer

The prostatoscope analyzer scanning imaging unit is a portable rectal NIR scanning polarization imaging unit with an optical fiber-based rectal probe. The unit was designed to be capable of recording sets of 2D images of the prostate by scanning the illuminating beam on the prostate through rectum walls. The major optical and electronic components of the unit include laser diodes and their power supplies, miniature scanning Galvanometric mirrors with their electronic control boards (servo driver circuit boards) and the LabVIEW control software, illumination and imaging coherent optical fiber-bundles, and the rectal optical probe.

A photograph of the portable rectal NIR scanning polarization imaging prostatoscope analyzer unit is shown in Fig. 11.2a. Three diode lasers emitting at 635, 750, and 980 nm, respectively, are alternatively used as light sources. These wavelengths were selected to probe the native molecules such as Hb, HbO_2, and H_2O in tissues. The output beam from a laser diode is directed to the scanning galvanometric mirror system (ThorLab GVSM002

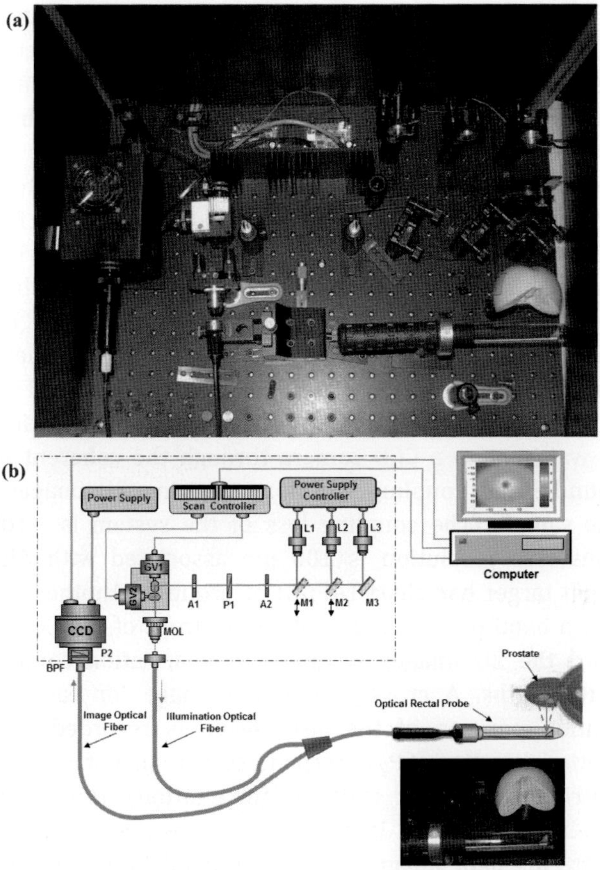

Figure 11.2 (a) A photograph of the Prostatoscope Analyzer: a portable rectal NIR scanning polarization imaging unit. The power supplies and computer system are not shown in the picture. (b) The schematic diagram of the portable rectal NIR scanning polarization imaging unit, where L-laser, M-mirror, A-aperture, P-polarizer, GV-galvanometric scanning mirror, BPF-band pass filter, MOL-microscopy objective lens.

with Dual Axis Galvo Mirrors) after passing through two pinholes and a polarizer (P_1). The beam can be scanned by two miniature galvanometric mirrors in the *x*- and *y*-directions, respectively. The beam reflected by the galvanometric mirrors is focused using a microscopy objective lens on a coherent fiber-bundle (Mytiad Fiber Imaging, 20–0826 Fiberscope Assy) used for illumination. The position of the microscopy objective lens can be adjusted in the *x*-, *y*- and *z*-directions for the beam-fiber coupling. The output beam from the illumination fiber is directed to a small reflection prism located inside the rectal probe. The beam reflected from the prism is used to illuminate a prostate sample. This illumination beam with a diameter of ~1 mm can be scanned in the *x*- and *y*-directions on the prostate sample.

The light backscattered (or emitted) from a prostate sample is first passing through the rectal wall and reflected from the prism inside the probe head. The diameter of the probe is ~2 cm and the length is ~12 cm. The beam is then collected by a lens into another coherent fiber-bundle (Mytiad Fiber Imaging, 20–0826 Fiberscope Assy) used for imaging. The diameter of a single fiber in the bundle is ~3.2 μm each with numerical aperture N.A. ≊ 0.4. The image information formed in the optical fiber bundle was sent to a CCD camera through the coherent imaging fiber bundle and coupling lens for recording 2D images of the prostate sample. The coupling loss of the system is ~10% and the transverse resolution is 100 μm associated with Air Force resolution target bar chart (AFBC) at group 3. Another polarizer (P_2) and a band pass filter are placed in front of the CCD camera to record the 2D images at different polarization configuration and wavelengths. A cross-polarization image for each scanned illumination position of the laser beam is recorded when the polarization direction of P_2 is perpendicular to that of P_1 to suppress the contribution of light scattered (or emitted) from the surface and sub-surfaces to the images of the prostate sample. For the backscattering light imaging, a narrow band filter corresponding to the illumination wavelength is used in front of the CCD camera to ensure that the recorded images are formed only by the light backscattered from the prostate sample. For the tissue emission and/or contrast agent emission light imaging, a long pass filter is used in front of the CCD camera to ensure that the recorded images are formed only by the light emitted from the prostate sample. When the illumination light beam is scanned in the *x-y* plane of the sample with $n \times n$ points, an array of n^2 2D images will be recorded.

The key part of the scanning imaging unit is the scan of the illumination beam on the surface of the prostate sample with adjustable scanning parameters such as scanning area, step, speed and the number of scanning points. Figure 11.2b schematically shows the layout of the major parts and the control boards for the imaging unit. This scanning system consists of (1) the Dual Axis Galvo mirrors (ThorLab GVS 002), (2) the two servo driver circuit boards and their power supply (ThorLab, GPS011), (3) the drive unit for sending voltage output to servo circuit boards (National Instruments, NI DAQmx USB-9263 USB DAQ—data acquisition), and (4) a PC with an installed LabVIEW software (LabVIEW Signal Express 2009) to power and send a command to DAQ USB 9263 to generate desired output voltage through a USB connection. In the scanning system, the drive unit NI-9263 powered by USB interface of the PC, is used to generate an analog voltage, which can be varied from −10 V to +10 V using the LabVIEW software. This analog voltage is sent to the two servo driver boards to drive the rotations of the Galvo Mirrors. For example, 1 V input to the servo board can make the mirror rotate 1°. One servo board controls the *X*-axis, and another is for *Y*-axis. Both servo driver boards are powered by the power supply of GPS011. The beam scanning, the image acquiring and recording, and the synchronization of scanning and imaging are controlled by the graphical user interface (GUI) software developed using LabVIEW. All parameters of the scanning imaging unit (the position of the original, scanning steps, step size, exposure time, and waiting time between two adjacent imaging acquiring,) can be adjusted through the GUI.

11.3.2 Test Model and Prostate Samples

The prostatoscope analyzer scanning imaging unit was tested for different types of tissue samples. The first sample used for scanning imaging was a black rubber absorber with the size of ~2.7 mm × ~3.0 mm × ~1.5 mm embedded in chicken breast tissue at different depth. The fresh chicken breast tissues were purchased at the local super market. The black rubber was covered by the chicken breast tissues at two different depths of ~5.1 mm and ~6.6 mm with a lateral dimension of 38 mm × 29 mm for scanning imaging measurements. The thickness of the tissue behind the object is ~8 mm. A transport length l_t = 1.1 mm and an absorption coefficient μ_a = 0.007 mm^{-1} were used for chicken breast tissue for imaging analysis [4].

The second study was performed to distinguish in vitro cancerous prostate tissue from surrounding normal prostate tissue. Human prostate tissue samples were obtained from the Co-operation Human Tissue Network (CHTN) and the National Disease Research Interchange (NDRI) under the approval of the Institutional Review Board (IRB) at CCNY. The cancerous and normal prostate tissue samples were diagnosed by the pathologist before the optical imaging experiments. Samples were neither chemically treated nor frozen prior to the experiments. The time elapsed between tissue resection and taking the scanning imaging measurements may vary for different sample sources. The longest elapsed time is about 30 hours. The sample consists of a small piece of cancerous prostate tissue (4 mm × 4 mm × 1.5 mm) embedded inside a large piece of normal prostate tissue at the depth of z = 3.0 mm from the front surface. The thickness of the whole tissue sample is 10 mm. For imaging analysis, the absorption and reduced scattering coefficients for normal prostate tissue (the host medium) were taken as $\mu_a \approx 0.026$ mm^{-1} and $\mu_s' \approx 0.53$ mm^{-1}. The cancerous prostate tissue absorbs much less light with $\mu_a \approx 0.0025$ mm^{-1} and scatters less with $\mu_s' \approx 0.44$ mm^{-1} at the probing wavelength of 635 nm estimated from our earlier spectroscopic investigations of ex vivo prostate tissues [22]. The piece of cancerous prostate tissue embedded behaves predominantly as an absorption inhomogeneity in homogeneous normal prostate tissue. Here, we simply assume that the optical parameters of normal tissue are macroscopically homogeneous throughout the tissue volume. Although this is not a rigorous description of tissue, for many cases of interest, it will be sufficient. OPTICA is applied here to extract different independent components by treating cancerous tissue as inhomogeneity embedded inside a "homogeneous" normal tissues.

The third study was designed to characterize cancerous prostate tissue after enhancing imaging contrast using extrinsic fluorescent marker, a smart receptor-targeted contrast agent, namely Cytate. Cytate was synthesized by Achilefu's group at the Washington University School of Medicine at St. Louis. This contrast agent is mainly composed of ICG the only FDA approved florescence dye in the NIR range, and the somatostatin receptor ligand, which delivers ICG to the corresponding somatostatin receptors over-expressed in the tumor [23, 24]. The prostate tissue samples for the scanning imaging study were prepared following this protocol: (1) parts of cancerous and normal prostate tissue samples were cut into very tiny size, which is less than ~1 mm (in length, width and thickness) pieces. Each pair of cancerous tissue and the corresponding normal tissue used as a control sample was obtained from same patient; (2) both small pieces of prostate tissue samples were soaked in a same Cytate solution with a concentration of ~3.2×10^{-6} M for ~10 minutes; (3) Cytate-stained tissue samples were put into sodium phosphate buffer (Sigma-Aldrich) to wash off and consequently reduce the amount of un-bound Cytate [24]; and (4) the Cytate-stained small pieces of cancerous and normal prostate tissues were covered by a large piece of normal prostate tissue at two different depths of ~2.5 mm and ~3.7 mm for scanning imaging measurements.

The orthogonal co-ordinate system to locate 3D positions of the target in the scanning polarized image measurements is schematically shown in Fig. 11.3. The samples to be tested were embedded inside large slices of host tissue at different depths. The illuminating laser beam was scanned along the x- and y-directions on the front surface of the sample. The 2D images formed by the light backscattered from the sample in the normal direction are recorded. In the first and second studies, only one target (black rubber or cancerous prostate tissue) in each case is involved and the sample was illuminated by a collimated laser beam with λ = 635 nm. The elastic backscattering from the samples was collected through a narrow band filter (NF). In the third study for contrast agents' fluorescence, the Cytate-stained tiny cancerous and normal prostate tissue samples were embedded inside large pieces of normal prostate tissue. The sample was illuminated by a laser beam with λ = 635 nm in the direction close to the normal to the surface. A narrow band-pass filter at 830 nm was placed in front of a CCD camera to record emission images based on the emission peak of Cytate at ~837 nm [23] in the first optical window [24].

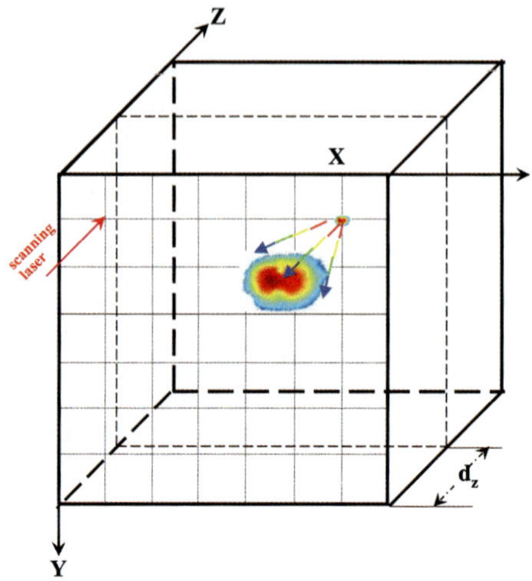

Figure 11.3 A screenshot of the graphical user interface (GUI) after completing an n × n scan. The GUI was developed using LabVIEW for operation of the prostatescope analyzer. All parameters of the scanning imaging (the position of the original, scanning steps, step size, exposure time, and waiting time between two adjacent imaging acquiring, e.g.) can be adjusted through the GUI.

11.4 Experimental Results

To illustrate the procedures and improvement of OPTICA using prostatoscope analyzer unit particularly developed for backscattering geometry, a set of scanning optical polarized images of a small piece of cancerous prostate tissue embedded in normal prostate tissue were acquired by the prostatoscope analyzer. Figure 11.4 shows the typical scanning images chosen from the total 8 × 8 polarized images recorded by scanning the incident beam on the *x*–*y* plane of the sample. The images shown on the four corners [images (1, 1), (1, 8), (8, 1), (8, 8)] are chosen by the *x*–*y* positions of the incident light far from the embedded object while other images in the central area [images (3, 5), (3, 6), (4, 5), (4, 6)] are chosen by the *x*–*y* positions of the incident light close to the embedded cancerous tissue. There is no big difference of the light intensity distribution pattern among these recorded images except the position of maximum intensity. This is because of the sharp peak of light intensity scattered from the surface and subsurface in the backscattered direction [3], which suppresses the perturbation caused by μ_a and/or μ_s' from the embedded cancerous tissue. To overcome this difficulty, a CBI needs to be synthesized. In "clean" background image synthesis, the incident light position of each of 8 × 8 images is set to origin. All array images are then shifted to the origin and the size of each image is cropped at the boundary while incident light distribution reaches the noise level. The images minimally perturbed by the embedded targets are selected for synthesizing the "clean" image.

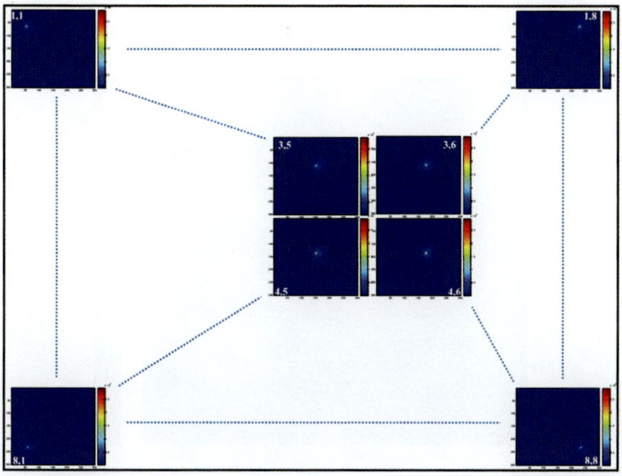

Figure 11.4 A set of 64 (8 × 8) images recorded by scanning the incident beam on the *x–y* plane of the sample (a small piece of cancerous prostate tissue hidden in the host normal tissue).

After the "clean" background image is obtained, the perturbation 2D (*x–y*) images are generated by extracting the "clean" background image from the recorded images. In the perturbation 2D images, the perturbation caused by the embedded cancerous tissue was highlighted. ICA is then performed upon the perturbation images to recognize leading ICs. Each target is associated with one independent component (IC). As an example, Fig. 11.5 shows the OPTICA-generated independent intensity distributions of a cancerous prostate tissue embedded in normal prostate tissue. Figure 11.5a is for the leading independent component, and Fig. 11.5b is for residual (noise) component. It is clear that the existence of target (the cancerous prostate tissue) can be discerned from Fig. 11.5a. The *x*– and *y*-locations of the cancer tissue can be obtained from the position of maximum intensity of Fig. 11.5a. The signal strength of Fig. 11.5a is much stronger than that of Fig. 11.5b, indicating independent intensity distribution in Fig. 11.5a generated by OPTICA is the leading IC. The OPTICA-generated intensity distributions of the leading independent component(*s*) on the detector plane can be used to locate the *z*-position of the cancerous tissue by numerical matching the target to the surface of the medium until matching the retrieved IC [6]. The result of $z = 3.1$ mm obtained from the OPTICA analysis is in good agreement with the actual depth of ~3 mm.

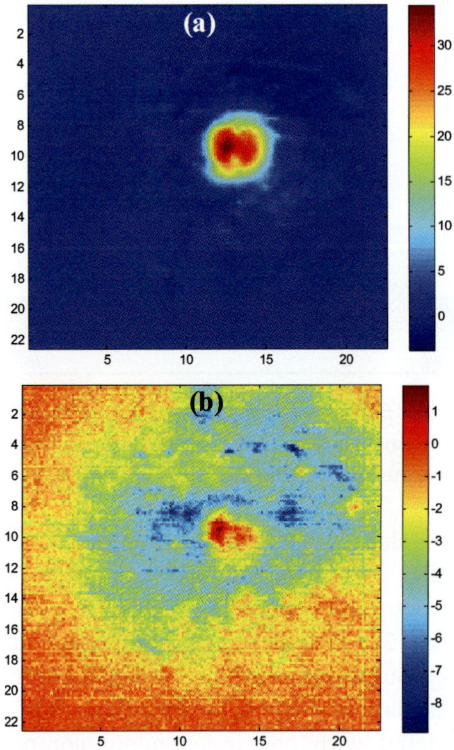

Figure 11.5 OPTICA-generated intensity distributions of the independent components on the detector plane for the sample consisting of a small piece of cancerous prostate tissue embedded in normal prostate tissues. (a) is the leading IC and (b) is the residual (noise) component.

The OPTICA-generated independent components using data of the scanning fluorescent images of Cytate-stained cancerous and normal prostate tissues embedded in large normal prostate tissue are shown in Figs. 11.6(a–f). Figures 11.6a,b show the first leading IC standing for Cytate-stained cancerous tissue location; Figs. 11.6c,d show the second IC indicating Cytate-stained normal tissue location; Figs. 11.6e,f show residual noise. The 3D locations of the Cytate-stained cancerous prostate tissues were obtained using similar methods described above.

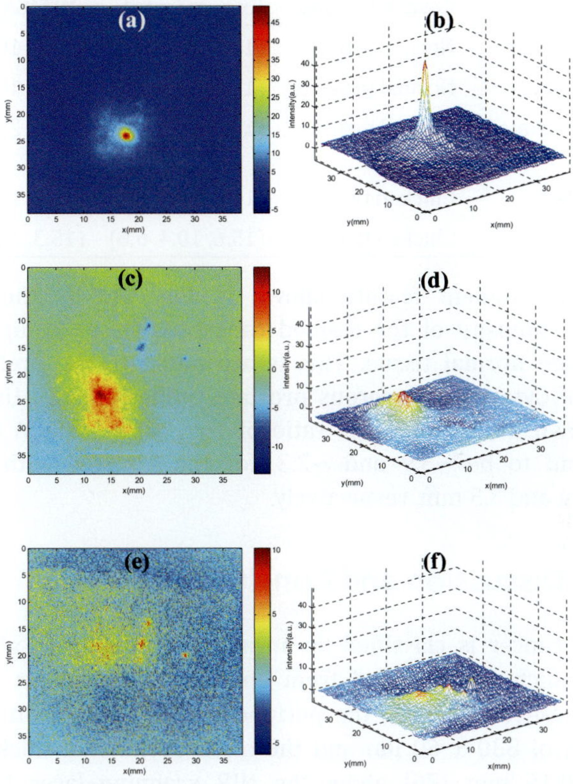

Figure 11.6 OPTICA-generated intensity distributions of the independent components on the detector plane for the sample consisting of the Cytate-stained cancerous and normal prostate tissues embedded in normal prostate tissue. (a) and (b) show the first leading IC indicating for Cytate-stained cancerous prostate tissue; (c) and (d) show the second leading IC standing for Cytate-stained normal prostate tissue; and (e) and (f) show the residual (noise) component.

OPTICA-generated results are summarized in Table 11.2 under different conditions using backscattering polarized imaging and OPTICA, which lists the OPTICA-determined positions for different objects of the different experiments mentioned above in comparison with their known 3D locations.

Table 11.2 Comparison of the known and OPTICA-determined positions of embedded objects

	NP		
Object	**Covered tissue**	**Known (mm)**	**OPTICA (mm)**
Prostate cancer	Normal prostate	(13.5, 10.1, 3.0)	(13.8, 10.3, 3.1)
Prostate cancer	Normal prostate	(17.8, 14.6, 2.5)	(17.9, 14.4, 2.3)
Stained prostate cancer	Normal prostate	(18, 24, 3.5)	(17.6, 23.8, 3.6)
Back rubber	Chicken tissue	(15.6, 10.4, 5.1)	(15.2, 10.2, 4.7)
Back rubber	Chicken tissue	(15.6, 10.4, 6.6)	(15.3, 10.8, 6.2)

Another salient feature shown in Fig. 11.6 is the higher emission intensity of the stained cancerous tissue compared to the stained normal tissue. This is attributed to the preferential uptake of Cytate by cancerous prostate tissue [23, 25]. Using the data shown in Fig. 11.6, the ratio of $\bar{I}^c/I_n$ after OPTICA analysis was found to be ~3.9 and ~2.3, for the samples with depths of 2.5 mm and 3.5 mm, respectively.

11.5 Discussion and Conclusion

Prostate cancer is classified as an adenocarcinoma or glandular cancer which is developed from epithelial cells. About 70% of prostate cancer arises in the peripheral zone. The mean mucosa thickness of 830 ± 60 μm and the mean rectal wall thickness of 2.57 ± 0.15 mm [26] allow the NIR scanning laser beam to penetrate the rectal wall to reach prostate for optical imaging through rectum [27]. By employing scanning NIR imaging, the prostatoscope analyzer can be used to characterize cancerous prostate tissue up to ~3 mm without contrast agents. With the enhancement of the receptor-target contrast agent, prostatoscope analyzer unit (PAU) can detect cancer located at much deeper tissue areas. Previous spectral polarized imaging experiments performed in IUSL show that ICG-stained small object hidden inside the host prostate tissues in the rectum-membrane-prostate structures at depths of 7.5 mm can be imaged and identified using fluorescence imaging methods [27]. ICG is the fluorescent part of Cytate. It is reasonable to assume that PAU can detect cancer in deep prostate tissue layer more than 7.5 mm under optimal condition with the enhancement of the receptor-target contrast agent. The PAU will achieve diagnostic function as a screening tool other than transrectal ultrasound (TRUS). It is well known that TRUS alone cannot be considered as a screening tool for prostate cancer detection. It is used as location map when numbers of biopsy samples were taken randomly from the prostate using biopsy needles of ~1.2 mm in diameter. The number of biopsies that should be performed is debated. Twelve or eighteen biopsies suggested by different protocols were reported. The oversampling and overtreatment of prostate cancer is now a great concerning in prostate cancer research [12] not only because of the increased numbers of sampling will cause bleeding and pain but also because the penetration and bleeding may cause the metastasis of prostate cancer to other organ. In contrast to TRUS, prostatoscope analyzer can locate the 3D positions of cancerous prostate sites. Therefore, it may function as a detecting and screening instrument. The prostatoscope analyzer, an optical scanning imaging system, may have positive impact in clinic applications to reduce significant pain and amount of bleeding for patients in prostate cancer detection.

In conclusion, a portable rectal NIR scanning polarization imaging unit with an optical fiber-based rectal probe, named prostatoscope analyzer, was designed, developed, assembled, and tested on ex vivo tissues at IUSL in CCNY. This instrument combined with OPTICA was achieved to characterize and 3D-locate cancerous prostate tissue embedded in normal prostate tissue in backscattered geometry. The retrieved 3D locations are in good agreement with the known positions of the embedded objects. This technique was able to sense weak/small absorptive, scattering, or fluorescent inhomogeneities. This new developed prostatoscope analyzer instrument provides an opportunity for alternative optical NIR imaging technique for prostate cancer detection, which is accurate, of higher spatial resolution and non-or-less invasive, and can be applied in near-real-time for in vivo detection.

Acknowledgments

This research was supported by U. S. Army Medical Research and Material Command (USAMRMC) grant of W81XWH-08-1-0717 (CUNY RF # 47170-00-01) and W81XWH-11-1-0335 (CUNY RF # 47204-00-01). The authors acknowledge the contributions from our collaborator: Prof. S. Achilefu at Washington University School of Medicine at St. Louis, and the help of CHTN and NDRI for providing normal and cancerous prostate tissue samples for the measurements.

References

1. Benaron, D. A. (2002) The future of cancer imaging, *Cancer and Metast. Rev.*, **21**, pp. 45–78.
2. Wang, L., Ho, P. P., Liu, C., Zhang, G., and Alfano, R. R. (1991) Ballistic 2-D imaging through scattering wall using an ultrafast Kerr gate, *Science,* **253**, pp. 769–771.
3. Pu, Y., Wang, W. B., Xu, M., Eastham, J. A., Tang, G. C., and Alfano, R. R. (2012) Characterization and three-dimensional localization of cancerous prostate tissue using backscattering scanning polarization imaging and independent component analysis, *J. Biomed. Opt.*, **17**, pp. 081419-1-8.
4. Cheong, W.-F., Prahl, S. A., and Welch, A. J. (1990) A review of the optical properties of biological tissues, *IEEE J. Quantum Electron.*, **26**, pp. 2166–2185.
5. Pu, Y., Wang, W. B., Tang, G. C., Budansky, Y., Sharonov, M., Xu, M., Achilefu, S., Eastham, J. A., and Alfano, R. R. (2012) Screening prostate cancer using a portable near infrared scanning imaging unit with an optical fiber-based rectal probe, *Proc. SPIE,* **8220**, pp. 822002 (in U.S.A.).
6. Xu, M., Alrubaiee, M., Gayen, S. K., and Alfano, R. R. (2005) Optical imaging of turbid media using independent component analysis: Theory and simulation. *J. Biomed. Opt.*, **10**, pp. 051705-1-12.
7. Wood, D., Feke, G., Vizard, D., and Papineni, R. (2008) Refining epifluorescence imaging and analysis with automated multiple-band flat-field correction, *Nat. Methods,* **5**, pp. i–ii.
8. Vandelest, C., Versteeg, E., Veerkamp, J., and Vankuppevelt, T. (1995) Elimination of autofluorescence in immunofluorescence microscopy with digital image processing, *J. Histochem. Cytochem.*, **43**, pp. 727–730.

9. Montcuquet, A., Herve, L., Navarro, F., Dinten, J., and Mars, J. (2010) Nonnegative matrix factorization: A blind spectra separation method for in vivo fluorescent optical imaging, *J. Biomed. Opt.*, **15**, pp. 056009-1-14.

10. Pu, Y., Wang, W. B., Tang, G. C., Budansky, Y., Alfano, R. R., and Xu, M. (2011) Backscattering scanning optical imaging with independent component analysis: Three-dimensional localization of objects in tissue, *Proc. SPIE*, **7895**, pp. 78950K-1-5 (in U.S.A.).

11. Xu, M., Alrubaiee, M., Gayen, S. K., and Alfano, R. R. (2005) Three-dimensional localization and optical imaging of objects in turbid media with independent component analysis, *Appl. Opt.*, **44**, pp. 1889–1897.

12. Pu, Y., Wang, W. B., Xu, M., Tang, G. C., Budansky, Y., Sharanov, M., Achilefu, S., Eastham, J. A., and Alfano, R. R. (2011) Near infrared photonic finger imager for prostate cancer screening, *Technol. Cancer Res. Treat.*, **10**, pp. 507–517.

13. Xu, M., Alrubaiee, M., Gayen, S. K., and Alfano, R. R. (2008). Optical diffuse imaging of an ex vivo model cancerous human breast using independent component analysis, *IEEE J. Sel. Top. Quantum Electron.*, **14**, pp. 43–49.

14. Comon, P. (1994) Independent component analysis—a new concept? *Signal Process.*, **36**, pp. 287–314.

15. Konecky, S. D., Mazhar, A., Cuccia, D., Durkin, A., Schotland, J., and Tromberg, B. J. (2009) Quantitative optical tomography of sub-surface heterogeneities using spatially modulated structured light, *Opt. Express*, **17**, pp. 14780–14790.

16. Xu, M., Cai, W., Lax, M., and Alfano, R. R. (2001) A photon transport forward model for imaging in turbid media, *Opt. Lett.*, **26**, pp. 1066–1068.

17. Ripoll, J., and Ntziachristos, V. (2006) From finite to infinite volumes: Removal of boundaries in diffuse wave imaging, *Phys. Rev. Lett.*, **96**, p. 173903.

18. O'Leary, M., Boas, D. A., Chance, B., and Yodh, A. G. (1992) Refraction of diffuse photon density waves, *Phys. Rev. Lett.*, **69**, pp. 2658–2661.

19. Case, K. M. (1957) Transfer problems and the reciprocity principle, *Rev. Mod. Phys.*, **29**, pp. 651–663.

20. Cai, W., Lax, M., and Alfano, R. R. (2000) Analytical solution of the elastic Boltzmann transport equation in an infinite uniform medium using cumulant expansion. *J. Phys. Chem. B*, **104**, pp. 3996–4000.

21. Wang, L. H., Jacques, S. L., and Zheng, L. Q. (1995) MCML—Monte Carlo modeling of photon transport in multi-layered tissues. *Comput. Meth. Prog. Bio.*, **47**, pp. 131–146.
22. Pu, Y. (2010) Fractal dimensional parameters and optical coefficients of cancerous and normal prostate tissues, in *Time-Resolved Spectroscopy and Near Infrared Imaging for Prostate Cancer Detection: Receptor-targeted and Native Biomarker,* (A dissertation thesis of the City University of New York, New York), Chapter 3, pp. 48–76.
23. Pu, Y., Wang, W. B., Das, B. B., Achilefu, S., and Alfano, R. R. (2008) Time-resolved fluorescence polarization dynamics and optical Imaging of Cytate: A prostate cancer receptor-targeted contrast agent, *Appl. Opt.*, **47**, pp. 2281–2289.
24. Sordillo, L. A., Pu, Y., Pratavieira, S., Budansky, Y., and Alfano, R. R. (2014) Deep optical imaging of tissue using the second and third near-infrared spectral windows, *J. Biomed. Opt.*, **19**, pp. 056004-1-6.
25. Pu, Y., Wang, W. B., Achilefu, S., and Alfano, R. R. (2008) Study of rotational dynamics of receptor-targeted contrast agents in cancerous and normal prostate tissues using time-resolved picosecond emission spectroscopy, *Appl. Opt.*, **50**, pp. 1312–1322.
26. Huh, C. H., Bhutani, M. S., Farfan, E. B., and Bolch, W. E. (2003) Individual variations in mucosa and total wall thickness in the stomach and rectum assessed via endoscopic ultrasound, *Physiol. Meas.*, **24**, pp. N15–N22.
27. Pu, Y., Wang, W. B., Tang, G. C., Zeng, F., Achilefu, S., Vitenson, J. H., Sawczuk, I., Peters, S., Lombardo, J. M., and Alfano, R. R. (2005) Spectral polarization imaging of human prostate cancer tissue using near-infrared receptor-targeted contrast agent, *Technol. Cancer Res. Treat.*, **4**, pp. 429–436.

Chapter 12

Terahertz Propagation in Tissues and Its Thickness Limitation

Burcu Karagoz and Hakan Altan

Department of Physics, Middle East Technical University, Cankaya, Ankara 06800, Turkey

haltan@metu.edu.tr

The terahertz region offers a multitude of possibilities for new diagnostic techniques in biology. Recent advancements in terahertz sources and detectors have allowed such systems to be employed in investigations of biological tissues. The main interaction with such media is evident in the water content, which can be detrimental in certain instances such as deep tissue diagnostics and also be beneficial in others such as the detection of certain types of cancers. The field has developed to deal with such challenges, and the improvement in overall signal to noise of especially pulsed diagnostic methods has led to the achievement of penetration depths on the order of millimeters, which has opened up a variety of new investigations. These investigations are described with special emphasis on diagnostics on hard tissues such as teeth.

Deep Imaging in Tissue and Biomedical Materials: Using Linear and Nonlinear Optical Methods
Edited by Lingyan Shi and Robert R. Alfano

ISBN 978-981-4745-88-8 (Hardcover), 978-1-315-20655-4 (eBook)
www.panstanford.com

12.1 Introduction

Despite the limited science and technology compared to the neighboring spectral bands, there has been an increasing interest in terahertz (THz) radiation due to the many interesting biological phenomena that can be investigated in this frequency range—typically accepted to be from 0.1 to 10 THz [1], [2]. The rotational and vibrational transitions of many molecules show strong absorption and dispersion at THz frequencies [3]. The location and amplitude of these resonances which are specific to the molecules structure provide a means for spectroscopic identification. Besides this, the longer wavelengths of THz radiation compared to visible and IR waves prevent high scattering and THz waves can easily penetrate into many dry materials such as plastic, paper, textile, and wood, which are opaque at optical wavelengths. Likewise, an enhanced contrast can also be obtained in THz imaging methods through low scattering. On the other hand, the high absorbance of water in THz frequencies limits the penetration of THz waves into water-containing materials. This presents a challenge when investigating tissues using instruments designed to work in the THz range. Furthermore, the limited development of sources and detectors as compared to the other regions of the electromagnetic spectrum have constrained the development of applications with tissue analysis where measurements are typically done in vitro and in laboratory settings as opposed to clinical ones. Nonetheless, the ability to identify species as well as provide a higher spatial resolution as compared to microwaves due to its shorter wavelengths has spurred a growing interest in the development and application of technologies for investigations in this field. While the focus of this study is to assess the ability of THz radiation to penetrate hard and soft biological tissue matter with an emphasis on human teeth, there are a number of noteworthy reviews that address the applicability of instruments working in the THz frequency range on biological media [4], [5], [6], [7]. The majority of studies on biological tissues have been done using pulsed THz spectroscopy and/or imaging systems with a few notable exceptions where high average power continuous wave (CW) sources such as quantum cascade lasers were used to study certain types of soft tissues [8]. The power levels involved using pulsed THz instrumentation is typically quite low, thus the THz radiation cannot present any harm to biological tissues [9]. This is aided by the fact that THz waves have limited penetration in soft tissues such as skin. In the same way, the low photon energy of THz waves cannot cause photoionization in living tissues.

12.2 THz Generation and Detection

There are various methods to generate and detect THz waves. THz sources are generally categorized into two groups according to operating frequency and emission mode: CW and pulsed (time-domain) sources [10]. CW systems have typically been developed from the electrical side of the electromagnetic spectrum by the up-conversion of frequency, whereas most of the pulsed systems have been developed from the optical side. Pulsed systems rely on the generation and detection of the electromagnetic transitions using ultrafast femtosecond (fs) lasers. CW systems tend to be narrowband and provide continuous emission at a single frequency. Compared to CW systems, pulsed systems are broadband sources and emission is not continuous.

THz spectroscopy and imaging are two main applications of THz wave technologies. In time-domain systems, short pulses composed of many frequencies with sub to few picosecond (ps) duration are generated and detected. Therefore, pulsed systems are ideal in many dynamical applications because they provide picosecond resolution [11]. When the source is a pulsed laser, the used spectroscopy technique is commonly referred to as THz time-domain spectroscopy (THz-TDS) [12] and the imaging technique is named as THz pulsed imaging (TPI). The advent of time-domain spectroscopy utilizing ultrafast pulsed lasers has created interest in THz science by enabling new spectroscopic information and imaging techniques in the submillimeter wavelength range. Auston et al. reported the first use of photoconductive antennas for the generation of THz pulses from ultrafast pulses in 1984 [13]. These devices have been widely preferred in the generation and detection of THz pulses. In the same year, an electro-optic (EO) crystal (lithium tantalate, $LiTaO_3$) was shown that it could be used to rectify laser pulses for THz generation [14]. In 1989, Martin van Exter et al. laid the foundation of THz time-domain spectroscopy technique [15]. Following this, the research into terahertz generation and detection methods has accelerated and some simple studies oriented to parameter extraction of materials using THz-TDS method had begun. The first two-dimensional (2D) THz imaging system, now named THz pulsed imaging, was demonstrated in 1995 by Hu and Nuss [16]. In this study, plant and animal tissues were investigated using a system similar to the THz-TDS method. The comparison of transmission images of a fresh leaf sample and that after a drying process of 48 h indicated the usability of water as a contrast mechanism and was a precursor in the study of biological samples using the THz frequency range. A difference in the THz images of muscle and fat tissues derived from porcine was also observed in the same study. These images added momentum to biomedical investigations. An electro-optic method for sampling the resulting pulse was presented in 1996 by Wu et al. and since been subsequently used for imaging applications instead of photoconductive antennas [17]. The first publication of a three-dimensional (3D) THz image using a TPI system was in 1997 where the inside of a conventional floppy disk consisting of a series of layers was characterized using a reflection-based system [18].

12.2.1 THz-TDS Method

Terahertz time-domain spectroscopy is the most versatile technique in the biological applications of THz imaging [19]. It is based on the generation and coherent detection of broadband terahertz pulses using ultrafast pulsed lasers enabling both spectroscopic investigation and imaging.

In a typical THz-TDS system, the pulse from an ultrafast laser is split into pump and probe beams using a beam splitter. The pump beam is used to excite the THz emitter and generate picosecond terahertz pulses, while the probe beam is used to gate the THz detector and sample the THz electric field after its interaction with the sample. An optical delay line which is often located in the pump beam arm adjusts the path lengths between the generation and detection beams and changes the relative arrival time of the THz and probe pulses to the detector. By scanning this time delay, the THz waveform is recorded. In THz-TDS, the electric field of the THz pulse is measured. Since the electric field has both amplitude and phase information, the absorption coefficient and refractive index of the sample can be obtained by Fourier transform of the electric field [20]. Thus, TDS provides the complex permittivity of the sample without using the Kramers–Kronig analysis [11]. The delay line and data acquisition from using phase sensitive techniques are controlled by a computer. Then, the comparison of the THz pulses obtained with and without a sample provides information native to the sample such as its THz absorption and dispersion profile.

Terahertz systems can be either constructed in transmission or reflection geometry according to sample properties under investigation. The typical schematic layouts of THz-TDS systems operating in transmission and reflection modes are illustrated in Fig. 12.1. In both systems, a pair of off-axis mirrors are typically used to focus the generated THz pulse to the sample located at the focus of THz pulse and to deliver the transmitted or reflected pulse after the interaction with the sample to the detector. In transmission-based THz-TDS, the incident THz pulse passes through the sample and the transmitted pulse from the sample is sampled by the detector. In reflection geometry however, the generated THz beam by the emitter is reflected from the sample at normal incidence or at a small incidence angle and the reflected pulses are measured by the detector.

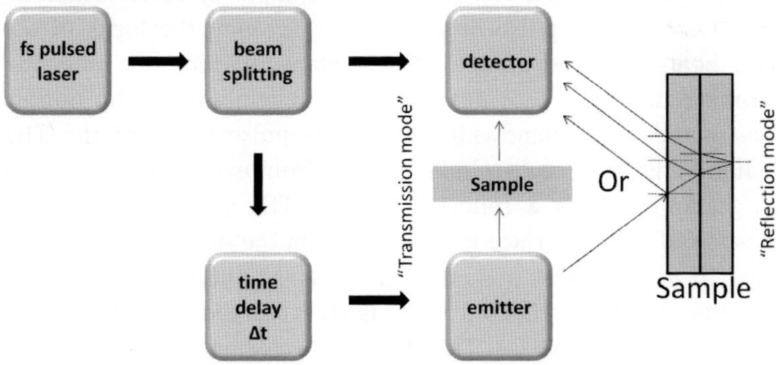

Figure 12.1 General applied form of THz time-domain spectroscopy systems.

In THz-TDS, the majority of the THz spectroscopy studies that have been done have utilized transmission geometry due to easier alignment and data analysis. Transmission mode THz-TDS is advantageous for samples having moderate absorption and low dispersion. However, samples such as biological tissues which can be thicker than the penetration depth are opaque in the THz range due to the high water content. Therefore, the preferential geometry in the investigation of such samples that are strongly absorbing, reflecting and dispersive is using the reflection geometry. The measurements of biological samples with a thickness of around 500 μm or more is unlikely using transmission geometry because of the high attenuation of THz pulse by the water content [21]. The signal to noise ratio (SNR) of the generated THz pulse is the determining factor for this thickness limit. The strong absorption of water restricts the transmission measurements on lyophilized or thin tissues. However, lyophilization and tissue sectioning processes can also affect the optical properties measured by THz. Too thin samples can also cause unwanted etalon effects. Reflection-based approaches seem to be more advantageous in the measurement of these samples. Therefore, reflection geometry is the preferred approach for in vivo and ex vivo measurements of biological samples [22].

THz time-domain spectroscopy method can also be used to generate two- and three-dimensional images. Terahertz pulsed imaging is a continuation of THz-TDS. It involves the mapping of the detected pulses after the transmission through the sample or the reflection from the sample by raster scanning of either the THz pulse or the sample through the focus of the THz beam. The advantage of a system working in reflection geometry is that it provides the depth information of the sample due to the broad bandwidth and short pulse width of the THz radiation. Each refractive index change from each interface within the sample causes a reflection of the THz pulse. Because the location of each interface is different from the detector, reflections from these interfaces arrive the detector at different delay times and the resulting reflected THz signal from the sample consists of a train of THz pulses. The thickness information of the layers within the sample is obtained by the time delay between each reflected THz pulse since each peak corresponds to an interface. This allows one to monitor the inside of the samples which is one of the reasons why this technique has become so popular. Thus, it gives an opportunity of the diagnosis of the extent of a disease if found for in vitro or in vivo measurements.

In a typical THz pulsed imaging system, the covered frequency range is 0.1–3 THz [2]. The strong absorption of water in this range [23] makes TPI method applicable in biomedical measurements. While the water content degree of biological tissues can be used as a contrast mechanism in medical imaging, the penetration depth of THz radiation is limited into tissues due to the high absorption losses resulting from the high water content. The drawback of transmission-based imaging through thick tissues is on the account of this high loss. The attenuation of THz radiation in fat containing tissues is lower than in water and skin consequently [24], [25] such that THz radiation can only penetrate a few tens of micrometers into tissue having high water content, while the penetration depth into fatty tissues is in the range of a few millimeters [26], [27]. For example, the penetration of THz radiation up to 1 mm inside of skin and 6 mm inside fat tissue were achieved using a commercial terahertz imaging system (TeraView TPI Imaga 1000) [7]. The SNR of the constructed system also determines the penetration depth. THz pulsed imaging systems are likely to have high SNR due to coherent detection schemes involved [28], [29]. This high SNR enables the detection of small refractive index changes. However, the acquisition speed of the imaging process and the high absorption coefficients of biological tissues reduce the overall SNR.

The other important factor in medical imaging is spatial resolution. The bandwidth of the THz imaging systems affects the resolution and a spatial resolution range of 100 μm to 1 mm is achievable due to the broad bandwidth of the THz radiation available in THz-TDS [30].

12.2.2 Principles of THz-TDS

In the mid- and far-infrared regions, THz-TDS and Fourier transform infrared (FTIR) are two complimentary spectroscopic methods. While the light source for THz pulses is generated by a fs laser in THz-TDS, it is a thermal source in the FTIR technique. They both use broadband sources, but differences between these spectroscopic techniques are evident. FTIR typically provides information in a broader frequency spectrum compared to THz-TDS. Signal-to-noise in THz-TDS is far better than FTIR for frequencies below 3 THz. However, FTIR has better SNR for frequencies over 5 THz [31].

In THz-TDS, temporal resolution is in the picosecond scale which makes this technique suitable in time resolved measurements. The other advantage of THz-TDS is that the electric field of THz pulses is measured directly. Since both the amplitude and phase information are obtained through electric field measurement, absorption coefficient and complex refractive index of samples can be determined. In FTIR, intensity of light is measured and the amplitude information can only be obtained. Therefore, Kramers–Kronig relations are necessary to extract the refractive index of samples.

To better understand the analysis of samples using THz-TDS the transmission of the THz electric field through the sample has to be investigated. The electric field of the temporally sampled THz waveform $E(t)$ can be transformed into frequency-domain via Fourier transform. The spectral distribution of electric field in frequency-domain is calculated as

$$\widetilde{E}(\omega) = A(\omega)e^{-i\varphi(\omega)} = \int dt E(t)e^{-i\omega t}, \tag{12.1}$$

where $\widetilde{E}(\omega)$ is the frequency dependent complex electric field, $A(\omega)$ is the electric field amplitude in frequency-domain, $\varphi(\omega)$ is the phase of electric field and ω is the angular frequency where $\omega = 2\pi f$ in which f is the frequency. Thus, the amplitude and phase information can be extracted. Measurements either in transmission or reflection geometries can be used to calculate the complex refractive index of the sample, $\tilde{n}(\omega)$. The real and imaginary components of refractive index can be represented as

$$\tilde{n}(\omega) = n(\omega) + i\kappa(\omega), \tag{12.2}$$

where $n(\omega)$ is the real valued refractive index and $\kappa(\omega)$ is the extinction coefficient which describes the attenuation of the electric field. The complex refractive index is defined as

$$\tilde{n}^2 = \varepsilon_r \mu_r = \frac{\varepsilon}{\varepsilon_0}\frac{\mu}{\mu_0}, \tag{12.3}$$

where ε_r is the relative permittivity of the medium, μ_r is the relative permeability of the medium, ε is the medium permittivity, ε_0 is the vacuum permittivity, μ is the medium permeability and μ_0 is the vacuum permeability. By using the assumption $\mu = \mu_0$ for biological media, the complex refractive index has a form given as $\tilde{n} = \sqrt{\varepsilon_r}$ and the relation between the extinction coefficient $\kappa(\omega)$ and absorption coefficient $\alpha(\omega)$ is expressed by the formula

$$\kappa(\omega) = \frac{\alpha(\omega)c}{2\omega}, \quad (12.4)$$

where c is the speed of light in vacuum. The amplitude of the electric field is attenuated during its propagation in the medium. The absorption coefficient is a fraction of this light intensity loss per unit length of the travelled distance. For linear interactions, the attenuation is described with the Beer–Lambert law which can be written as

$$I = I_0 e^{-\alpha(\omega)d}, \quad (12.5)$$

where I is the intensity at distance d and I_0 is the incident intensity. In transmission or reflection measurements, reflections can be observed in time-domain waveforms due to both components used for generation and detection of the THz pulse as well as boundaries of the samples under study. These reflections appear as multiple waveforms which are replicas of the main pulse and their attenuation as well as their delay depends on the refractive index and thickness of the sample. The refractive index of samples can then be calculated using time difference between these consecutive pulses emerging from the reflection of the THz pulse or with respect to the original pulse or reference pulse when no sample is present.

For more complicated layered geometries, the transmission and reflection coefficients of the medium can be extracted by Fresnel's equations to describe the fraction of the incident pulse to be transmitted through the sample and reflected from the surface. In Fig. 12.2, the transmitted and reflected signals during the propagation of THz pulse through a flat layer are illustrated.

The normal incident THz pulse, $E_0(\omega)$, which includes the amplitude and phase information of the pulse is a form of incident field obtained by the Fourier transform of time-domain THz pulse. If multiple reflections in the sample are ignored, the transmitted and reflected fields are then defined as

$$E_t(\omega) = E_0(\omega)t'_{12}t'_{23}e^{-\alpha d/2}e^{in\omega d/c}, \tag{12.6}$$

$$E_r(\omega) = E_0(\omega)r'_{12}, \tag{12.7}$$

$$E'_r(\omega) = E_0(\omega)\, t'_{12}r'_{23}t'_{21}e^{-\alpha d/2}e^{in\omega d/c}. \tag{12.8}$$

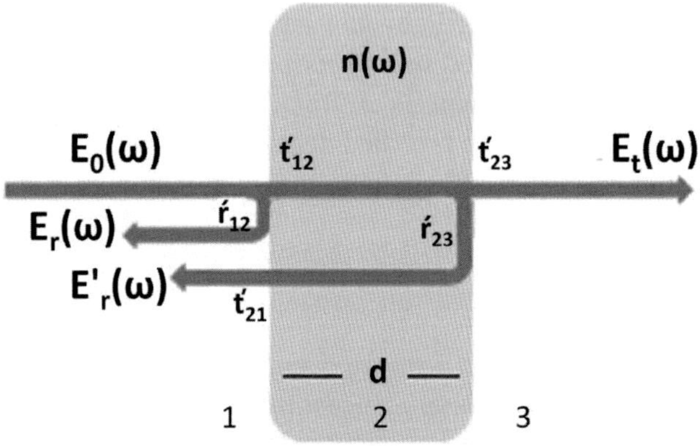

Figure 12.2 Transmission and reflection of a normal incident THz pulse passing through a single layer sample where $E_0(\omega)$ is the incident THz field, $E_r(\omega)$ and $E'_r(\omega)$ are the reflected THz field from the entrance and exit surfaces of the sample, respectively; and $E_t(\omega)$ is the transmitted THz field. r'_{12} and r'_{23} are the complex Fresnel reflection coefficients from the first and second interfaces of the sample, respectively. t'_{12}, t'_{21} and t'_{23} are the complex Fresnel transmission coefficients from region 1 to 2, region 2 to 1 and region 2 to 3, respectively. n is the refractive index and d is the thickness of the sample.

Detection geometry of THz spectrometers is configured according to sample properties and constraints on account of its geometry. THz radiation can penetrate many materials, so the response of such samples can be measured in transmission geometry. Some dry, non-polar, and non-metallic matters are transparent to THz radiation, whereas some materials such as metals are opaque to THz radiation [28]. In investigation of such non-transparent materials, THz-TDS systems operating in reflection geometry are used. Dynamic range (DR) of THz-TDS systems which limits the maximum measurable absorption coefficient of a sample is frequency dependent. Therefore, highly absorbing samples can exceed the dynamic range of transmission mode systems and this prevents the determination of optical parameters. Reflection measurements can be performed when more dynamic range is required. Reflection geometry is more suitable for highly absorbing, reflective and dispersive samples. Transmission spectroscopy is not also appropriate samples with large sample thickness. However, reflection spectroscopy enables measurement without thickness limitation.

Transmission geometry allows more uncomplicated data analysis, whereas it is more difficult in reflection measurements. Even still, suitable methods have been developed and are continuously improved. Noting that parameter extraction principles differ in both geometries, the fundamental approach employed in both geometries is to obtain the response of a sample and compare it to the reference and signal spectra. Reference spectrum is measured without a sample in transmission or reflection and can be described as a single pulse. Signal spectrum is obtained by measuring either the transmitted or reflected pulses from the sample and can consist of multiple pulses presenting themselves as near echoes coming from the different interfaces within the sample. The analyses in time-domain and in frequency-domain after the calculation of complex transmission and reflection coefficients of samples are done to extract the optical parameters.

12.2.2.1 Measurements in transmission geometry

THz time-domain spectroscopy in transmission geometry is typically the common method used to determine the optical parameters of the sample. The coherent detection of both the amplitude and phase of THz wave allows one to extract the optical constants. Figure 12.3 shows a THz waveform through a 200 μm-thick silicon wafer measured using a home-built THz-TDS spectrometer [32]. The echoes in Fig. 12.3 are well separated because silicon is optically thick; however, these would be superimposed if the sample was thinner. The refractive index of the sample can be calculated by the time difference between the THz peak in air and the first THz peak in silicon because thickness of the sample is known. It can also be obtained by the time delay between the first and second THz peaks or between the second and third THz peaks in sample time-domain spectrum.

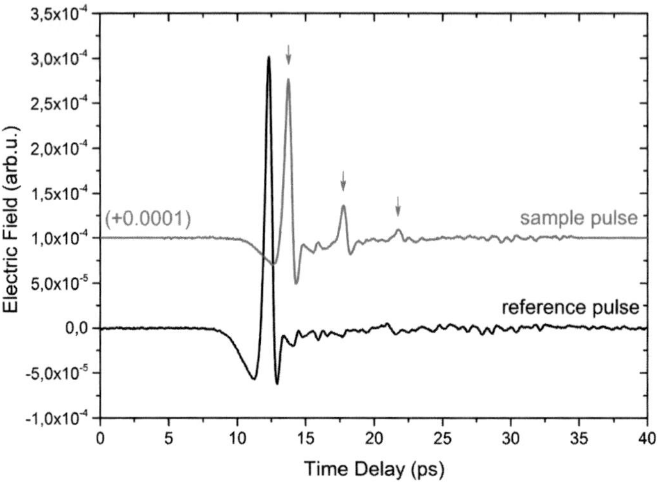

Figure 12.3 Example of a reference and single layer sample signal in the time-domain measured using a transmission geometry THz-TDS system.

If the analysis is restricted to samples having low absorption coefficients, then the Fresnel transmission coefficients will be real-valued and refractive index and absorption coefficient of the sample at normal incidence are given as

$$n(\omega) = 1 + \frac{\phi_T(\omega)c}{\omega d} = 1 + \frac{(\phi_{\text{sam},T}(\omega) - \phi_{\text{ref}}(\omega))c}{\omega d}, \tag{12.9}$$

$$\alpha(\omega) = -\frac{2}{d}\, ln\left(\frac{(n(\omega)+1)^2}{4n(\omega)}\,\tau(\omega)\right), \tag{12.10}$$

where $\phi_{\text{sam},T}(\omega)$ and $\phi_{\text{ref}}(\omega)$ are the phases of the transmitted and reference THz pulses, respectively $E_{\text{sam},T}(\omega)$. In frequency-domain, no meaningful measurements can be obtained above a specific frequency value which is called the maximal cutoff frequency. Equations (12.9) and (12.10) are only valid for samples having low absorption and infinite thickness. However, additional phase shifts can come from finite absorption due to complex Fresnel transmission coefficients. In thin samples, multiple reflections occur inside the sample and these are observed as a sequence of echoes following the main pulse in time-domain signal. For such a sample, the transmission coefficient $T(\omega)$ is the amplitude ratio of the transmitted signal $E_{\text{sam},T}(\omega)$ and reference THz signal $E_{\text{ref}}(\omega)$ given as

$$T(\omega) = \frac{E_{\text{sam},T}(\omega)}{E_{\text{ref}}(\omega)} = \frac{4\tilde{n}}{(\tilde{n}+1)^2} e^{-\alpha d} e^{\frac{in\omega d}{c}} \text{FP}(\omega), \tag{12.11}$$

where FP(ω) describes the Fabry–Perot modes and is defined as

$$\text{FP}(\omega) = \frac{1}{1 - \dfrac{\tilde{n}-1}{\tilde{n}+1} e^{-\alpha d} e^{2in\omega d/c}}. \tag{12.12}$$

Fabry–Perot echoes, multiple reflections in the sample, are well separated for thick samples, whereas they are overlaid for thin samples in the time-domain profile of the transmitted pulse.

12.2.2.2 Measurements in reflection geometry

THz-TDS in reflection geometry can also be used to obtain spectroscopic information. In this mode, the reference signal is usually obtained by removing sample from THz focus and placing a flat metallic or dielectric mirror instead of it. These mirrors have high reflectivity at THz frequencies [33]. However, readjustment of the sample placement can lead to some phase delay errors due to failure in the alignment of sample position which has to be same with the reference mirror position and this can lead to wrong parameter extraction [33], [34].

A window in contact with the sample can also be used for reference measurements. Reflection from the window surface is used as reference signal while the reflected signal from the back surface which is in contact with the sample is the sample signal.

Determination of refractive index is one of the simplest methods to classify samples. This can be measured precisely using transmission mode THz-TDS systems [35], [36]. Many of the reflection THz-TDS systems have been constructed with non-normal incidence angles [37], and even though it is more complicated, normal incidence geometries in the reflection mode can also be employed.

In this geometry, the refractive index can be calculated from the pulse travel time within the sample. The path of THz pulse in a system working in reflection geometry with an incidence angle to the normal is shown in Fig. 12.4. The optical path ($x_{\mathrm{opt},s}$) of THz pulse within the sample is calculated as

$$x_{\mathrm{opt},s} = \frac{2nd}{\cos\beta}. \tag{12.13}$$

The optical path difference between the reflected pulses from the surface of the sample and from an interface inside or the lower part of it is then defined as

$$c\Delta t = \frac{2d}{\cos\beta}(n - \sin\beta\sin\alpha) = 2nd\cos\beta. \tag{12.14}$$

Using Eq. (12.14), if the refractive index of the sample is known, its thickness can be calculated or if the sample thickness is known, its refractive index can be calculated without a reference measurement. Reflected THz signal from a sample with a series of interfaces is composed of multiple pulses in sequence which have different amplitude, polarity and temporal delay. Spectroscopic parameters about the first interface are obtained by analyzing the first reflected pulse in the waveform. By combining these parameters with the second pulse, spectroscopic information about the second layer can be determined.

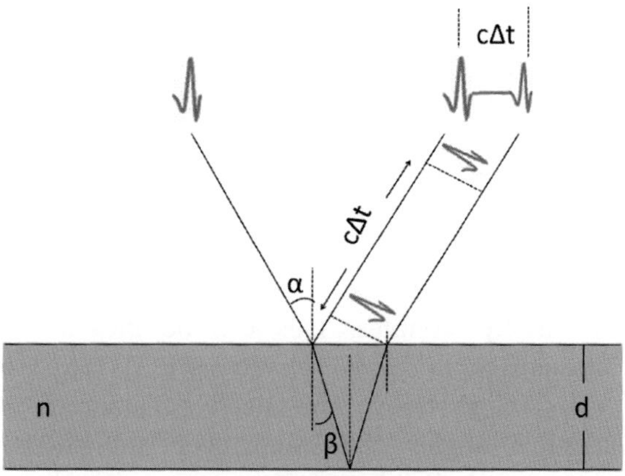

Figure 12.4 THz-TDS reflection mode measurement principles where α and β are the incident and refraction angles, respectively. n is the refractive index of sample, d is the sample thickness and Δt is time delay difference.

When a reference measurement is taken by using a metallic mirror; and it is assumed that the refractive index of air is ~1 and the reflectivity of this mirror is high at THz frequencies, ratio of the reflected and reference THz pulses at normal incidence is given as

$$R(\omega) = \frac{E_{\text{sam},R}(\omega)}{E_{\text{ref}}(\omega)} = r(\omega)e^{i\phi_R(\omega)} = \frac{\tilde{n}(\omega) - 1}{\tilde{n}(\omega) + 1} = \frac{n(\omega) + i\kappa(\omega) - 1}{n(\omega) + i\kappa(\omega) + 1}, \quad (12.15)$$

where $R(\omega)$ is the complex reflection coefficient of sample, $E_{\text{sam},R}(\omega)$ is the amplitude of the reflected THz pulse, $r(\omega)$ is the amplitude ratio and $\phi_R(\omega)$ is the phase difference of reflected and reference pulses. This expression involves the refractive index and absorption coefficient of sample defined as

$$n(\omega) = \frac{1 - r(\omega)^2}{1 + r(\omega)^2 - 2r(\omega)\cos\phi_{\text{R}}(\omega)}, \quad (12.16)$$

$$\alpha(\omega) = \frac{2\omega}{c}\frac{2r(\omega)\sin\phi_{\text{R}}(\omega)}{1 + r(\omega)^2 - 2r(\omega)\cos\phi_{\text{R}}(\omega)}. \quad (12.17)$$

Complex reflection coefficients $\tilde{R}_{\text{p}}(\omega)$ and $\tilde{R}_{\text{s}}(\omega)$, where subscripts p and s denote parallel and perpendicular polarized non-normal incidence THz beam, respectively, include refractive index and absorption coefficient information of the medium. The complex reflectivities for two polarization states through the Fresnel formulas are expressed as

$$\tilde{R}_{\text{p}}(\omega) = -\frac{\tilde{n}^2(\omega)\cos(\theta) - \sqrt{\tilde{n}^2(\omega) - \sin^2(\theta)}}{\tilde{n}^2(\omega)\cos(\theta) + \sqrt{\tilde{n}^2(\omega) - \sin^2(\theta)}} \quad (12.18)$$

$$\tilde{R}_{\text{s}}(\omega) = -\frac{\cos(\theta) - \sqrt{\tilde{n}^2(\omega) - \sin^2(\theta)}}{\cos(\theta) - \sqrt{\tilde{n}^2(\omega) - \sin^2(\theta)}} \quad (12.19)$$

where θ is the incident THz radiation angle. Measurement of the reflectivity of differently polarized THz radiation can be performed by turning both the emitter and detector antennas 90°. Complex reflectivities $\tilde{R}_{\text{p}}$ and $\tilde{R}_{\text{s}}$ are characterized by amplitudes R_p and R_s and phases φ_p and φ_s as $\widetilde{R}_p = R_p e^{i\varphi_p}$ and $\widetilde{R}_s = R_s e^{i\varphi_s}$, respectively. If the incidence angle θ is close to Brewster angle $\theta_{Br} = \tan^{-1}(n)$, resonance lines will become more pronounced in reflection spectrum. When the incidence angle θ is equal to Brewster angle θ_{Br}, amplitude values tend to zero and the extraction of the material parameters becomes a more complicated task.

12.3 Soft Tissues

The interaction of THz pulses with soft tissues is highly dependent on the water content of the tissue and whether there are scattering features which can become important as their feature size scales up toward the investigating wavelength. Tissues studied by THz spectroscopy can be characterized into the following forms: fresh tissues, frozen tissues, formalin fixed paraffin-embedded tissues and more recently penetration enhancement agent embedded tissue samples [38], [39]. Investigations into various tissue samples have shown that if the reflection and/or transmission is not strongly frequency dependent, then analysis of the time-domain waveforms can produce superior images [38]. In the case of archived (formalin-fixed, alcohol-dehydrated and paraffin-mounted) tissue samples the time delay, for example, is found to depend little on frequency. Often a practical approach towards classification consists in studying the time delay information and forming the image based on the fluctuations in the THz electric field amplitude. This approach avoids artifacts which may arise upon Fourier transformation.

Except for skin measurements, the majority of in vitro measurements of soft tissues such as muscle or adipose have been performed using porcine samples. Depending on the signal to noise ratio of the system the penetration depth for biological samples during in vivo measurements is typically limited to thicknesses less than 1 mm [40]. These limits are much higher for adipose tissue where penetration depths can exceed 5 mm [7]. These criteria change in reflection mode where it is preferential to have a thicker tissue samples so as to block any reflections. For many tissue types this requirement means that the sample should be at least 1 mm thick, but for samples with a high fat content such as adipose tissue this limit will be much larger [21].

12.3.1 Brain Tissue

THz-TDS is not only sensitive to water content in tissue samples. Brain tumors can have any number of abnormalities compared to healthy tissue such as nuclear atypia, mitotic activity, florid microvascular proliferation, and necrosis which can affect the transmission of THz pulses. In one study, the complex refractive index of paraffin embedded brain glioma and normal brain tissues were measured using terahertz time-domain spectroscopy. Compared with normal brain tissue, paraffin-embedded brain gliomas were shown to have a higher refractive index, absorption coefficient, and dielectric constant [41]. Recently, THz absorption and index of refraction of brain tissues from a mouse model of Alzheimer's disease (AD) and a control wild-type (normal) mouse were compared using THz time-domain spectroscopy [42]. Absorption peaks associated to torsional–vibrational modes were observed in AD tissue, close to the peaks of free tryptophan molecules which suggest that these modes could be used as biomarkers for detection of certain diseases.

Preparation of samples can also affect the transmission of THz pulses. Freezing tissues rapidly as in snap freezing which employ liquid nitrogen can result in reduced of ice crystals as compared to regular slow freezing processes. This can reduce scattering losses, which can enhance differences between healthy and diseased brain tissue [43].

12.3.2 Skin, Muscle, and Adipose Tissue

Tissues such as skin offer the opportunity to perform both in vivo as well as in vitro measurements. Again measurements in the time-domain can provide better information for distinguishing between cancerous and non-cancerous tissue samples. Prevalent skin cancers such as Basal cell carcinoma show different transmissive characteristics compared to inflammation and scar tissue as well as healthy tissue [44]. The stratum corneum, which is the outermost layer of the epidermis, has varying thicknesses and consists mainly of dead cells. Even so this layer contains water which attenuates the transmission of the THz pulse. In vivo studies on this layer show that its thickness can range from a few microns to hundreds of microns on various parts of the body [45], [46]. The variation of the thickness of this layer can impact the analysis of studies which look toward identifying certain types of skin cancers based on water content. Water content in the skin also can aid in determining the hydration status of patients undergoing hemodialysis [47]. Muscle and adipose tissue also show significant variations in water content. Measurements done with porcine tissue samples show that striated muscle and adipose tissue could be clearly distinguished in time-domain spectra owing to the difference in absorbance values and refractive indices [48], [49]. The refractive index of muscle, adipose, and frozen/thawed skin have been shown to exhibit comparable refractive index values ranging between 2.5 and 2.0, whereas freshly harvested skin showed values between 1.8 and 2.0 [22]. Since water is the key absorbing constituent the penetration depth can be enhanced by using a biocompatible material agent which can be absorbed by the tissue.

Using glycerol as a THz-peak enhancement agent (THz-PEA), the peak value of the THz signal which was transmitted through the fresh tissue was almost double compared to that of tissue without glycerol [39]. Such material agents can also adversely affect the characteristics of the tissue and care must be taken in their use. For example, formalin fixing is used to preserve tissues for histopathological diagnosis. However, along with sample variability, the sample distortion caused by formalin fixing can alter the DNA. In THz studies, this fixing process can displace the water molecules and introduce new intermolecular interactions between the sample and the formalin causing changes in both the time-dependent and frequency-dependent THz spectra [50].

12.3.3 Other Cancerous Tissue and Tumors

Other investigations have focused on determining differences between certain types of cancerous and healthy tissues. Typically such measurements were performed on histological sections either in transmission or reflection mode. Spectral differences can be seen between cancerous and healthy breast tissue samples where samples containing cancer had a higher refractive index and absorption coefficient [51]. However, the authors noted that the difference in absorption between fibrous tissue and cancer was not so clear most likely due to the added attenuation being a combination of both absorption and scattering. The breast cancer tissue showed higher differences in refractive index near the low end of the THz spectrum as compared to the high end. Similar differences in absorption were observed for liver cells. Most tumors and serious inflamed livers had lower absorption than normal tissues in the measurement range of the THz system [52]. While all these studies involve excision of the sample, one area has shown promise that further developments in THz technology could potentially lead to in vivo measurements. Several frozen oral tissues resected from patients showed differences in THz transmission between normal mucosa and cancerous tissue [53].

12.4 Hard Tissues

The study of osseous tissues and teeth are important from a variety of applications from medicine to criminal investigations. THz radiation is sensitive to structural changes and can be used to understand the rigidity or structural integrity of the hard tissue samples such as bone [54]. Furthermore, the hard tissue structure allows for investigations using much higher power sources such as high power gas lasers and QCLs [55]. The non-uniform surface of these samples limited the penetration depths in these studies to a few mm or less.

12.4.1 Bone

There have not been many publications about the study of human bones using THz radiation up to now most likely due to the non-uniform surface and sample geometry with respect to instrument design and sample placement. In one study, properties of cortical bone derived from human femur were investigated [54]. While the refractive index of samples subject to ambient drying was found as 2.49, this value decreased to 2.32 after a hot-air drying process. THz transmission through the bone which was higher for samples subject to hot-air drying was very significant below 1.2 THz. A linear relationship was found between the transmitted

THz radiation from dried human cortical bone and sample density in the frequency range of 0.1–1.25 THz. However, the measured THz parameters could not be correlated with the values of Young's modulus. The attenuation of THz radiation was found higher in dried spongy human bone than the compact one [55]. The THz imaging of human bone became difficult above 1 THz on account of inorganic components of sample. In 0.3–1.9 THz frequency range, they found that the refractive index of compact bone was 1.92–1.97. Inhomogeneous structures inside the samples can alter these measurements and their effects need to be addressed.

12.4.2 Cartilage

Cartilage is a heterogeneous tissue and its molecular composition changes in depth. Jung et al. observed the gradual decreasing of absorption coefficient from the articular surface to the subchondral bone while the change of refractive index was relatively low along the depth of cartilage [56]. They attributed the reason of this variation to the molecular composition and predominantly to water content which might have particular depth profile in cartilage. The characterization of water distribution in depth is critical in the medical diagnosis of osteoarthritis, a widely observed disease in elderly people, because water content alteration of the structure is associated with the degeneration degree of the cartilage [57]. In another study by the same group [58], normal and diseased human articular cartilage tissues with 500 μm in thickness were distinguishable mainly due to the differences in water content. Different structures in the cartilage sample could be separated in the transmission images of normal and diseased dehydrated tissues. However, no structural distinction was observed in the THz images of normal and diseased fresh samples.

In the diagnosis of osteoarthritis, thickness information of the cartilage is crucial to understand the spread of the disease if any. The degeneration of cartilage due to osteoarthritis causes the hyaline and calcified cartilage to get thinner [59]. Kan et al. used a reflection mode terahertz pulsed imaging system to measure the thicknesses of formalin fixed cartilage samples with osteoarthritis [60]. The optical delays which gave the thickness information of the layers showed that the sample from the immobilized leg was found to be thinner than the mobilized one.

12.4.3 Teeth

Teeth comprise the first studied diseased tissues using THz imaging systems [61]. It has been shown that THz is capable of distinguishing the different layers of a human tooth [27]. Sim et al. found the refractive index of enamel to be higher than that of dentine and these values were constant along the 0.1 to 1 THz frequency range [62]. As the frequency increased, the power absorption of enamel and dentine also increased, but remained close. While the power absorptions of the wet and dried enamels were different, there was no significant difference between the wet and dried dentines due to the water content in the layers. Nazarov et al. measured the refractive index of dried enamel to be between 3.0–3.2 and of dried dentine to be between 2.2 and 2.4 [63]. In the same study, the investigation of the microtubules direction and concentration in dentine showed that refractive index increases with frequency when samples are cut along the tubules at frequencies lower than 1.5 THz, but it decreases with frequency or stays constant when this cut is made across the tubules. These initial spectroscopic measurements quickly developed into imaging applications. Crawley et al. demonstrated a pulsed THz system that could obtain images and gather 3D information from tooth samples [64].

To better understand the interaction of THz pulses with different portions of the tooth, various tooth samples were obtained with an ethical approval from the Faculty of Dentistry of Ankara University, Ankara, Turkey. These samples are permanent teeth slices sectioned mesiodistally in parallel. In segmentation, a water-cooled diamond saw operating at low speed was used, thereby layered structure of teeth and dental caries can be monitored by THz pulse. These sections were then inspected using a stereomicroscope (×10) (Stemi 2000; Carl Zeiss, Jena, Germany). Slices stored in distilled water were dried in air before being examined in reflection geometry home-built THz pulsed imaging system. Figure 12.5 shows the visible images of the tooth slices studied.

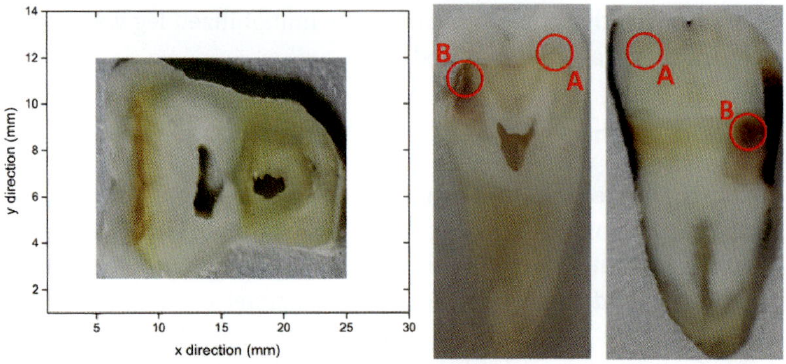

Figure 12.5 Sliced (a) tooth sample 1 with scale (b) tooth sample 2 (c) tooth sample 3 where (A) and (B) indicate healthy and caries points, respectively.

Figures 12.6 and 12.7 show two-dimensional THz images of tooth slice presented in Fig. 12.5a where various sections of tooth, enamel and dentine (outer) and pulp (center), are represented. These THz images are obtained by keeping the delay axis at fixed points separately as that these time delays correspond to peak THz amplitude reflected from top surface and bottom surface of a predefined pixel respectively and scanning the sample along THz focus. In this way, structural changes of the sample became evident.

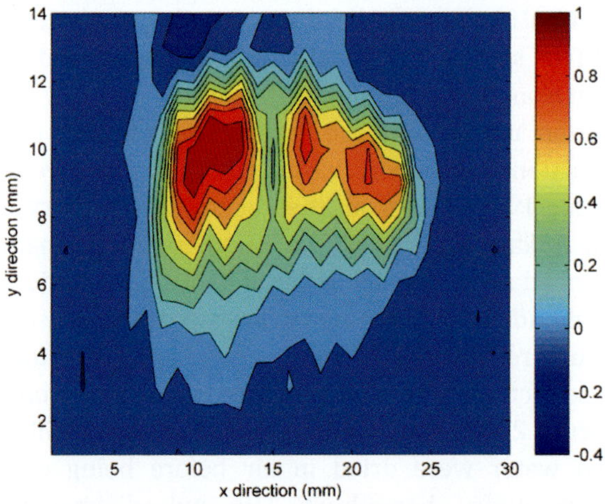

Figure 12.6 Terahertz reflection image of sliced tooth given in Fig. 12.5a to scale at the peak of THz signal according to top surface of tooth slice.

In analysis of the images obtained, since the refractive index of enamel (3.06 ± 0.09) is higher than that of dentine (2.57 ± 0.05) in 0.5–1.5 THz spectral range [40], the reflection from enamel part is expected to be higher and from dentine part is expected to be lower. Additionally, the absorption coefficient of enamel (62 ± 7 cm^{-1}) is lower than the absorption coefficient of dentine (70 ± 7 cm^{-1}) in spectral range of 0.5–1.5 THz [40]. Consequently, the THz pulse is less absorbed in enamel and reflected signal is also higher from this region. The coloring of Figs. 12.6 and 12.7 in which the red color shows high reflectance is consistent with these assertions.

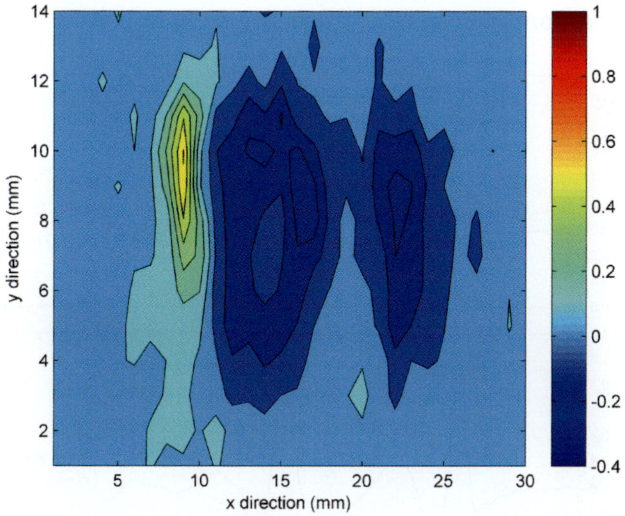

Figure 12.7 Terahertz reflection image of sliced tooth given in Fig. 12.5a to scale at the peak of THz signal according to bottom surface of tooth slice.

Time-domain profiles also provide information which can be helpful in understanding the properties of the tooth sample. Normalized time-domain THz waveforms taken from a healthy point and a caries point of the second tooth sample given in Fig. 12.5b are shown in Fig. 12.8. The primary (A1) and (B1) peaks represent partial reflections from top surfaces of healthy point, A, and unhealthy point, B, respectively. The secondary (A2) and (B2) peaks signify reflections from bottom surfaces of healthy point, A, and caries point, B, respectively.

Because refractive index of tooth is higher than that of air, these reflections are observed as positive peaks. Regions between the primary and secondary partial reflection peaks in reflected THz waveforms from healthy and caries points provide information about the internal structures of these areas. Even in the time-domain profiles differences between healthy and unhealthy regions can be observed.

The frequency-domain analysis of the healthy and caries points on the same tooth sample is shown in Fig. 12.9. Spectral amplitude of healthy point was extended up to 0.4 THz; however, this range was up to 0.5 THz for the caries point. This result suggests that characteristic differences of healthy and caries points are more evident in the frequency-domain.

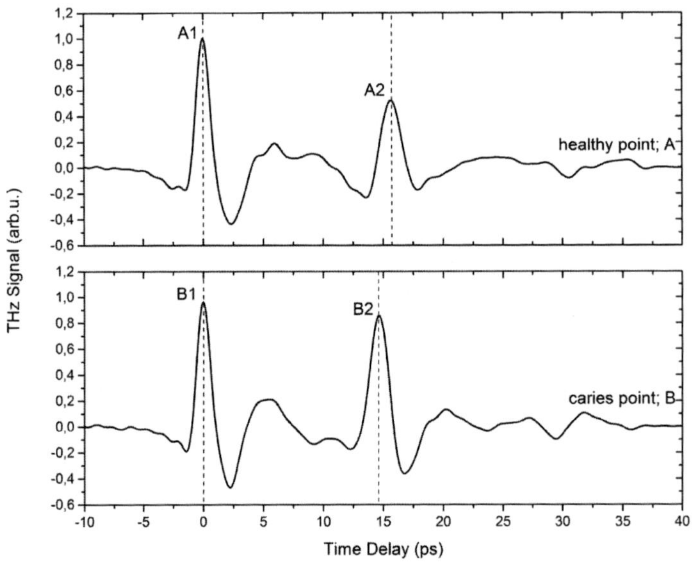

Figure 12.8 Temporal waveforms obtained from a single point THz measurement of a healthy point (A) and caries point (B) of tooth sample 2, respectively.

Measured THz waveforms from a single point THz measurement of a healthy and an unhealthy point on the third tooth sample shown in Fig. 12.5c in the time-domain are displayed in Fig. 12.10. Similarly, the primary (A1) and secondary (A2) peaks show THz reflections from the top and bottom surface of healthy point, A, respectively; while the primary (B1) and secondary (B2) peaks indicate reflections from the top and bottom surface of unhealthy point, B, respectively.

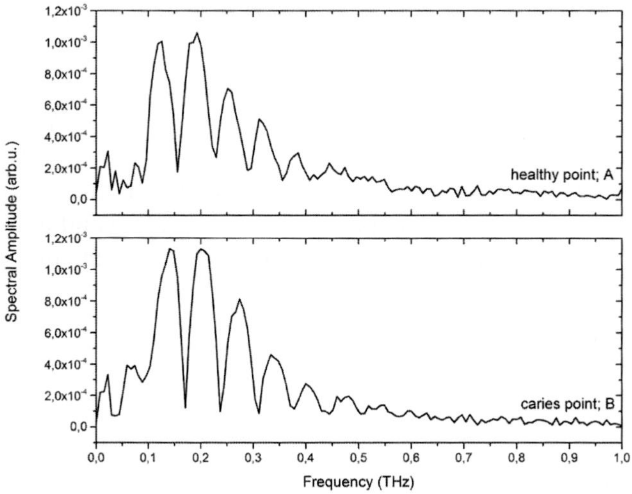

Figure 12.9 Amplitude of reflected THz pulse from healthy (A) and caries (B) points of tooth sample 2 in frequency-domain, respectively.

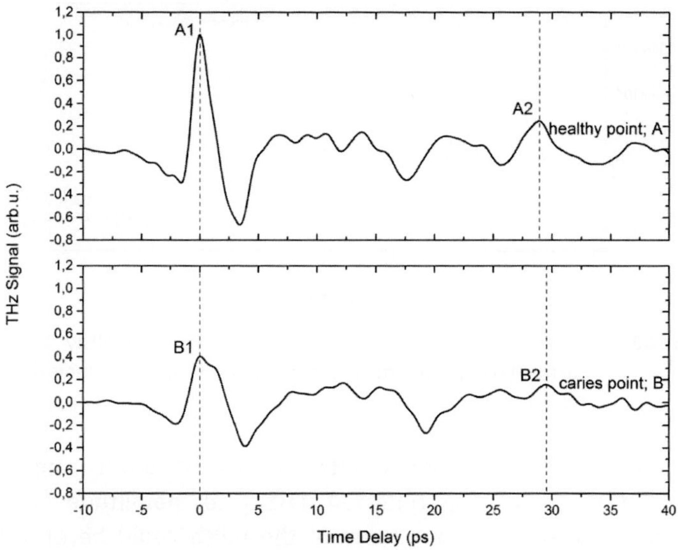

Figure 12.10 Temporal waveforms obtained from a single point THz measurement of a healthy point (A) and caries point (B) of tooth sample 3, respectively.

Although reflected signal from the bottom surface of these points were weak in Fig. 12.10, differences in time-domain signals were observed at delay times coming after the upper surface reflection as a result of the exposure of these points to THz radiation.

Frequency-domain waveforms obtained through a fast Fourier transformation of these healthy and unhealthy points for tooth sample 3 are given in Fig. 12.11. Amplitude range of the healthy point was up to 0.4 THz; however, frequencies up to 0.3 THz were observed for the caries point. Healthy and unhealthy zones were likewise distinguishable even though the cutoff frequency ranges were different.

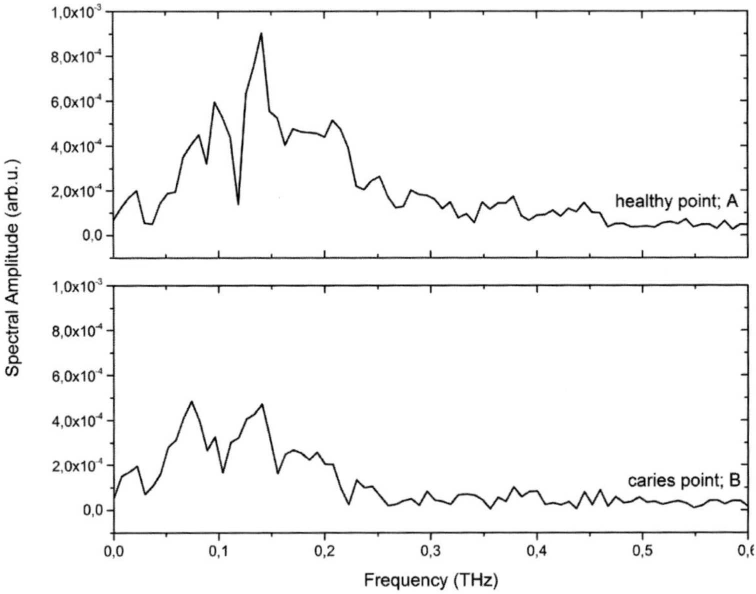

Figure 12.11 Amplitude of reflected THz pulse from healthy (A) and caries (B) points of tooth sample 3 in frequency-domain, respectively.

Finally, two-dimensional THz images of tooth sample 1 obtained by measuring reflected THz peak amplitude showed that differences at various parts of the tooth could be observed by the TPI method. Dental caries were also investigated in time-domain and frequency-domain and differences between the healthy and unhealthy regions were detected in both domains, but distinctions were more explicit in the frequency-domain [65], [66]. These images display the sensitivity of the THz pulse to structural changes both on surface and inside the tooth samples. The potential of TPI to diagnose such changes is promising to detect early dental caries.

12.5 Discussion and Conclusion

Overall the attenuation of the sample and signal to noise ratio of the constructed THz measurement system determine maximum detectable penetration depth in the analysis of tissue samples. For example, the generated THz pulse can only penetrate up to 1.5 mm depth in the tooth samples for our home-built system. Therefore, a better SNR is required which will increase the overall observable bandwidth because penetration depth increases with SNR. Since the usable bandwidth of our system is about 0.7 THz, the coherence length is then calculated to be approximately 70 μm assuming roughly $n = 3$ for the refractive index of tooth. Thus, it is necessary to increase the bandwidth of THz pulse for smaller features to be resolved. However, further problems can arise since the absorption coefficient of teeth or any tissue sample for that matter increases with increasing frequency.

The ability for THz waves to distinguish between cancerous and non-cancerous tissues is promising and further developments in instrumentation with regard to improvement of SNR can aid in the detection of such diseases. So far the penetration depth in these studies are limited to the type of tissues where the cancer is prevalent, and by far the best applicability seems to be in the measurements of the stratum corneum and layers beneath up to 1 mm [40], [45]. In vitro studies with excised tissue samples benefit from freezing or snap-freezing processes to ascertain differences between cancerous and healthy tissues [41]. Another promising field is the use of PEAs to enhance the penetration depths, and so far certain biocompatible materials have yielded results which have doubled typical depth values [39]. Terahertz pulsed measurement techniques have also proven useful in measuring other tissues such as muscle and adipose, where the largest penetration depths up to few mm has been observed [22].

The development of innovative measurement schemes using heterodyne or homodyne Schottky mixers [67] as well as THz microscopy techniques can improve the measurement and enhance detection limits by finding new way to channel the THz light into the samples [68].

Dental caries, a serious prevalent disease, is the main underlying cause of tooth losses and the potential for diagnosis of such diseases without the harmful side effects of such technologies that use x-rays continue to drive the research in this field. This disease both destructs tooth structure and affects ultimately surrounding tissues and pulp. Its formation is a dynamic process and the detection of lesions is difficult at early stages due to deficiency of methods to identify the presence of tooth decay precisely on the entire tooth structure. Therefore, there is a requirement for new methods to monitor whole structure and characterize defects. Transverse microradiography (TMR) is one of the methods to investigate mineral content of lesions [69]. However, this technique requires a destructive sample preparation phase since tooth samples have to be cut into thin slices before microradiography. X-ray microtomography (XMT) is an alternative method to TMR [70]. Because this method uses ionizing radiation, detrimental effects of this method on human health can be significant [71], [72]. On the other hand, TPI is by far one of the more successful applications of THz biomedical applications and so far has been extensively employed due to its non-destructive and non-ionizing nature. There have also been some studies on the identification of natural dental caries [73] and artificial lesions [74] and detection of remineralization of artificially induced lesions [75] using TPI. These investigations and others sustain the usability of TPI in studies on identification of demineralization.

Acknowledgments

The authors acknowledge the support and contributions of Dr. Kivanc Kamburoglu at the Faculty of Dentistry of Ankara University, Ankara, Turkey, in the dental studies section of this chapter. We also would like to acknowledge support from Middle East Technical University Research Office METU-BAP#01052014002 and METU-BAP#08112013023 research grants.

References

1. G.-S. Park, Y. H. Kim, H. Han, J. K. Han, J. Ahn, J.-H. Son, W.-Y. Park, and Y. U. Jeong, eds., *Convergence of Terahertz Sciences in Biomedical Systems*, Springer Netherlands, 2012.
2. X.-C. Zhang, and J. Xu, *Introduction to THz Wave Photonics*, Springer Science + Business Media LLC, 2010.
3. B. Yu, F. Zeng, Y. Yang, Q. Xing, A. Chechin, X. Xin, I. Zeylikovich, and R. R. Alfano, Torsional vibrational modes of tryptophan studied by terahertz time-domain spectroscopy, *Biophys. J.*, vol. 86, no. 3, pp. 1649–1654, 2004.
4. M. Nazarov, A. Shkrurinov, V. V. Tuchin, and X.-C. Zhang, Terahertz Tissue spectroscopy and imaging, *Handbook Photonics Biomed. Sci.*, vol. 30, no. 8, pp. 519–617, 2010.
5. X. Yin, B. W.-H. Ng, and D. Abbott, *Terahertz Imaging for Biomedical Applications*, New York: Springer-Verlag, 2012.
6. S. W. Smye, J. M. Chamberlain, A. J. Fitzgerald, and E. Berry, The interaction between Terahertz radiation and biological tissue, *Phys. Med. Biol.*, vol. 46, no. 9, pp. R101–R112, 2001.
7. E. Pickwell and V. P. Wallace, Biomedical applications of terahertz technology, *J. Phys. D Appl. Phys.*, vol. 39, no. 17, pp. R301–R310, 2006.
8. S. M. Kim, F. Hatami, J. S. Harris, A. W. Kurian, J. Ford, D. King, G. Scalari, M. Giovannini, N. Hoyler, J. Faist, and G. Harris, Biomedical terahertz imaging with a quantum cascade laser, *Appl. Phys. Lett.*, vol. 88, p. 153903, 2006.
9. M. Saviz, O. Spathmann, J. Streckert, V. Hansen, M. Clemens, and R. Faraji-Dana, Theoretical estimations of safety thresholds for terahertz exposure of surface tissues, *IEEE Trans. Terahertz Sci. Technol.*, vol. 3, no. 5, pp. 635–640, 2013.
10. K.-E. Peiponen, A. Zeitler, and M. Kuwata-Gonokami, eds., *Terahertz Spectroscopy and Imaging*, Berlin: Springer-Verlag, 2013.
11. Y. Sun, M. Y. Sy, Y. X. Wang, A. T. Ahuja, Y. T. Zhang, and E. Pickwell-MacPherson, A promising diagnostic method: Terahertz pulsed imaging and spectroscopy, *World J. Radiol.*, vol. 3, no. 3, pp. 55–65, 2011.
12. C. A. Schmuttenmaer, Exploring dynamics in the far-infrared with terahertz spectroscopy, *Chem. Rev.*, vol. 104, no. 4, pp. 1759–1780, 2004.

13. D. H. Auston, K. P. Cheung, and P. R. Smith, Picosecond photoconducting Hertzian dipoles, *Appl. Phys. Lett.*, vol. 45, pp. 284–286, 1984.
14. D. H. Auston, K. P. Cheung, J. A. Valdmanis, and D. A. Kleinman, Cherenkov radiation from femtosecond optical pulses in electro-optic media, *Phys. Rev. Lett.*, vol. 53, p. 1555, 1984.
15. M. van Exter, Ch. Fattinger, and D. Grischkowsky, High-brightness terahertz beams characterized with an ultrafast detector, *Appl. Phys. Lett.*, vol. 55, no. 4, pp. 337–339, 1989.
16. B. B. Hu and M. C. Nuss, Imaging with terahertz waves, *Opt. Lett.*, vol. 20, no. 16, pp. 1716–1718, 1995.
17. Q. Wu, T. D. Hewitt, and X.-C. Zhang, Two-dimensional electo-optic imaging of THz beams, *Appl. Phys. Lett.*, vol. 69, no. 8, pp. 1026–1028, 1996.
18. D. M. Mittleman, S. Hunsche, L. Boivin, and M. C. Nuss, T-ray tomography, *Opt. Lett.*, vol. 22, no. 12, pp. 904–906, 1997.
19. J.-H. Son, Terahertz electromagnetic interactions with biological matter and their applications, *J. Appl. Phys.*, vol. 105, no. 10, p. 102033, 2009.
20. M. M. Nazarov, A. P. Shkurinov, E. A. Kuleshov, and V. V. Tuchin, Terahertz time-domain spectroscopy of biological tissues, *Quantum Elec.*, vol. 38, no. 7, pp. 647–654, 2008.
21. E. Pickwell-MacPherson and V. P. Wallace, Terahertz pulsed imaging—A potential medical imaging modality?, *Photodiagnosis Photodynamic Ther.*, vol. 6, no. 2, pp. 128–134, 2009.
22. G. J. Wilmink, B. L. Ibey, T. Tongue, B. Schulkin, N. Laman, X. G. Peralta, C. C. Roth, C. Z. Cerna, B. D. Rivest, J. E. Grundt, and W. P. Roach, Development of a compact terahertz time-domain spectrometer for the measurement of the optical properties of biological tissues, *J. Biomed. Opt.*, vol. 16, no. 4, p. 047006, 2011.
23. K. Humphreys, J. P. Loughran, M. Gradziel, W. Lanigan, T. Ward, J. A. Murphy, and C. O'Sullivan, Medical applications of terahertz imaging: A review of current technology and potential applications in biomedical engineering, in *Conf Proc IEEE Eng Med Biol Soc*, San Francisco, 2004.
24. M. He, A. K. Azad, S. Ye, and W. Zhang, Far-infrared signature of animal tissues characterized by terahertz time-domain spectroscopy, *Opt. Commun.*, vol. 259, no. 1, pp. 389–392, 2006.
25. E. Pickwell-MacPherson, Practical considerations for in vivo THz imaging, *IEEE Trans. Terahertz Sci. Technol.*, vol. 3, no. 4, pp. 163–171, 2010.

26. S. Fan, Y. He, B. Ung, and E. Pickwell-MacPherson, The growth of biomedical terahertz research, *J. Phys. D Appl. Phys.*, vol. 47, no. 37, p. 374009, 2014.

27. D. D. Arnone, C. M. Ciesla, A. Corchia, S. Egusa, M. Pepper, J. M. Chamberlain, C. Bezant, E. H. Linfield, R. Clothier, and N. Khammo, Applications of terahertz (THz) technology to medical imaging, in *Proc. SPIE 3828, Terahertz Spectroscopy and Applications II*, Munich, 1999.

28. D. M. Mittleman, R. H. Jacobsen, and M. C. Nuss, T-ray imaging, *IEEE J. Sel. Top. Quantum Elec.*, vol. 2, no. 3, pp. 679–692, 1996.

29. X.-C. Zhang, Terahertz wave imaging: Horizons and hurdles, *Phys. Med. Biol.*, vol. 47, no. 21, pp. 3667–3677, 2002.

30. J. B. Baxter and G. W. Guglietta, Terahertz spectroscopy, *Anal. Chem.*, vol. 83, no. 12, pp. 4342–4368, 2011.

31. P. Y. Han, M. Tani, M. Usami, S. Kono, R. Kersting, and X.-C. Zhang, A direct comparison between terahertz time-domain spectroscopy and far-infrared Fourier transform spectroscopy, *J. Appl. Phys.*, vol. 89, no. 4, pp. 2357–2359, 2001.

32. Z. Ozer, S. Gok, H. Altan, and F. Severcan, Concentration-based measurement studies of L-tryptophan using terahertz time-domain spectroscopy (THz-TDS), *Appl. Spectrosc.*, vol. 68, no. 1, pp. 95–100, 2014.

33. T.-I. Jeon and D. Grischkowsky, Characterization of optically dense, doped semiconductors by reflection THz time-domain spectroscopy, *Appl. Phys. Lett.*, vol. 72, no. 23, pp. 3032–3034, 1998.

34. A. Pashkin, M. Kempa, H. Nemec, F. Kadlec, and P. Kuzel, Phase-sensitive time-domain terahertz reflection spectroscopy, *Rev. Sci. Instrumentation*, vol. 74, no. 11, pp. 4711–4717, 2003.

35. M. van Exter and D. Grischkowsky, Optical and electronic properties of doped silicon from 0.1 to 2 THz, *Appl. Phys. Lett.*, vol. 56, no. 17, pp. 1694–1696, 1990.

36. M. Herrman, M. Tani, K. Sakai, and R. Fukasawa, Terahertz imaging of silicon wafers, *J. Appl. Phys.*, vol. 91, no. 3, pp. 1247–1250, 2002.

37. M. Khazan, R. Meissner, and I. Wilke, Convertible transmission-reflection time-domain terahertz spectrometer, *Rev. Sci. Instrum.*, vol. 72, no. 8, pp. 3427–3430, 2001.

38. T. Loffler, K. Siebert, S. Czasch, T. Bauer, and H. G. Roskos, Visualization and classification in biomedical terahertz pulsed imaging, *Phys. Med. Biol.*, vol. 47, no. 21, pp. 3847–3852, 2002.

39. S. J. Oh, S.-H. Kim, K. Jeong, Y. Park, Y.-M. Huh, J.-H. Son, and J.-S. Suh, Measurement depth enhancement in terahertz imaging of biological tissues, *Opt. Express*, vol. 21, no. 18, pp. 21299–21305, 2013.
40. E. Berry, A. J. Fitzgerald, N. N. Zinov'ev, G. C. Walker, S. Homer-Vanniasinkam, C. D. Sudworth, R. E. Miles, J. M. Chamberlain, and M. A. Smith, Optical properties of tissue measured using terahertz-pulsed imaging, in *Proc. SPIE 5030, Medical Imaging 2003: Physics of Medical Imaging*, San Diego, 2003.
41. K. Meng, T.-N. Chen, T. Chen, L.-G. Zhu, Q. Liu, Z. Li, F. Li, S.-C. Zhong, Z.-R. Li, H. Feng, and J.-H. Zhao, Terahertz pulsed spectroscopy of paraffin-embedded brain glioma, *J. Biomed. Opt.*, vol. 19, no. 7, p. 077001, 2014.
42. L. Shi, P. Shumyatsky, A. Rodriguez-Contreras, and R. R. Alfano, Terahertz spectroscopy of brain tissue from a mouse model of Alzheimer's disease, *J. Biomed. Opt.*, vol. 21, no. 1, p. 015014, 2016.
43. G. M. Png, R. Flook, B. W.-H. Ng, and D. Abbott, Terahertz spectroscopy of snap-frozen human brain tissue: An initial study, *Electron. Lett.*, vol. 45, no. 7, pp. 343–345, 2009.
44. R. M. Woodward, B. E. Cole, V. P. Wallace, R. J. Pye, D. D. Arnone, E. H. Linfield, and M. Pepper, Terahertz pulse imaging in reflection geometry of human skin cancer and skin tissue, *Phys. Med. Biol.*, vol. 47, no. 21, pp. 3853–3863, 2002.
45. E. Pickwell, B. E. Cole, A. J. Fitzgerald, M. Pepper, and V. P. Wallace, In vivo study of human skin using pulsed terahertz radiation, *Phys. Med. Biol.*, vol. 49, no. 9, pp. 1595–1607, 2004.
46. R. M. Woodward, V. P. Wallace, D. D. Arnone, E. H. Linfield, and M. Pepper, Terahertz pulsed imaging of skin cancer in the time and frequency domain, *J. Biolog. Phys.*, vol. 29, no. 2–3, pp. 257–261, 2003.
47. F. Kadlec, M. Berta, P. Kuzel, F. Lopot, and V. Polakovic, Assessing skin hydration status in haemodialysis patients using terahertz spectroscopy: A pilot/feasibility study, *Phys. Med. Biol.*, vol. 53, no. 24, pp. 7063–7071, 2008.
48. H. Hoshina, A. Hayashi, N. Miyoshi, F. Miyamaru, and C. Otani, Terahertz pulsed imaging of frozen biological tissues, *Appl. Phys. Lett.*, vol. 94, no. 12, p. 123901, 2009.
49. A. J. Fitzgerald, E. Berry, N. N. Zinov'ev, S. Homer-Vanniasinkam, R. E. Miles, J. M. Chamberlain, and M. A. Smith, Catalogue of human tissue optical properties at terahertz frequencies, *J. Biolog. Phys.*, vol. 29, no. 2, pp. 123–128, 2003.

50. Y. Sun, B. M. Fischer, and E. Pickwell-MacPherson, Effects of formalin fixing on the terahertz properties of biological tissues, *J. Biomed. Opt.*, vol. 14, no. 6, p. 064017, 2009.

51. P. C. Ashworth, E. Pickwell-MacPherson, E. Provenzano, S. E. Pinder, A. D. Purushotham, M. Pepper, and V. P. Wallace, Terahertz pulsed spectroscopy of freshly excised human breast cancer, *Opt. Express*, vol. 17, no. 15, pp. 12444–12454, 2009.

52. C. H. Zhang, G. F. Zhao, B. B. Jin, Y. Y. Hou, H. H. Jia, J. Chen, and P. H. Wu, Terahertz imaging on subcutaneous tissues and liver inflamed by liver cancer cells, *IEEE Trans. Terahertz Sci. Technol.*, vol. 5, no. 3, pp. 114–123, 2012.

53. Y. C. Sim, J. Y. Park, K.-M. Ahn, C. Park, and J.-H. Son, Terahertz imaging of excised oral cancer at frozen temperature, *Biomed. Opt. Express*, vol. 4, no. 8, pp. 1413–1421, 2013.

54. M. R. Stringer, D. N. Lund, A. P. Foulds, A. Uddin, E. Berry, R. E. Miles, and A. G. Davies, The analysis of human cortical bone by terahertz time-domain spectroscopy, *Phys. Med. Biol.*, vol. 50, no. 14, pp. 3211–3219, 2005.

55. M. Bessou, B. Chassagne, J.-P. Caumes, C. Pradere, P. Maire, M. Tondusson, and E. Abraham, Three-dimensional terahertz computed tomography of human bones, *Appl. Opt.*, vol. 51, no. 28, pp. 6738–6744, 2012.

56. E. Jung, H. J. Choi, M. Lim, H. Kang, H. Park, H. Han, B.-H. Min, S. Kim, I. Park, and H. Lim, Quantitative analysis of water distribution in human articular cartilage using terahertz time-domain spectroscopy, *Biomed. Opt. Express*, vol. 3, no. 5, pp. 1110–1115, 2012.

57. C.-B. James and T. L. Uhl, A review of articular cartilage pathology and the use of glucosamine sulfate, *J. Athl. Train.*, vol. 36, no. 4, pp. 413–419, 2001.

58. E. Jung, H. Park, K. Moon, M. Lim, Y. Do, H. Han, H. J. Choi, B.-H. Min, S. Kim, I. Park, and H. Lim, THz time-domain spectroscopic imaging of human articular cartilage, *J. Infrared Millimeter Terahertz Waves*, vol. 33, no. 6, pp. 593–598, 2012.

59. W. C. Kan, W.-S. Lee, W.-H. Cheung, and E. Pickwell-MacPherson, A pilot study of terahertz pulsed imaging of osteoarthritis, in *IRMMW-THz 2008, 33rd International Conference*, Pasadena, 2008.

60. W.-C. Kan, W.-S. Lee, W.-H. Cheung, V. P. Wallace, and E. Pickwell-MacPherson, Terahertz pulsed imaging of knee cartilage, *Biomed. Opt. Express*, vol. 1, no. 3, pp. 967–974, 2010.

61. C. M. Ciesla, D. D. Arnone, A. Corchia, D. A. Crawley, C. Longbottom, E. H. Linfield, and M. Pepper, Biomedical applications of terahertz pulse imaging, in *Proc. SPIE 3934, Commercial and Biomedical Applications of Ultrafast Lasers II*, San Jose, 2000.
62. Y. C. Sim, I. Maeng, J.-H. Son, Frequency-dependent characteristics of terahertz radiation on the enamel and dentin of human tooth, *Curr. Appl. Phys.*, vol. 9, pp. 946–949, 2009.
63. M. M. Nazarov, A. P. Shkurinov, and V. V. Tuchin, Tooth study by terahertz time-domain spectroscopy, in *Proc. SPIE 6791, Saratov Fall Meeting 2007: Optical Technologies in Biophysics and Medicine IX*, Saratov, 2008.
64. D. Crawley, C. Longbottom, V. P. Wallace, B. Cole, D. Arnone, and M. Pepper, Three-dimensional terahertz pulse imaging of dental tissue, *J. Biomed. Opt.*, vol. 8, no. 2, pp. 303–307, 2003.
65. B. Karagoz, H. Altan, and K. Kamburoglu, Terahertz pulsed imaging study of dental caries, in *Proc. SPIE 9542, Medical Laser Applications and Laser-Tissue Interactions VII*, Munich, 2015.
66. K. Kamburoglu, N. O. Yetimoglu, and H. Altan, Characterization of primary and permanent teeth using terahertz spectroscopy, *Dentomaxillofac. Radiol.*, vol. 43, no. 6, p. 20130404, 2014.
67. S. Barbieri, J. Alton, H. E. Beere, E. H. Linfield, D. A. Ritchie, S. Withington, G. Scalari, L. Ajili, and J. Faist, Heterodyne mixing of two far-infrared quantum cascade lasers by use of a point-contact Schottky diode, *Opt. Lett.*, vol. 29, no. 4, pp. 1632–1634, 2004.
68. J. Cunningham, M. Byrne, P. Upadhya, M. Lachab, E. H. Linfield, and A. G. Davies, Terahertz evanescent field microscopy of dielectric materials using on-chip waveguides, *Appl. Phys. Lett.*, vol. 92, no. 3, p. 032903, 2008.
69. M. C. Huysmans and C. Longbottom, The challenges of validating diagnostic methods and selecting appropriate gold standards, *J. Dent. Res.*, vol. 83, pp. C48–C52, 2004.
70. S. E. Dowker, J. C. Elliott, G. R. Davis, and H. S. Wassif, Longitudinal study of the three-dimensional development of subsurface enamel lesions during in vitro demineralisation, *Caries Res.*, vol. 37, no. 4, pp. 237–245, 2003.
71. F. Angelieri, V. Carlin, D. M. Saez, R. Pozzi, and D. A. Ribeiro, Mutagenicity and cytotoxicity assessment in patients undergoing orthodontic radiographs, *Dentomaxillofac. Radiol.*, vol. 39, no. 7, pp. 437–440, 2010.

72. W. T. J. Longstreth, L. E. Phillips, M. Drangsholt, T. D. Koepsell, B. S. Custer, J. A. Gehrels, and G. van Belle, Dental X-rays and the risk of intracranial meningioma: A population-based case-control study, *Cancer*, vol. 100, no. 5, pp. 1026–1034, 2004.

73. D. A. Crawley, C. Longbottom, B. E. Cole, C. M. Ciesla, D. D. Arnone, V. P. Wallace, and M. Pepper, Terahertz pulse imaging: A pilot study of potential applications in dentistry, *Caries Res.*, vol. 37, no. 5, pp. 352–359, 2003.

74. E. Pickwell, V. P. Wallace, B. E. Cole, S. Ali, C. Longbottom, R. J. M. Lynch, and M. Pepper, A comparison of terahertz pulsed imaging with transmission microradiography for depth measurement of enamel demineralisation in vitro, *Caries Res.*, vol. 41, pp. 49–55, 2007.

75. D. Churchley, R. J. Lynch, F. Lippert, J. S. Eder, J. Alton, and C. Gonzales-Cabezas, Terahertz pulsed imaging study to assess remineralization of artificial caries lesions, *J. Biomed. Opt.*, vol. 16, no. 2, p. 026001, 2011.

Chapter 13

Detection of Brain Tumors Using Stimulated Raman Scattering Microscopy

Spencer Lewis and Daniel Orringer

Department of Neurosurgery, University of Michigan Medical School,
1500 E Medical Center Drive, Suite 3552
Ann Arbor, Michigan 48109, USA

dorringe@med.umich.edu

One of the greatest challenges in brain tumor surgery is the differentiation of neoplastic tissue from surrounding brain. This difficulty contributes to historically poor outcomes in brain tumor surgery. Stimulated Raman scattering (SRS) microscopy is a label-free, non-destructive modality which images tissue microstructure with chemical specificity. Features captured by SRS not only provide potentially diagnostic information but also permit the rapid, accurate, and automatic characterization of tissue as either normal or tumor-infiltrated.

13.1 Introduction

There is a need for improved structural visualization in brain tumor surgery, particularly at tumor margins where such

Deep Imaging in Tissue and Biomedical Materials: Using Linear and Nonlinear Optical Methods
Edited by Lingyan Shi and Robert R. Alfano

ISBN 978-981-4745-88-8 (Hardcover), 978-1-315-20655-4 (eBook)
www.panstanford.com

technology could guide surgical resection. There are several optical methods with the potential to identify neoplastic tissue using the electronic states and vibrational states of biomolecules within tissue. Point spectroscopy techniques, such as infrared microscopy and spontaneous Raman point spectroscopy, lack spatial detail and thus diagnose pathology purely by spectral character. While Raman spectroscopy has been successfully used in the operative setting, it requires custom light sources and is thus impractical to implement. Two-photon excitation fluorescence (TPEF), second-harmonic generation and third-harmonic generation (SHG and THG, respectively) yield spatial information; however, THG is without chemical specificity, and both SHG and TPEF highlight primarily collagen, limiting their utility in the brain. Stimulated Raman scattering (SRS) is a technique which allows for a substantial increase in Raman gain, such that signal can be acquired under ambient light conditions. In this chapter, the focus lies with the use of stimulated Raman scattering to image brain tissue, allowing for accurate and spatially precise tumor detection.

13.2 Background

Outcomes in brain tumor surgery correlate closely with the fraction of tumor mass removed, termed the extent of resection (EOR) [1, 2]. Yet maximizing EOR with preservation of contiguous healthy tissue presents a challenge: tumor and surrounding tissue may be difficult or impossible to differentiate grossly [3]. As a result, the majority of neoplasms in the CNS recur locally, often within 1 cm of the original tumor margin [3, 4]. A number of technologies have been developed to improve neurosurgical outcomes by refining the surgeon's ability to differentiate tumor from normal brain.

Frameless stereotaxy, for one, has been used to map instruments within the operative field onto an image space defined by preoperative tomography, facilitating resection of tissue with MR or CT signal changes characteristic of tumor. While stereotactic neuronavigation has become an invaluable part of the neurosurgical workflow, its effect on recurrence and mortality has been modest at best. This can be largely attributed to its reliance on static preoperative imaging, which fails to capture soft tissue shifts (sometimes on the order of a centimeter or more) within the brain due to perioperative edema and tumor debulking [5]. Intraoperative MRI (iMRI) was developed to address this issue, facilitating the acquisition of updated images depicting residual tumor throughout the surgery and improving outcomes in glioma resection. Yet despite the substantial cost of an iMRI suite (approximately $10 million) and two hour increase in surgical duration, little survival benefit has been demonstrated [6, 7].

The orally administered fluorescent tag 5-aminolevulinic acid (5-ALA) has also been shown to improve EOR, particularly among high-grade glioma, and provides real-time operative guidance unaffected by brain shift [8]. However, its low sensitivity for tumor detection and minimal uptake in low-grade glioma [9] have hampered its wide adoption and FDA approval. More broadly, extrinsic tumor labels have been plagued by issues of heterogeneous label uptake and inconsistent sensitivity across a range of neoplasms [10].

Optical techniques which rely on both intrinsic and extrinsic biomarkers in tissue provide another potential avenue for achieving both sensitivity and specificity in intraoperative brain tumor detection. Raman spectroscopy, a method which relies on the vibrational states of molecules within a sample to determine its chemical composition based on how molecular bonds interact with incident light, has shown promise for this application [11]. It is rapid, label free and non-destructive [11]. The spontaneous Raman method is weak (about 10 Raman photons per incident pump photon). Briefly, light of one wavelength (commonly referred to as the "pump wavelength" or "pump beam") is used to illuminate a sample. This light is scattered upon interaction with chemical bonds within the sample, each of which initially lies in a vibrational ground state. These bonds are excited to a high-energy "virtual state" before re-emitting a photon and relaxing back to a low-energy state. If the bond relaxes back to its ground state, the emitted photon is of the pump frequency (this is known as "elastic scattering"). However, the bond can also relax to either a higher or lower energy state than its initial state ("inelastic scattering"), with the difference in energy conserved through the emission of either a longer (Stokes) or shorter wavelength (anti-Stokes) of emitted photon, respectively. Photons emitted at a longer wavelength (due to a net excitation of the bond) are referred to as "Stokes scattering," while those emitted at a shorter wavelength (due to net bond relaxation) are referred to as "Anti-Stokes." The difference in wavelength between the pump and Stokes photons, known as the Raman shift, is specific to each chemical bond, and thus the spectrum of light inelastically scattered represents a linear combination of the moieties it contains [12].

Spontaneous Raman scattering is an extremely weak phenomenon [11]. Stimulated Raman scattering is a coherent modality which boosts Raman signal by illuminating the sample with both a pump beam and a second beam at a chosen Stokes wavelength. This stabilizes the virtual excitation state and allows coherent amplification of Raman signal at the corresponding shift in much the same way as stimulated emission of laser light. The energy difference between the pump and Stokes beams is tuned to the vibrational Raman shift of a moiety of interest, with inelastic scattering at that frequency manifesting as intensity loss in the pump beam (known as Raman loss) [12]. In Stimulated Raman Gain, two beams are incident on to the sample: a pump laser beam and Stokes Raman probe beam. The difference of the frequencies between the Pump and Raman probe beams is the vibrations frequency, such as CH_3 and CH_2 bonds (2930 and 2845 cm^{-1}). The signal of the Raman beam is enhanced from the transfer to the Raman probe beam giving larger signals at the site of the vibrations by over 10- to 10000-fold. This makes allows for SRS signals vastly exceeding endogenous fluorescence. This loss and gain are detected by the SRS microscope, whose schematic is shown in Fig. 13.1a. Stokes (1064 nm) and pump (690–990 nm) beams (power >40 mW) with 7 ps pulse width from a optical parametric oscillator were used. Because SRS microscopy is a self-heterodyning process, a modulation-demodulation method should be used to detect the signal. Our system modulates the Stokes intensity at a high frequency (10 MHz) with an EOM and detected the weak SRS signal over the large pump intensity ($\Delta I/I < 1 \times 10^{-4}$) using a fast home-built demodulator.

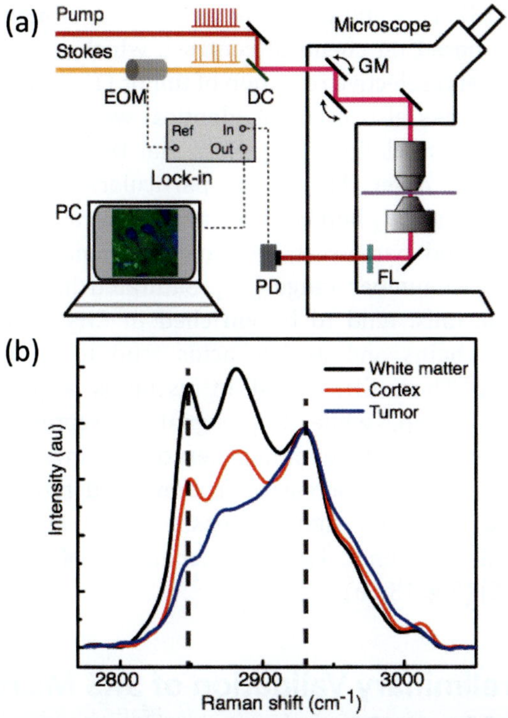

Figure 13.1 SRS microscopy workflow and spectra. (a) Experimental setup of SRS microscopy: the Stokes beam was modulated at high frequency (10 MHz), and the weak stimulated Raman loss signal was demodulated by a lock-in amplifier. A transmission mode detection scheme was used for ex vivo imaging on fresh tissues. DC, dichroic mirror; EOM, electro-optical modulator; FL, optical filter; PD, photodiode; GM, galvanometer mirror. (b) Raman spectra from fresh sections of human glioblastoma multiforme (GBM) biopsy show white matter, cortex, and tumor. The marked frequencies (dashed lines) at 2845 and 2930 cm-1 were chosen for two-color SRS imaging (from Ji et al., 2015).

Like spontaneous Raman, SRS is a non-destructive and label-free modality from the native vibrations which requires no tissue processing and can be performed in situ [14]. However, because the SRS signal is four orders of magnitude more powerful than spontaneous Raman, it facilitates the rapid acquisition of images based on the distribution of a chemical moiety within a sample. Further, images can be acquired under ambient light conditions [13]. This allows for the acquisition of high-resolution micrographs based on chemical contrast, which can assist in the identification and selective resection of tumor [13, 14].

Spectral peaks also occur at identical wavelengths for given moieties in both modalities [13]. Note that the spectra of normal brain and tumor differ (Fig. 13.1b), particularly at wavenumbers related to CH_3 and CH_2 bonds (2930 and 2845 cm^{-1}, respectively), the strongest and narrowest spontaneous Raman spectral lines in the high-wavenumber range [13]. Saturated lipids, possessing long aliphatic tails, tend to be enriched in CH_2 moieties, while protein side chains and nucleic acids tend to be enriched in CH_3 moieties. Thus, CH_2 signal in tissue is a proxy for the abundance of lipid, while CH_3 signal is represents protein content. Hypercellularity, with degeneration of myelinated axonal processes, leads to an increased protein: lipid ration (and thus an increased CH_3: CH_2 signal intensity ratio) in homogenized samples of dense tumor, largely independent of specific tumor histology [15] (Fig. 13.1b).

13.3 Preliminary Validation of SRS Microscopy in Mouse Models

Freudiger et al. documented the first use of SRS microscopy to interrogate ex vivo brain tissue samples from mouse in 2012 [15]. In their study, SRS images were acquired at wavelengths corresponding to CH_3 and CH_2 bond stretch (810.5 and 716.7 nm), as well as a third wavelength to stimulate two-photon absorption in hemoglobin (700 nm), and overlayed to produce virtual color micrographs depicting lipid in green, protein in blue, and hemoglobin in red. In such a color map, normal gray matter appears as sparse, round, blue nuclei interspersed throughout dark green neuropil, while white matter demonstrates tracts of slender bright green axonal processes* with few small blue oligodendrocyte nuclei interspersed. Note that, unlike brain tissue stained with hematoxylin and eosin (H&E), neuropil visualized through SRS is free of vacuolations due to the fact that neither fixatives nor freezing are used in tissue preparation. Within nuclei, chromatin and nucleoli can be visualized. Capillaries can be seen as well, with erythrocytes appearing in the hemoglobin two-photon absorption channel. This study further surveyed a broad range of murine tissues, including lung, liver, ovary, kidney, muscle, and skin. However, the presence of complex lipid-rich architecture in neural tissue makes SRS uniquely suited for examination of CNS pathology.

Their work also examined mouse models of both glioma and metastatic breast cancer using SRS imaging. In doing so, they noted that certain histoarchitectural features easily identified with SRS were common to disparate neoplasms within the CNS. Both tumor models revealed a qualitative increase in cellularity, as well as a decrease in the total length of axon within a given area, or axonal density, in Dense tumor. However, it was observed that the hypercellularity and abnormal cells of the breast cancer metastasis abruptly stopped at the tumor margin, while atypical cells of the glioma continued to infiltrate axonal tracts beyond the primary tumor mass, in line with the infiltrative nature of these tumors.

Meanwhile, resonance Raman studies investigating the spectral characteristics of gliomas had been conducted, covering the 500 cm^{-1} to 4000 cm^{-1} range. The strongest and narrowest Raman lines were found at 2896 cm^{-1} (CH_2), 2938 cm^{-1} (CH_3), and 1647 cm^{-1} [16]. These vibrational modes were hypothesized to be candidate wavelengths for highlighting tumor structure.

Building upon these findings, Ji and Orringer et al. (also of the Xie group) used the same schema to identify glioma in a mouse model both ex vivo and in vivo [14]. Imaging full coronal sections of mouse brain, they were able to examine variations in normal microanatomy more broadly, achieving cell-to-cell correlation with H&E imaging. In doing so, they surveyed regional differences in chemical composition using the ratio of protein: lipid signals, here measured at the microscale, demonstrating normal ranges for these signals in both grey and white matter (grey matter being more protein rich, and white matter being more lipid-rich).

The bulk of this study focused on using a glioblastoma multiforme (GBM) xenograft mouse model to explore how SRS could be integrated into the neurosurgical workflow. Currently, intraoperative analysis of tissue histopathology is achieved through H&E staining of frozen tissue sections. Ji et al. constructed a survey demonstrating exceptionally high correlation between pathologists' classification of paired frozen sections stained with H&E and imaged with SRS as either normal/minimally hypercellular, infiltrating tumor or dense tumor.

They further used epi-illumination SRS system to examine the margins of a GBM en face, both ex vivo and in simulated tumor resections. The chemical contrast generated by differences in CH_3: CH_2 ratio was demonstrated to effectively demarcate the tumor margin, even when in vivo brightfield microscopy showed no discernible difference between tumor and surrounding tissue. As the tumor resection progressed, SRS was able to appreciate the normal architecture underlying the tumor mass. In doing so, they demonstrated that SRS imaging was possible in a surgical setting, movement artifact and the presence of blood and dissected tissue in the operative field notwithstanding.

13.4 Preliminary Validation of SRS Microscopy in Human Tissue

13.4.1 Qualitative Histology

While Ji et al. went on to examine a small number of human glioma biopsies with SRS, a more comprehensive survey of the qualitative appearance of human brain tumor came in a 2015 work by Ji and Lewis et al. [17] in which ex vivo biopsies from 19 patients with primary and secondary CNS neoplasms were imaged with both SRS traditional microscopy. Three specimens of histologically normal brain were also collected from patients undergoing anterior temporal lobe resection for intractable epilepsy. Since this study, our group has imaged a larger and more diverse case series with SRS, currently containing tissue from 56 additional patients.

The appearance of normal human brain with SRS microscopy is detailed in Fig. 13.2, with normal histoarchitecture verified by H&E staining. The soma of pyramidal neurons appear as protein-rich triangular bodies speckled with bright green lipofuscin granules. These cells are seated in a lipid-rich, unvacuolated neuropil with short, intracortical myelinated processes scattered throughout. Efferent axons form fascicles of lipid-rich, linear processes, streaming perpendicular to the gray-white junction and merging into large, dense tracts within the lipid-rich white matter. The white matter is largely acellular save a few small, faint oligodendrocytes. At lower magnification, the cortex and white matter are clearly differentiable based on protein:lipid signal ratio.

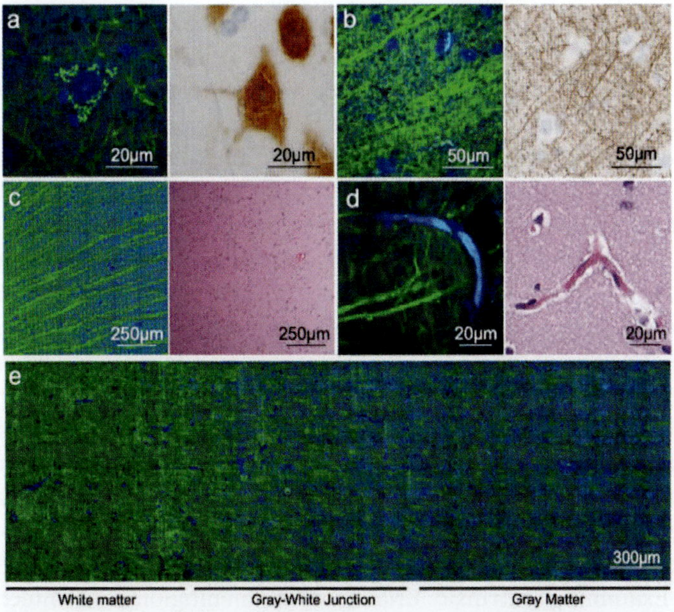

Figure 13.2 SRS Imaging of normal brain. (a) SRS imaging of normal gray matter at high magnification shows neuronal soma with pyramidal architecture filled with lipofuscin-rich granules (left), NeuN immunohistochemical stains the neuronal cell body (right). (b) SRS imaging of white matter (left) demonstrates individual axons appearing as linear, lipid-rich structures that correspond well with neurofilament immunohistochemical staining (right). (c) An SRS image of the gray-white junction (left) demonstrates parallel bundles of lipid-rich white matter tracts that are not visible with H&E staining (right). (d) Capillaries filled with protein-rich erythrocytes appear blue on SRS imaging (left) and eosinophilic on H&E stained section (right). (e) at low magnification, the biochemical differences between protein-rich gray matter (blue; right) and myelinated white matter (green; left) are apparent (from Ji et al., 2015).

Imaging of primary CNS neoplasms revealed many characteristic histopathologic findings seen on H&E (Fig. 13.3).

Pediatric medulloblastoma, another high-grade neoplasm, is distinctive for densely packed small, round, blue cells and brisk mitosis, with cells often organized into pseudorosettes. Hypercellularity and marked nuclear atypia can be seen GBM, with regions of microvascular proliferation and multiple mitotic figures visible beside regions of pseudopalisading necrosis. Dense GBM demonstrates a complete obliteration of normal axonal architecture, with extremely high cellularity and occasional nuclear molding. Contrast with the appearance of a low-grade oligodendroglioma, characterized by round cells with protein-rich cytoplasm and delicate capillary architecture. Unlike in GBM, axonal architecture is somewhat preserved even in dense oligodendroglioma, with perineuronal satellitosis visible. Of note, tissue response at the periphery of these neoplasms differs. Neuropil surrounding the GBM appears to have fewer interspersed axonal segments than tissue abutting low-grade glioma. Importantly, imaging at tumor margins was able to demonstrate a graded normalization of tissue histoarchitecture with increasing distance from the primary tumor in both high- and low-grade glioma. The ability of SRS to capture this gradation is important, as glioma cells adjacent to the primary tumor are thought to be responsible for 90% of recurrences [4].

Examining extrinsic CNS tumors, meningioma architecture demonstrates a fascicular growth pattern with collagen whorls and calcifications known as psammoma bodies. In some specimens, collagen deposition completely obscures the underlying cellular architecture. Metastatic squamous cell carcinoma of the lung shows anaplastic cells (some of which possess multiple nuclei) in a bland matrix, with broad areas of necrosis. Imaging of spinal schwannoma depicts sheets of fusiform nuclei with prominent nucleoli and occasional hemosiderin-laden macrophages.

To assess the ability of pathologists to interpret SRS images, a survey similar to that performed by Ji et al. was used to verify correlation between paired SRS and H&E fields of view representing minimally hypercellular, infiltrating tumor and dense tumor. This survey was administered to three neuropathologists and demonstrated near-perfect detection of tumor infiltration, as well as near-perfect correlation between SRS and H&E ratings of tumor density.

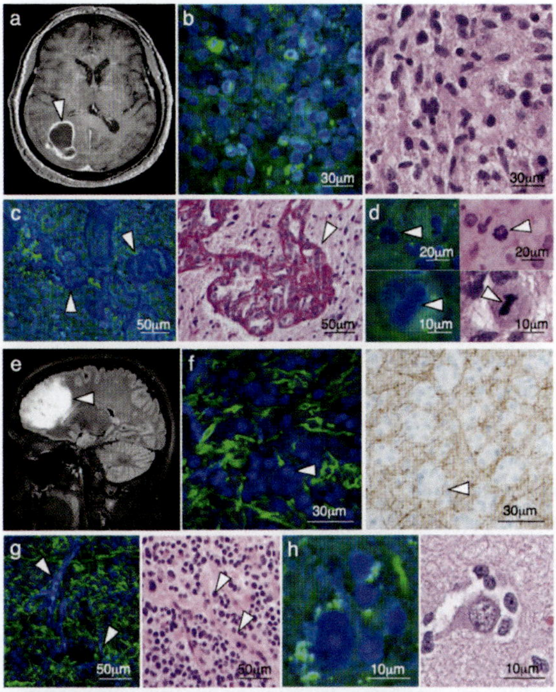

Figure 13.3 SRS and traditional microscopy of intrinsic brain tumors. (a) SRS imaging of a GBM (arrowhead) demonstrating ring enhancement on MRI. (b) Hypercellularity and nuclear atypia of viable tumor is apparent on both SRS (left) and H&E (right) microscopy. (c) Microvascular proliferation creates tortuous vascular complexes evident on SRS microscopy (left, arrowheads) and highlighted with periodic acid Schiff staining (right, arrowhead). (d) Mitotic figures are also visible (arrowheads) with SRS microscopy (left) and H&E staining (right). (e) A non-enhancing, low-grade oligodendroglioma (arrowhead) consists of (f) hypercellular tissue with nests of "fried-egg" morphology (arrowheads) causing minimal axonal disruption on SRS imaging (left), as confirmed through neurofilament immunostaining (right). (g) "Chicken wire" blood vessels.

13.4.2 Quantitative Analysis of Tumor Infiltration with SRS

As in Ji et al., the protein:lipid ratio was measured across each ex vivo human biopsy and shown to monotonically increase from minimally hypercellular tissue to dense or necrotic tumor. However, differences in the chemical composition of various tumor subtypes, as well as differences in the protein:lipid ratio of surrounding tissue for tumors growing in disparate brain regions meant that this measurement alone had poor predictive value for tumor infiltration in our data set. Mouse models of glioma, while more infiltrative than mouse models of metastatic breast cancer, are less infiltrative than human glioma arising in situ. As a result, the changes in protein:lipid ratio approaching the tumor edge can be so gradual as to be overwhelmed by intrinsic variations in tissue composition between brain regions. Further, some CNS neoplasms, most notably low-grade glioma, demonstrate a less dramatic disruption of normal histoarchitecture. Viable tumor can thus exist among dense axon tracts, yielding a protein: lipid ratio higher than normal white matter, but lower than the more protein-rich cortex.

Yet in order for SRS to be effectively integrated into the neurosurgical workflow, it must be amenable to rapid and unambiguous interpretation. Micrographs, while incredibly detailed, require the training of a skilled pathologist to interpret, and are not amenable to analysis in the operative environment. Thus, a more detailed quantification was undertaken using image segmentation, calculating the axonal density and cellularity (as well as the protein:lipid ratio) for each 300 × 300 μm^2 field of view (FOV).

As expected, an inverse relationship was seen between axonal density and cellularity, persisting across tumor subtypes. Yet while axonal density decreased to nearly zero in dense and necrotic high-grade tumor, as well as in extrinsic CNS neoplasms, the persistence of axonal structure was verified in dense, viable low-grade glioma. And while cellularity was highest in dense, viable high-grade tumor, cellularity dropped to zero in regions of necrosis. Thus no single metric is sufficiently sensitive and specific to detect tumor in all cases observed.

However, each neoplasm had characteristic deviations from normality in at least one of these metrics. We quantified and plotted cellularity, axonal density, and protein: lipid ratio of 1477 300 × 300 μm^2 FOVs from 51 fresh tissue biopsies from 18 patients (3 epilepsy patients, 15 brain and spine tumors encompassing 8 distinct histologic subtypes). A three-dimensional plot of averaged values for each of the 51 biopsies reveals the variability in each attribute among the tissues analyzed in the study, with clustering of normal to minimally hypercellular biopsies apparent.

One approach to generating a binary tumor versus normal decision from these data is to create a classifier function based on the measured parameters. We divided our dataset randomly into a training set and a test set of equal size and used the training set to fit a generalized additive model (GAM) with a quasi-likelihood approach, which was then tested on the other half of the data. The classifier function takes into account not only the parameters quantified in our segmentation, but their two-way interactions as well. It outputs a value between 0 and 1, 0 corresponding to the lowest possible likelihood that an FOV contains tumor, and 1 corresponding to the greatest possible certainty.

Using this approach, we were able to differentiate tumor from normal/minimally hypercellular brain with 97.5% sensitivity and 98.5% specificity across all CNS neoplasms tested. Current work focuses on validating this approach on a larger series of biopsies, as well as determining whether segmentation of additional morphologic features could help improve the speed, accuracy and versatility of this schema.

While this approach has yet to be used in the operating room, it has been validated using a unique cadaveric specimen acquired from a GBM patient who expired prior to treatment. A coronal slice was made through the brain and samples were collected from the necrotic tumor core, the viable margin and at 5 mm intervals from the margin up to 25 mm. These samples were each imaged with SRS and stained with H&E, as well as neurofilament, and EGFR immunostains (depicting axons and tumor cells, respectively). Examination of viable, dense tumor demonstrated hypercellularity with EGFR positivity and nearly complete obliteration of the neuropil. Images taken ≥10 mm from the margin demonstrated moderate hypercellularity with partial preservation of axonal architecture and focal expression of EGFR, while those ≤15 mm from the apparent tumor margin demonstrated normal cellularity and axonal density with no evidence of neoplastic infiltration. Quantification of 295 FOVs from the cadaveric specimens demonstrated decreasing protein:

lipid ratio and cellularity, and increased axonal density with increasing distance from the tumor. Notably, the quasi-likelihood GAM classifier accurately predicted tumor infiltration up to 10 mm from the tumor margin in these specimens.

13.5 Other Intraoperative Microscopy Techniques in Neurosurgery

Other optical techniques are also being investigated for intraoperative guidance in tumor surgery. Raman spectroscopy provides a means of differentiating tumor from normal brain based on chemical content. Unlike SRS, Raman does not produce an image, but rather a local chemical fingerprint constituting a weighted average of the Raman-active compounds in a sample. Jermyn et al. [11] have developed a handheld Raman probe which is able to acquire tissue spectra in an operative setting. Using a machine learning approach, they were able to identify a subset of peaks which differentiated normal tissue from tumor with 93% sensitivity and 91% specificity. However, intraoperative Raman cannot be carried out under ambient light condition and does not allow for visualization of tissue architecture.

Optical coherence tomography (OCT) is based on low coherence interferometry and generates images based on differences in optical attenuation though depth scans. Kut et al. [18] demonstrated that OCT could use the differential attenuation properties of tumor and normal tissue, achieving 92–100% sensitivity and 80–100% specificity, with particular sensitivity for infiltrating tumor. However, OCT does not give any information as to the chemical content or microarchitecture on tissue and thus does not yield any diagnostic or prognostic information.

Broadband coherent anti-Stokes Raman Scattering (BCARS) microscopy is the most similar technology to SRS being investigated for neurosurgical application. Like SRS, it is a coherent Raman imaging modality, and is thus able to generate images of tissue histoarchitecture based on chemical composition [19]. However, unlike SRS, BCARS is able to image the entire biologically relevant Raman window with spectral resolution >10 cm^{-1}. This makes BCARS well-suited to image chemical signatures specific to cancers. Yet because BCARS is acquiring full spectra at each pixel, its scan times are longer and spatial resolution poorer than SRS, making SRS potentially better suited for integration into a surgical workflow [19].

13.6 Clinical Implementation of SRS Microscopy and Future Work

The clinical adoption of SRS microscopy had been hindered by the need for a robust, tunable laser with an ultrafast, dual-wavelength source. Until recently, a synchronously pumped picosecond optical parametric oscillator (OPO) has been the gold standard source for SRS microscopy. Such a laser is both costly and highly sensitive to vibration (due to its reliance on a free space cavity), and is thus ill-suited for a conventional operating room. Yet a new fiber laser design by Freudiger et al. has greatly reduced laser alignment issues and demonstrated stability over a long-term continuous use, facilitating construction of the first SRS microscope designed specifically for intraoperative use [20].

With this advance, we have begun to explore integration of SRS microscopy into the operative workflow. At this point, human imaging has been performed exclusively ex vivo. Yet examination of tissue resected from the operative margins can still guide operative decision making. As with any microscopic imaging modality, the FOV is several orders of magnitude smaller than the tumor cavity, making imaging of the full margin impractical. Adaptation of existing sampling protocols for intraoperative microscopy of both squamous cell carcinoma of the head and neck and breast cancer may guide the use of ex vivo SRS imaging. Further, we are integrating SRS microscopy with neuronavigation technologies with the aim of imaging stereotactically registered biopsy samples in the OR. This will allow the mapping of histoarchitectural features to co-registered tomographic images, revealing regions of the tumor cavity with probable residual tumor. We also hope to use SRS-assisted neuronavigation in the intraoperative MRI to verify the presence of infiltrative tumor outside the enhancing margins, as was seen in our cadaveric specimen. Further work will also focus on direct comparison of SRS and 5-ALA, seeking to assess tumor burden in peripheral tissues without visible fluorescence in both high and low-grade glioma.

Ex vivo microscopy allows for transmission mode imaging, which allows for greater signal-to-noise ratio and clearer images. Yet previous work in mouse has demonstrated that SRS signal intensity is sufficient for in vivo, reflectance mode imaging [14]. This is possible even in the presence of blood, coagulated tissue, and cardiorespiratory cycle motion. With the construction of a handheld probe, it would be possible to interrogate elegant brain or tissue deemed too dangerous to remove surgically, guiding prognosis and the utility of adjuvant therapies.

The utility of SRS may extend beyond CNS neoplasms to other cancers where local recurrence is common, or proximity of tumor cells to the margins of resection holds prognostic value. One such neoplasm is squamous cell carcinoma of the head and neck. Yet SRS is limited in its depth of penetration (≤100 μm) [13], meaning that it may not be well suited for neoplasms such as breast cancer, where depth of invasion commonly exceeds 100 μm.

While our implementation generates contrast based exclusively on CH_2 and CH_3 signal intensity, SRS is capable of imaging additional moieties in the high-wavenumber range. We are currently focused on whether additional Raman-active compounds with peaks in this range can be detected in subsets of brain cancers. This would allow for the identification of tumor infiltration not only through structural changes in tissue, but through specific biochemical changes in the neoplastic cells themselves.

In summary, SRS microscopy is able of detecting brain tumor through both chemical and structural changes in fresh tissue without labels or fixatives. Research in mouse has demonstrated that imaging can be conducted both in vivo and ex vivo and correlates well with traditional microscopy, while simultaneously detecting tumor not visible by brightfield microscopy in an operative setting. Imaging of human tissues with SRS redemonstrates key histopathologic features of a wide array of CNS neoplasms qualitatively. Quantitative analysis of these images with image segmentation reveals that hypercellularity, axonal degradation and increased protein:lipid ratio tend to follow increasing tumor density. Statistical modeling allows the construction of a classifier function based on quantitative analysis of a wide range of CNS neoplasms which can predict the presence of tumor infiltration with high accuracy. Recent development of a fiber laser source has reduced the cost and environmental sensitivity of SRS and made it more attractive as an intraoperative modality. Present work is focused on the validation of image quantification and tumor detection in a large patient series, as well as optimizing the integration of SRS into the operative workflow. Future work will attempt to image Raman-active aberrant metabolites, effectively allowing the specific identification of tumor cells through virtual immunohistochemistry.

References

1. Sanai, N., and M. S. Berger (2008). Glioma extent of resection and its impact on patient outcomes. *Neurosurgery* **62**(4): 753–766.
2. Smith, J. S., et al. (2008). Role of extent of resection in the long-term outcome of low-grade hemispheric gliomas. *J. Clin. Oncol.* **26**(8): 1338–1345.
3. Albert, F. K., et al. (1994). Early postoperative magnetic resonance imaging after resection of malignant glioma: Objective evaluation of residual tumor and its influence on regrowth and prognosis. *Neurosurgery* **34**(1): 45–61.
4. Hochberg, F. H., and A. Pruitt (1980). Assumptions in the radiotherapy of glioblastoma. *Neurology* **30**(9): 907–911.
5. Berkels, B., et al. (2014). Co-registration of intra-operative brain surface photographs and pre-operative MR images. *Int. J. Comput. Assisted Radiol. Surg.* **9**(3): 387–400.
6. Makary, M., et al. (2011). Clinical and economic outcomes of low-field intraoperative MRI-guided tumor resection neurosurgery. *J. Magn. Reson. Imaging* **34**(5): 1022–1030.
7. Senft, C., et al. (2011). Intraoperative MRI guidance and extent of resection in glioma surgery: A randomised, controlled trial. *Lancet Oncol.* **12**(11): 997–1003.
8. Stummer, W., et al. (2006). Fluorescence-guided surgery with 5-aminolevulinic acid for resection of malignant glioma: A randomised controlled multicentre phase III trial. *Lancet Oncol.* **7**(5): 392–401.
9. Stummer, W., et al. (2014). 5-Aminolevulinic acid-derived tumor fluorescence: The diagnostic accuracy of visible fluorescence qualities as corroborated by spectrometry and histology and postoperative imaging. *Neurosurgery* **74**(3): 310.
10. Ewelt, C., et al. (2015). Fluorescence in neurosurgery: Its diagnostic and therapeutic use. Review of the literature. *J. Photochem. Photobiol. B: Biol.* **148**: 302–309.

11. Jermyn, M., et al. (2015). Intraoperative brain cancer detection with Raman spectroscopy in humans. *Sci. Transl. Med.* **7**(274): 274ra219.
12. Raymer, M. G., and J. Mostowski (1981). Stimulated Raman scattering: Unified treatment of spontaneous initiation and spatial propagation. *Phys. Rev. A* **24**(4): 1980–1993.
13. Freudiger, C. W., et al. (2008). Label-free biomedical imaging with high sensitivity by stimulated Raman scattering microscopy. *Science* **322**(5909): 1857–1861.
14. Ji, M., et al. (2013). Rapid, label-free detection of brain tumors with stimulated Raman scattering microscopy. *Sci. Transl. Med.* **5**(201): 201ra119.
15. Freudiger, C. W., et al. (2012). Multicolored stain-free histopathology with coherent Raman imaging. *Lab Invest.* **92**(10): 1492–1502.
16. Zhou Y., Liu C., Sun Y., et al. (2012). Human brain cancer studied by resonance Raman spectroscopy. *J. Biomed. Opt.* 17(11): 116021. doi:10.1117/1.JBO.17.11.116021.
17. Ji, M., et al. (2015). Detection of human brain infiltration with quantitative stimulated Raman scattering microscopy. *Sci. Transl. Med.* **7**(309): 309ra163.
18. Kut, C., et al. (2015). Detection of human brain cancer infiltration ex vivo and in vivo using quantitative optical coherence tomography. *Sci. Transl. Med.* **7**(292): 292ra100.
19. Camp, C. H., Jr., et al. (2014). High-speed coherent Raman fingerprint imaging of biological tissues. *Nat. Photonics* **8**: 627–634.
20. Freudiger, C. W., et al. (2014). Stimulated Raman scattering microscopy with a robust fibre laser source. *Nat. Photonics* **8**(2): 153–159.

Chapter 14

Chemical and Molecular Imaging of Deep Tissue through Photoacoustic Detection of Chemical Bond Vibrations

Yingchun Cao[a] and Ji-Xin Cheng[a,b]

[a]*Weldon School of Biomedical Engineering, West Lafayette, IN 47907, USA*
206 S Martin Jischke Dr., West Lafayette, IN 47907, USA
[b]*Department of Chemistry, Purdue University, 560 Oval Dr., West Lafayette, IN 47907, USA*

jcheng@purdue.edu

Imaging deep in a biological tissue in vivo with macroscopic field of view and biochemistry and molecular information is important for early diagnosis of human disease and accurate assessment of a therapy. In this chapter, we introduce a technology that allows vibrational imaging of chemical content in a deep tissue through photoacoustic detection of vibrational overtone and combination absorption of molecules.

14.1 Introduction

Optical imaging of biological tissue presents a number of advantages, including high resolution, noninvasive to tissue,

Deep Imaging in Tissue and Biomedical Materials: Using Linear and Nonlinear Optical Methods
Edited by Lingyan Shi and Robert R. Alfano

ISBN 978-981-4745-88-8 (Hardcover), 978-1-315-20655-4 (eBook)
www.panstanford.com

label free, and chemical composition selectivity. Optical imaging modalities, including fluorescence microscopy [1, 2], particularly two-photon excitation fluorescence microscopy [3–6], optical coherent tomography [7–9], near-infrared spectroscopy [10, 11], harmonic generation [12], and Raman scattering, which includes spontaneous Raman scattering [13, 14], coherent anti-Stokes Raman scattering [15–18], and stimulated Raman scattering [19–22], have been intensively investigated for biomedicine applications with high sensitivity and three-dimensional (3D) imaging capability. However, due to the strong scattering in biological tissue, the penetration depths of these modalities are limited to sub-millimeter range, which is far from enough for clinical settings of depth tissue imaging in many cases.

In order to go deep inside the tissue, diffused photons other than ballistic photons are employed for biomedical study [23, 24]. In this case, the photons received at the detector have undergone multiple scatterings blurring images and can only carry the spatially averaged information after a tissue thickness of millimeter to centimeter scale. Therefore, the spatial resolution is highly reduced. Moreover, it is difficult to differentiate the optical absorption from scattering during the optical excitation or data processing algorithms. One needs to use longer wavelengths in the near infrared region from 700 to 2500 nm to obtain reduced scattering and deeper penetration depths [25]. The golden window from about 1550 to 1880 nm offers the deepest penetration for tissue [25].

Photoacoustic imaging (PAI), or photoacoustic tomography (PAT), which converts the optical absorption into acoustic wave that is detectable with an ultrasound transducer, brings new opportunities to the biological imaging area [26–29]. In this method, all the absorbed photons, including transmitted and scattered, contribute to the photoacoustically generated ultrasonic signal. Moreover, the ultrasonic wave detected forward or backward shows a much lower scattering coefficient compared with optical wave. Thus, imaging based on this modality can reach an imaging depth up to several centimeters [29].

Vibrational photoacoustic imaging (VPAI), or vibrational photoacoustic tomography (VPAT), which inherits all the advantages of PAI [30], can provide chemical and molecular information by spectroscopically detecting the vibrational absorption happened to the specific chemical bond through overtone or combination transitions. With proper data processing algorithms, the density distribution of each composition can be resolved to achieve a 3D image with high sensitivity and high spatial resolution. The hybrid system of VPAI uses molecular mechanical vibrations, optics, and sound to image.

14.2 Fundaments of Vibrational Photoacoustic Imaging

The basic principle of VPAI will be introduced in this section. In addition, the absorption spectra of common contrast chemicals and molecular fingerprints will be provided to show optimal spectral regions for different label-free applications with specific contrast mechanism. A useful modality of multispectral photoacoustic imaging (MS-PAI) is presented as well for differentiation of chemical compositions in tissues.

14.2.1 Principle of Vibrational Photoacoustic Imaging

The principle of VPAI can be illustrated by Fig. 14.1. A pulsed laser beam illuminates a biological specimen containing chemical compositions with strong optical absorption at the laser wavelength. The optical absorption in the tissue will result in jumping of molecules to a higher vibrational energy level and then dropping to the initial energy level through nonradiative process due to perturbations such as molecular collision. During this process, a localized temperature increase is induced and subsequently transformed to pressure wave propagating in the tissue due to the thermoelastic effect. This process is known as photoacoustic effect, which was discovered by Alexander Bell in 1880 [31]. The generated acoustic wave with ultrasonic frequency due to a nanosecond laser pulse excitation can be detected by a sensitive ultrasound transducer (Fig. 14.1a).

The amplitude of the detected pressure wave on the transducer is determined by the optical absorption and can be expressed as

$$P \propto E\alpha C\beta\left(\frac{K}{\rho C_p}\right), \tag{14.1}$$

where E is the laser radiation fluence, α is absorption coefficient, β is the thermal coefficient of volume expansion, C is the concentration of absorbers, K is the bulk modulus, ρ is the mass density, and C_p is the specific heat capacity at constant pressure.

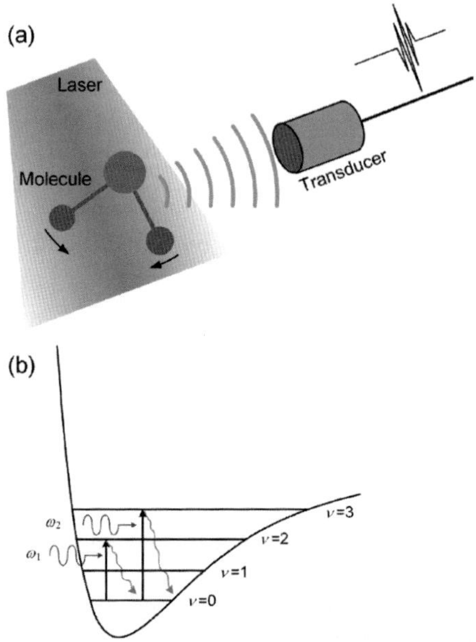

Figure 14.1 (a) Principle of VPAI; (b) vibrational energy levels of a chemical bond and examples of the first and second overtone transitions.

Optical absorption in VPAI is based on the vibrational overtone or combination transitions. Figure 14.1b shows the molecular vibrational energy levels determined by the anharmonic vibrations of a specific chemical bond [32]. The frequency of an overtone transition is described by

$$\Omega = \Omega_0 \nu - \chi \Omega_0 (\nu + \nu^2), \tag{14.2}$$

where Ω_0 is the vibration frequency of chemical bond, χ is the anharmonicity constant, and ν = 2, 3, ... represents the first, second, and so on, overtone vibrations. When the frequency of the incident laser equals to that of an overtone transition, the photons will be absorbed to generate a pressure wave photoacoustically. It should be noted that the vibrational energy levels are not evenly separated. A higher order overtone transition requires a laser pulse with a larger frequency, or a shorter wavelength.

14.2.2 Overtone Absorption Spectra of Molecular Vibration

The most important advantage of VPAI is bond-selective absorption of target chemical composition in tissue as the contrast mechanism. Although overtone photoacoustic spectroscopy was reported over 30 years ago [33–36], the early application of photoacoustic spectroscopy was mainly focused on the spectroscopic study of simple liquid or gaseous analytes such as H_2O, D_2O, and benzene. Water is the main molecule in tissue. The application of overtone photoacoustic spectroscopy on biological imaging has not been exploited until very recently [30, 37–41]. In a pioneering study of VPAI, H.-W. Wang et al. demonstrated the feasibility of photoacoustic microscopy based on the second vibrational overtone absorption of CH_2 bond [30]. With this method, they successfully achieved 3D imaging of biological tissue with several millimeter penetration depth and high spatial resolution. The use of the Golden window in tissue offers deep penetration [25].

Photoacoustic tomography has been focused on virtualizing the vessel structure deep inside biological tissues [26, 42–44]. This is based on the large absorption coefficient of hemoglobin, the dominant absorber in blood, in an optical window ranging from visible to near-infrared region (i.e., 0.4–1.0 μm) as can be seen from Fig. 14.2a. As the spectral region exceeds 1 μm, the water absorption increases quickly, which prevents the water-based imaging in biological samples because the abundant water absorption will attenuate the optical fluent quickly to go deep inside a tissue. Therefore, water is not a suitable target for bond-selective photoacoustic imaging in live samples. As an important composition in tissue, lipid is widely selected as contrast mechanism in biological imaging in recent years. The absorption curve of lipid is also presented in Fig. 14.2a. The spectral regions labeled with I and II represent the second and first overtone absorption bands of lipid.

As an important and universal chemical component in biological tissue, C–H bonds, not like water, are only abundant in certain types of tissue compositions, such as CH_2-rich lipid bodies, CH_3-rich collagens. It means in C–H bond-selected spectral regions, light can go deeper into the tissue while maintaining good contrast.

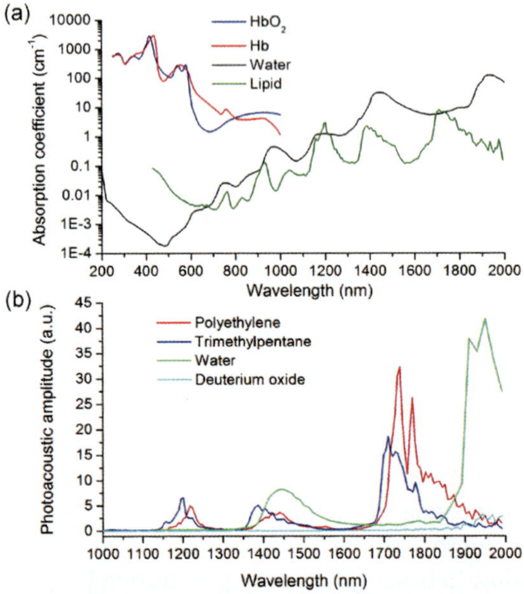

Figure 14.2 (a) Absorption coefficients of whole blood, water and lipid in the infrared region; (b) absorption curves of polyethylene, trimethylpentane, water and deuterium oxide obtained by photoacoustic approach. I, II represent two optical windows for the second and first overtone of C–H bond, respectively.

A detailed study of overtone absorption of C–H bonds can be found in P. Wang's paper [38]. Polyethylene film was selected as the target to provide a typical absorption of CH_2 group, while trimethylpentane provides a characteristic absorption profile of CH_3 group. Laser wavelength was scanned across the near-infrared range from 1000 to 2000 nm to cover the whole overtone and combination absorption region. The signals were photoacoustically detected by an ultrasound transducer. The absorption magnitude of CH_2 and CH_3 groups in terms of photoacoustic amplitude were shown in Fig. 14.2b. For CH_2 group, the first overtone absorption (2ν CH_2) happens around 1730 nm. Peak at 1730 nm is thought to be a combination band of symmetric and asymmetric stretching

($\nu_a + \nu_s$), while the 1760 nm peak is assigned to the first overtone stretching mode of CH_2 [45]. The peak located at 1210 nm is assigned to be the second overtone stretching of CH_2 (3ν CH_2), while the peak between 1350 and 1500 nm is thought to be the second combination band of CH_2 ($2\nu+\delta$) through harmonic stretching and non-stretching modes, such as bending, rocking and twisting [45]. While for CH_3 bond, the absorption peaks, i.e., the first overtone (2ν CH_3) at 1700 nm, the second overtone (3ν CH_3) at 1195 nm and the second combination band ($2\nu + \delta$) around 1380 nm, are found to be very similar to that of CH_2 bond. It should be noted that the first and second overtone of C–H bonds reside in spectral regions that correspond to relatively low absorption bands of water as labeled as I and II in Fig. 14.2, and CH_2 and CH_3 groups exhibit distinguishable absorption profiles at these two overtones, which indicates these two overtone bands can be used for multispectral photoacoustic imaging to differentiate CH_2-rich lipids from CH_3-rich collagen, for example, in biological tissues. Moreover, the first overtone of CH_2 bond shows a ~6.3 times stronger absorption than that of the second overtone of CH_2, indicating an optimal spectral band for CH_2-rich lipid imaging in the near-infrared range. For comparison, the absorption curves of H_2O and deuterium oxide (D_2O) were investigated as well in this range as shown in Fig. 14.2b [38]. As expected, the strong absorption peaks of H_2O, i.e., $\nu_1 + \nu_3$ combination band at 1450 nm and $\nu_2 + \nu_3$ combination band at 1940 nm, are well separated from the first and second overtones of C–H bonds. D_2O shows a negligible absorption in the entire range from 1000 to 2000 nm, indicating that D_2O can be served as a perfect medium for acoustic coupling in ex vivo study during VPAI.

14.2.3 Multispectral Photoacoustic Imaging

MS-PAI is a powerful modality that allows for highly specific chemical mapping in a biological tissue through multispectral/multiwavelength scanning of the excitation laser across the characteristic absorption spectral region of interested molecules [41, 46–49]. As mentioned above, the distinct absorption curves of CH_2 and CH_3 in their first and second vibrational overtone bands provide a mean to spectroscopically differentiate tissue compositions that contain these two different groups. For example, P. Wang et al. designed a proof-of-concept experiment to successfully distinguish CH_2-rich fat and CH_3-rich collagen by MS-PAI in the optical window from 1650 to 1850 nm, where the first overtone absorption of C–H bonds reside [39]. A research group in Netherlands demonstrated a method to detect and differentiate different types of lipids in atherosclerotic human coronaries by taking advantage of the slightly different absorption profiles of C–H bonds in the second vibrational overtone band centered on 1200 nm [41, 50].

Among the quantitative analysis methods of MS-PAI, the most widely used is multivariate curve resolution alternating least squares (MCR-ALS) method [39, 51]. In this method, the concentration distributions of interested chemicals are resolved by least squares fitting of multispectral images by the known absorption profiles of each pure component. Therefore, the advantages of MCR-ALS lie in two aspects: (1) by applying the principle component analysis and multivariate curve resolution, one can easily determine the main chemical compositions in a sample and map their concentration distributions in the imaging area, without a priori knowledge of chemical components containing in the specimen and their spectral absorption profiles, which indicates MCR-ALS can be a potential tool for real tissue imaging where the major compositions and their absorption is unknown; (2) when dealing with a deep tissue problem, the complex scattering and absorption in deep tissue prevents the accuracy of a simple linear inversion method. For MCR, however, the absorption profiles used for initial iterative optimization come from the pure chemical compositions. Once the self-optimization process reaches a convergence, the resolved absorption curve for each component matches the spectral profile of the real composition in the deep tissue.

A detailed description of MCR-ALS applied for MS-PAI can be found in Ref. [39]. Assuming multispectral method with Ω wavelength is applied to a B-scan section imaging with pixels of $X \times Z$. Thus, the image shows a row data of 3D matrix ($X \times Z \times \Omega$) as shown in Fig. 14.3a. The first step is to reform the matrix to a 2D matrix D with a size of $(X \times Z) \times \Omega$, where the row number $X \times Z$ represents the total pixels, and the data for each row shows the spectral absorption at each pixel. The MCR-ALS method is to fit matrix D by a bilinear model to produce two matrices, C and S^T, plus an error matrix of E as

$$D = C \cdot S^T + E. \tag{14.3}$$

Each column in C, with a size of $X \times Z \times q$, denotes the concentration distribution of one of the q components, where each row in S^T, with a size of $q \times \Omega$, represents the absorption profiles for one of the q components at the measurement spectral range. During the ALS process, the spectral matrix S^T and distribution matrix C are iterated to approach an optimal solution with certain constrains, such as non-negativity, unimodality, closure, and so on, according to the chemical properties and origins of mathematics to minimize the error matrix E [52, 53]. Eventually, each column of the distribution matrix C is reformed as an $X \times Z$ size matrix to represent the distribution of the resolved components as displayed in Fig. 14.3a.

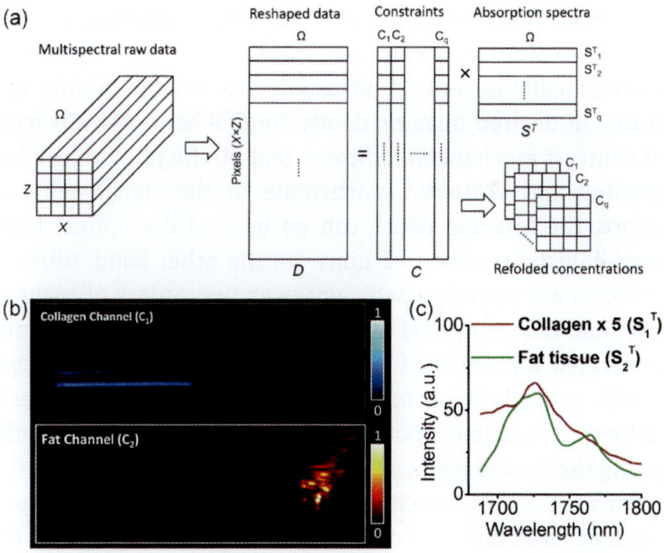

Figure 14.3 MS-PAI of tissue phantom containing rat-tail tendon and fat with MCR-ALS analysis. (a) Flow chart of MCR-ALS; (b) reconstructed collagen and fat distributions; (c) recovered absorption profiles of collagen and fat from MCR-ALS analysis. Images in B and C are reproduced from Ref. [31] © 2012 SPIE.

P. Wang et al. designed an experiment of MS-PAI to prove the concept of MCR-ALS. The tissue phantom containing rat-tail tendon (rich in collagen) and fat (rich in lipids) are served as the imaging target. 60 wavelengths in the spectral range from 1650 to 1800 nm, where the first overtone of C–H bonds reside, were adopted to perform MS-PAI. For each wavelength, a B-scan image was recorded to include the preassigned tendon and fat. All the multispectral images were then analyzed with MCR-ALS method. The resolved distributions of tendon and lipids were displayed in Fig. 14.3b and the recovered spectral absorption in this region were plotted in Fig. 14.3c. The experiment results agree well the phantom design, indicating the validity of MCR-ALS method for multispectral images analysis.

14.3 Modalities of Vibrational Photoacoustic Imaging

As a hybrid imaging modality, PAI takes advantages of the optical absorption as contrast mechanism, and detects the photoacoustically generated ultrasonic wave with scalable spatial resolution at desired imaging depth. On one hand, the absorption-based contrast mechanism indicates that all the photons, including transmitted and scattered, participate in the signal generation. Therefore, the imaging depth can go beyond the optical ballistic or quasi-ballistic regime (~1 mm). On the other hand, ultrasound wave shows a scattering coefficient over two orders of magnitude smaller than that of optical wave in biological tissue, which means the generated ultrasound wave can go much deeper than optical wave with a negligible propagation loss. This unique property of PAI allows a scalable spatial resolution at different depth by managing the optical beam size inside the tissue.

According to its specific application areas, PAI is typically operated at three different modes to realize 3D imaging of a tissue as depicted in Fig. 14.4. The first and most straightforward one is based on focused photons or phonons, and raster scanning of the optical beam or tissue as shown in Fig. 14.4a. This modality is known as photoacoustic microscopy (PAM). In PAM, focused ultrasound transducer can largely enhance the collecting efficiency of photoacoustic signal with an improved spatial resolution less than 100 μm. A tightly focused optical beam can further push the lateral resolution to the optical resolution regime, which is also known as optical-resolution PAM, with the sacrifice of penetration depth less than 1 mm. By assuming the sound velocity in tissue to be ~1.54 mm/μs, the time-of-flight ultrasound signal is converted into depth information in tissue. The 3D image is then resolved by combining the time-of-flight depth signal with the raster scanning information.

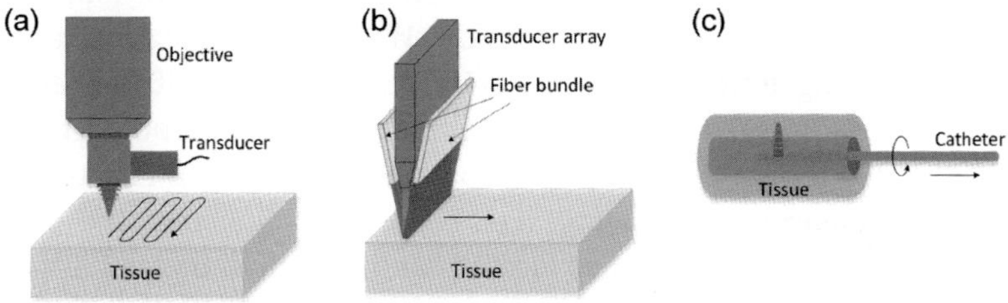

Figure 14.4 Operation modes of PAI to implement 3D imaging. (a) PAM with focused photons or phonons and 2D raster scanning; (b) PACT with linear scanning with a probe array; (c) PAE with rotational and pullback scanning of the catheter.

The second mode is based on diffused photons and scanning of an unfocused transducer or transducer array, which is called photoacoustic computed tomography (PACT) or simply PAT [27, 54]. Figure 14.4b depicted a representative example of PACT by scanning an optical delivery fiber bundle and a linear transducer array. In PACT, the image is reconstructed by inversely solving the photoacoustic equations. Due to the parallel acquisition of a number of A-lines, this method allows high frame rates as high as 50 Hz [42, 54–57].

The third operation mode is based on a rotational scanning and combined linear pullback of the imaging probe. This modality is typically used in catheter-based photoacoustic endomicroscopy (PAE) [58–62]. In this approach, a catheter probe is designed to illuminate the tissue through side view, and the echo ultrasound signal generated by the photoacoustic effect is detected by a transducer embedded in the probe. Each A-line in a pulse duration forms the time-of-flight depth information along the axial direction, and each rotational scanning will result in a B-scan cross-sectional image. 3D imaging is implemented by introducing the linear pullback system.

The spatial resolution and penetration depth of VPAI is scalable with the optical focus and ultrasonic frequency. With a tightly focused optical beam, PAM can be operated at optical resolution configuration with lateral resolution down to 5 μm and imaging depth >0.7 mm. Based on this method, capillaries, the smallest blood vessels, in a mouse ear was visualized in vivo with contrast from the optical absorption of hemoglobin (shown in Fig. 14.5a) [43]. In another work, P. Wang et al. took advantage of the strong first overtone of C–H bond around 1730 nm and successfully imaged the lipid deposition in atherosclerotic artery with a depth over 3 mm under blood layer as shown in Fig. 14.5b [38]. Photoacoustic imaging has been also demonstrated for identification of even deeper located sentinel lymph nodes (SLNs) [63]. SLNs have a mean depth of 12 ± 5 mm in human tissue and are usually surrounded by fatty tissue. After methylene blue injected, a dye with ten-fold stronger optical absorption than hemoglobin at 677 nm, SLN under a depth of 31 mm has been identified in vivo in a rat with a 3.5 MHz ultrasound transducer.

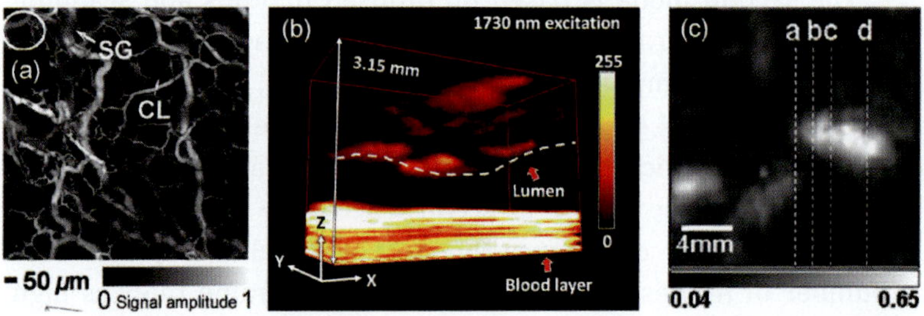

Figure 14.5 Scalable photoacoustic imaging for (a) capillaries, (b) lipid, and (c) sentinel lymph nodes. Images in A, B, and C are reproduced from Ref. [35] © 2008 OSA, Ref. [30] © 2012 Wiley, and [55] © 2008 SPIE.

14.4 Applications of Vibrational Photoacoustic Imaging

As an emerging tool for biological imaging, VPAI shows a number of advantages over other imaging modalities including scalable deep tissue imaging beyond the optical ballistic or quasi-ballistic regime, bond-selective chemical composition imaging, easily co-registered with ultrasound imaging as a hybrid imaging modality. Therefore, after its adventure, VPAI has drawn extensive attentions and been applied to a variety of biomedical areas, such as atherosclerotic lipid deposition detection, breast cancer diagnosis [64], white matter assessment in spinal cord [65], imaging of peripheral nerve [66], and so on. In this section, we focus on the two outstanding applications in breast cancer diagnosis and atherosclerotic plaque identification based on bond-selective VPAI.

14.4.1 Breast Tumor Margin Assessment

In breast cancer treatment, knowing how close of the tumor to the removed tissue edge to determine whether a further treatment is needed or not has practical importance in breast cancer surgery [67–69]. For a positive tumor margin, typically defined as the tumor margin with a distance between the tumor and the excised specimen edge small than 2 mm, additional surgery is needed to reduce the chance of cancer recurrence [67, 70]. However, current imaging modalities either fail to distinguish the tumor tissue from the normal tissue or cannot provide enough imaging depth to visualize the detailed compositions of the tumor margin.

R. Li et al. proposed and demonstrated a multispectral VPAT based imaging modality to detect breast cancer margin [64]. The schematic setup for their imaging system is shown in Fig. 14.6a. Nd:YAG pumped optical parametric oscillator (OPO) with a 10 Hz repetition rate and 5 ns pulse width is used as the excitation source. The wavelength of the OPO was tuned from 1100 to 1250 nm, which includes the strong absorption of hemoglobin and second overtone absorption band of lipid, with 10 nm step size to perform multispectral photoacoustic imaging. Optical energy was delivered to the specimen by a fiber bundle with two rectangular distal terminals, which is stick to each side of an ultrasound transducer array in parallel. The generated photoacoustic signals were detected by ultrasound transduce array with 128 elements and sent to a high-frequency ultrasound imaging system for image reconstruction.

The multispectral VPAT system was performed for human breast tumor sample imaging. First, the boundary of the tissue was defined with the help of the ultrasound image. Then the multispectral photoacoustic imaging was employed to mapping the areas with hemoglobin and lipid, respectively. The region containing only lipid was characterized as normal tissue, while the ones with hemoglobin was considered as tumor. The imaging results for a typical positive and negative breast tissue margin are shown in Figs. 14.6b,c, with corresponding histology showing in Figs. 14.6d,e. It is shown the imaging results agree well with histology, indicating the validity of the demonstrated method for breast cancer margin determination.

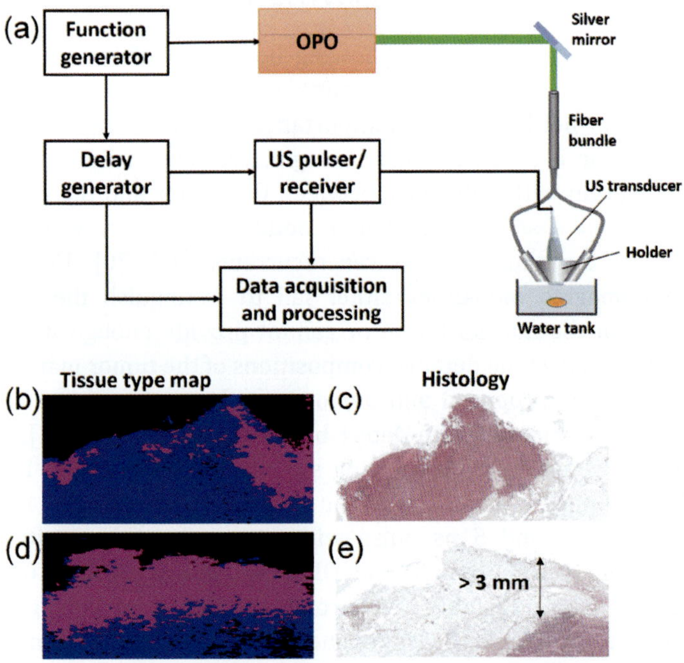

Figure 14.6 Multispectral photoacoustic tomography for breast cancer margin detection. (a) Experimental setup; (b, c) resolved photoacoustic images for positive and negative cancer margins; (d, e) histology results of the cancer margins in (b) and (c), respectively. Magenta color in (b) and (c) represents normal tissue while blue color shows tumor regions. Image a is redrawn and b–e are reproduced from Ref. [56]. © 2015 OSA.

14.4.2 Intravascular Imaging of Atherosclerotic Plaque

Another important application of VPAI is intravascular photoacoustic (IVPA) imaging of atherosclerotic plaque, which is belonged to the category of PAE. The rupture of atherosclerotic plaque can directly lead to fatal acute cardiovascular disease or even death [71–75]. Therefore, accurate detection and characterization of atherosclerotic plaque has a significant impact in the diagnosis of atherosclerosis for a cardiologist. A vulnerable plaque is generally composed by lipid core and a thin layer of fibrous cap, and located around the luminal-intima surface [74, 75]. The vulnerability of a plaque is characterized by the lumen area, the size of lipid core and the thickness of fibrous cap. Intravascular ultrasound (IVUS) imaging can show the profile of the lumen structure, but cannot provide chemical information [76, 77]. Other pure optical imaging modalities, such as optical coherent tomography, have a limited penetration depth (<1 mm) due to strong light scattering effect in tissue [78]. Thus they failed to provide depth information in a vascular tissue. Early PAI of atherosclerotic plaque using wavelengths in the near infrared region from external illumination cannot effectively differentiate lipid from other chemical components [61, 79].

As mentioned before, VPAI utilizes the optical absorption as contrast and converted the optical absorption into ultrasound signal via photoacoustic effect. Thus, it can go deep into biological tissue while providing chemical information by bond-selectively exciting vibrational overtone absorption. For example, CH_2-rich lipid has shown strong absorptions in their first (~1730 nm) and second (1210 nm) overtone peaks, which can be used for intravascular lipid detection.

A number of works have been done based on the vibrational photoacoustic detection of intravascular lipid deposition [37, 40, 50, 62, 80–85]. For example, K. Jansen et al. demonstrated IVPA imaging of human coronary atherosclerotic plaque ex vivo by using a 1.25 mm diameter catheter [62]. The catheter comprises an angle-polished optical fiber and a 30 MHz ultrasound transducer. By targeting the second overtone absorption region of CH_2 bond between 1180 and 1230 nm, they performed IVPA/IVUS imaging for a human coronary artery with advanced lesion as shown in Fig. 14.7. Figure 14.7a presents the histology result of the imaging section, Fig. 14.7b shows the IVUS result, and Figs. 14.7c,d display the IVPA images at 1210 nm (high lipid absorption) and 1230 nm (low lipid absorption), respectively. They are found to be in accordance with each other. Later, researchers at the University of Texas at Austin group reported in vivo IVPA imaging of a coronary stent deployed in a rabbit's thoracic aorta with the presence of luminal blood [80]. Subsequently, they presented IVPA imaging of lipid deposition in atherosclerotic plaque at 1720 nm, where the strong first overtone absorption of CH_2 bond resides [37]. P. Wang et al. reported a high-speed IVPA system based on a 2 kHz barium nitrite ($Ba(NO_3)$) Raman laser and a ring-shaped ultrasound transducer, which improves the imaging speed by 50 times compared with traditional Nd:YAG laser based IVPA systems [82]. Y. Cao et al. developed a collinear IVPA catheter design, which enables high sensitivity intravascular imaging up to 6 mm depth with a clinically relevant catheter size [83]. Most recent researches have achieved high-speed IVPA imaging at 1.7 μm to target the strong first overtone band of CH_2 bond in lipid [84–86].

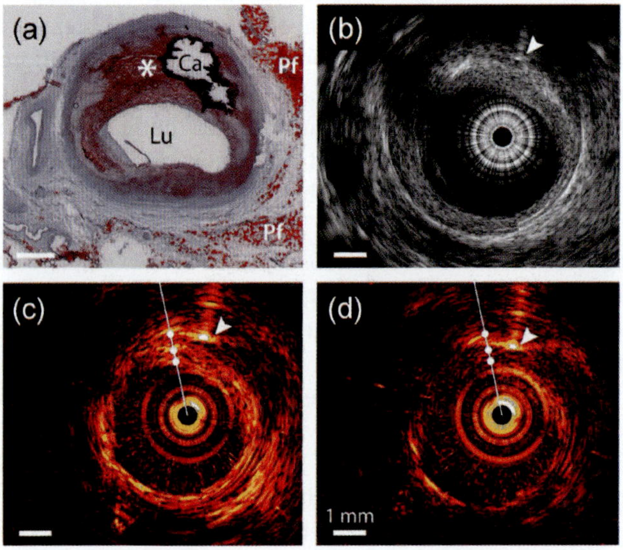

Figure 14.7 IVPA/IVUS imaging of a human coronary artery. (a) Oil Red O based histology result; (b) IVUS image; (c) and (d) IVPA images at 1210 nm and 1230 nm, respectively. Reproduced with permission from Ref. [54]. © 2011 OSA.

14.5 Conclusions and Perspective

In this chapter, we have introduced an emerging biological imaging method named VPAI for deep tissue imaging. This method takes the near-infrared vibrational overtone absorption as the imaging contrast, and converts the periodical optical absorption in tissue into localized heat production, which induces the mechanical vibration of tissue via thermoelastic effect and generates ultrasonic wave that is detectable by a sensitive ultrasound transducer. The most promising properties of VPAI are chemical selectivity and deep tissue accessibility due to a much smaller scattering coefficient of ultrasound in tissue.

By controlling the optical beam size in tissue and selecting different transducer frequency, the imaging depth and spatial resolution of VPAI can be scaled to image biological tissues from organelles to organs. Several typical examples of scalable imaging depth based on VPAI have been presented.

Optical absorptions for different characteristic chemical compositions or bonds, including hemoglobin, CH_2, CH_3, OH bonds, provide a useful guidance to select specific excitation laser wavelength for target tissue composition detection. With MS-PAI modality, different compositions in the tissue can be differentiated and quantified with MCR-ALS analysis. Finally, two important applications, i.e., breast cancer margin detection and intravascular imaging of atherosclerotic plaque, based on the bond-selective photoacoustic detection of lipid have been reviewed.

VPAI has shown its powerful ability for deep tissue imaging. However, it is still in its early stage. Future development of VPAI may be focused on the optical beam management to reduce the optical scattering in deep tissue, excitation laser with specific wavelength, high repetition rate and high output power pulsed laser to realize high-sensitivity video-rate imaging, and the miniaturization of imaging device, for example catheter in IVPA system, to access deep body tissues with minimal invasion and reliable performance. This exciting technology is anticipated to bring great opportunity to biomedical engineering and eventually benefit the health of entire human beings.

References

1. Herman, B. (1998). *Fluorescence Microscopy*. 2nd ed., Oxford, UK: Bios Scientific Publishers.
2. Alfano, R. R., Tata, D. B., Cordero, J., Tomashefsky, P., Longo, F. W., and Alfano, M. A. (1984). Laser-induced fluorescence spectroscopy from native cancerous and normal tissue, *IEEE J. Quantum Electron.*, **20**, pp. 1507–1511.
3. Albota, M. A., Xu, C., and Webb, W. W. (1998). Two-photon fluorescence excitation cross sections of biomolecular probes from 690 to 960 nm, *Appl. Opt.*, **37**, pp. 7352–7356.
4. Sanchez, E. J., Novotny, L., and Xie, X. S. (1999). Near-field fluorescence microscopy based on two-photon excitation with metal tips, *Phys. Rev. Lett.*, **82**, pp. 4014–4017.
5. So, P. T. C., Dong, C. Y., Masters, B. R., and Berland, K. M. (2000). Two-photon excitation fluorescence microscopy, *Annu. Rev. Biomed. Eng.*, **2**, pp. 399–429.
6. Diaspro, A., Chirico, G., and Collini, M. (2005). Two-photon fluorescence excitation and related techniques in biological microscopy, *Q. Rev. Biophys.*, **38**, pp. 97–166.
7. Gelikonov, V. M., Gelikonov, G. V., Kuranov, R. V., Pravdenko, K. I., Sergeev, A. M., Feldshtein, F. I., Khanin, Y. I., Shabanov, D. V., Gladkova, N. D., Nikulin, N. K., Petrova, G. A., and Pochinko, V. V. (1995). Coherent optical tomography of microscopic inhomogeneities in biological tissues, *Jetp. Lett.*, **61**, pp. 158–162.
8. Kuranov, R. V., Sapozhnikova, V. V., Shakhova, H. M., Gelikonov, V. M., Zagainova, E. V., and Petrova, S. A. (2002). Combined application of optical methods to increase the information content of optical coherent tomography in diagnostics of neoplastic processes, *Quantum Electron.*, **32**, pp. 993–998.
9. Knight, O. J., Chang, R. T., Feuer, W. J., and Budenz, D. L. (2009). Comparison of retinal nerve fiber layer measurements using time domain and spectral domain optical coherent tomography, *Ophthalmology*, **116**, pp. 1271–1277.
10. Kleinschmidt, A., Obrig, H., Requardt, M., Merboldt, K. D., Dirnagl, U., Villringer, A., and Frahm, J. (1996). Simultaneous recording of cerebral blood oxygenation changes during human brain activation by magnetic resonance imaging and near-infrared spectroscopy, *J. Cerebr. Blood F. Metab.*, **16**, pp. 817–826.

11. Reich, G. (2005). Near-infrared spectroscopy and imaging: Basic principles and pharmaceutical applications, *Adv. Drug Deliv. Rev.*, **57**, pp. 1109–1143.

12. Fine, S., and Hansen, W. P. (1971). Optical second harmonic generation in biological systems, *Appl. Opt.*, **10**, pp. 2350–2353.

13. Abraham, J. L., and Etz, E. S. (1979). Molecular microanalysis of pathological specimens in situ with a laser-Raman microprobe, *Science*, **206**, pp. 716–718.

14. Alfano, R. R., Liu, C. H., Sha, W. L., Zhu, H. R., Akins, D. L., Cleary, J., Prudente, R., and Cellmer, E. (1991). Human breast tissues studied by IR Fourier transform Raman spectroscopy, *Lasers Life Sci.*, **4**, pp. 23–28.

15. Zumbusch, A., Holtom, G. R., and Xie, X. S. (1999). Three-dimensional vibrational imaging by coherent anti-Stokes Raman scattering, *Phys. Rev. Lett.*, **82**, pp. 4142–4145.

16. Volkmer, A., Cheng, J. X., and Xie, X. S. (2001). Vibrational imaging with high sensitivity via epidetected coherent anti-Stokes Raman scattering microscopy, *Phys. Rev. Lett.*, **87**, pp. 023901.

17. Evans, C. L., Potma, E. O., Puoris'haag, M., Cote, D., Lin, C. P., and Xie, X. S. (2005). Chemical imaging of tissue in vivo with video-rate coherent anti-Stokes Raman scattering microscopy, *P. Natl. Acad. Sci. U. S. A.*, **102**, pp. 16807–16812.

18. Cheng, J. X., and Xie, S. S. (2012). *Coherent Raman Scattering Microscopy*, New York: Taylor & Francis.

19. Freudiger, C. W., Min, W., Saar, B. G., Lu, S., Holtom, G. R., He, C. W., Tsai, J. C., Kang, J. X., and Xie, X. S. (2008). Label-free biomedical imaging with high sensitivity by stimulated Raman scattering microscopy, *Science*, **322**, pp. 1857–1861.

20. Saar, B. G., Freudiger, C. W., Reichman, J., Stanley, C. M., Holtom, G. R., and Xie, X. S. (2010). Video-rate molecular imaging in vivo with stimulated Raman scattering, *Science*, **330**, pp. 1368–1370.

21. Zhang, D. L., Wang, P., Slipchenko, M. N., and Cheng, J. X. (2014). Fast vibrational imaging of single cells and tissues by stimulated Raman scattering microscopy, *Acc. Chem. Res.*, **47**, pp. 2282–2290.

22. Liao, C. S., Slipchenko, M. N., Wang, P., Li, J., Lee, S. Y., Oglesbee, R. A., and Cheng, J. X. (2015). Microsecond scale vibrational spectroscopic imaging by multiplex stimulated Raman scattering microscopy, *Light Sci. Appl.*, **4**, p. e265.

23. Wang, L. V., and Wu, H.-I. (2007). *Biomedical Optics: Principles and Imaging*, Hoboken, New Jersey: Wiley.
24. Wang, L., Ho, P. P., Liu, C., Zhang, G., and Alfano, R. R. (1991). Ballistic 2-D imaging through scattering walls using an ultrafast optical Kerr gate, *Science*, **253**, pp. 769–771.
25. Shi, L., Sordillo, L. A., Rodríguez-Contreras, A., and Alfano, R. (2015). Transmission in near-infrared optical windows for deep brain imaging, *J. Biophotonics*, **9**, pp. 38–43.
26. Xu, M. H., and Wang, L. H. V. (2006). Photoacoustic imaging in biomedicine, *Rev. Sci. Instrum.*, **77**, p. 041101.
27. Wang, L. V. (2009). Multiscale photoacoustic microscopy and computed tomography, *Nat. Photonics*, **3**, pp. 503–509.
28. Beard, P. (2011). Biomedical photoacoustic imaging, *Interface Focus*, **1**, pp. 602–631.
29. Wang, L. V., and Hu, S. (2012). Photoacoustic tomography: In vivo imaging from organelles to organs, *Science*, **335**, pp. 1458–1462.
30. Wang, H. W., Chai, N., Wang, P., Hu, S., Dou, W., Umulis, D., Wang, L. V., Sturek, M., Lucht, R., and Cheng, J. X. (2011). Label-free bond-selective imaging by listening to vibrationally excited molecules, *Phys. Rev. Lett.*, **106**, p. 238106.
31. Bell, A. G. (1880). On the production and reproduction of sound by light, *Am. J. Sci.*, **20**, p. 20.
32. Siesler, H. W., Ozaki, Y., Kawata, S., and Heise, H. M. (2008). *Near-Infrared Spectroscopy: Principles, Instruments, Applications*, Wiley.
33. Patel, C. K. N., and Tam, A. C. (1979). Optoacoustic spectroscopy of liquids, *Appl. Phys. Lett.*, **34**, pp. 467–470.
34. Patel, C. K. N., and Tam, A. C. (1979). Optical-absorption coefficients of water, *Nature*, **280**, pp. 302–304.
35. Patel, C. K. N., Tam, A. C., and Kerl, R. J. (1979). High overtones of C–H stretch in liquid benzene measured by laser optoacoustic spectroscopy, *J. Chem. Phys.*, **71**, pp. 1470–1474.
36. Tam, A. C., and Patel, C. K. N. (1979). Optical absorptions of light and heavy-water by laser optoacoustic spectroscopy, *Appl. Opt.*, **18**, pp. 3348–3358.
37. Wang, B., Karpiouk, A., Yeager, D., Amirian, J., Litovsky, S., Smalling, R., and Emelianov, S. (2012). In vivo intravascular ultrasound-guided photoacoustic imaging of lipid in plaques using an animal model of atherosclerosis, *Ultrasound Med. Biol.*, **38**, pp. 2098–2103.

38. Wang, P., Wang, H. W., Sturek, M., and Cheng, J. X. (2012). Bond-selective imaging of deep tissue through the optical window between 1600 and 1850 nm, *J. Biophotonics*, **5**, pp. 25–32.
39. Wang, P., Wang, P., Wang, H. W., and Cheng, J. X. (2012). Mapping lipid and collagen by multispectral photoacoustic imaging of chemical bond vibration, *J. Biomed. Opt.*, **17**, pp. 96010–96011.
40. Jansen, K., Wu, M., van der Steen, A. F. W., and van Soest, G. (2013). Lipid detection in atherosclerotic human coronaries by spectroscopic intravascular photoacoustic imaging, *Opt. Express*, **21**, pp. 21472–21484.
41. Wang, P., Rajian, J. R., and Cheng, J. X. (2013). Spectroscopic imaging of deep tissue through photoacoustic detection of molecular vibration, *J. Phys. Chem. Lett.*, **4**, pp. 2177–2185.
42. Wang, X. D., Pang, Y. J., Ku, G., Xie, X. Y., Stoica, G., and Wang, L. H. V. (2003). Noninvasive laser-induced photoacoustic tomography for structural and functional in vivo imaging of the brain, *Nat. Biotechnol.*, **21**, pp. 803–806.
43. Maslov, K., Zhang, H. F., Hu, S., and Wang, L. V. (2008). Optical-resolution photoacoustic microscopy for in vivo imaging of single capillaries, *Opt. Lett.*, **33**, pp. 929–931.
44. Hu, S., Maslov, K., and Wang, L. V. (2011). Second-generation optical-resolution photoacoustic microscopy with improved sensitivity and speed, *Opt. Lett.*, **36**, pp. 1134–1136.
45. Workman, J., and Weyer, L. (2007). *Practical Guide to Interpretive Near-Infrared Spectroscopy*, CRC Press.
46. Ntziachristos, V., and Razansky, D. (2010). Molecular imaging by means of multispectral optoacoustic tomography (MSOT), *Chem. Rev.*, **110**, pp. 2783–2794.
47. Razansky, D., Buehler, A., and Ntziachristos, V. (2011). Volumetric real-time multispectral optoacoustic tomography of biomarkers, *Nat. Protoc.*, **6**, pp. 1121–1129.
48. Razansky, D. (2012). Multispectral optoacoustic tomography-volumetric color hearing in real time, *Selected Top. Quantum Electron. IEEE J.*, **18**, pp. 1234–1243.
49. Joseph, J., Tomaszewski, M., Morgan, F. J. E., and Bohndiek, S. E. (2015). Evaluation of multispectral optoacoustic tomography (MSOT) performance in phantoms and in vivo. In *Photons Plus Ultrasound: Imaging and Sensing 2015*. Proceedings of the SPIE.

50. Wu, M., Jansen, K., van der Steen, A. F. W., and van Soest, G. (2015). Specific imaging of atherosclerotic plaque lipids with two-wavelength intravascular photoacoustics, *Biomed. Opt. Express*, **6**, pp. 3276–3286.

51. Jaumot, J., Gargallo, R., de Juan, A., and Tauler, R. (2005). A graphical user-friendly interface for MCR-ALS: A new tool for multivariate curve resolution in MATLAB, *Chemometr. Int. Lab.*, **76**, pp. 101–110.

52. Bro, R., and DeJong, S. (1997). A fast non-negativity-constrained least squares algorithm, *J. Chemometr.*, **11**, pp. 393–401.

53. Bro, R., and Sidiropoulos, N. D. (1998). Least squares algorithms under unimodality and non-negativity constraints, *J. Chemometr.*, **12**, pp. 223–247.

54. Xu, M., and Wang, L. V. (2005). Universal back-projection algorithm for photoacoustic computed tomography, *Phys. Rev. E Stat. Nonlin. Soft Matter Phys.*, **71**, p. 016706.

55. Kruger, R. A., Liu, P., Fang, Y. R., and Appledorn, C. R. (1995). Photoacoustic ultrasound (PAUS)—reconstruction tomography, *Med. Phys.*, **22**, pp. 1605–1609.

56. Haltmeier, M., Scherzer, O., Burgholzer, P., and Paltauf, G. (2004). Thermoacoustic computed tomography with large planar receivers, *Inverse Probl.*, **20**, p. 1663.

57. Mark, A. A., Jin, Z., Dimple, M., and Patrick, J. L. R. (2007). Application of inverse source concepts to photoacoustic tomography, *Inverse Probl.*, **23**, p. S21.

58. Yang, J. M., Maslov, K., Yang, H. C., Zhou, Q. F., Shung, K. K., and Wang, L. H. V. (2009). Photoacoustic endoscopy, *Opt. Lett.*, **34**, pp. 1591–1593.

59. Yang, J. M., Favazza, C., Chen, R. M., Yao, J. J., Cai, X., Maslov, K., Zhou, Q. F., Shung, K. K., and Wang, L. H. V. (2012). Simultaneous functional photoacoustic and ultrasonic endoscopy of internal organs in vivo, *Nat. Med.*, **18**, pp. 1297–1303.

60. Yang, J. M., Li, C. Y., Chen, R. M., Rao, B., Yao, J. J., Yeh, C. H., Danielli, A., Maslov, K., Zhou, Q. F., Shung, K. K., and Wang, L. V. (2015). Optical-resolution photoacoustic endomicroscopy in vivo, *Biomed. Opt. Express*, **6**, pp. 918–932.

61. Sethuraman, S., Amirian, J. H., Litovsky, S. H., Smalling, R. W., and Emelianov, S. Y. (2008). Spectroscopic intravascular photoacoustic imaging to differentiate atherosclerotic plaques, *Opt. Express*, **16**, pp. 3362–3367.

62. Jansen, K., van der Steen, A. F. W., van Beusekom, H. M. M., Oosterhuis, J. W., and van Soest, G. (2011). Intravascular photoacoustic imaging of human coronary atherosclerosis, *Opt. Lett.*, **36**, pp. 597–599.

63. Song, K. H., Stein, E. W., Margenthaler, J. A., and Wang, L. V. (2008). Noninvasive photoacoustic identification of sentinel lymph nodes containing methylene blue in vivo in a rat model, *J. Biomed. Opt.*, **13**, pp. 054033-6.

64. Li, R., Wang, P., Lan, L., Lloyd, F. P., Goergen, C. J., Chen, S. X., and Cheng, J. X. (2015). Assessing breast tumor margin by multispectral photoacoustic tomography, *Biomed. Opt. Express*, **6**, pp. 1273–1281.

65. Wu, W., Wang, P., Cheng, J. X., and Xu, X. M. (2014). Assessment of white matter loss using bond-selective photoacoustic imaging in a rat model of contusive spinal cord injury, *J. Neurotraum*, **31**, pp. 1998–2002.

66. Li, R., Phillips, E., Wang, P., Goergen, C. J., and Cheng, J.-X. (2015). Label-free in vivo imaging of peripheral nerve by multispectral photoacoustic tomography, *J. Biophotonics*, **8**, pp. 1–5.

67. Aziz, D., Rawlinson, E., Narod, S. A., Sun, P., Lickley, H. L. A., McCready, D. R., and Holloway, C. M. B. (2006). The role of reexcision for positive margins in optimizing local disease control after breast-conserving surgery for cancer, *Breast J.*, **12**, pp. 331–337.

68. Cèfaro, G., Genovesi, D., Marchese, R., Ursini, L., Cianchetti, E., Ballone, E., and Nicola, M. (2006). Predictors of local recurrence after conservative surgery and whole-breast irradiation, *Breast Cancer Res. Treat.*, **98**, pp. 329–335.

69. Cabioglu, N., Hunt, K., Sahin, A., Kuerer, H., Babiera, G., Singletary, S. E., Whitman, G., Ross, M., Ames, F., Feig, B., Buchholz, T., and Meric-Bernstam, F. (2007). Role for intraoperative margin assessment in patients undergoing breast-conserving surgery, *Ann. Surg. Oncol.*, **14**, pp. 1458–1471.

70. Smitt, M. C., Nowels, K., Carlson, R. W., and Jeffrey, S. S. (2003). Predictors of reexcision findings and recurrence after breast conservation, *Int. J. Radiat. Oncol. Biol. Phys.*, **57**, pp. 979–985.

71. Buja, L. M., and Willerson, J. T. (1994). Role of inflammation in coronary plaque disruption, *Circulation*, **89**, pp. 503–505.

72. Farb, A., Burke, A. P., Tang, A. L., Liang, Y. H., Mannan, P., Smialek, J., and Virmani, R. (1996). Coronary plaque erosion without rupture into a lipid core-A frequent cause of coronary thrombosis in sudden coronary death, *Circulation*, **93**, pp. 1354–1363.

73. Depre, C., Wijns, W., Robert, A. M., Renkin, J. P., and Havaux, X. (1997). Pathology of unstable plaque: Correlation with the clinical severity of acute coronary syndromes, *J. Am. Coll. Cardiol.*, **30**, pp. 694–702.

74. Vallabhajosula, S., and Fuster, V. (1997). Atherosclerosis: Imaging techniques and the evolving role of nuclear medicine, *J. Nucl. Med.*, **38**, pp. 1788–1796.

75. Lusis, A. J. (2000). Atherosclerosis, *Nature*, **407**, pp. 233–241.

76. Mintz, G. S., Popma, J. J., Pichard, A. D., Kent, K. M., Satler, L. F., Wong, S. C., Hong, M. K., Kovach, J. A., and Leon, M. B. (1996). Arterial remodeling after coronary angioplasty-A serial intravascular ultrasound study, *Circulation*, **94**, pp. 35–43.

77. Vakharia, K. T., Mishell, J. M., Yeghiazarians, Y., Ports, T. A., and Michaels, A. D. (2005). Determination of adequate stent expansion: A comparison of quantitative coronary angiography, intravascular ultrasound, and novel Stent Boost X-ray imaging, *Am. J. Cardiol.*, **96**, pp. 195H–195H.

78. Jang, I. K., Bouma, B. E., Kang, D. H., Park, S. J., Park, S. W., Seung, K. B., Choi, K. B., Shishkov, M., Schlendorf, K., Pomerantsev, E., Houser, S. L., Aretz, H. T., and Tearney, G. J. (2002). Visualization of coronary atherosclerotic plaques in patients using optical coherence tomography: Comparison with intravascular ultrasound, *J. Am. Coll. Cardiol.*, **39**, pp. 604–609.

79. Sethuraman, S., Amirian, J. H., Litovsky, S. H., Smalling, R. W., and Emelianov, S. Y. (2007). Ex vivo characterization of atherosclerosis using intravascular photoacoustic imaging, *Opt. Express*, **15**, pp. 16657–16666.

80. Karpiouk, A. B., Wang, B., Amirian, J., Smalling, R. W., and Emelianov, S. Y. (2012). Feasibility of in vivo intravascular photoacoustic imaging using integrated ultrasound and photoacoustic imaging catheter, *J. Biomed. Opt.*, **17**, p. 096008.

81. Jansen, K., Wu, M., van der Steen, A. F. W., and van Soest, G. (2014). Photoacoustic imaging of human coronary atherosclerosis in two spectral bands, *Photoacoustics*, **2**, pp. 12–20.

82. Wang, P., Ma, T., Slipchenko, M. N., Liang, S., Hui, J., Shung, K. K., Roy, S., Sturek, M., Zhou, Q., Chen, Z., and Cheng, J. X. (2014). High-speed intravascular photoacoustic imaging of lipid-laden atherosclerotic plaque enabled by a 2 kHz barium nitrite Raman laser, *Sci. Rep.*, **4**, p. 6889.

83. Cao, Y., Hui, J., Kole, A., Wang, P., Yu, Q., Chen, W., Sturek, M., and Cheng, J.-X. (2016). High-sensitivity intravascular photoacoustic imaging of lipid–laden plaque with a collinear catheter design, *Sci. Rep.*, **6**, p. 25236.

84. Piao, Z., Ma, T., Li, J., Wiedmann, M. T., Huang, S., Yu, M., Kirk Shung, K., Zhou, Q., Kim, C.-S., and Chen, Z. (2015). High speed intravascular photoacoustic imaging with fast optical parametric oscillator laser at 1.7 μm, *Appl. Phys. Lett.*, **107**, p. 083701.

85. Li, Y., Gong, X., Liu, C., Lin, R., Hau, W., Bai, X., and Song, L. (2015). High-speed intravascular spectroscopic photoacoustic imaging at 1000 A-lines per second with a 0.9 mm diameter catheter, *J. Biomed. Opt.*, **20**, pp. 065006–065006.

86. Hui, J., Yu, Q., Ma, T., Wang, P., Cao, Y., Bruning, R. S., Qu, Y., Chen, Z., Zhou, Q., Sturek, M., Cheng, J.-X., and Chen, W. (2015). High-speed intravascular photoacoustic imaging at 1.7 μm with a KTP-based OPO, *Biomed. Opt. Express*, **6**, pp. 4557–4566.

Chapter 15

Deep Tissue Imaging: Acoustic and Thermal Wave Propagation and Light Interactions in Tissue

Idan Steinberg,[a,b] Asaf Shoval,[a] Michal Balberg,[c] Adi Sheinfeld,[b] Michal Tepper,[a] and Israel Gannot[a,d]

[a]*Department of Biomedical Engineering, Tel-Aviv University, Haim Levanon St., Tel Aviv 6997801, Israel*
[b]*School of Electrical Engineering, Tel-Aviv University, Haim Levanon St., Tel Aviv 6997801, Israel*
[c]*Faculty of Engineering, Holon Institute of Technology, 52 Golomb St., Holon, 5810201, Israel*
[d]*Department of Electrical & Computer Engineering, Johns Hopkins University, Baltimore, Maryland 21218, USA*

Idanstei@gmail.com

Deep tissue imaging in combination with treatment is a powerful tool for theranostics applications. This chapter describes a set of tools developed by the authors for functional and structural imaging with localized treatment under feedback. Photothermal, photoacoustic, radiotherapy under thermal imaging for real time assessment of treatment are described and discussed in this chapter. The conclusion of this approach is that a synergetic approach has higher chanced for a successful detection and treatment methods.

Deep Imaging in Tissue and Biomedical Materials: Using Linear and Nonlinear Optical Methods
Edited by Lingyan Shi and Robert R. Alfano

ISBN 978-981-4745-88-8 (Hardcover), 978-1-315-20655-4 (eBook)
www.panstanford.com

15.1 Introduction

Understanding light interaction with tissue enables scientists and engineers to develop powerful tools for diagnostics and treatment. It has the potential of being low cost, portable, and safe, so it can be used at the point of care or near patient with minimal damage associated. Adding other components to the scheme of operation may lead to a much more efficient and useful interaction. In order to detect deeper structures and functionality, we should make use of additional signals that are created by the light tissue interaction. This group of authors has developed a few methods that utilized more than optical signal only with no exogenous components. The works that are presented in this chapter mainly consisted of photoacoustic imaging to quantify blood flow rate through the Doppler effect and the thermal diffusion effect. A second project utilized the optical absorption with acoustic signal propagation for early diagnosis of osteoporosis. This method has the potential to replace both two-wavelength x-ray (DXA) for composition changes and ultrasound for mechanical properties changes. Both parameters as a whole have higher probability of early detection of osteoporosis in comparison to each of them separately. Another strong synergetic combination is thermal imaging. It can be used for deep assessment of tissue functionality through local oxygen saturation measured with photothermal techniques. Thermal imaging is also very powerful in assessment of tumor size and aggressiveness and response to treatment. It can also be used for real-time feedback of treatment using other modalities such as embedded radioactive wires (DaRT) or external radiation therapy.

15.2 Photoacoustics Quantitative Assessment of Deep Tissue Functionality

15.2.1 Introduction

Photoacoustic (PA) imaging is a relatively new, non-ionizing, biomedical imaging modality, based on light-generated ultrasound. Although the PA effect was first discovered by Alexander Graham Bell in the 19th century [1], sensing applications of PA were not developed significantly until the 1970s, after the invention of the laser. In the 1970s and 1980s the research focused mainly on spectroscopy of gases and solids [2, 3] and it was not until the mid-1990s when the first research of this technique for biomedical applications began [4, 5]. In PA imaging or sensing, pulsed or modulated laser light in the visible or near-infrared spectral range is launched into a medium of interest [6–8]. Endogenous or exogenous molecules found in tissue absorb the light and generate a local thermo-elastic expansion and accompanying pressure wave that can be detected by ultrasound transducers [9]. The PA technique can provide high-resolution tomography at larger penetration depths compared to pure optical imaging methods [7], due to the lower attenuation of ultrasound waves relative to light and since the resolution can be controlled acoustically rather optically.

Despite the similarities in detection methods, ultrasound (US) and PA imaging are based on different contrast sources. US contrast comes from acoustic impedance mismatch between different tissues [10] and therefore US imaging provide mainly structural information on the tissue content. In PA imaging, the signal amplitude is proportional to the optical absorption. Thus, an obvious advantage is improved contrast due to the high optical absorption of tissue chromophores—most importantly oxy- and deoxy-hemoglobin, but also water, lipids, and melanin [7]. In addition, there is a variety of biocompatible exogenous contrast agents that absorb at the visible and NIR spectral range and may be used as a source for spectrally selective PA contrast. Among them are organic dyes, like indocyanine green and methylene blue, and nanoparticles, at various geometries. The last can be bound to various antibodies in order to target cell receptors associated with specific diseases, such as cancer [11, 12]. These contrast agents often provide higher contrast than endogenous agents and their absorption wavelength can be engineered.

While PA imaging was demonstrated in vivo for various applications, quantitative PA measurements are more challenging [13]. The PA signal is proportional to the optical absorption; however, extracting the absolute absorption in vivo is a highly complicated task, even in the simple scenario when the target includes only a single absorber at the excitation wavelength.

First, it requires accurate estimation of the optical fluence at the target, which in turn requires modeling the light propagation in the tissue until it reaches the target. Second, it requires compensation for the ultrasonic detector response. And third, the efficiency of converting the absorbed energy to ultrasound depends on the mechanical and thermal characteristics of the target as well as on the local temperature and is not always accurately known in advance.

The last two factors can be eliminated by using differential measurement of the same target with two or more different excitation wavelengths. This measurement cannot provide the absolute absorption; however, in blood it can yield the oxygen saturation (sO_2), which is a highly important metabolic parameter, based on the different absorption spectrum of oxy- and deoxy-hemoglobin [14]. The same principle of multi-spectral excitation will be demonstrated in Section 15.1.3 for the estimation of the ratio of fat and blood in bones. It should be noted that accurate multi-spectral measurements in depth still require correction for the wavelength dependence of the local fluence.

Another example of quantitative PA is measuring blood velocity or flow, which will be presented in the next section. In this case, the measured value is derived from changes in the PA signal in the frequency domain, either Doppler frequency shift or sidebands resulting from thermal effects. These changes result directly from the velocity regardless of the optical fluence.

15.2.2 Deep Tissue Photoacoustic Flow Measurements

The most common non-invasive technique today for continuously measuring blood flow is Doppler ultrasound (DUS), which can be used over a wide range of velocities and in large penetration depths. However, there are several advantages for using optical techniques for vascular flowmetry, and specifically for using photoacoustics. The first is the option to combine blood flow and sO_2 measurements simultaneously in order to obtain the important metabolic parameter of oxygen consumption. An additional advantage of PA, especially in small blood vessels and capillaries, is reducing background noise since the blood is the main optical absorber, while DUS is based on scattering.

Several techniques have been proposed for implementing PA flowmetry. The first family of techniques is similar to the concept of DUS and measures the change generated by the movement of an inhomogeneous fluid either in the time [14–16] or the frequency domain [17–20]. DUS and optical techniques such as laser Doppler velocimetry [21] and Doppler optical coherence tomography [22] are based on the Doppler effect, in which a wave reflected by a moving object, in this case a red blood cell (RBC), exhibits a frequency shift proportional to the object's velocity.

In light of the similarities of photoacoustics and ultrasound, the Doppler effect was a natural basis to develop a photoacoustic flowmetry method termed photoacoustic Doppler (PAD) [18, 19]. In this technique, the optical excitation was generated by laser diodes (LDs), which are relatively low-cost and compact, compared to the pulsed lasers commonly used in PA imaging. The use of LDs allowed tailoring a unique excitation pattern: a series of coherent tone-bursts at a carrier frequency which matches the transducer's central frequency. Time-frequency analysis of the detected ultrasonic signals using short-time-Fourier-transform (STFT) yielded simultaneous measurement of velocity and axial location of the target, as demonstrated in Fig. 15.1. The first was derived from the Doppler shift in the frequency domain and the second from the time shift of the acoustic signal relative to the excitation time.

The ability to control the excitation in both the time and the frequency domains was used also for simultaneous measurements of Doppler photoacoustic and Doppler US with similar excitation pattern and similar signal analysis, by separating the US and PA excitations in the frequency domain rather than in time. PAD was demonstrated in vitro in phantom vessels filled with flowing suspensions of carbon particles and later in blood with and without contrast agent (indocyanine green—ICG) [23].

Though PAD signal was observed in whole blood in vitro, it was still rather low and could not be measured clearly at high velocities. The Doppler signal is generated from the sum of incoherent contributions of the moving particles. In high particle concentrations, as is the case of RBCs in blood, the particles' locations become closer and therefore more correlated and their contributions start canceling each other, leading to reduced SNR. The same behavior can be observed in Doppler ultrasound; however, the sensitivity and permissible power levels of this modality are such that a high enough SNR can be obtained despite this limitation. Other studies of similar PA flowmetry methods achieved in vivo measurement only in the optical-resolution mode, where the optical beam was focused to approximately the size of a single RBC, therefore eliminating the aforementioned problem [14]. In vivo implementation of PAD requires SNR improvement, on either the detection or the excitation side.

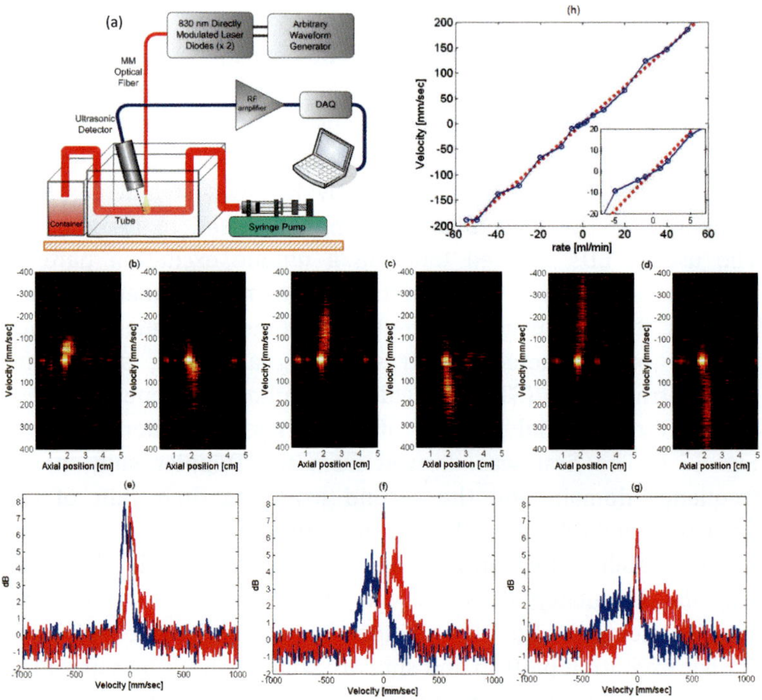

Figure 15.1 (a) Laser diode setup (at 830 nm) with a water-immersed phantom-vessel filled with blood with\without ICG or with carbon particles suspension. Optional ultrasonic transmitter was added for simultaneous US and PA measurements. (b–d) Spectrograms of velocity vs. axial position for withdrawal (left spectrograms) and infusion (right spectrograms) at rates of 10, 30, and 50 ml/min for a suspension of carbon particles. (e-g) Cross section at distance of 2 cm, location of maximum Doppler signal, infusion (red) and withdrawal (blue) for the three different rates respectively. (h) Velocity vs. flow-rate: estimated from Doppler shift (blue with markers at measurement points) and theoretic, based on rate and tube dimensions (dotted red). Inset includes zoom-in on the low velocities section.

This challenge of PAD motivated the development of a different PA flowmetry technique, termed PA thermal diffusion flowmetry (PA-TDF), based on measuring the heat clearance rate from the target volume, which is directly related to the flow [20]. In this technique, a local thermal inhomogeneity is generated in the blood vessel, and therefore it is not based on the heterogeneity of the flowing medium like PAD. The rate of heat clearance from the target is monitored using photoacoustics, based on the near-linear dependence of the PA generation efficiency in blood on the local temperature. This dependence results from the increase in thermal expansion with increased temperature.

PA-TDF was implemented using two LDs for excitation in an experimental setup similar to PAD. The intensity of the first LD was modulated harmonically at low frequencies between 0.1 and 20 Hz to induce temperature variations in the medium. The second LD was modulated either harmonically or with tone-bursts at a frequency of 0.5 MHz to generate a photoacoustic signal. The local temperature variations induced amplitude modulation on the PA signal, expressed in the frequency domain as side-bands around the PA modulation frequency.

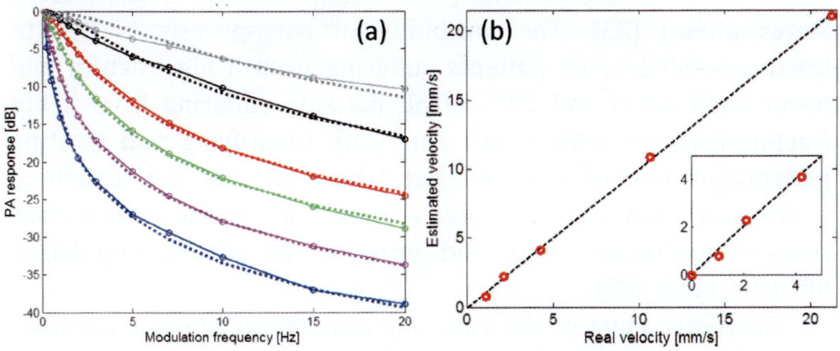

Figure 15.2 (a) Normalized PT-PA modulation frequency responses for stationary (blue) and velocities of 1.06 (pink), 2.1 (green), 4.2 (red), 10.6 (black) and 21.2 (gray) mm/s in a blood vessel phantom immersed in water. Solid lines with markers are experimental data. Dotted lines are theoretical Lorentzian fitting. (b) Estimated velocity vs. real velocity: Estimation for the 5 velocities (red circles). Dotted black line indicates the 45° line. Inset: zoom-in on the range of low velocities.

In order to estimate the flow from the photothermal-modulated PA signal, the PA modulation frequency response (MFR) was measured from these sidebands by scanning the photothermal (PT) modulation frequency. The MFR was fitted to a simplified thermal model resembling an RC circuit. This model could be expressed as exponential decay of the temperature in the time domain, or a Lorentzian function with a single time constant t_{eff} in the frequency domain. t_{eff} represents the time it takes for the temperature to return to its steady state value, after a short heating pulse and is composed of a conduction term and a convection term, the second inversely proportional to the velocity. Accurate estimations of blood velocity in vessel phantoms immersed in water [20] or embedded in scattering tissue-mimicking phantoms [24] were demonstrated with this method (Fig. 15.2).

15.2.3 Photoacoustic Characterization of Bone Pathologies for Early Detection of Osteoporosis

Osteoporosis is considered a major public health problem worldwide due to the number of fractures and the overwhelming related healthcare costs. For example, in the EU alone there are 19 million osteoporosis patients with annual health care costs of about 50 billion dollars (~20 billion dollars in the United States alone) [25]. The morbidity of osteoporosis is due to fractures—50% of all patients suffering from a hip fracture will never walk again and 20% of all patients suffering from a hip fracture will die within one year [26]. Identifying and treating patients at risk of fracture, but who have not yet sustained a fracture, can substantially reduce the fracture incidence, reduced healthcare costs and improve the elderly population life expectancy [27].

Currently, most of the clinically available diagnostic methods are based on the use of either X-ray (DXA method) or quantitative ultrasound (QUS) [27, 28]. Those methods were shown to measure bone mass density (BMD) as well as other mechanical parameters [30]. However, these methods have a relatively low sensitivity to molecular and functional changes in the tissue. Such sensitivity is of great importance. It was shown, for example, that biochemical changes in the bone marrow fat can serve as an early indication of osteoporosis prior to degradation of the BMD [31]. To address these issues, hybrid photoacoustic methods were recently proposed by several groups. PA has significant advantages for assessing osteoporosis, as it provides both molecular information and bone mechanical status.

To date, only a handful of experimental works have studied the use of PA for evaluating bone status. Thus, the potential of PA for bone evaluation extends far beyond the current available techniques and is largely unexploited. In recent years, Petro et al. demonstrated single wavelength photoacoustic excitation and detection on bone samples coated with gelatin [20]. Low-frequency ultrasound (50 kHz) was used to investigate the slow fundamental flexural guided wave alone. This method measures a single QUS parameter. Lashkari et al. have used a dual backscattered ultrasound and PA radar system for the assessment of bone structure and density variations [21]. Changes in the PA and US signatures due to induced osteoporotic process were demonstrated. Yuan and Jiang developed a finite element–based photoacoustic tomographic method for reconstruction of spatially varying optical and acoustic properties [22]. This method was used for the imaging of finger joints in 3D for the early detection of joint diseases such as osteoarthritis [23] and recently achieved high spatial resolution comparable to that of MRI [24]. Excellent results were achieved for tomographic imaging of joints. Finally, Feng et al. recently explored the feasibility of thermal photoacoustic measurement [25] and photoacoustic spectral analysis [26, 27] for the ex vivo assessment of bone microarchitecture in rat models. The in vivo implementation of the thermal PA method is expected to be challenging due to the precise control of the bone temperature required over a large range of temperatures. The photoacoustic spectral analysis seems quite promising and is expected to yield better results in the near future.

Steinberg et al. have demonstrated, for the first time, the potential of multispectral PA quantitative sensing to provide both molecular information from the bone absorption spectrum and bone mechanical status from clinically acceptable acoustical parameters as shown in Fig. 15.3 [32].

Those parameters include the speed of sound (SOS) and the broadband ultrasound attenuation (BUA). Thus, it has a significant advantage over both pure optical or ultrasonic techniques as well as over the other PA methods. The quantitative predications of a low-cost laser diode-based PA technique were compared to those of QUS under osteoporosis mimicking conditions. Such comparison was never performed before and is valuable in order to assess the performance of the PA method based on clinical measures used in practice.

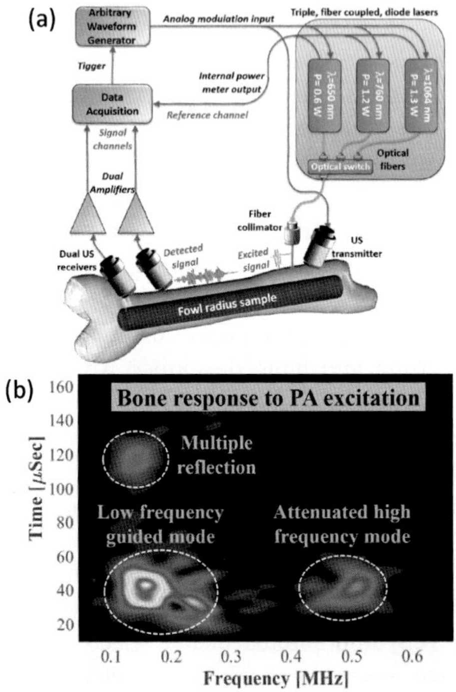

Figure 15.3 (a) Experimental setup–Differential, multispectral PA and QUS estimation of fowl bone parameters. (b) Spectrogram showing the different acoustic modes propagating along the bone shaft. From [32], with permission.

Recently, Steinberg et al. have developed a low-cost and compact dual-wavelength photoacoustic system based on pulsed laser diodes. Fiber-coupled, pulsed laser diodes offer a great improvement over the CW laser diodes used in earlier stages of the project. They are low-cost and allow for much shorter excitation duration (with sufficient energy per pulse) that enables efficient generation of the PA signal. Moreover, since the pulse repetition rate can be controlled, even for high peak power (tens of watts), the overall excitation can be made skin safe, eye safe and allow the laser to be operated everywhere and without specialized laser protection equipment. Demonstrated here are advantages of the PA system over QUS. It is shown, ex vivo that the PA method is sensitive and able to quantitatively estimate the blood/fat ratio in the bone marrow while the QUS is insensitive to that parameter. Such sensitivity is of great importance.

To demonstrate the sensitivity to optical parameters, a fresh tibia samples from a 4-month-old, Sinclair-type miniature swine made into a vile-like container by removing the marrow. Pure silicone oil and pure blood were placed inside the marrow intermittently six times (i.e. oil, blood, oil, etc.) and excited by a pulsed laser diode (wavelength 910 nm peak power 25 W) to generate PA signals as shown in Fig. 15.4a. Silicone oil was chosen since it is biologically inherent and has very similar acoustic impedance to that of blood (Oil—1.55MRayl, Blood—1.6MRayl). However, it has very different optical absorption in the visible and near IR regimes shown in Fig. 15.4b. Thus, mixtures are very similar in their acoustic parameters but not in their optical parameters. One can appreciate from Fig. 15.4c, that although measurements were performed in intermittently, the oil signal and blood signal are very distinct and demonstrates a stable behavior. The same experiment was repeated but with US excitation by acoustic transducer. As seen in the figure, although the signal power is 40 times higher, there is no clear distinction between the blood and oil. Thus the PA signal is sensitive to the marrow parameters behind the 5 mm highly scattering cortical wall, whereas the US signal is much less sensitive.

Finally, first-in-man experiments on healthy human volunteers were performed. Presented here are in vivo measurements of these parameters taken with both QUS and PA. To that end a bi-modal prototype apparatus was assembled (Fig. 15.4a). It facilitated implementation of both methods in the same site. Dual laser diodes (wavelengths 910 nm and 975 nm, peak power 25 W and 30 W, respectively) were alternatively connected to the optical switch with pulse durations of 400 ns. Laser safety calculations, categorized both lasers diode (in the above mentioned configuration) as skin safe, eye safe and class 3b—meaning only protective eye wear was needed. The apparatus, together with an optical switch, allows optically irradiating different points along the subject leg above the tibia, while detecting with a single US transducer, as seen in Fig. 15.5a.

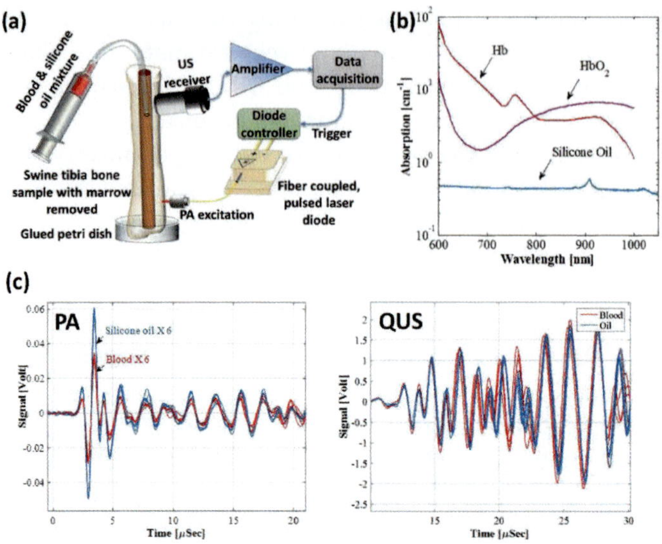

Figure 15.4 (a) Experimental setup. (b) Optical absorption spectrum of blood and silicone oil. (c) (Left) PA signals from pure oil and blood measured in alternation. (Right) The same with US excitation.

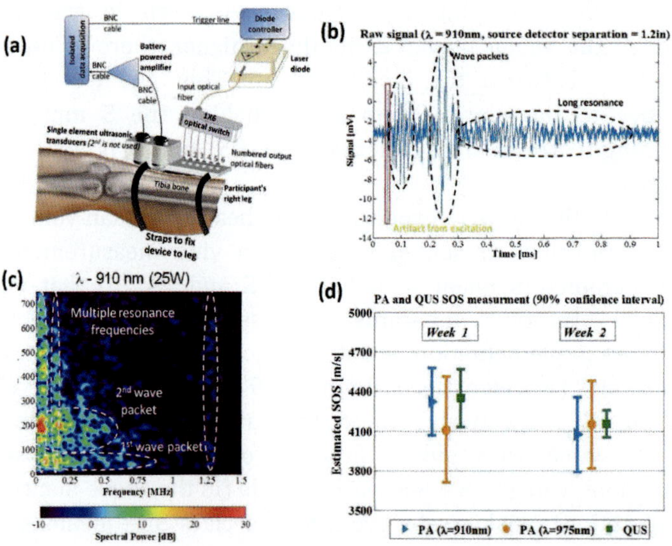

Figure 15.5 (a) First-in-man experimental setup. (b) Time signal showing the different wavepackets and resonances. (c) time-frequency representation of the same signal, (d) compression between PA signal at multiple wavelengths and QUS.

Figure 15.5b shows typical signal from human leg. The fastest wave-packet is associated with propagation in the bone (rather than soft tissue) since the SOS in bones is ~2.5 times higher than its value in soft tissues. Also shown are US resonances related to the bone structure. These are even more apparent in time/frequency representation shown in Fig. 15.5c. Finally, Fig. 15.5d shows repeated measurement performed on the same subject week after week. It demonstrates the repeatability of the PA method and its agreement with QUS.

In the long run, this method can expand and complement QUS as well as to open avenues for bedside early detection and monitoring of disease progression and response to therapies.

15.3 Acousto-Optic Imaging in Deep Tissue

15.3.1 Introduction

The use of near-infrared light to monitor and even image hemodynamic and physiologic changes within brain tissue relies on measuring the intensity of light, at several wavelengths, that escapes cerebral tissue as a function of distance from the light source or sources. Spatially resolved spectroscopy, whether using constant wave, time or frequency varying sources, is able to track variations in the oxygen saturation of hemoglobin within cerebral tissue, but is also affected by variations of oxygen saturation and blood flow through extra-cerebral, superficial tissue [33]. Diffuse correlation spectroscopy, that measures blood flow velocity, using temporal correlation of coherent light signals, can track changes in cerebral blood flow, but is also affected by extra-cerebral perfusion [34].

Acousto-optic modulation of coherent, near-infrared light signals within scattering biological tissue enables to better localize the optical path of photons within the tissue and thus reduce the effect of superficial contamination. Moreover, since coherent light is used for illumination, both spectral and temporal properties of the tissue can be investigated simultaneously.

15.3.2 Acousto-Optics in Turbid Media

When coherent light is used to probe deep tissue, the light escaping the surface exhibits random fluctuations, called speckles [35] due to the turbid nature of the tissue. Acousto-optics is based on locally modulating coherent light with a localized low power ultrasound beam [36, 37]. The ultrasound waves induce a local, artificial modulation in the detected random speckle intensity. This artificial modulation is temporally correlated with the pattern of the ultrasound wave as a function of the delayed time of propagation of the ultrasound wave through the media.

Mahan [36] and Leutz [38] introduced the theory of frequency modulation of light waves by ultrasonic waves, and Wang [37, 39] and Sfez et al. expanded the theory into practical applications in vitro [40–42] and in vivo [43]. Ramaz and colleagues [44, 45] demonstrated various configurations for improving the signal to noise of ultrasound-modulated light, using photorefractive and other complex detection schemes.

Several reviews of the field were published during the past decade, summarizing the benefits of acousto-optic imaging in biological tissue [46, 47]. Most demonstrations rely on scanning a focused ultrasound beam while illuminating with a coherent light source, and detecting the modulated optical speckle pattern via a CCD camera or a single photodetector.

15.3.3 Acousto-Optic Measurement of Blood Flow

Movement of scattering particles (such as blood cells) within the tissue causes phase shifts in the scattered light and thus modulates the random speckle pattern, producing temporal fluctuations in the speckle pattern. The temporal fluctuations caused by movement of the scattering centers are not correlated with the generated ultrasound pattern, and as a result an increase in blood flow results in poorer correlation between the light signal and the ultrasound wave. By measuring the degree of local correlation over the random fluctuations, acousto-optics is able to measure changes in flow in the microcirculation in deeper and larger (about 0.1–0.5 cm^3) tissue volumes compared to laser Doppler–based technologies [48, 49].

Monte Carlo simulations of acousto-optic-modulated light [50–52] demonstrate the modulation of the phase of the light by the acoustic field, and its effect on the acquired signal. These papers examined the amplitude of the modulation of the acquired signal by artificially separating the modulated and

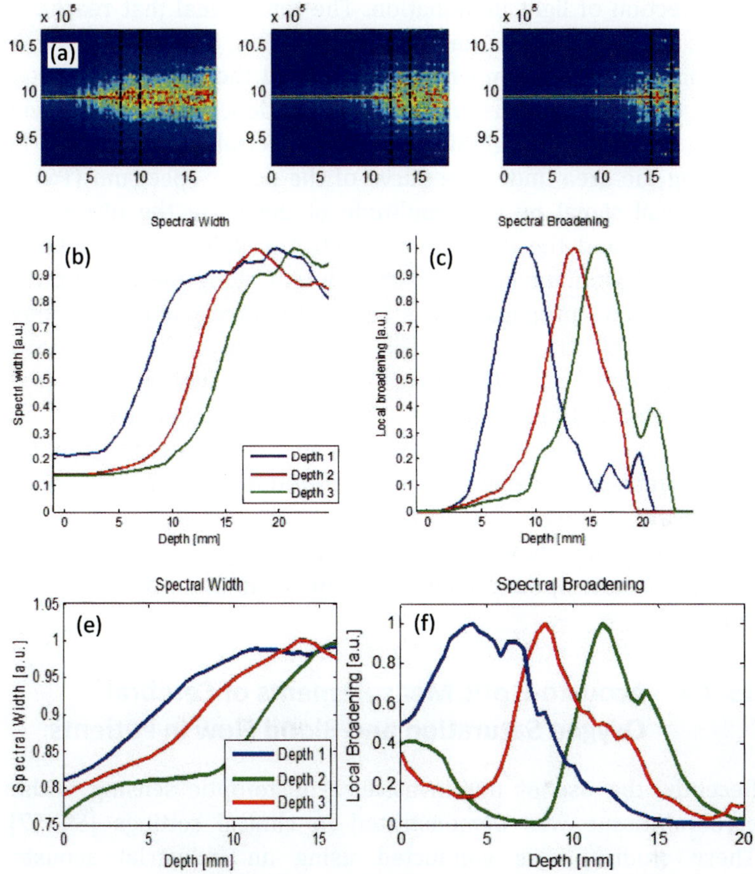

Figure 15.6 (a–c) Simulation results for three different depths of the 1 mm-wide flow layer. Spectrograms (where dotted lines show location of the flow channel) (a), spectral width (b) and spectral broadening (c) curves exhibit apparent depth discrimination ability. (d–e) Experimental results obtained for flow in channels at three different depths. Spectral width (d) and local broadening (e) both indicate a distinct depth discrimination. From [53], with permission.

non-modulated signals, where the ultrasound beam was either homogenous or perpendicular to the direction of light. These simulations did not take into account coherent effects within the turbid media that are affected by movement induced by blood flow. Tsalach et al. [53] simulated a more realistic configuration where an ultrasound pulse is propagated in a direction parallel to the direction of light illumination. The total signal that reaches a detector, in a reflection based light detection, is simulated, while summing both the number of photons and the phases associated with their paths, generating a time-dependent heterodyne signal sensed by a single detector. The spectral width (defined by dividing the area under the curve of the power spectrum (PS) of the optical signal by the amplitude of the PS at the ultrasound frequency), and the differential spectral broadening, (defined as the spatial gradient of the spectral width along the direction of the ultrasound propagation) is shown to correspond to the flow velocity within the channel. The simulation was compared to in vitro sensing of variable flow in flow channels with a diameter of 1 millimeter at different depths. The work demonstrates the ability to selectively monitor changes in deep (>1 cm deep) channels, and the ability to separate the contribution from flow channels at different depths.

Figure 15.6 shows a comparison between simulation and in vitro acousto-optic measurements in channels at three different depths.

15.3.4 Acousto-Optic Measurements of Cerebral Oxygen Saturation and Blood Flow in Patients

Recently, the use of non-invasive, acousto-optic sensing within cerebral tissue was demonstrated in clinical settings [54–57]. These studies were conducted using an industrial acousto-optic monitor[1] that emits coherent, near-infrared light, at three wavelengths between 780 and 830 nm, through an optical fiber, and collects the backscattered light through an optical fiber to an avalanche photodiode. The acousto-optic modulation is conducted by sonification of the tissue by an ultrasound wave, with a central frequency around 1 MHz that is able to penetrate the skull, with relatively low attenuation. By analyzing the cross-correlation between the detected light signal and the generated, coded, ultrasound wave, the system is able to measure a "virtual cross section" of the light distribution between a source and a detecting optical fiber [58]. Analysis of the spatial decay of these signals enables to extract the effective attenuation coefficient of light propagating in the tissue, at selective depths. The absorption of

[1]Developed by Ornim Medical, Israel.

different wavelengths of light by oxygenated and deoxygenated hemoglobin can be extracted, and the regional oxygen saturation of blood in the microvasculature of the brain can be determined. A clinical study on traumatic brain injured patient [54] demonstrated that the correlation between the oxygen saturation values obtained by the non-invasive acousto-optic signals and invasive measurements using a catheter placed in the jugular vein bulb was high ($r = 0.6$) and significant ($p < 0.001$). The correlation was high only for the non-invasive measurement obtained from the same hemisphere, where the invasive catheter was placed [54]. This demonstrates the volume/hemispheric localization of the acousto-optic measurement, within cerebral tissue at least 1 cm deep below the skin.

As explained above, movement of blood cells through the microvasculature affects the spectral width, and the degree of correlation between the light signal and the ultrasound signal. By analyzing the temporal properties of the modulated light signals, a blood flow index is determined. A good correlation between the blood flow index underneath the sensor, and regional cerebral blood flow measurement using 133Xe SPECT imaging was demonstrated following administration of acetazolamide, a drug known to affect cerebral blood flow [56]. Measurements with CW NIRS with ICG as contrast agent, during similar manipulations, could not demonstrate a significant correlation between the NIRS signal and the SPECT measurement [59].

15.4 Thermography for Assessment of Deep Tissue Tumor Volumes and Monitoring of Diffused Alpha Radiation Therapy

15.4.1 Introduction

Another method that allows us to gain necessary information regarding processes occurring in the depths of the tissue is thermography. Thermography, based on the detection of mid-IR radiation inertly emitted from the surface of a measured object, is an imaging modality that enables accurate, high resolution surface temperature measurements [60]. This method has several advantages such as rapidness, non-invasiveness as well as non-ionizing and low cost nature. Thermal imaging has already been investigated thoroughly and utilized for numerous medical applications: peripheral vascular disorders [61–63] inflammatory diseases [64] and cancer detection, especially breast cancer [65–72]. Since malignant tumors are characterized by abnormal metabolic and perfusion rates [73–75], they are expected to show abnormal temperature distribution compared to their healthy surroundings [65, 76]. This characteristic can be used to detect the tumor and evaluate its properties [77, 78], although improving the accuracy of automated algorithms in vivo is a great challenge.

15.4.2 Estimating Tumor Sizes from Thermographic Imaging

One such work investigates the correlation between tumor-healthy tissue temperature differences and tumor area in DA3 breast cancer carcinoma tumors [79]. Balb/c mice were each injected with breast carcinoma cells and treated with Diffused alpha Radiation Therapy (DaRT) [80–82] wires. Initially, thermal images of three mice were analyzed in order to evaluate the ability to estimate tumor areas and to observe changes in the temperature of tumors and their surroundings. In this initial experiment, thermal images of the three mice were obtained and analyzed to create an optimal basis for comparison in order to better observe the tumor and the undergoing processes. For each image, the tumor and its boundaries were identified manually and were represented by an elliptical shape. In order to show thermal changes in the tumor, the temperature scales of all the images were standardized by reducing the healthy tissue temperature for each day. The healthy tissue temperature was determined by examining the image and manually selecting an adjacent tissue with representing temperature. Figure 15.7 depicts the average and maximal temperature differences as a function of tumor area (Mouse #1 and Mouse #2 were injected with 60 kBq wires while Mouse #3 was injected with an inert wire as a control).

The general behavior of the temperature difference as a function of the area seemed to be the same: The tumor temperature decreases as its area increases. This can be explained by the necrosis at the center of the tumor and its extruding shape, typical to this type of transplantable tumors. It was also cautiously observed that the slopes for all mice are in the same order of magnitude. For example, a growth of 10 mm^2 in tumor area seemed to be linked to a reduction of 0.2–0.36° in the average tumor temperature. Thus, indicating that the generated tumors were similar in their biological and physical characteristics. The same can be concluded regarding their heat transfer patterns in various areas. The small changes between line slopes in the graphs were hypothesized to be indicative of internal biological changes.

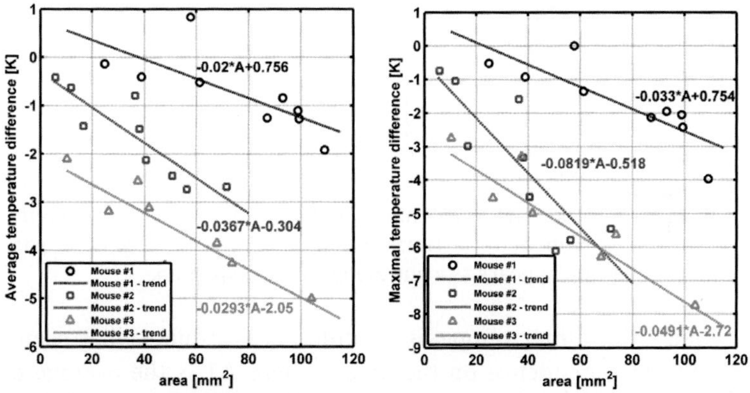

Figure 15.7 Average (left) and maximal (right) temperature difference as a function of estimated tumor area. Mouse #1 and Mouse #2 were injected with 60 kBq wires while Mouse #3 was injected with an inert wire as a control A linear fit for each mouse appears as a line along with its equation.

15.4.3 Comparison of DART vs. Inert via Thermography

In order to test these hypotheses, a second experiment was conducted, including nine mice. On average, mice with DaRT treatment increased their area by 78%, compared with mice with inert wires which increased their area by 165%. Next, the average and maximal temperature differences between the tumor and the healthy skin tissue temperature were calculated and presented in Fig. 15.8. As the figures show, all tumors exhibited a pattern similar to the one observed in the first experiment: the tumors were colder than their surroundings and their temperature further reduced as their area increased. The calculated slopes for the DaRT treated tumors were larger than the slopes calculated for the tumors with inert wires. Both slopes were relatively similar to each other and were only slightly higher than the slopes calculated in the first experiment.

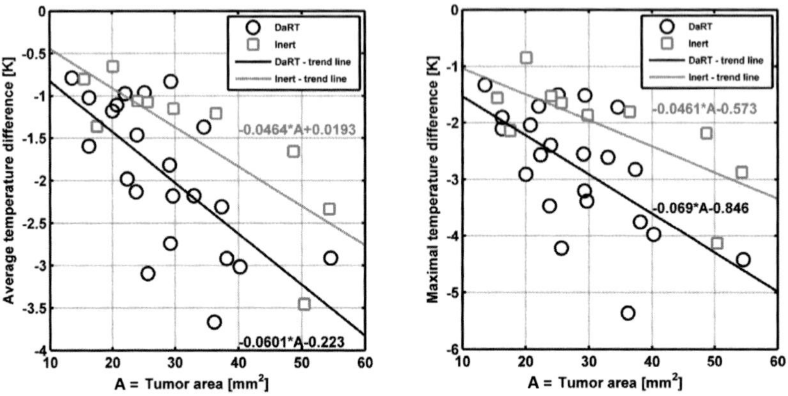

Figure 15.8 Average (left) and maximal (right) temperature difference as a function of estimated tumor area (A). A linear fit for each group appears as a line along with its equation.

Since the tumor temperature differences exhibit what seems as a linear dependence on the area, (where ΔT is the average or maximal temperature difference and A is the tumor area), it is expected that, on average, the ratio between the temperature difference change and the area change between measurements will be similar to the slopes calculated, as demonstrated by Eq. 15.1:

$$\frac{\Delta T_n - \Delta T_m}{A_n - A_m} = \frac{(a \cdot A_n + b) - (a \cdot A_m + b)}{A_n - A_m} = a, \tag{15.1}$$

where the indices n and m indicate the corresponding values for days n and m, respectively. For each mouse, the ratio described in Eq. 15.1 was calculated between all consecutive measurements (a total of three ratios for each mouse). Figures 15.9 shows the calculated ratios for the DaRT and inert wire treatment groups, between the average and maximal temperature difference changes and the area changes, averaged for each group. The error bars represent the range of all averages.

As shown in those figures, DaRT treated tumors have a larger ratio on average (−0.27 [K/mm^2] and −0.313 [K/mm^2]) for the average and maximal temperature differences, respectively) than the inert treated tumors (−5 × 10^{-4} [K/mm^2] and 0.0335 [K/mm^2]).

These values mean that the temperature of DaRT treated tumors decreases at an increased rate when compared to tumors with inert wires. This finding possibly indicates reduced metabolic activity due to the destruction processes caused by the treatment, leading to reduced heat production. Both ratio groups seem to be almost separated and distinct, although they have a small overlap. These findings were compared and found to be in accordance with previous experimental results found by other research groups [83, 84].

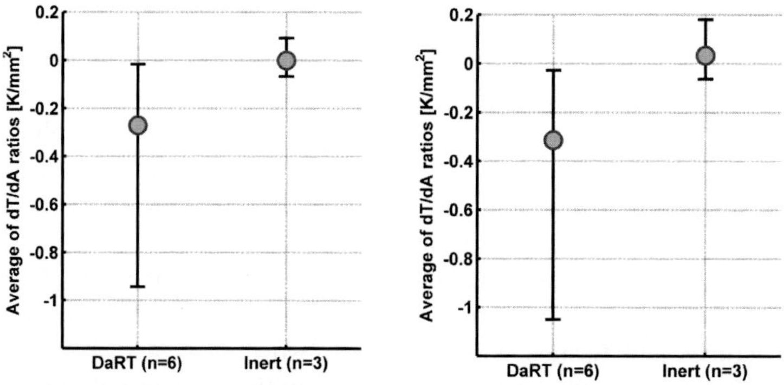

Figure 15.9 Ratio of change in average (left) and maximal (right) tumor temperature difference to change in tumor area, averaged on all mice during the second experiment.

Overall, all mice depicted similar behavior for this correlation. More specifically, treated mice were expected to show higher slopes, as treatment should induce more necrosis and therefore larger thermal differences. This method can provide a useful tool for tumor development assessment and demonstrate how tumor identification could benefit from improved integration of both visual and thermal images.

15.5 Photothermal Techniques for Estimation of Superficial and Deep Tissue Functionality

15.5.1 Introduction

Unlike the previous study which was based on heat generated by the tissue, photothermal spectroscopy allows to utilize the intrinsic absorption contrast of tissues for assessing tissue functionality. This is the same contrast utilized by the photoacoustic method. The basic principle of the photothermal method is the measurement of temperature changes resulting from illumination on the sample and extracting from them the material content of the sample. In one of the first research works [85], the samples (potassium sulfate powders) were illuminated by a CO_2 laser in order to estimate their spectra. The illumination was selected to be pulsatile, causing increased and fast heating of the surface layers, followed by a gradual heating of the lower layers and the entire sample. The algorithm is based on the assumption that the temperature increase is proportional to the absorption coefficient, and therefore uses the pulsatile pattern to deduce the spectrum, thus reducing the effect of lower layers on the spectral measurement. Researchers tried to avoid reaching "photothermal saturation," in which the heat from lower layers reaches the surface of the specimen, causing the temperature increase to depend not only on the upper layer and lowering the sensitivity to depth distribution. The theory of the photothermal effects dependence on tissue structure was later developed to include structures and tissues with different absorbing layers. Assuming the tissue has bulk optical properties, with a single sub-surface layer with higher absorption, the temperature measurement can indicate the depth of the absorbing layer. A connection is described between the depth of the absorbing layer and the time form pulse emission to measurement of temperature peak [86].

15.5.2 Photothermal Monitoring of Port-Wine-Stain (PWS) Lesions

Depth profiling of absorbing layers in biological tissues was thoroughly investigated for port-wine-stain (PWS) therapy. The treatment includes illumination with laser on the PWS, causing thermal damage to the abnormal tissue followed by a wound-healing response, clearing the PWS. The physician in such treatments should be aware of the optical and structural properties of the tissue in order to avoid causing thermal damage to other absorbing layers in the skin. A photothermal algorithm was developed to assess the depth and width of the PWS layer (in which the main absorber is hemoglobin) [87]. Figure 15.10 shows a comparison of the typical response to a pulse of normal tissue and PWS tissue. The normal tissue has mainly one absorbing layer, the epidermis which is effectively the highest layer (if we ignore the stratum corneum) and therefore exhibits immediate temperature rise, followed by a slow decrease. The PWS skin, however, shows the epidermal response, followed by another temperature increase, caused by the light absorbed in the PWS.

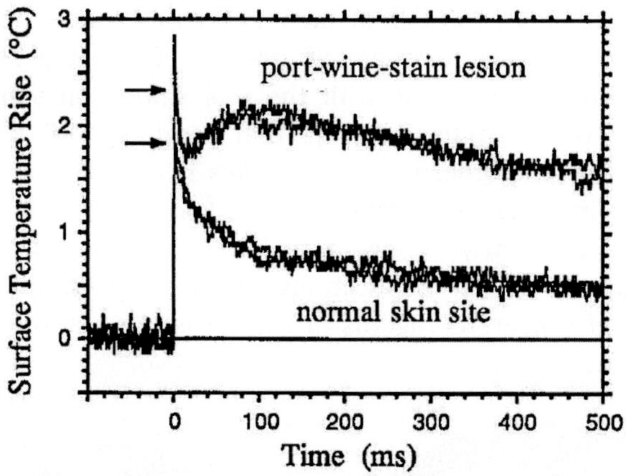

Figure 15.10 Comparison of photothermal response of normal and PWS tissue.

15.5.3 Photothermal Estimation of the Oxygenation Level of Deep-Tissue

Some studies have explored the possibilities of utilizing photothermal techniques to acquire information about the tissue deep below the surface. One such study was performed by Tepper et al. [88] in order to develop a minimal invasive thermal imaging method to determine the oxygenation level of an internal tissue. Detection of oxygen saturation in general, and minimally invasive detection specifically, is an important task. Measuring oxygen saturation is significant in many procedures, including tumor detection, cancer treatment adjustment, and ischemia monitoring during medical procedures.

In this method, the tissue was illuminated using an optical fiber by several wavelengths in the visible and near-IR range. Each wavelength is absorbed by the tissue and thus causes increase in its temperature. The temperature increase was observed by a coherent waveguide bundle in the mid-IR range (Fig. 15.11). The thermal imaging of the tissue is done using a thermal camera through the coherent bundle. Analyzing the temperature rise allows estimating the tissue composition in general, and specifically the oxygenation level. A curve-fitting algorithm was used to find the most suitable saturation value affecting the temperature function. The algorithm was tested on a theoretical tissue model with various parameters, implemented for this study. Later it was tested on agar phantom models and the calculated saturation values were in agreement with the real saturation values [89]. Results of those studies indicate that combining IR waveguide bundles with photothermography makes it a good candidate for saturation imaging in internal cavities. For example, it could be used to detect early stages of colon cancer during a simple colonoscopy. External applications are also possible, such as detection of cancerous regions in the breast, or any other tumor close to the surface.

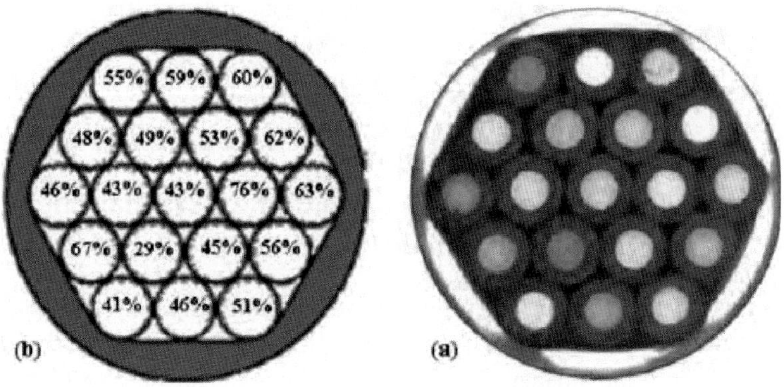

Figure 15.11 The IR camera image through bundle (a) and the estimate compounds ratio for that bundle (b).

References

1. A. G. Bell, LXVIII. Upon the production of sound by radiant energy, *Philos. Mag. Ser. 5*, vol. 11, no. 71, pp. 510–528, Jan. 1881.
2. A. Rosencwaig, Photoacoustic spectroscopy of solids, *Opt. Commun.*, vol. 7, no. 4, pp. 305–308, Apr. 1973.
3. Y.-H. Pao, *Optoacoustic Spectroscopy and Detection,* Academic Press, New-York, San-Francisco, London, 2012.

4. G. B. Christison and H. A. MacKenzie, Laser photoacoustic determination of physiological glucose concentrations in human whole blood, *Med. Biol. Eng. Comput.*, vol. 31, no. 3, pp. 284–290, May 1993.

5. R. A. Kruger, Photoacoustic ultrasound, *Med. Phys.*, vol. 21, no. 1, p. 127, 1999.

6. D. Ta, W. Wang, Y. Wang, L. H. Le, and Y. Zhou, Measurement of the dispersion and attenuation of cylindrical ultrasonic guided waves in long bone, *Ultrasound Med. Biol.*, vol. 35, no. 4, pp. 641–652, Apr. 2009.

7. P. Beard, Biomedical photoacoustic imaging, *Interface Focus*, vol. 1, no. 4, pp. 602–631, Aug. 2011.

8. M. Xu and L. V. Wang, Photoacoustic imaging in biomedicine, *Rev. Sci. Instrum.*, vol. 77, no 4, p. 041101–041122, 2006.

9. L. V. Wang and S. Hu, Photoacoustic tomography: In vivo imaging from organelles to organs, *Science*, vol. 335, no. 6075, pp. 1458–1462, 2012.

10. B. B. Goldberg, J.-B. Liu, and F. Forsberg, Ultrasound contrast agents: A review, *Ultrasound Med. Biol.*, vol. 20, no. 4, pp. 319–333, Jan. 1994.

11. J. Levi, S. R. Kothapalli, T.-J. Ma, K. Hartman, B. T. Khuri-Yakub, and S. S. Gambhir, Design, synthesis, and imaging of an activatable photoacoustic probe, *J. Am. Chem. Soc.*, vol. 132, no. 32, pp. 11264–11269, Aug. 2010.

12. J. Levi, S.-R. Kothapalli, S. Bohndiek, J.-K. Yoon, A. Dragulescu-Andrasi, C. Nielsen, A. Tisma, S. Bodapati, G. Gowrishankar, X. Yan, C. Chan, D. Starcevic, and S. S. Gambhir, Molecular photoacoustic imaging of follicular thyroid carcinoma, *Clin. Cancer Res.*, vol. 19, no. 6, pp. 1494–502, Mar. 2013.

13. B. Cox, J. G. Laufer, S. R. Arridge, and P. C. Beard, Quantitative spectroscopic photoacoustic imaging: A review Quantitative spectroscopic photoacoustic imaging, *J. Biomed. Opt.*, vol. 17, no. 6, p. 061202, 2012.

14. J. Yao, K. I. Maslov, Y. Zhang, Y. Xia, and L. V. Wang, Label-free oxygen-metabolic photoacoustic microscopy in vivo., *J. Biomed. Opt.*, vol. 16, no. 7, p. 076003, 2011.

15. J. Yao, K. I. Maslov, Y. Shi, L. A. Taber, and L. V. Wang, In vivo photoacoustic imaging of transverse blood flow by using Doppler broadening of bandwidth, *Opt. Lett.*, vol. 35, no. 9, p. 1419, 2010.

16. J. Brunker and P. Beard, Pulsed photoacoustic Doppler flowmetry using time-domain cross-correlation: Accuracy, resolution and scalability, *J. Acoust. Soc. Am.*, vol. 132, no. 3, p. 1780, 2012.

17. H. Fang, K. Maslov, and L. V. Wang, Photoacoustic Doppler effect from flowing small light-absorbing particles, *Phys. Rev. Lett.*, vol. 99, no. 18, pp. 184501-1–184501-4, 2007.

18. A. Sheinfeld, S. Gilead, and A. Eyal, Photoacoustic Doppler measurement of flow using tone burst excitation, *Opt. Express*, vol. 18, no. 5, pp. 4212–4221, Mar. 2010.

19. A. Sheinfeld, S. Gilead, and A. Eyal, Simultaneous spatial and spectral mapping of flow using photoacoustic doppler measurement, *J. Biomed. Opt.*, vol. 15, no. 6, p. 066010, 2010.

20. A. Sheinfeld and A. Eyal, Photoacoustic thermal diffusion flowmetry, *Biomed. Opt. Express*, vol. 3, no. 4, p. 800, 2012.

21. P. Oberg, Laser-Doppler flowmetry., *Crit. Rev. Biomed. Eng.*, vol. 18, no. 2, pp. 125–163, 1990.

22. Z. Chen, T. E. Milner, S. Srinivas, X. Wang, A. Malekafzali, M. J. van Gemert, and J. S. Nelson, Noninvasive imaging of in vivo blood flow velocity using optical Doppler tomography, *Opt. Lett.*, vol. 22, no. 14, pp. 1119–1121, 1997.

23. A. Sheinfeld, *Photoacoustic Flowmetry with Laser-Diodes Excitation*. University of Tel-Aviv, 2013.

24. A. Sheinfeld and A. Eyal, Photoacoustic thermal diffusion flowmetry in tissue-mimicking phantoms, in *SPIE BiOS*, 2013, p. 858110.

25. R. Lindsay, C. Christiansen, T. A. Einhorn, D. M. Hart, S. Ljunghall, C. A. Mautalen, P. J. Meunier, H. Morii, G. R. Mundy, S. A. Rapado, and J. Stevenson, Who are candidates for prevention and treatment for osteoporosis?, *Osteoporos. Int.*, vol. 7, no. 1, pp. 1–6, Jan. 1997.

26. National Osteoporosis Foundation [Online]. Available: http://nof.org/. [accessed: 24-Mar-2015].

27. R. Lindsay, S. Pack, and Z. Li, Longitudinal progression of fracture prevalence through a population of postmenopausal women with osteoporosis, *Osteoporos. Int.*, vol. 16, pp. 306–312, 2005.

28. J. A. Kanis, Assessment of fracture risk and its application to screening for postmenopausal osteoporosis: Synopsis of a WHO report, *Osteoporos. Int.*, vol. 4, no. 6, pp. 368–381, Nov. 1994.

29. M. Talmant, J. Foiret, J. Minonzio, E. Bossy, Q. Grimal, P. Laugier, A. Hosokawa, Y. Nagatani, M. Matsukawa, G. Haiat, G. Ha, R. Barkmann, and C. Gl, Bone Quantitative Ultrasound, Springer, Heidelberg, New York, pp. 47–71, 2011.

30. P. Pisani, M. D. Renna, F. Conversano, E. Casciaro, M. Muratore, E. Quarta, M. Di Paola, and S. Casciaro, Screening and early diagnosis

of osteoporosis through X-ray and ultrasound based techniques, *World J. Radiol.*, vol. 5, no. 11, pp. 398–410, 2013.

31. M. Kawai, F. J. A. de Paula, and C. J. Rosen, New insights into osteoporosis: The bone-fat connection, *J. Int. Med.*, vol. 272, no. 4, pp. 317–329, Oct. 2012.
32. I. Steinberg, N. Turko, O. Levi, I. Gannot, and E. Avishay, Quantitative study of optical and mechanical bone status using multispectral photoacoustics, *J. Biophotonics*, vol. 9, no. 9, pp. 924-933, Sep. 2016.
33. E. Ohmae, Y. Ouchi, M. Oda, T. Suzuki, S. Nobesawa, T. Kanno, E. Yoshikawa, M. Futatsubashi, Y. Ueda, H. Okada, and Y. Yamashita, Cerebral hemodynamics evaluation by near-infrared time-resolved spectroscopy: Correlation with simultaneous positron emission tomography measurements, *Neuroimage*, vol. 29, no. 3, pp. 697–705, 2006.
34. R. C. Mesquita, S. S. Schenkel, D. L. Minkoff, X. Lu, C. G. Favilla, P. M. Vora, D. R. Busch, M. Chandra, J. H. Greenberg, J. A. Detre, and A. G. Yodh, Influence of probe pressure on the diffuse correlation spectroscopy blood flow signal: Extra-cerebral contributions, *Biomed. Opt. Express*, vol. 4, no. 7, pp. 978–994, 2013.
35. D. A. Boas and A. K. Dunn, Laser speckle contrast imaging in biomedical optics, *J. Biomed. Opt.*, vol. 15, no. 1, p. 011109, 2010.
36. G. D. Mahan, W. E. Engler, J. J. Tiemann, and E. Uzgiris, Ultrasonic tagging of light: Theory, *Proc. Natl. Acad. Sci. U. S. A.*, vol. 95, no. November, pp. 14015–14019, 1998.
37. L. V. Wang, Mechanisms of ultrasonic modulation of multiply scattered coherent light: An analytic model, *Phys. Rev. Lett.*, vol. 87, no. 4, p. 043903, 2001.
38. W. Leutz and G. Maret, Ultrasonic modulation of multiply scattered light, *Phys. B Condens. Matter*, vol. 204, no. 1–4, pp. 14–19, 1995.
39. H. Li and L. V. Wang, Autocorrelation of scattered laser light for ultrasound-modulated optical tomography in dense turbid media, *Appl. Opt.*, vol. 41, no. 22, pp. 4739–4742, 2002.
40. E. Granot, A. Lev, Z. Kotler, B. G. Sfez, and H. Taitelbaum, Detection of inhomogeneities with ultrasound tagging of light, *J. Opt. Soc. Am. Opt. Image Sci. Vis.*, vol. 18, no. 8, pp. 1962–1967, 2001.
41. A. Lev and B. G. Sfez, Direct, noninvasive detection of photon density in turbid media, *Opt. Lett.*, vol. 27, no. 7, pp. 473–475, 2002.
42. A. Lev and B. G. Sfez, Pulsed ultrasound-modulated light tomography, *Opt. Lett.*, vol. 28, no. 17, pp. 1549–1551, 2003.

43. A. Lev, E. Rubanov, B. Sfez, S. Shany, and A. J. Foldes, Ultrasound-modulated light tomography assessment of osteoporosis, *Opt. Lett.*, vol. 30, no. 13, pp. 1692–1694, 2005.
44. F. Ramaz, B. C. Forget, M. Atlan, A. C. Boccara, M. Gross, P. Delaye, and G. Roosen, Photorefractive detection of tagged photons in ultrasound modulated optical tomography of thick biological tissues, *Opt. Express*, vol. 12, no. 22, pp. 5469–5474, 2004.
45. M. Lesaffre, S. Farahi, M. Gross, P. Delaye, C. Boccara, and F. Ramaz, Acousto-optical coherence tomography using random phase jumps on ultrasound and light, *Opt. Express*, vol. 17, no. 20, pp. 18211–18218, 2009.
46. D. S. Elson, R. Li, C. Dunsby, R. Eckersley, and M.-X. Tang, Ultrasound-mediated optical tomography: A review of current methods, *Interface Focus*, vol. 1, no. 4, pp. 632–648, 2011.
47. L. V Wang, Ultrasound-mediated biophotonic imaging: A review of acousto-optical tomography and photo-acoustic tomography, *Dis. Markers*, vol. 19, no. 2–3, pp. 123–138, 2004.
48. A. Ron, N. Racheli, I. Breskin, Y. Metzger, Z. Silman, M. Kamar, A. Nini, R. Shechter, and M. Balberg, Measuring tissue blood flow using ultrasound modulated diffused light, *Spie*, vol. 8223, p. 82232J–82232J-7, 2012.
49. N. Racheli, A. Ron, Y. Metzger, I. Breskin, G. Enden, M. Balberg, and R. Shechter, Non-invasive blood flow measurements using ultrasound modulated diffused light, *SPIE BiOS*, vol. 8223, p. 82232A–82232A-8, 2012.
50. L. V. Wang, Mechanisms of ultrasonic modulation of multiply scattered coherent light: A Monte Carlo model, *Opt. Lett.*, vol. 26, no. 15, pp. 1191–1193, 2001.
51. S. Sakadžić and L. V. Wang, Ultrasonic modulation of multiply scattered coherent light: An analytical model for anisotropically scattering media, *Phys. Rev. E—Stat. Nonlinear Soft Matter Phys.*, vol. 66, no. 2, pp. 14–19, 2002.
52. S. Powell and T. S. Leung, Highly parallel Monte-Carlo simulations of the acousto-optic effect in heterogeneous turbid media, *J. Biomed. Opt.*, vol. 17, no. 4, p. 045002, 2012.
53. A. Tsalach, Z. Schiffer, E. Ratner, I. Breskin, R. Zeitak, R. Shechter, and M. Balberg, Depth selective acousto-optic flow measurement, *Biomed. Opt. Express*, vol. 6, no. 12, p. 4871, 2015.

54. G. Rosenthal, A. Furmanov, E. Itshayek, Y. Shoshan, and V. Singh, Assessment of a noninvasive cerebral oxygenation monitor in patients with severe traumatic brain injury, *J. Neurosurg.*, vol. 120, no. 4, pp. 901–907, 2014.

55. J. M. Murkin, M. Kamar, Z. Silman, M. Balberg, and S. J. Adams, Intraoperative cerebral autoregulation assessment using ultrasound-tagged near-infrared-based cerebral blood flow in comparison to transcranial Doppler cerebral flow velocity: A pilot study, *J. Cardiothorac. Vasc. Anesth.*, vol. 29, no. 5, pp. 1187–1193, 2015.

56. H. W. Schytz, S. Guo, L. T. Jensen, M. Kamar, A. Nini, D. R. Gress, and M. Ashina, A new technology for detecting cerebral blood flow: A comparative study of ultrasound tagged NIRS and 133Xe-SPECT, *Neurocrit. Care*, vol. 17, no. 1, pp. 139–145, 2012.

57. D. Hori, C. W. Hogue, A. Shah, C. Brown, K. J. Neufeld, J. V. Conte, J. Price, C. Sciortino, L. Max, A. Laflam, H. Adachi, D. E. Cameron, and K. Mandal, Cerebral autoregulation monitoring with ultrasound-tagged near-infrared spectroscopy in cardiac surgery patients, *Anesth. Analg.*, vol. 121, no. 5, pp. 1187–1193, 2015.

58. A. Ron, N. Racheli, I. Breskin, and R. Shechter, A tissue mimicking phantom model for applications combining light and ultrasound, in *Proc. SPIE 8583, Design and Performance Validation of Phantoms Used in Conjunction with Optical Measurement of Tissue V*, 2013, p. 858307.

59. H. W. Schytz, T. Wienecke, L. T. Jensen, J. Selb, D. A. Boas, and M. Ashina, Changes in cerebral blood flow after acetazolamide: An experimental study comparing near-infrared spectroscopy and SPECT, *Eur. J. Neurol.*, vol. 16, no. 4, pp. 461–467, 2009.

60. M. Diakides, J. D. Bronzino, and D. R. Peterson, *Medical Infrared Imaging—Principles and Practices*. Boca Raton, FL.: CRC Press, 2013.

61. S. Bagavathiappan, T. Saravanan, J. Philip, T. Jayakumar, B. Raj, R. Karunanithi, T. M. R. Panicker, M. P. Korath, and K. Jagadeesan, Infrared thermal imaging for detection of peripheral vascular disorders, *J. Med. Phys.*, vol. 34, no. 1, pp. 43–7, Jan. 2009.

62. M. Banić, D. Kolarić, N. Borojević, Željko Ferenčić, S. Pleško, L. Petričušić, T. Božin, and S. Antonini, Thermography in patients with inflammatory bowel disease and colorectal cancer: Evidence and review of the method, *Period. Biol.*, vol. 113, no. 4, pp. 439–444, Nov. 2011.

63. T. J. Love, Thermography as an indicator of blood perfusion, *Ann. N. Y. Acad. Sci.*, vol. 335, pp. 429–37, Jan. 1980.

64. P. Bacon, A. Collins, F. Ring, and J. Cosh, Thermography in the assessment of inflammatory arthritis, *Clin. Rheum. Dis.*, vol. 2, pp. 51–65, Jan. 1976.

65. W. C. Amalu, W. B. Hobbins, J. F. Head, and R. L. Elliot, Infrared imaging of the breast—An overview, in *The Biomedical Engineering Handbook Medical Devices and Systems*, Boca Raton, FL: CRC press, 2006.

66. M. Gautherie and C. M. Gros, Breast thermography and cancer risk prediction, *Cancer*, vol. 45, no. 1, pp. 51–56, Jan. 1980.

67. M. Gautherie, Thermobiological assessment of benign and malignant breast diseases, *Am. J. Obstet. Gynecol.*, vol. 147, no. 8, pp. 861–869, Dec. 1983.

68. M. Gautherie, A. Kotewicz, and P. Gueblez, Accurate and objective evaluation of breast thermograms: Basic principles and new advances with special reference to an improved computer-assisted scoring system, in *Thermal Assessment of Breast Health*, 1983, pp. 72–97.

69. M. Mital and E. P. Scott, Thermal detection of embedded tumors using infrared imaging, *J. Biomech. Eng.*, vol. 129, no. 1, p. 33, Feb. 2007.

70. A. L. Shada, L. T. Dengel, G. R. Petroni, M. E. Smolkin, S. Acton, and C. L. Slingluff, Infrared thermography of cutaneous melanoma metastases, *J. Surg. Res.*, vol. 182, no. 1, pp. e9–e14, Jun. 2013.

71. J. P. Agnelli, A. A. Barrea, and C. V. Turner, Tumor location and parameter estimation by thermography, *Math. Comput. Model.*, vol. 53, no. 7–8, pp. 1527–1534, Apr. 2011.

72. S. C. Fok, E. Y. K. Ng, and K. Tai, Early detection and visualization of breast tumor with thermogram and neural network, *J. Mech. Med. Biol.*, vol. 02, no. 02, pp. 185–195, Jun. 2002.

73. P. Vaupel, F. Kallinowski, and P. Okunieff, Blood flow, oxygen and nutrient supply, and metabolic microenvironment of human tumors: A review, *Cancer Res.*, vol. 49, no. 23, pp. 6449–6465, Dec. 1989.

74. F. Kallinowski, K. H. Schlenger, S. Runkel, M. Kloes, M. Stohrer, P. Okunieff, and P. Vaupel, Blood flow, metabolism, cellular microenvironment, and growth rate of human tumor xenografts, *Cancer Res.*, vol. 49, no. 14, pp. 3759–3764, Jul. 1989.

75. R. A. Macbeth and J. G. Bekesi, Oxygen consumption and anaerobic glycolysis of human malignant and normal tissue, *Cancer Res.*, vol. 22, pp. 244–248, Feb. 1962.

76. T. Yahara, T. Koga, S. Yoshida, S. Nakagawa, H. Deguchi, and K. Shirouzu, Relationship between microvessel density and thermographic hot areas in breast cancer, *Surg. Today*, vol. 33, no. 4, pp. 243–248, Jan. 2003.

77. J. Manuel Luna, R. Romero-Mendez, A. Hernandez-Guerrero, and F. Elizalde-Blancas, Procedure to estimate thermophysical and geometrical parameters of embedded cancerous lesions using thermography, *J. Biomech. Eng.*, vol. 134, no. 3, p. 031008, Mar. 2012.

78. V. Umadevi, S. V. Raghavan, and S. Jaipurkar, Framework for estimating tumour parameters using thermal imaging, *Indian J. Med. Res.*, vol. 134, no. 5, pp. 725–731, Nov. 2011.

79. M. Tepper, A. Shoval, O. Hoffer, H. Confino, M. Schmidt, I. Kelson, Y. Keisari, and I. Gannot, Thermographic investigation of tumor size, and its correlation to tumor relative temperature, in mice with transplantable solid breast carcinoma, *J. Biomed. Opt.*, vol. 18, no. 11, p. 111410, Nov. 2013.

80. L. Arazi, T. Cooks, M. Schmidt, Y. Keisari, and I. Kelson, Treatment of solid tumors by interstitial release of recoiling short-lived alpha emitters, *Phys. Med. Biol.*, vol. 52, no. 16, pp. 5025–5042, Aug. 2007.

81. T. Cooks, L. Arazi, M. Schmidt, G. Marshak, I. Kelson, and Y. Keisari, Growth retardation and destruction of experimental squamous cell carcinoma by interstitial radioactive wires releasing diffusing alpha-emitting atoms, *Int. J. Cancer*, vol. 122, no. 7, pp. 1657–1664, Dec. 2007.

82. E. Lazarov, L. Arazi, M. Efrati, T. Cooks, M. Schmidt, Y. Keisari, and I. Kelson, Comparative in vitro microdosimetric study of murine- and human-derived cancer cells exposed to alpha particles, *Radiat. Res.*, vol. 177, no. 3, pp. 280–287, Mar. 2012.

83. C. Song, V. Appleyard, K. Murray, T. Frank, W. Sibbett, A. Cuschieri, and A. Thompson, Thermographic assessment of tumor growth in mouse xenografts, *Int. J. Cancer*, vol. 121, no. 5, pp. 1055–1058, Sep. 2007.

84. W. Xie, P. McCahon, K. Jakobsen, and C. Parish, Evaluation of the ability of digital infrared imaging to detect vascular changes in experimental animal tumours, *Int. J. Cancer*, vol. 108, no. 5, pp. 790–794, Feb. 2004.

85. P.-E. Nordal and S. O. Kanstad, Photothermal radiometry, *Phys. Scr.*, vol. 20, no. 5–6, pp. 659–662, Nov. 1979.

86. F. H. Long, R. R. Anderson, and T. F. Deutsch, Pulsed photothermal radiometry for depth profiling of layered media, *Appl. Phys. Lett.*, vol. 51, no. 25, p. 2076, Dec. 1987.
87. S. L. Jacques, J. S. Nelson, W. H. Wright, and T. E. Milner, Pulsed photothermal radiometry of port-wine-stain lesions, *Appl. Opt.*, vol. 32, no. 13, pp. 2439–2446, May 1993.
88. M. Tepper, R. Neeman, Y. Milstein, M. Ben David, and I. Gannot, Thermal imaging method for estimating oxygen saturation, *J. Biomed. Opt.*, vol. 14, no. 5, p. 054048, 2010.
89. Y. Milstein, M. Tepper, M. Ben David, J. A. Harrington, and I. Gannot, Photothermal bundle measurement of phantoms and blood as a proof of concept for oxygenation saturation measurement, *J. Biophotonics*, vol. 4, no. 4, pp. 219–223, 2011.

Chapter 16

Using the Transmission Matrix to Image Disordered Media

Matthieu Davy,[a] **Sylvain Gigan,**[b] **and Azriel Z. Genack**[c]

[a]*Institut d'Electronique et de Télécommunications de Rennes, University of Rennes 1, Rennes 35042, France*
[b]*Laboratoire Kastler Brossel, ENS-PSL Research University, CNRS, UPMC-Sorbonne Universites, College de France, 24 rue Lhomond, F-75005 Paris, France*
[c]*Department of Physics, Queens College and Graduate Center of the City University of New York, Flushing, New York 11367, USA*

matthieu.davy@univ-rennes1.fr, sylvain.gigan@lkb.ens.fr, genack@qc.edu

Waves that are scattered many times within a static disordered medium form a random field pattern in transmission which is uncorrelated with the incident waveform. Nevertheless, knowledge of the matrix of field transmission coefficients could allow a degree of control over the transmitted field if the incident wavefront could be shaped. This holds exciting prospects for imaging in disordered media. In this chapter, we review approaches based on the measurement of the transmission matrix (TM). First, we discuss the properties of the TM in terms of the transmission eigenchannels that make it possible not only to control the net transmission through random media but also to selectively deliver radiation deep into the medium. In addition, intensity can be

Deep Imaging in Tissue and Biomedical Materials: Using Linear and Nonlinear Optical Methods
Edited by Lingyan Shi and Robert R. Alfano

ISBN 978-981-4745-88-8 (Hardcover), 978-1-315-20655-4 (eBook)
www.panstanford.com

greatly enhanced at a focal spot by phase-conjugating elements of the TM for monochromatic radiation or time-reversing scattered waves utilizing the spectral components of the TM. This provides the basis for the application of the TM to imaging in biomedical optics. We discuss the potential of the photoacoustic TM for imaging deep inside the medium and explore the use of the TM for endoscopic imaging with multimode optical fiber.

16.1 Introduction

Although the output field pattern of an optical or acoustic wave transmitted through a disordered sample is uncorrelated with the incident wave, the transmitted field can be controlled by manipulating the incident wave. The nature and extent of this control is determined by the properties of the transmission matrix (TM) t which relates the outgoing and incident fields. The TM was first introduced to characterize the scaling and statistics of electronic conductance [1–10]. It has subsequently proved to be a powerful tool for describing the statistics of local and integrated transmission of classical waves [9–12]. An important difference in the application of the TM to electronic and classical waves is that electrons entering a sample derive from a thermal reservoir while coherent sources are available for classical waves.

Control of scattered waves was demonstrated in acoustic [13–15] and microwave [16, 17] experiments in which a pulse emerging from a random sample is time reversed and refocused on a source within a scattering medium. In recent years, there has been a surge of interest in the TM because of the possibility of controlling optical transmission by manipulating the incident waveform with use of a spatial light modulator (SLM) [18–29]. The possibilities for biomedical applications abound.

Transmitted waves can be focused to within a wavelength even when the random medium is illuminated by an optical system with a low numerical aperture. Light may be focused at any point in the transmitted wave by bringing the field of all the incident channels into phase at the selected focal point [30]. The focused intensity can be enhanced over the average intensity for plane-wave illumination by a factor of the number of orthogonal incident channels which can be controlled. Linear combinations of the incident channels with associated output channels form orthogonal transmission eigenchannels with a wide range of transmission values τ. These are the transmission eigenvalues of the matrix product $tt^{\dagger}$. Transmission eigenvalues in samples larger than the transport mean free path, $L > \ell$, in which the average transmission is ℓ/L, range from unity, giving complete transmission, to exponentially small in the ratio l/ℓ, so that in principle the total transmission can be controlled. The dwell time of a wave in a channel increases with τ [31]. In addition, the intensity profile for a wave inside a medium varies systematically with τ [32–35]; the position of the peak intensity moves towards the center of the sample as τ increases and the peak achievable energy density, reached at the center of the sample for the completely transmitting eigenchannel, increases linearly with sample thickness [35].

In the next section, we discuss the properties of a complete TM and the impact of incompleteness in the measurement of the TM upon the control of transmission. In Section 16.3, we describe the variation with τ of the average profiles of energy density of transmission eigenchannels within the sample. We discuss wave-front shaping via phase conjugation and spatiotemporal focusing in Section 16.4. Applications to biomedical imaging, including the use of the photoacoustic TM and endoscopic imaging in multimodal optical fiber are discussed in Section 16.5.

16.2 Distribution of Transmission Eigenvalues

16.2.1 Predictions

The TM was first introduced in the early 1980s to explain electronic conductance in samples in which the wavefunction is temporally coherent. Fisher and Lee generalized the two-port one-dimensional (1D) Landauer relation equating the conductance in disordered samples in units of the quantum of conductance, (e^2/h), and the transmission to multiple channel systems, $g = G/(e^2/h) = T$, where T is the conductance [36]. The transmittance can be expressed as the sum of the N^2 flux transmission coefficients $T_{ab} = |t_{ab}|^2$ between all incident and outgoing channels, a and b, respectively, where the t_{ba} are the field transmission coefficients, which are the elements of the TM. The transmittance is also given by the sum of N transmission eigenvalues, $g = T = \sum_{a,b=1}^{N} T_{ba} = \sum_{n=1}^{N} \tau_n$. The channels are the propagating transverse modes in the ideal leads attached to the sample. The total number of channels is approximately $2\pi A/\lambda^2$, including two spin states for electrons and two orthogonal polarizations for electromagnetic waves. Here A is the cross-sectional area of the sample and λ is the wavelength in the space surrounding the sample. The transmission eigenvalues can be obtained via the singular value decomposition of the TM, $t = U\Lambda V^{\dagger}$, where U and V are unitary matrices with columns which are the singular vectors of the TM and Λ is the diagonal matrix whose elements are the singular values $\sqrt{\tau_n}$. In the absence of absorption, eigenchannels of the reflection matrix are the same as the transmission eigenchannels and have eigenvalues $(1 - \tau_n)$.

Dorokhov showed that the variation of conductance of an N-channel wire with sample length can be described in terms of the scaling of the individual channels [1]. Instead of a single localization length, there are N distinct auxiliary localization lengths, ξ_n, which characterize the scaling of the averages of each of the transmission eigenvalues via the relation [2–8],

$$\tau_n = 1/\cosh^2(L/\xi_n) \tag{16.1}$$

The inverse of these localization lengths given by $1/\xi_n = (2n - 1)/2\xi$ for $N < N/2$ so that they are equally spaced, $1/\xi_{n+1} - 1/\xi_n = 1/\xi$, with spacing equal to the inverse of the localization length. Dorokhov showed that electronic conduction through disordered conductors is dominated by $\xi/L = g$ eigenvalues with auxiliary localization lengths longer than the sample length, $\xi_n > L$ and transmission eigenvalues $\tau_n \geq 1/e$. Such channels are known as "open" channels, in contrast to "closed" channels with low transmission [2, 3]. The uniform spacing of L/ξ_n leads to a bimodal distribution of the transmission eigenchannels τ, with peaks as τ approaches unity and at exponentially small values of τ,

$$\rho(\tau) = \frac{\xi}{2L}\frac{1}{\tau\sqrt{1-\tau}} \tag{16.2}$$

The bimodal distribution is shown by the copper curve in Fig. 16.1.

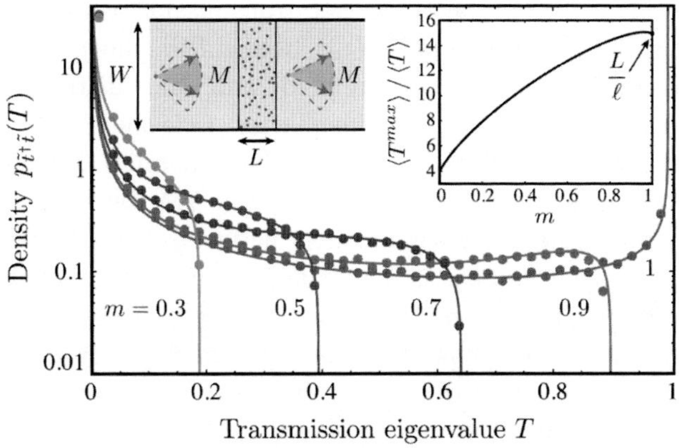

Figure 16.1 Transmission eigenvalue density of a disordered slab placed in a waveguide with $N = 485$ channels for different fractions of controlled channels $m = M/N$. Numerical results (dots) are compared to theoretical prediction (solid curves). Inset shows the maximal transmission enhancement possible for a given m. Reproduced from [37].

In practice, it is very challenging to make a perfect measurement of the TM. The completeness of optical measurements is limited by the numerical aperture of the illumination and collection optics, which restricts the number of channels that can be controlled. In addition, the angular or translational resolution of the sources and detectors is limited in optical, acoustic and microwave experiments. The impact of incomplete control of the TM was studied by Goetschy and Stone using free probability theory [37]. The probability distribution of a filtered matrix with missing columns and rows that reduces the fraction of channels on both the input and output sides of the sample to a fraction m is shown in Fig. 16.1. The theoretical results shown by the curves are in accord with simulations given by the dots in the figure. The same distributions are found in the slab geometry as in the quasi-1D geometry for all values of m. The equivalence of the eigenvalue distribution in a slab to the distribution in the quasi-1D geometry was found previously be Nazarov for $m = 1$ [7]. As shown in Fig. 16.1, even a small drop in m from 1 to 0.9 completely suppresses the high-transmission peak. However, the value of peak transmission at the upper cutoff of the distribution of τ for moderate values of m is still much larger than the average transmission T/N so that the net transmission can be greatly enhanced. As the value of m falls below T/N, the transmission eigenchannels become uncorrelated and the distribution of singular values $\sqrt{\tau}$ approaches the quarter circle law [38].

The transmission and reflection eigenchannels are identical in the absence of absorption for the full TM. Once absorption is introduced, the correspondence between transmission and reflection eigenchannels is lost since absorption destroys flux conservation. The maximum transmission eigenvalue is then no longer unity and the peak of $\rho(\tau)$ at high values of τ is suppressed [39–41]. Perfect control of the incident wavefront, however, may open the door to complete absorption of the incident energy [42].

For a sample with the same internal reflection at the input and output surfaces, perfect transmission persists for any level of internal reflection [43]. However, a transition occurs at a critical value of the difference between the reflectivities at the two sample boundaries. At the critical value the singularity in $\rho(\tau)$ at $\tau = 1$ changes from $(1 - \tau)^{-1/2}$, as given in Eq. (16.2), to $(1 - \tau)^{-1/3}$ [44]. Above the critical value of the difference in reflectivity, τ is strictly less than unity, $\tau < 1$ and its maximal value decreases as the difference in reflectivities increases. The critical difference in the internal reflectivity corresponds to the point at which the difference in the extrapolation lengths on the output side of the sample equals the sample length, $|z_b - z_{b'}| = L$ [45]. The extrapolation length is the length beyond the boundary of a diffusive sample at which the average intensity within the sample due to a remote source extrapolates to zero.

16.2.2 Measurements

The enhancement of total transmission with wavefront shaping was first demonstrated in the context of focusing by Vellekoop and Mosk [19]. The phase of two polarizations of a complex incident wave was shaped with an SLM while the transmitted intensity pattern was measured with a CCD camera. When the intensity at an arbitrarily selected point in transmission through a slab of ZnO particles is optimized, the total transmission relative to transmission for plane wave illumination also increased. This reflects the preferential coupling to high transmission eigenchannels in the phase-conjugation process of optimal focusing. Measurements of the total transmission and of the focused intensity over a wide range of the degree of control of the incident wavefront are in accord with random matrix theory. Optimal focusing will be discussed in Section 16.4.

The TM was subsequently measured using an SLM to control the incident field while the transmitted field was detected on a CCD via interference with a coherent wave derived from the incident laser [21]. The singular values of the TM, $\tau_n^{1/2}$, are calculated and normalized by the total intensity. When alternate elements are removed to suppress inter-element correlation, the distribution is a quarter circle as shown in Fig. 16.2a. The quarter circle law is predicted for random matrices without correlation [38]. This distribution indicates that there is no correlation between the elements as would be expected here since only a small number of elements of the TM are measured.

The TM was measured for microwave radiation transmitted through a copper tube filled with randomly positioned dielectric spheres embedded in a low-index shell in samples of different length and in different frequency ranges so that transmission of both diffusive and localized waves could be studied [46]. The field is measured by a source and detector that are moved on a grid on the input and output surfaces of the sample. This enables the measurement of the TM, so that its properties can be characterized. When the wave is localized, calculations of the transmission eigenchannels from the measured matrix give $\tau_1 >> \tau_2$, so that the transmitted field is dominated by the first transmission eigenchannel and the transmitted field does not change with position of the source. Since the transmission is essentially equal to τ_1, the transmission in the highest transmission eigenchannels is enhanced over the average transmission by a factor nearly equal to the number of channels of ~30. The enhancement for diffusive waves is less than an order of magnitude. In this sample, the degree of control is approximately 0.7 so that the bimodal distribution is not observed. Nonetheless, strong correlation of the transmission eigenvalues is observed with values of τ_n falling exponentially with n, as shown in Fig. 16.2b. The constant spacing between the averages of $\ln \tau_n$ in various random ensembles is proportional to the inverse of the bare conductance $\xi/(L + 2z_b)$.

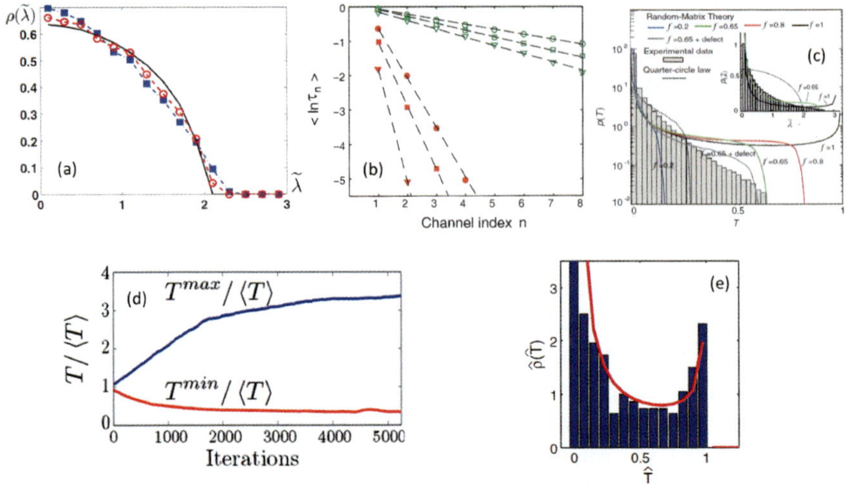

Figure 16.2 (a) Singular value distribution of the TM obtained in optics when only a small part of the TM is measured [21]. (b) Variation of <ln τ_n> with channel index for diffusive (green) and localized (red) samples measured in microwave experiments in random waveguides [46]. (c) Transmission eigenvalue distribution of the measured TM [26]. The lines represent simulations with full information on a fraction f = 0.65 of the incident and transmitted modes and an additional fraction of 0.24 to account for the transverse spread of the light. The dashed line represents the quarter-circle law distribution. (Inset) Normalized singular values distribution of the measured TM. (d) Measured total transmission of light through random medium versus the optimization step for transmission controlled for maximum (blue curve) and minimum (red curve) transmission [27]. (e) Bimodal distribution of eigenvalues of the TM obtained from measurements of elastic waves propagating in a disordered waveguide corrected for instrumental noise [47]. Adapted from [21], [46], [26], [27], and [47], respectively.

In optical studies, Yu et al. used a rotating galvanometer mirror to sequentially illuminate an 18 × 18 μm^2 area of a random slab with two polarizations with an angular resolution great enough to determine the channels that fell within the numerical aperture (NA) of their optical system [26]. The transmitted signal interferes with a portion of the incident field on a CCD camera. The completeness of the measured TM was limited only by the NA of the illumination and by the expansion of the transverse dimensions of the light on the output surface. The measured distribution of singular values of the TM is decidedly different from the quarter circle law with larger values of $\sqrt{\tau_n}$ and the distribution of transmission eigenvalues in agreement with simulations in which a fraction 0.65 of the channels are missing and the spread of the light at the output results in an additional 24% loss of collected channels, as shown in Fig. 16.2c. These results are consistent with calculations in [37] of filtered random matrices.

Coherent control over transmission was demonstrated by Popoff et al. in measurements of the maximum total transmission, $T^{\max}$ due to a wavefront shaped by two SLMs that modified the phase of two polarizations of light illuminating a random slab [27]. The transmitted and reflected energies were fully collected with photodiodes. Consequently, the completeness of the TM in these measurements was only reduced by the fraction of controlled channels m_1 of the incident illumination. The enhancement of the average of $T^{\max}$ relative to the average total transmission, $<T^{\max}>/<T>$, as a function of the fraction of controlled input channels, m_1, and as a function of the thickness of the random slab are in excellent agreement with analytic theory for a filtered random matrix [27, 37]. The degree to which transmission is enhanced depends only on the average transmission, which can be measured directly, and m_1, which is equal to the ratio of the variance of the transmission eigenvalues $\tilde{\tau}$ of the measured TM and the same ratio for the full TM, $m_1 = \mathrm{var}(\tilde{\tau})/\mathrm{var}(\tau)$ [27, 37]. The maximum and minimum values of transmission after increasing numbers of iterations towards the optimized values are shown in Fig. 16.2d. The corresponding results for the maximum and minimum values of reflection agree closely with the reflection R expressed as $(1 - T)$.

Evidence of the bimodal distribution has been found in measurements by Gérardin et al. [46] of the scattering matrix of elastic waves in a segment of an aluminum alloy plate in which small holes are drilled along the length of a longer aluminum plate. Elastic waves are generated at the longitudinal boundaries of the random region via the thermoelastic effect by focusing a high-energy 20 ns laser pulse at a point. The resulting wave is detected along the plate using a heterodyne interferometer. Measurement noise makes it difficult to accurately determine the TM. However, the noise can be suppressed by relying on the unitarity of the S-matrix, which requires that its eigenvalues lie on the unit circle. A modified S-matrix is found by adjusting the amplitude of the elements of the S-matrix but not the phase to place all eigenvalues on the unit circle. The TM found from this modified S-matrix exhibits the bimodal distribution as shown in Fig. 16.2e. Simulations of propagation in this sample were carried out with and without added noise. Modifying the S-matrix, in the way described above for the sample with added noise gave a distribution of transmission eigenvalues close to that for the matrix without added noise.

The challenges of making a perfect measurement of the TM and accounting for incompleteness in the measurement have stimulated the search for enhanced control of transmission and reflection and improved dynamic range in focusing. The spatial distribution within the sample and their dynamics [28, 48, 49] as well as new opportunities for imaging are also under intense investigation.

16.3 Eigenchannel Intensity Profiles

The TM does not provide any information regarding the spatial distribution of energy in different transmission eigenchannels. Only the average of the spatial distribution of the energy is known. For diffusive samples, this is a linear decrease of the energy density governed by Fick's law for particle diffusion.

The variation of the distribution of the energy density for transmission eigenchannels with a specific transmission eigenvalue has recently been explored in homogeneously disordered samples. Simulations [32–34, 50, 51] in single realizations of random samples show that a peak of the energy density profile increases and moves towards the center of the sample as increases. This trend has been confirmed in measurements with elastic waves propagating in a disordered waveguide (see Fig. 16.3) [47].

An expression for the ensemble average of the spatial distribution of the energy density for eigenchannels with a specified transmission eigenvalue τ is found in [35]. Numerical simulations based on the recursive Green's function method in 2D [52, 53] make it possible to find not only the TM but also the matrix of field values at depth x into the sample due to the incident field, $e(x)$. The product of $e(x)$ and the singular vector v_n of the TM gives the field due to the nth incoming eigenchannel of transmission τ_n at depth x within the surface. The square of the field integrated across the cross section at this depth for transmission eigenchannels with transmission τ gives the longitudinal energy density profile.

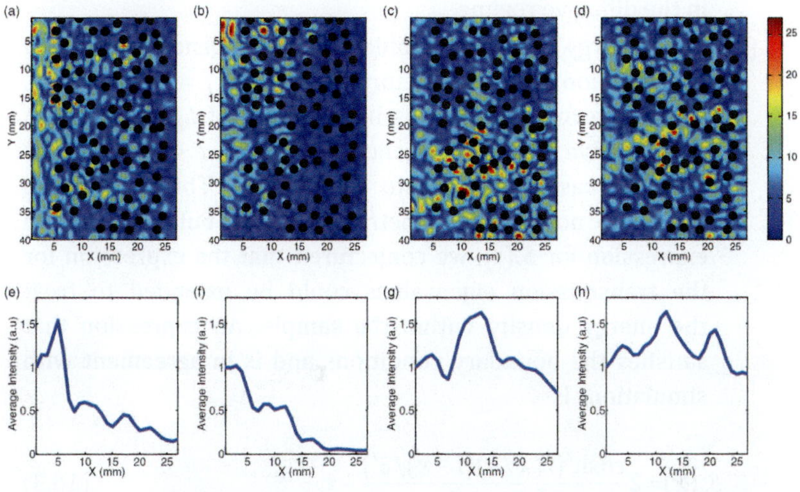

Figure 16.3 Absolute value of the field in the scattering medium associated with (a) an incident plane wave, (b) a closed eigenchannel, (c,d) an open eigenchannel. The corresponding intensities averaged over the waveguide cross section are shown with depth x in panels (e–h). Reproduced from [47].

The profiles of the normalized longitudinal energy density profile averaged over 500 random configurations, $W_\tau(x)$, for eigenchannels of index matched samples with transmission $\tau = 1$, $\tau = 0.5$, $\tau = 0.1$ and $\tau = 0.001$ are presented in Fig. 16.4 for two diffusive samples with $L/\xi = 0.05$. The width and length of the sample in (*d*) is half the width and length of the sample in (*a*, *b*). The profiles are normalized so that $W_\tau(x = L) = \tau$. With this normalization, $W_\tau(x = 0) = 2 - \tau$. Fluctuations $\delta\varepsilon(x, y)$ in the position-dependent dielectric constant, $\varepsilon(x, y) = 1 + \delta\varepsilon(x, y)$, are drawn from a rectangular distribution. In Fig. 16.1c, the sum of profiles for all eigenchannels is shown to decrease linearly with depth into the sample, as expected from the diffusion equation. Moreover, we find the following:

(1) The energy density of the transmitting eigenchannel with $\tau = 1$ is the sum of a spatially uniform background and the probability density that scattered waves will arrive at and return to the cross section at given depth within the sample. This return probability can be found from the generalized diffusion equation and gives $W_{\tau=1}(x) = 1 + \pi x\,(L - x)/2L\ell$ in the diffusive regime.

(2) The energy density profile of an eigenchannel with transmission τ can be factorized as, $W_\tau(x) = S_\tau(x)W_{\tau=1}(x)$, where the function $S_\tau(x)$ is independent of L/ξ and L/ℓ in the diffusive limit. For eigenchannels with $\tau < 1$, $S_\tau(x)$ falls with increasing depth into the sample. This is because waves do not readily penetrate into the bulk. To find an expression for $S_\tau(x)$, we conjectured that the expression for the transmission eigenvalues could be extended to treat the energy density within the sample. An expression that satisfies the boundary conditions and is in agreement with simulations is

$$S_\tau(x) = 2\frac{\cosh^2(h(x/L)(L-x)/\xi')}{\cosh^2(h(x/L)L/\xi')} - \tau, \qquad (16.3)$$

where $h(x/L)$ is an empirical function. It is obtained for a single value of τ in one diffusive sample and is found to give good agreement for all τ in all diffusive samples, $L\xi < 1$. As seen in Fig. 16.4, the expressions for $W_{\tau=1}(x)$ and $S_\tau(x)$ are in excellent agreement with the profiles found in simulations for diffusive waves.

These results extend the knowledge of the average of the disposition of energy from the sample boundary into the bulk. It is in general difficult to probe the energy density within the medium. Integrating $W_\tau(x)$ over the depth is proportional to the contribution of the eigenchannel to the density of states (DOS) of the sample, the eigenchannel density of states (EDOS)

[31]. The sum of the EDOS over the N eigenchannels gives the DOS. The EDOS can also be found from measurements of spectra of the TM. It is given by the derivative with frequency of the composite phase of the transmission eigenchannel [31, 54]. This composite phase derivative is equal to the dwell time of this eigenchannel. The relation between eigenchannel dwell time and the DOS is an extension to multichannel systems of the equality between the DOS and the transmission delay time in 1D systems [54].

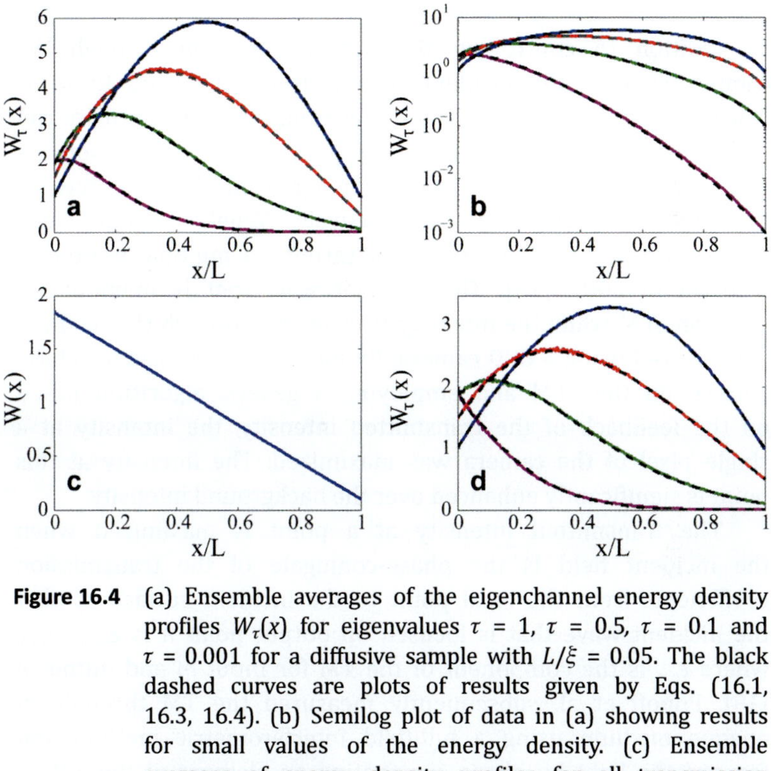

Figure 16.4 (a) Ensemble averages of the eigenchannel energy density profiles $W_\tau(x)$ for eigenvalues τ = 1, τ = 0.5, τ = 0.1 and τ = 0.001 for a diffusive sample with L/ξ = 0.05. The black dashed curves are plots of results given by Eqs. (16.1, 16.3, 16.4). (b) Semilog plot of data in (a) showing results for small values of the energy density. (c) Ensemble averages of energy density profiles for all transmission eigenchannels with the eigenchannel indices n from 1 to N, $W(x) = N^{-1}\sum_{n=1}^{N} W_{\tau_n}(x)$. The linear falloff of the average of the energy density over all eigenchannels is in accord with diffusion theory. (d) $W_\tau(x)$ for a sample with half the width and length as the sample in (a). Note that, though the value of L/ξ is the same as for (a), the form of $W_\tau(x)$ differs with lower values of the peak energy density in the shorter sample. Reproduced from [35].

Shaping the incident waveform therefore also allows for control of the distribution of excitation inside a random medium by selecting the transmission eigenchannels. This makes it possible to deliver radiation deep into multiply scattered media for imaging, depth profiling, or energy deposition.

16.4 Wavefront Shaping for Imaging through Turbid Media

In addition to controlling the net transmission through the sample, shaping the incident wavefront can be used to focus waves. Using the linearity of the scattering process, intensity can indeed be strongly enhanced at any point by phasing all sources of the wave so that the fields arriving from these sources at the selected point add constructively. Vellekoop and Mosk demonstrated the focusing of light through a multiple scattering medium in 2007 [18]. The incident wavefront is manipulated using an SLM while the intensity transmitted through the medium is measured using a CCD camera. By modifying the incident phase pattern on the SLM and employing a genetic algorithm based on the feedback of the transmitted intensity, the intensity at a single pixel of the camera was maximized. The intensity at this point is significantly enhanced over the background intensity.

The transmitted intensity at a point is maximized when the incident field is the phase-conjugate of the transmission coefficients from the focal point to the incident surface so that the incident wave that is focused on output point n is $E_m = t^*_{\mathrm{nm}}$, where t_{nm} is the component of the TM for input m and output n [19]. Popoff et al. subsequently measured the TM through an opaque medium using a full-field interferometric method and focused the light by phase conjugating an element of the TM as shown in Fig. 16.5 [21].

The contrast μ between the intensity at the focus and the averaged background intensity is related to the number of degrees of freedom that can be controlled [24]. For monochromatic radiation, the contrast is equal to the effective number of independent eigenchannels which contribute to the transmitted field, which is the participation number of the eigenvalues of the

TM, $M = (\sum_{n=1}^{N} \tau_n)^2 / \sum_{n=1}^{N} \tau_n^2$ [56, 57]. When the number of measured incoming channels N' is much smaller than the dimensionless conductance g, the components of the TM are statistically independent and $\mu = N'$. However, the contrast saturates for $N' > g$. When the measured TM is complete, the distribution of transmission eigenchannels is bimodal and $M \sim 3g/2$. This saturation of the contrast for optimal focusing reflects the stronger coupling of waves to high transmission eigenchannels in a phase-conjugation process. This can be seen in the enhancement of total transmission. The background intensity is given by $C_{4,2} = \sum_n \tau_n^2 / \sum_n \tau_n = g/M$ instead of the background of g/N for random illumination [19].

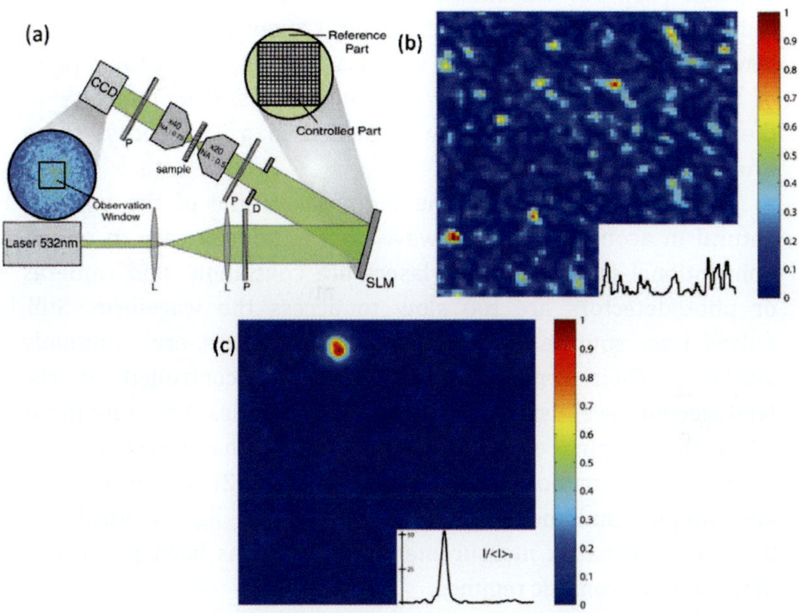

Figure 16.5 (a) Schematic of the apparatus used to measure the TM in optics. (b) Initial speckle pattern. (c) Example of focusing on one point by phase-conjugating an element of the TM. Adapted from [21].

Phase conjugation is analogous to time reversal (TR) for broadband pulsed signals [14] that allows spatio-temporal focusing of the waves. TR has been demonstrated to focus ultrasound and electromagnetic radiation within reverberant cavities [58, 59] or through random media [13, 59]. The phase-conjugation of spectra of the TM is equivalent to reversing the impulse response in the time domain. Acoustic measurements have demonstrated the richness of the interplay between temporal and spatial degrees of freedom of complex media [15]. The number of spatiotemporal degrees of freedom, which is the product of the number of spatial and spectral degrees of freedom, M and N_f, respectively. N_f is linked to the number of correlation frequencies within the bandwidth of the pulse and can be expressed as $N_f = \Delta\omega/\delta\omega$. Here $\Delta\omega$ is the bandwidth of the signal and $\delta\omega$ is the width of the spectral field-correlation function [61]. For transmission through a random medium, $\delta\omega$ is inversely proportional to the typical time in which a photon traverses the medium and so varies as the inverse of the square of the sample thickness for diffusing waves. The intensity at the focus is therefore greatly enhanced in a TR process for a broadband field.

While manipulation of the temporal profile of the wave is natural in acoustics or microwave, this is not the case in optics. Conventional monochromatic lasers are continuous and cameras or photodetectors are too slow to access the waveform. Still, pulsed laser sources, such as femtosecond lasers, are commonly available. Their temporal profile can be controlled at the femtosecond level using pulse shaping techniques. These methods allow the shaping of a pulse and can be characterized using interferometry or non-linear measurements [62]. These methods are complex and most of the early work on light control and transmission matrix measurements [19, 21] has been performed in the monochromatic regime.

Temporal focusing of a polychromatic wave that has traversed a complex medium could be accomplished by controlling spatial degrees of freedom. Using either time-gated [23] or a nonlinear optical signals [63], the input wavefront was optimized to focus a short pulse in time after it was stretched by the medium (see Fig. 16.6). In parallel, and in close analogy with TR experiments, a transform-limited pulse was produced from a stretched pulse without spatial control [64].

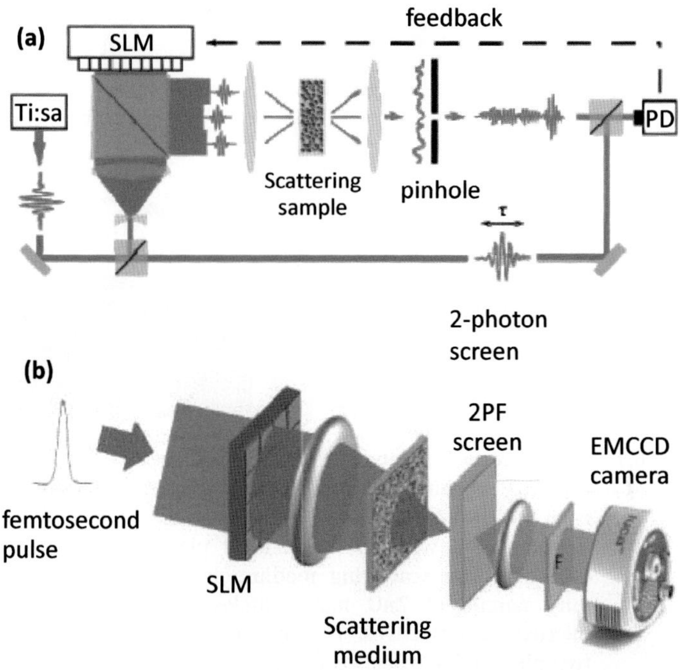

Figure 16.6 Two experiments for spatiotemporal focusing by optimization of the input wavefront. (a) Time-gated measurements are performed via interferometry [23] and (b) temporal refocusing is achieved using nonlinear fluorescence optimization [63]. Reproduced from [23] and [63].

Beyond optimization, full control of the spatiotemporal profile requires the knowledge of the spectral variation of the TM. Using a tunable-frequency laser, it was shown in [65] that the method of [21] could be readily extended to measure the multispectral TM (MSTM). In order to have sufficient information to fully describe the spectral behavior, the spectral domain must be sampled with sufficient resolution to measure $N_f = \Delta\omega/\delta\omega$ different frequencies [60]. In [65], the reference wave used to record the MSTM was itself a speckle pattern so that the phase-relation between the spectral components of the MSTM was unknown. Mono- or multispectral focusing was demonstrated, but temporal control remained elusive. As shown in Fig. 16.7, this was achieved in [66] using a common reference to measure

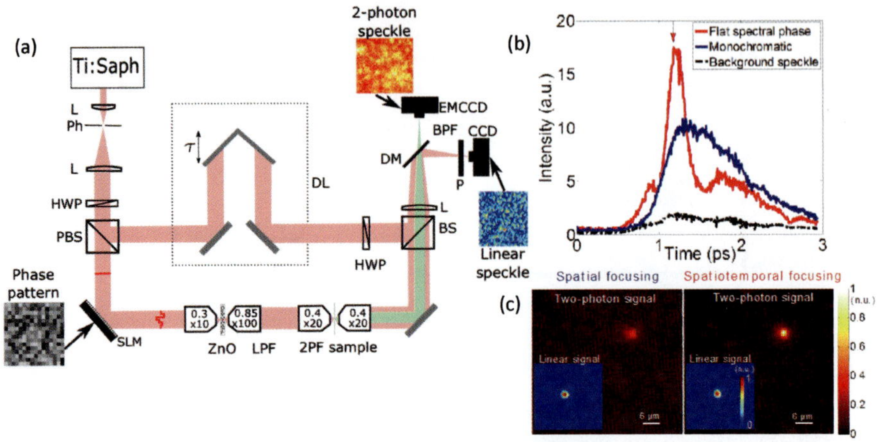

Figure 16.7 Spatiotemporal control with the multispectral TM (MSTM) [66]. (a) Experimental setup. The laser (ultrashort pulse 110 fs at FWHM) is expanded and illuminates the SLM which is conjugated with the back focal plane of a microscope objective. The scattering medium is a thick (around 100 μm) sample of ZnO nanoparticles, placed between two microscope objectives. The output plane is imaged on the two-photon fluorescence (2PF) sample. The long-pass filter (LPF) allows one to eliminate the autofluorescence of the ZnO. A dichroic mirror (DM) separates the two-photon fluorescence signal (imaged with an EMCCD) from the linear signal (imaged with a CCD). Lens (L), pinhole (Ph), half plate wave (HPW), polarized beam-splitter (PBS), delay line (DL), beam splitter (BS), polarizer (P), band pass filter (BPF). (b) Result for temporal focusing obtained by setting all the spectral phases at a given position at the output equal (red curve) compared with a monochromatic focus that is dispersed in time (blue curve). (c) Comparison of linear and two-photon fluorescence image, showing that, for the same linear intensity, the two-photon signal is enhanced approximately two-fold when temporally compressing the pulse. Reproduced from [66].

the spectral phase relation within the MSTM. Thanks to the knowledge of the MSTM, not only is temporal refocusing possible, but arbitrary temporal shapes can be generated experimentally. Two pulses with a controllable delay have been produced. This is an important case which will facilitate coherent control of optical experiments.

16.5 Applications to Biomedical Imaging

While the TM relates the field on the two sides of a complex medium, in biomedical optics, where the goal is to achieve imaging in the depth of a sample, this input-output relation is of little use. This requires modifying the TM approach to image within the medium.

16.5.1 Photoacoustic Transmission Matrix

The first approach is to access the field inside the medium. This is the so-called "guide-star" problem which can be tackled via a variety of techniques, either all-optical such as time-gating or nonlinear fluorescence, or in combination with acoustic methods [69]. A particularly interesting modality is photoacoustics, in which optical intensity in an absorber within the sample generates local heating. This in turn generates an acoustic wave that can be measured and from which the position of the absorber can be determined. This is possible because, in contrast to optical waves, ultrasound propagates ballistically in a homogeneous environment in biological tissue. However, the strength of the acoustic signal is proportional not only to the absorption coefficient but also to the optical intensity; therefore, the acoustic signal is an indirect measure of the intensity in the interior of the sample. The TM method was extended in [67], where a photoacoustic TM was measured, linking output modes at the input of the medium (controlled using an SLM) to a set of input positions along a line inside the medium. This result was rapidly extended to 2D in [68]. Once this matrix is known, it is then possible to focus on any position at which an absorber is located (see Fig. 16.8).

Since there is no signal originating from regions without absorption, it is not possible to focus everywhere. Another strong difference from conventional TMs is that the resolution of the target output positions is dictated by the acoustic resolution, which is in the 100 micron range for conventional high-frequency ultrasound. This limits the resolution of optical focusing as well as its efficiency. While other guide-star methods also allow focusing in the depth of the sample [69] by optimizating phase-conjugation, scanning is still difficult with these methods [67] and remains to-date the only matrix method able to focus within the sample.

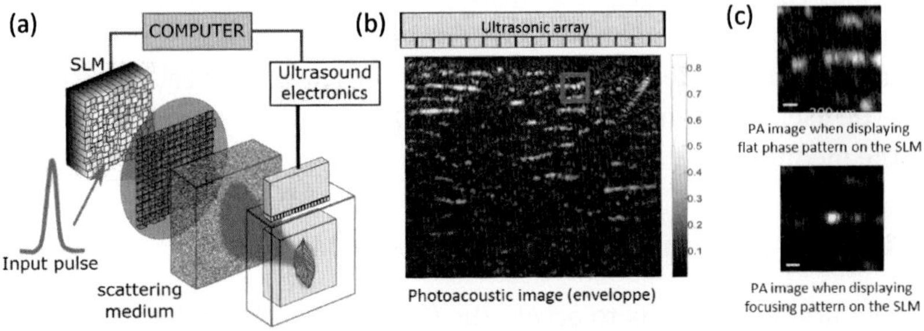

Figure 16.8 Photoacoustic TM. (a) Experimental setup [67], the wavefront of a nanosecond laser pulse is controlled using an SLM, and the measurement is obtained from photoacoustic images using an acoustic transducer. (b) Typical photoacoustic image of the absorbing structure under speckle illumination [67]. (c) Close-up on the red square in previous image, with a flat SLM pattern (top) or for an SLM pattern focusing in the center (bottom).

16.5.2 Endoscopic Imaging in Multimode Optical Fiber

Endoscopy is another strategy for focusing or imaging in depth in vivo which circumvents the problem of movement of the scattering tissue and absorption. This is usually done by inserting a dense array of single mode fibers, each of them constituting a pixel. Unfortunately, this approach is strategy invasive and the resolution is limited due to the inter-core separation required to avoid crosstalk. An interesting solution that has been proposed is to replace the fiber bundle with a multimode fiber. This drastically reduces the size and invasiveness of the endoscope, for the same number of pixels and greatly improves the resolution since it is only limited by the numerical aperture of the fiber. However, the multimode fiber itself becomes a complex medium to be compensated. Following early work showing that optimization or digital optical phase conjugation of the wavefront could restore a focus through a multimode fiber as shown in Fig. 16.9 [70–72], the TM of a multimode fiber was measured and exploited [70, 73]. This was used to obtain images from the distal tip of the fiber with use of a wide set of modalities, including bright and dark field microscopy, fluorescence microscopy [70], photoacoustics [74], confocal microscopy [75] and two-photon microscopy [76]. An important practical question for TM imaging in multimode fiber is the sensitivity of the TM to bending of the fiber. Several strategies for dynamic compensation have been proposed [77, 78], but this remains a major challenge. A major advance has been recently put forward in [79] where the TM and the impact of bending of a fiber were shown to be predictable. The rapid progress in multimode fiber imaging, as well as the numerous advantages of this method, makes it a promising domain for future biomedical applications.

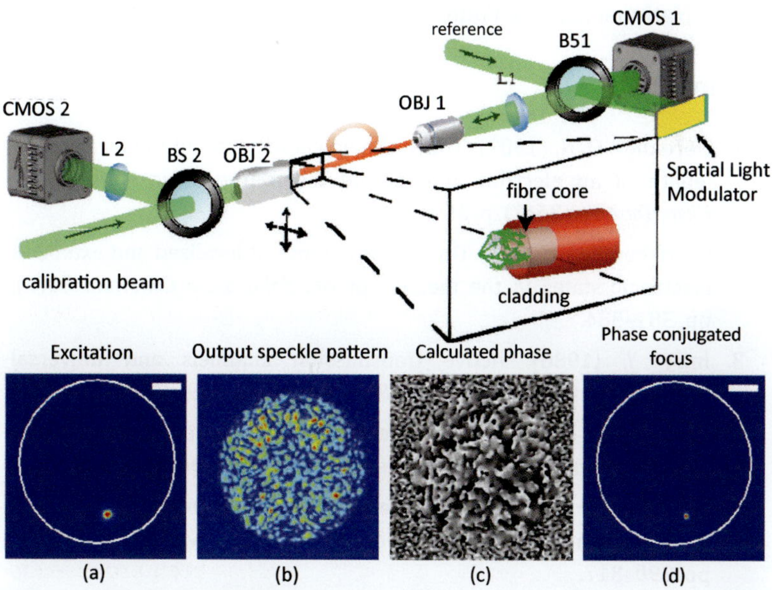

Figure 16.9 (Top) Experimental setup for multimode fiber wavefront shaping experiment. (Bottom) Example of wavefront shaping focusing. By injecting (a) on one side of the multimode fiber, one can record a complex pattern on the distal end (b), corresponding to a column of the TM of the fiber. By phase conjugating this pattern (c) one can refocus light on the original position (d) Reproduced from [71, 80].

16.6 Conclusion

Measurements of the TM show that it is possible to focus waves through disordered media and to control the net transmission by shaping the incident wavefront. Organizing light transmitted through scattering media holds great promise for imaging complex media for a wide range of applications, especially for optical imaging in biomedical tissue.

Acknowledgement

We thank Zhou Shi, Chushun Tian, Arthur Goetschy, and Doug Stone for stimulating discussions.

The research of S.G. was supported by the European Research Council (Grant No. 278025). The research of A.Z.G. was supported by the National Science Foundation grant DMR/-BSF-1609218.

References

1. Dorokhov, O. N. (1982). Transmission coefficient and the localization length of an electron in n bound disordered chains. *Pis'ma Zh. Eksp. Teor. Fiz.* **36**(7), p. 259.
2. Dorokhov, O. N. (1984). On the coexistence of localized and extended electronic states in the metallic phase. *Solid State Commun.* **51**(6), pp. 381–384.
3. Imry, Y. (1986). Active transmission channels and universal conductance fluctuations. *Euro. Phys. Lett.* **1**, pp. 249–256.
4. Mello, P. A. (1988). Macroscopic approach to universal conductance fluctuations in disordered metals. *Phys. Rev. Lett.* **60**(11), p. 1089.
5. Mello, P., Pereyra, P., and Kumar, N. (1988). Macroscopic approach to multichannel disordered conductors. *Ann. Phys.* (N.Y.). **181**(2), pp. 290–317.
6. Pichard, J.-L., Zanon, N., Imry, Y., and Douglas Stone, A. (1990). Theory of random multiplicative transfer matrices and its implications for quantum transport *J. Phys. France.* **51**(1990) p. 22.
7. Nazarov, Y. V. (1994). Limits of universality in disordered conductors. *Phys. Rev. Lett.* **73**(1), pp. 134–137.
8. Beenakker, C. W. J. (1997). Random-matrix theory of quantum transport *Rev. Mod. Phys.* **69**(3), p. 731.
9. van Rossum, M. C. W., and Nieuwenhuizen, T. M. (1999). Multiple scattering of classical waves: Microscopy, mesoscopy, and diffusion. *Rev. Mod. Phys.* **71**(1), p. 313.

10. Shi, Z., Wang, J., and Genack, A. Z. (2014). Microwave conductance in random waveguides in the cross-over to Anderson localization and single-parameter scaling. *Proc. Natl. Acad. Sci.* p. 201319704.
11. Nieuwenhuizen, T. M., and van Rossum, M. C. W. (1995). Intensity distributions of waves transmitted through a multiple scattering medium. *Phys. Rev. Lett.* **74**(14), p. 2674.
12. Kogan, E., and Kaveh, M. (1995). Random-matrix-theory approach to the intensity distributions of waves propagating in a random medium. *Phys. Rev. B.* **52**(6), p. R3813.
13. Derode, A., Roux, P., and Fink, M. (1995). Robust acoustic time reversal with high-order multiple scattering. *Phys. Rev. Lett.* **75**(23), p. 4206.
14. Fink, M. (1997). Time reversed acoustics. *Phys. Today.* **50**(3), pp. 34–40.
15. Lemoult, F., Lerosey, G., de Rosny, J., and Fink, M. (2009). Manipulating spatiotemporal degrees of freedom of waves in random media. *Phys. Rev. Lett.* **103**(17), p. 173902.
16. Lerosey, G., de Rosny, J., Tourin, A., Derode, A., Montaldo, G., and Fink, M. (2004). Time reversal of electromagnetic waves. *Phys. Rev. Lett.* **92**(19), p. 193904.
17. Lemoult, F., Lerosey, G., de Rosny, J., and Fink, M. (2010). Resonant metalenses for breaking the diffraction barrier. *Phys. Rev. Lett.* **104**(20), p. 203901.
18. Vellekoop, I. M., and Mosk, A. P. (2007). Focusing coherent light through opaque strongly scattering media. *Opt. Lett.* **32**(16), pp. 2309–2311.
19. Vellekoop, I. M., and Mosk, A. P. (2008). Universal optimal transmission of light through disordered materials. *Phys. Rev. Lett.* **101**(12), pp. 120601–120604.
20. Vellekoop, I. M., Lagendijk, A., and Mosk, A. P. (2010). Exploiting disorder for perfect focusing. *Nat. Photon.* **4**(5), pp. 320–322.
21. Popoff, S. M., Lerosey, G., Carminati, R., Fink, M., Boccara, A. C., and Gigan, S. (2010). Measuring the transmission matrix in optics: An approach to the study and control of light propagation in disordered media. *Phys. Rev. Lett.* **104**(10), p. 100601.
22. Choi, Y., Yang, T. D., Fang-Yen, C., Kang, P., Lee, K. J., Dasari, R. R., Feld, M. S., and Choi, W. (2011). Overcoming the diffraction limit using multiple light scattering in a highly disordered medium. *Phys. Rev. Lett.* **107**(2), p. 023902.

23. Aulbach, J., Gjonaj, B., Johnson, P. M., Mosk, A. P., and Lagendijk, A. (2011). Control of light transmission through opaque scattering media in space and time. *Phys. Rev. Lett.* **106**(10), p. 103901.
24. Mosk, A. P., Lagendijk, A., Lerosey, G., and Fink, M. (2012). Controlling waves in space and time for imaging and focusing in complex media. *Nat. Photon.* **6**(5), pp. 283–292.
25. Kim, M., Choi, Y., Yoon, C., Choi, W., Kim, J., Park, Q.-H., and Choi, W. (2012). Maximal energy transport through disordered media with the implementation of transmission eigenchannels. *Nat. Photon.* **6**(9), pp. 583–587.
26. Yu, H., Hillman, T. R., Choi, W., Lee, J. O., Feld, M. S., Dasari, R. R., and Park, Y. K. (2013). Measuring large optical transmission matrices of disordered media. *Phys. Rev. Lett.* **111**(15), p. 153902.
27. Popoff, S. M., Goetschy, A., Liew, S. F., Stone, A. D., and Cao, H. (2014). Coherent control of total transmission of light through disordered media. *Phys. Rev. Lett.* **112**(13), p. 133903.
28. Hsu, C. W., Goetschy, A., Bromberg, Y., Stone, A. D., and Cao, H. (2015). Broadband coherent enhancement of transmission and absorption in disordered media. *Phys. Rev. Lett.* **115**(22), p. 223901.
29. Sarma, R., Yamilov, A. G., Petrenko, S., Bromberg, Y., and Cao, H. (2016). Control of energy density inside a disordered medium by coupling to open or closed channels. *Phys. Rev. Lett.* **117**(8), p. 086803.
30. Vellekoop, I. M., and Mosk, A. P. (2008). Universal optimal transmission of light through disordered materials. *Phys. Rev. Lett.* **101**(12), p. 120601.
31. Davy, M., Shi, Z., Wang, J., Cheng, X., and Genack, A. Z. (2015). Transmission eigenchannels and the densities of states of random media. *Phys. Rev. Lett.* **114**(3), p. 033901.
32. Choi, W., Mosk, A. P., Park, Q. H., and Choi, W. (2011). Transmission eigenchannels in a disordered medium. *Phys. Rev. B.* **83**(13), p. 134207.
33. Choi, W., Park, Q. H., and Choi, W. (2012). Perfect transmission through Anderson localized systems mediated by a cluster of localized modes. *Opt. Express.* **20**(18), pp. 20721–20729.
34. Liew, S. F., Popoff, S. M., Mosk, A. P., Vos, W. L., and Cao, H. (2014). Transmission channels for light in absorbing random media: From diffusive to ballistic-like transport. *Phys. Rev. B.* **89**(22), p. 224202.
35. Davy, M., Shi, Z., Park, J., Tian, C., and Genack, A. Z. (2015). Universal structure of transmission eigenchannels inside opaque media. *Nat. Commun.* **6**.

36. Fisher, D. S., and Lee, P. A. (1981). Relation between conductivity and transmission matrix. *Phys. Rev. B.* **23**(12), pp. 6851–6854.

37. Goetschy, A., and Stone, A. D. (2013). Filtering random matrices: The effect of incomplete channel control in multiple scattering. *Phys. Rev. Lett.* **111**(6), p. 063901.

38. Marčenko, V. A., and Pastur, L. A. (1967). Distribution of eigenvalues for some sets of random matrices. *Math. USSR Sbornik.* **1**, pp. 457–483.

39. Brouwer, P. W. (1998). Transmission through a many-channel random waveguide with absorption. *Phys. Rev. B.* **57**(17), p. 10526.

40. Beenakker, C. W. J., Paasschens, J. C. J., and Brouwer, P. W. (1996). Probability of reflection by a random laser. *Phys. Rev. Lett.* **76**(8), pp. 1368–1371.

41. Yamilov, A., Petrenko, S., Sarma, R., and Cao, H. (2016). Shape dependence of transmission, reflection, and absorption eigenvalue densities in disordered waveguides with dissipation. *Phys. Rev. B.* **93**(10), p. 100201.

42. Chong, Y. D., Ge, L., Cao, H., and Stone, A. D. (2010). Coherent perfect absorbers: Time-reversed lasers. *Phys. Rev. Lett.* **105**(5), p. 053901.

43. Cheng, X., Tian, C., and Genack, A. Z. (2013). Transmission eigenvalues in random media with surface reflection. *Phys. Rev. B.* **88**(9), p. 094202.

44. Zhao, L., Tian, C., Bliokh, Y. P., and Freilikher, V. (2015). Controlling transmission eigenchannels in random media by edge reflection. *Phys. Rev. B.* **92**(9), p. 094203.

45. Li, J. H., Lisyansky, A. A., Cheung, T. D., Livdan, D., and Genack, A. Z. (1993). Transmission and surface intensity profiles in random media. *Euro. Phys. Lett.* **22**, p. 675.

46. Shi, Z., and Genack, A. Z. (2012). Transmission eigenvalues and the bare conductance in the crossover to Anderson localization. *Phys. Rev. Lett.* **108**(4), p. 043901.

47. Gérardin, B., Laurent, J., Derode, A., Prada, C., and Aubry, A. (2014). Full transmission and reflection of waves propagating through a maze of disorder. *Phys. Rev. Lett.* **113**(17), p. 173901.

48. Hsu, C. W., Goetschy, A., Bromberg, Y., Stone, A. D., and Cao, H. (2015). Broadband coherent enhancement of transmission and absorption in disordered media. *Phys. Rev. Lett.* **115**(22), p. 223901.

49. Shi, Z., and Genack, A. Z. (2015). Dynamic and spectral properties of transmission eigenchannels in random media. *Phys. Rev. B,* **92**(18), p. 184202.

50. Rotter, S., Ambichl, P., and Libisch, F. (2011). Generating particlelike scattering states in wave transport. *Phys. Rev. Lett.* **106**(12), p. 120602.

51. Peña, A., Girschik, A., Libisch, F., Rotter, S., and Chabanov, A. (2014). The single-channel regime of transport through random media. *Nature Commun.* **5**.

52. Baranger, H. U., DiVincenzo, D. P., Jalabert, R. A., and Stone, A. D. (1991). Classical and quantum ballistic-transport anomalies in microjunctions. *Phys. Rev. B.* **44**(19), p. 10637.

53. Metalidis, G., and Bruno, P. (2005). Green's function technique for studying electron flow in two-dimensional mesoscopic samples. *Phys. Rev. B.* **72**(23), p. 235304.

54. Brandbyge, M., and Tsukada, M. (1998). Local density of states from transmission amplitudes in multichannel systems. *Phys. Rev. B.* **57**(24), pp. R15088–R15091.

55. Avishai, Y., and Band, Y. B. (1985). One-dimensional density of states and the phase of the transmission amplitude. *Phys. Rev. B.* **32**(4), pp. 2674–2676.

56. Davy, M., Shi, Z., and Genack, A. Z. (2012). Focusing through random media: Eigenchannel participation number and intensity correlation. *Phys. Rev. B.* **85**(3), p. 035105.

57. Davy, M., Shi, Z., Wang, J., and Genack, A. Z. (2013). Transmission statistics and focusing in single disordered samples. *Opt. Express.* **21**(8), pp. 10367–10375.

58. Draeger, C., and Fink, M. (1997). One-channel time reversal of elastic waves in a chaotic 2d-silicon cavity. *Phys. Rev. Lett.* **79**(3), p. 407.

59. Lerosey, G., de Rosny, J., Tourin, A., Derode, A., Montaldo, G., and Fink, M. (2004). Time reversal of electromagnetic waves. *Phys. Rev. Lett.* **92**(19), p. 193904.

60. Derode, A., Tourin, A., and Fink, M. (2001). Random multiple scattering of ultrasound. II. Is time reversal a self-averaging process? *Phys. Rev. E.* **64**(3), p. 036606.

61. Genack, A., and Drake, J. (1990). Relationship between optical intensity, fluctuations and pulse propagation in random media. *Euro. Phys. Lett.* **11**(4), p. 331.

62. Monmayrant, A., Weber, S., and Chatel, B. (2010). A newcomer's guide to ultrashort pulse shaping and characterization. *J. Phys. B.* **43**(10), p. 103001.

63. Katz, O., Small, E., Bromberg, Y., and Silberberg, Y. (2011). Focusing and compression of ultrashort pulses through scattering media. *Nat. Photon.* **5**(6), pp. 372–377.

64. McCabe, D. J., Tajalli, A., Austin, D. R., Bondareff, P., Walmsley, I. A., Gigan, S., and Chatel, B. (2011). Spatio-temporal focusing of an ultrafast pulse through a multiply scattering medium. *Nat. Commun.* **2**, p. 447.

65. Andreoli, D., Volpe, G., Popoff, S., Katz, O., Grésillon, S., and Gigan, S. (2015). Deterministic control of broadband light through a multiply scattering medium via the multispectral transmission matrix. *Sci. Rep.* **5**, p. 10347.

66. Mounaix, M., Andreoli, D., Defienne, H., Volpe, G., Katz, O., Grésillon, S., and Gigan, S. (2016). Spatiotemporal coherent control of light through a multiple scattering medium with the multispectral transmission matrix. *Phys. Rev. Lett.* **116**, p. 253901.

67. Chaigne, T., Katz, O., Boccara, A. C., Fink, M., Bossy, E., and Gigan, S. (2014). Controlling light in scattering media non-invasively using the photoacoustic transmission matrix. *Nat. Photon.* **8**(1), pp. 58–64.

68. Chaigne, T., Gateau, J., Katz, O., Bossy, E., and Gigan, S. (2014). Light focusing and two-dimensional imaging through scattering media using the photoacoustic transmission matrix with an ultrasound array. *Opt. Lett.* **39**(9), pp. 2664–2667.

69. Horstmeyer, R., Ruan, H., and Yang, C. (2015). Guidestar-assisted wavefront-shaping methods for focusing light into biological tissue. *Nat. Photon.* **9**(9), pp. 563–571.

70. Cizmar, T., and Dholakia, K. (2012). Exploiting multimode waveguides for pure fibre-based imaging. *Nat. Commun.* **3**, p. 1027.

71. Papadopoulos, I. N., Farahi, S., Moser, C., and Psaltis, D. (2012). Focusing and scanning light through a multimode optical fiber using digital phase conjugation. *Opt. Express.* **20**(10), pp. 10583–10590.

72. Di Leonardo, R., and Bianchi, S. (2011). Hologram transmission through multi-mode optical fibers. *Opt. Express.* **19**(1), pp. 247–254.

73. Choi, Y., Yoon, C., Kim, M., Yang, T. D., Fang-Yen, C., Dasari, R. R., Lee, K. J., and Choi, W. (2012). Scanner-free and wide-field endoscopic imaging by using a single multimode optical fiber. *Phys. Rev. Lett.* **109**(20), p. 203901.

74. Papadopoulos, I. N., Simandoux, O., Farahi, S., Huignard, J. P., Bossy, E., Psaltis, D., and Moser, C. (2013). Optical-resolution photoacoustic

microscopy by use of a multimode fiber. *Appl. Phys. Lett.* **102**(21), p. 211106.

75. Loterie, D., Farahi, S., Papadopoulos, I., Goy, A., Psaltis, D., and Moser, C. (2015). Digital confocal microscopy through a multimode fiber. *Opt. Express.* **23**(18), pp. 23845–23858.
76. Morales-Delgado, E. E., Psaltis, D., and Moser, C. (2015). Two-photon imaging through a multimode fiber. *Opt. Express.* **23**(25), pp. 32158–32170.
77. Caravaca-Aguirre, A. M., Niv, E., Conkey, D. B., and Piestun, R. (2013). Real-time resilient focusing through a bending multimode fiber. *Opt. Express.* **21**(10), pp. 12881–12887.
78. Farahi, S., Ziegler, D., Papadopoulos, I. N., Psaltis, D., and Moser, C. (2013). Dynamic bending compensation while focusing through a multimode fiber. *Opt. Express.* **21**(19), pp. 22504–22514.
79. Plöschner, M., Tyc, T., and Čižmár, T. (2015). Seeing through chaos in multimode fibres. *Nat. Photon.* **9**(8), pp. 529–535.
80. Papadopoulos, I., Farahi S., Moser C., and Psaltis D. (2013). High-resolution, lensless endoscope based on digital scanning through a multimode optical fiber. *Biom. Opt. Express,* **4**(2), pp. 260–270.

Index